MODERN INORGANIC PHARMACEUTICAL CHEMISTRY

Second Edition

MODERN INORGANIC PHARMACEUTICAL CHEMISTRY

Second Edition

Clarence A. Discher
Professor Emeritus
Rutgers University, Piscataway, New Jersey

Thomas Medwick
Professor of Pharmaceutical Chemistry
Rutgers University, Piscataway, New Jersey

Leonard C. Bailey
Associate Professor of Pharmaceutical Chemistry
Rutgers University, Piscataway, New Jersey

WAVELAND PRESS, INC.
Prospect Heights, Illinois

For information about this book, write or call:
Waveland Press, Inc.
P.O. Box 400
Prospect Heights, Illinois 60070
(847) 634-0081

ISBN 0-88133-121-X

Printed in the United States of America

7 6 5 4

PREFACE

The organization and style of presentation used in this textbook have been developed because of a desire to bring a fresh perspective and new vitality to the teaching of inorganic pharmaceutical chemistry—an area of the pharmaceutical curriculum which has made little progress in the immediate past. To some extent this may be ascribed to the fact that too often inorganic medicinals has become just another chemistry course, with some pharmaceutical overtones, and frequently having the additional handicap of outmoded presentations. If inorganic pharmaceutical chemistry is to progress as a vital component of pharmaceutical education and the profession as a whole, certain changes are necessary. The two more important of these are: One, the modern theoretical approach and the newer thinking which have been developed in inorganic chemistry must be incorporated into and applied to the presentation and elucidation of the phenomena constituting this field. Two, a presentation is needed which ties the subject matter of inorganic pharmaceutical chemistry more closely to the health sciences.

Modern interpretation of phenomena based on the latest theories has been used throughout this textbook. Whenever possible the remarkable advances made in the over-all field of inorganic chemistry during the past two or three decades have been incorporated into the presentations. This has frequently necessitated a break from the traditional interpretations and presentations of inorganic pharmaceutical chemistry, as, for instance, the treatment afforded antacids. A modern presentation presupposes an adequate understanding by the reader of the underlying principles, laws, and nomenclature. Part I of this textbook provides such a foundation for those students requiring it. Basic chemical theory is reviewed and in many cases expanded somewhat beyond that traditional for general chemistry and qualitative analysis.

In order to tie more closely the subject matter of inorganic pharmaceutical chemistry to pharmacy and the health sciences, this textbook breaks with the traditional order of presentation—that based on families of the periodic table. Such a traditional presentation provides little or no emphasis of those properties which make a given compound suitable for a given application. Here a new organization, closely tied to the ultimate role of a given inorganic substance in pharmacy and medicine, has been adopted. All substances with similar applications are discussed as a group in a single section; for example, all antacids and the essentials of antacid action are considered in Chapter 15. The advantages of this presentation are many, but most important it is possible to emphasize precisely those chemical and physical properties essential if a compound is to act in a given pharmaceutical or medicinal capacity. Furthermore, the individual roles of the cation and anion of ionic substances may be clearly differentiated. This approach gives to inorganic pharmaceutical chemistry a real meaning and reason for existence, for it is precisely here that its presentation must differ from the ordinary course in inorganic chemistry.

This approach, based on use, requires a two-part presentation of inorganic pharmaceutical chemistry—one dealing with those inorganics of importance to the pharmacist in his efforts to produce a suitable dosage form, the other dealing with those inorganic substances employed for their medicinal activity. The first group, studied in Part II of this textbook, covers topics essential for an understanding of the chemistry of the techniques constituting the practice of pharmacy: the common solvent water, those acids and bases suitable for pharmaceutical neutralizations, pharmaceutical buffers, the solubility behavior in inorganic pharmaceutical systems and inorganic preservatives, and those inorganics important in a variety of physical operations (e.g., distributing agents, fillers, filter aids). The second group of substances, the inorganic medicinal agents, are discussed in Part III. One or more chapters are devoted to each, the trace minerals and the ions of the body fluids, the antacids, the germicides, astringents, and protectives, and the radiochemicals, together with a chapter on miscellaneous medicinal applications of inorganics. This arrangement not only makes it possible to investigate the essential chemistry of the substances studied, but also allows the elucidation of their mode of action and an emphasis of the chemical and physical properties essential in a compound or group of compounds to be employed for a given use. Here, amazingly enough, the major classes of medicinal substances fall into distinct regions of the periodic table. (Antacids are the only major group not doing so.)

Because of the intensive development of a sound understanding of the basic theory, some of the more traditional material, such as extensive recapitulations of the physical properties of and the analysis procedures for compounds studied, is not included unless a particular point of chemistry is illustrated. This information is available to the student in his compendia. Furthermore, inorganic pharmaceutical chemistry should not be used for the extensive memorization of such information. Rather, the course should provide the student with the theoretical and practical background necessary to understand and intelligently use the compendia and other source material available.

Ideally a textbook should stimulate a student to further reading and study. With this in mind the discussions of some topics have been purposely abbreviated, and from time to time terms have been introduced without providing a suitable definition. Some leading questions, somewhat beyond the scope of the text material, have also been included in the study aids. By these techniques the student will be encouraged to consult other source material; suitable references are provided as an aid in locating such supplementary reading. Study aids consisting of chapter outlines, terms, questions, and equation writing drill conclude each chapter of Parts II and III. These provide a cross section of the material covered in each. These study aids will help the student in achieving mastery of inorganic pharmaceutical chemistry by providing an opportunity to apply the material studied to appropriate situations.

Because of the wide range and diverse nature of the material covered, especially in substances having several applications, numerous cross references have been provided. For the same reason and because of a plethora of trivial names an extensive index is provided. Since modern inorganic nomenclature has been adopted almost entirely in this textbook, older chemical names are extensively cross indexed.

I am deeply indebted to my colleagues for the encouragement given me both before and during the writing of this textbook. I am especially indebted to Doctors N. Lordi, T. Medwick, and J. Shinkai for reading the manuscript and for the many valuable suggestions offered. Finally, credit must be given to Mrs. Discher, the other half of the team, for the countless hours spent over atrocious script, the typewriter, and copy in progress. Without her encouragement and continuing effort this work would have been extremely difficult.

C. A. D.

Belleville, N.J.
February 1964

SECOND EDITION PREFACE

In this new edition the arrangement and general chemical background have remained essentially unchanged. Where necessary, chemical advances and changes in practice have been introduced, e.g., Ewens-Bassett nomenclature. The student "home work" at the end of the chapters has been retained but answers are now included for some of the mathematical problems.

The pharmaceutical and medicinal content has been brought up-to-date by introducing advances which have occurred since the original edition was written. Newly introduced medicinals have been added. The chapter on radioisotopes in pharmacy and medicine has been rewritten. Compendial references have been made current.

Clarence A. Discher
Birnamwood, Wisconsin
Thomas Medwick
Piscataway, New Jersey
Leonard C. Bailey
Piscataway, New Jersey
June 27, 1984

CONTENTS

PART I • MODERN CONCEPTS OF CHEMICAL REACTIONS

The essential features of modern atomic and molecular theory, together with a study of reaction mechanisms, are reviewed in four of the following chapters. Although largely descriptive, the presentation is adequate for the elucidation of the chemical and physical phenomena encountered in the applied body of technical information constituting the field of inorganic pharmaceutical chemistry.

Chapter 5 deals with inorganic nomenclature. In general, modern nomenclature should be used exclusively. However, trivial and older, frequently nondefinitive, chemical names remain in use and appear in the older literature. Of necessity they cannot be ignored.

ONE · ATOMIC STRUCTURE AND PERIODIC CLASSIFICATION

The pharmaceutical chemist, like all chemists, works with atoms as his building blocks. At first glance it appears that molecules are the ultimate particle with which he is concerned, but under closer scrutiny it becomes apparent that certain atoms, within the structure of these molecules, are responsible for the changes he produces during his work. This is true in all branches of this diverse division of chemical science. The dispensing pharmacist controls atoms when overcoming a chemical incompatibility in a prescription being filled. In product development, a worker in search of a better vehicle for the administration of a drug or interested in improving the stability of a drug by control of pH operates on specific atoms of certain molecules. In synthesis, the worker rearranges atoms when he tries to modify a known compound to make it more effective or to decrease its side effects. In analysis, it is ordinarily an atom, or a specific group of two or three atoms, within an unknown's structure which serves as the focus for a successful determination.

It is impossible to understand the chemistry involved in a study of actual pharmaceutical problems without a thorough knowledge of the molecule and its properties. To understand the molecule, it is of fundamental importance to understand the atom—its make-up and properties. Thus a study of the atom, especially certain aspects, is an essential prerequisite to work in inorganic pharmaceutical chemistry.

The theoretical chemist can describe a given element by two pure numbers: its *atomic number* and its *atomic weight*. Starting from these two numbers, he can develop, by the use of the proper working tools, the atomic structure of a given element and further determine its physical and chemical properties in the various oxidation states in which it appears. Atomic numbers also serve in construction of the periodic table.

ATOMIC WEIGHTS

Atomic weight is the ratio of the mass of an atom of a given element to the mass of the most abundant isotope of carbon, carbon-12, taken as representing twelve mass units. This does not imply that carbon-12 has an exact atomic weight of twelve. Rather, in order to establish a useful scale of atomic weights, carbon-12 has been arbitrarily assigned an atomic weight of twelve mass units. The atomic weights of all other elements are thus ratios—based on carbon-12 as a reference point.

Atomic weight scale. Naturally occurring carbon exists as a mixture of three isotopes: 99 per cent carbon-12, about 1 per cent carbon-13, and traces of carbon-14. Since this ratio is not constant, natural carbon cannot be used as a basis for the atomic weight table. Chemical atomic weight tables published prior to 1961 were based on natural oxygen, which consists of a nearly constant mixture of oxygen-16, oxygen-17, and oxygen-18. But with the development of newer, more exact techniques of measurement, this basis was proven unsatisfactory. The proportions of O^{17} and O^{18} are not constant, their relative proportion depending on whether the natural oxygen sample is obtained from the earth's atmosphere, the water of the ocean, or the rocks of the earth's crust. Still another atomic weight table, that of the physicist, was based on oxygen-16, the most abundant isotope of the element.

The International Union of Pure and Applied Chemistry (IUPAC) and the International Union of Pure and Applied Physics (IUPAP) have adopted carbon-12 as the standard and have issued a new table of atomic weights based on this isotope.[1] Table 1.1 gives the revised atomic weights for all the elements. Close examination will reveal only three significant changes from the old atomic weight scales: bromine (79.909 vs. 79.916), chlorine (35.453 vs. 35.457), and silver (107.870 vs. 107.88). The use of the new atomic weights will have a negligible effect on practical calculations.

For descriptive purposes the atomic weight of an element represents its nuclidic mass, which is the sum of the number of protons and the number of neutrons in the nucleus of the given atom (referred to as mass number). If this fact were used as a theoretical basis for calculation, then all atomic weights should be simple whole numbers. In reality, however, whole numbers do not occur for several reasons. Firstly, almost all naturally occurring elements are isotopic mixtures: only a very few (aluminum, fluorine, sodium, and phosphorus) are mononuclidic. This isotopic effect is well illustrated by natural carbon; despite

[1] *J. Chem. Educ.*, **38**, 625 (1961).

the fact that carbon-12 is the basis of the atomic weight scale, its experimental atomic weight is 12.01. Secondly, the masses of the neutron and the proton are not exactly unity; in addition, the electron has mass, albeit almost insignificant. Finally, there is the binding energy of the nucleus, formerly called the mass defect. Mass is lost in the form of energy ($E = mc^2$) when nucleons coalesce to form a nucleus.

Stoichiometry. The chemist in practice thinks of atomic weights in terms of gram units, rather than as pure numbers. This usage arises from the fact that in stoichiometry calculations a mass unit must be used, and the gram has been the scientific unit of choice. It should be remembered, however, that other mass units, i.e., pounds, tons, could equally well be assigned. In English-speaking countries, pounds are usually used for the calculation of the reacting masses in chemical manufacturing. When weight units are assigned to atomic weights for calculation purposes, in order to avoid confusion the specified unit of weight should be written as a prefix to the term atomic weight. In this text, gram-atomic weights will be used in calculations. Of course, a worker must consistently use only one weight unit during a given series of calculations, and if other units of weight appear, he must change these to the chosen one.

The molecular mass, or, more frequently, molecular weight, represents the sum of the atomic masses of each and every atom making up the molecule. Like atomic weight, molecular weight is a pure number. In this text the gram mass unit is assigned to the molecular weight for calculation purposes.

As indicated, the atomic weight of an element is important because it determines the stoichiometry of chemical reactions. Thus the equation

$$Fe + I_2 \rightarrow FeI_2$$

tells us that:

(1) 1 gram-atom of iron + 2 gram-atoms of iodine $\rightarrow$ 1 gram-molecule of iron(II) iodide, or

(2) 55.85 Gm. of iron + 2 × 126.9 Gm. of iodine $\rightarrow$ 309.65 Gm. of iron(II) iodide, or

(3) 11.17 lb. of iron + 50.76 lb. of iodine $\rightarrow$ 61.93 lb. of iron(II) iodide, or

(4) 6.02×10^{23} atoms of iron + 1.204×10^{24} atoms of iodine $\rightarrow$ 6.02×10^{23} molecules of iron(II) iodide . . . .

In addition to stoichiometry, the law of conservation of mass is illustrated. An infinitesimally small quantity of mass is lost as heat during this reaction. In general chemical practice, any lost mass is neglected because it is much less than that amount detectable by ordinary measuring instruments (e.g., the analytical balance). Note also that atomic and

TABLE 1.1 ATOMIC WEIGHTS

Element	Symbol	Atomic Number	Atomic Weight	Element	Symbol	Atomic Number	Atomic Weight
Actinium	Ac	89		Mercury	Hg	80	200.59
Aluminum	Al	13	26.98	Molybdenum	Mo	42	95.94
Americium	Am	95		Neodymium	Nd	60	144.24
Antimony	Sb	51	121.75	Neon	Ne	10	20.18
Argon	Ar	18	39.95	Neptunium	Np	93	
Arsenic	As	33	74.92	Nickel	Ni	28	58.71
Astatine	At	85		Niobium	Nb	41	92.91
Barium	Ba	56	137.34	Nitrogen	N	7	14.01
Berkelium	Bk	97		Nobelium	No	102	
Beryllium	Be	4	9.01	Osmium	Os	76	190.2
Bismuth	Bi	83	208.98	Oxygen	O	8	16.00*
Boron	B	5	10.81	Palladium	Pd	46	106.4
Bromine	Br	35	79.91	Phosphorus	P	15	30.97
Cadmium	Cd	48	112.40	Platinum	Pt	78	195.09
Calcium	Ca	20	40.08	Plutonium	Pu	94	
Californium	Cf	98		Polonium	Po	84	
Carbon	C	6	12.01	Potassium	K	19	39.10
Cerium	Ce	58	140.12	Praseodymium	Pr	59	140.91
Cesium	Cs	55	132.91	Promethium	Pm	61	
Chlorine	Cl	17	35.45	Protactinium	Pa	91	
Chromium	Cr	24	52.00	Radium	Ra	88	
Cobalt	Co	27	58.93	Radon	Rn	86	
Copper	Cu	29	63.54	Rhenium	Re	75	186.2
Curium	Cm	96		Rhodium	Rh	45	102.91
Dysprosium	Dy	66	162.50	Rubidium	Rb	37	85·47
Einsteinium	Es	99		Ruthenium	Ru	44	101.07
Erbium	Er	68	167.26	Samarium	Sm	62	150.35
Europium	Eu	63	151.96	Scandium	Sc	21	44.96

* 15.9994

Element	Symbol	Atomic Number	Atomic Weight	Element	Symbol	Atomic Number	Atomic Weight
Fermium	Fm	100		Selenium	Se	34	78.96
Fluorine	F	9	19.00	Silicon	Si	14	28.09
Francium	Fr	87		Silver	Ag	47	107.87
Gadolinium	Gd	64	157.25	Sodium	Na	11	22.99
Gallium	Ga	31	69.72	Strontium	Sr	38	87.62
Germanium	Ge	32	72.59	Sulfur	S	16	32.06
Gold	Au	79	196.97	Tantalum	Ta	73	180.95
Hafnium	Hf	72	178.49	Technetium	Tc	43	
Helium	He	2	4.00	Tellurium	Te	52	127.60
Holmium	Ho	67	164.93	Terbium	Tb	65	158.92
Hydrogen	H	1	1.008	Thallium	Tl	81	204.37
Indium	In	49	114.82	Thorium	Th	90	232.04
Iodine	I	53	126.90	Thulium	Tm	69	168.93
Iridium	Ir	77	192.2	Tin	Sn	50	118.69
Iron	Fe	26	55.85	Titanium	Ti	22	47.90
Krypton	Kr	36	83.80	Tungsten	W	74	183.85
Lanthanum	La	57	138.91	Uranium	U	92	238.03
Lawrencium	Lw	103		Vanadium	V	23	50.94
Lead	Pb	82	207.19	Xenon	Xe	54	131.30
Lithium	Li	3	6.94	Ytterbium	Yb	70	173.04
Lutetium	Lu	71	174.97	Yttrium	Y	39	88.91
Magnesium	Mg	12	24.31	Zinc	Zn	30	65.37
Manganese	Mn	25	54.94	Zirconium	Zr	40	91.22
Mendelevium	Md	101					

molecular weights represent a counting process for the number of atoms and molecules involved. There are 6.02×10^{23} atoms in one gram-atomic weight and 6.02×10^{23} molecules in one gram-molecular weight. This constant is known as Avogadro's number. (The number of molecules in a pound-molecular weight would obviously differ from 6.02×10^{23} by a factor of 454.)

Effect on atomic properties. The atomic weight and, through it, the molecular weight are useful in ways other than stoichiometry. The mass of a given atom is related to its size (or volume). And size is important in determining crystal structure. *Atomic radius*, which directly expresses the relative size of an atom, determines in many cases whether or not an atom of an element can be accommodated in a crystal or in a given series of compounds. The molecule ortho nitric acid, H_3NO_4,[2] corresponding to ortho phosphoric acid, H_3PO_4, does not exist because there is insufficient space around the small nitrogen atom to accommodate a fourth oxygen atom. Thus the nitrogen series of acids has as its final member HNO_3. Similarly, there is no carbon analogue of tetra-*o*-tolylsilane.

Size, together with electronic structure, is also important in determining the strength of the electrostatic (and magnetic) fields surrounding the atom. For a given series of atoms, differing only in size, the ability to hold a valence electron is inversely related to atomic radius. For example, an atom of sodium holds its single valence electron more securely than the much larger cesium atom. In this way chemical properties are mediated by size.

Further, the atomic radius plays a significant role in determining many physical properties of substances: e.g., melting and boiling points, malleability, ductility. Since the centers of smaller atoms can come closer to one another, the individual atoms are brought into and held more effectively within one another's electrostatic fields. For this reason, within a given family of elements, the smallest (i.e., the one with the lowest atomic weight) will generally have the highest melting point. Aluminum is a solid at room temperature, whereas gallium, a larger but otherwise similar element, liquefies when held in the hand.

ATOMIC NUMBERS AND EXTRANUCLEAR ELECTRONS

In addition to mass (or size), certain other fundamental atomic properties contribute to the properties of a given element. The most important factor is the electronic structure of an atom. Electronic structure, together with atomic radius, determines the strength and nature of the

[2]Not to be confused with peroxonitric acid, which has an identical formula but a different structure.

electrostatic field surrounding an atom. A knowledge of this electrostatic field is essential since it is the main determinant of the properties of the atom.

One of the problems that has intrigued the chemist and physicist for the past half century has been the role of the electron. The first comprehensive attempt to describe the arrangement of extranuclear particles—electrons—resulted in the Bohr-Sommerfeld model. In this model, the atom resembles a subminiature solar system, the nucleus representing the sun and the electrons playing the role of the planets. Although quantum mechanics (or wave mechanics) no longer permits the concept of such rigidly fixed orbits, the word orbit has survived in modern terminology as *orbital*. Interestingly enough, the electron orbit of the Bohr-Sommerfeld atom is the region of the highest probable electron distribution in the currently accepted electron arrangements.

The electronic structure of an atom may be deduced from its *atomic number*. The atomic number of an atom is equal to the number of protons found in its nucleus and thus gives the number of positive charges residing in the nucleus. An equal number of electrons must be present to preserve the electroneutrality of the atom in its ground state.[3] Knowing the allowed number of electrons and applying the *Aufbau principle* (or build-up principle, an application of quantum theory and Pauli's exclusion principle), the chemist can assign each electron to its proper energy level. This is of fundamental importance since it is the incompletely filled, outermost electron energy levels which ultimately determine the chemical properties of an element.

In this discussion of the electron arrangements of atoms, the over-all atomic structure will be simplified. The structure of the nucleus, as well as the nature of subatomic particles other than the proton, neutron, and electron, will not be developed. Electron arrangements and the underlying theory used in their elucidation will be presented from a descriptive or qualitative point of view. Since chemical reactions involve only the outermost electrons, it is essential to develop a sufficient understanding of the electron arrangements within the atom. The student interested in a more theoretical presentation might refer to a modern advanced textbook in which relativistic quantum mechanics is thoroughly discussed.[4]

Today, wave mechanics describes the wave properties of the electron. It requires that certain stationary wave patterns occur around the nucleus,

---

[3]The unreacted, unexcited condition of an atom.

[4]Cotton, F.A. and Wilkinson, G. *Advanced Inorganic Chemistry,* 4th Ed., Wiley, New York, 1980.

very much like the standing waves on a vibrating string whose profile is fixed. (This wave motion is not that from a wave moving out along the string, as in a whip, which would actually represent the escape of an electron during the ionization of an atom.) Such a standing wave can occur only when an integral number of wavelengths complete the orbital; otherwise the waves tend to be destroyed by interference. Thus for a given orbital only certain frequencies, or discrete energy values, are allowed, and quantum theory is thereby satisfied.

For purposes of this discussion, however, it is more useful to think of an electron probability distribution within specific zones around the nucleus. As a crude approximation we may say that since an electron has a certain energy and a given momentum and is spinning on its own axis either clockwise or counterclockwise, it will very probably be found within a restricted volume of extranuclear space. Because an electron moving around a nucleus cannot be located precisely (Heisenberg's uncertainty principle), its location must be defined as a closed region around the nucleus whose limits are determined in terms of the probability of finding the electron in the given volume at any given instant. This region may be so defined that the probability of finding the electron within it is 95% or, in a slightly larger region, 98%. The probability of finding the electron outside the region is not zero, but it is very small.

Quantum numbers, energy levels, and orbitals. Identification of any one of the electrons in the extranuclear space of the atom is best accomplished by means of four quantum numbers. By their use an electron can be fully characterized, and by extending the process to each of the available electrons of a given atom, an exact understanding of the chemical behavior of that atom may be achieved.

To be confined to a given volume of extranuclear space, an electron can have only discrete values of kinetic energy; that is, the electron exists in a discrete energy level. If the energy possessed by a given electron is increased (e.g., by absorption of heat), the electron can move out to a volume of space further removed from the nucleus. If its energy is reduced, (e.g., by emission of light), the electron falls back to a lower energy level, i.e., to a volume of space closer to the nucleus. The *principal quantum number, n,* indicates the relative position of an energy level with respect to other energy levels present and with respect to the nucleus. Thus n designates the distance of an electron shell from the nucleus; to put it in another way, n indicates the number of the breaks or discontinuities—energy levels—in the electron cloud of an atom. The principal quantum number may have values of $n = 1, 2, 3, \ldots$ An electron with $n = 1$ is in the first energy level, or volume of space closest to the nucleus; with $n = 2$, it is in the next volume of space beyond the nucleus; when $n = 3$, it is in a space still further removed from the nucleus; etc. When $n = \infty$, the electron has sufficient energy to escape from the influence of the nucleus, and an ion of the atom results. (The values 1, 2, 3, 4, 5, . . . , referring to energy levels, correspond to the K, L, M, N, O, . . . shells of older literature.)

Within a given energy level, different electron distributions are possible. For a given value of n (except for $n = 1$), electrons within that energy level are not all equivalent; the electrons exist in different orbitals. However, in the atomic state, orbitals of a given type are equivalent to one another.

The simplest of these orbitals is a spherical distribution, found, for instance, in hydrogen and helium atoms. Such a distribution is dependent only upon the radius of the distribution and is independent of direction in space. Electrons with spherical distributions are known as s electrons. $1s$ if in the first principal quantum group, $2s$ if $n = 2$, $3s$ if $n = 3$, etc.

Electron distributions other than spherical are also found. The origin of nonspherical distributions is related to the angular momentum of the electron. (It is zero for the s subshell.) The particular subgroup within a major quantum group or energy level to which an electron belongs is described by the *subsidiary quantum number* (or subquantum number), l. It may have values of 0, 1, 2, . . . $(n - 1)$. It is quickly evident that l cannot be equal to or greater than its principal quantum number. Thus if $n = 1$, $l = 0$ is the single allowed value; for $l = 0$ we have the spherical distribution of s electrons already discussed. If $n = 2$, $l = 0$ and 1; if $n = 3$, $l = 0$, 1, and 2; etc.

Like the principal quantum number, a subsidiary quantum number can have different quantized values. These can best be visualized by recalling that a moving electron constitutes an electrical current and is associated with a magnetic field. Since an extranuclear electron is also in motion, it must be associated with a magnetic moment, which will affect the angular momentum when in a magnetic field. The restraints imposed are given by the *magnetic quantum number*, m_l. It has permitted values of $-l, \ldots, 0, \ldots, +l$. Thus for $l = 1$, m_l may have values of -1, 0, and $+1$. One may best think of this quantum number as the magnetic contribution to the angular momentum of an electron.

Finally, an electron spins as it moves through its assigned volume of space. This gives rise to another magnetic component of the angular momentum. This component is identical in value for all electrons, but may be + or − depending upon whether the electron spins in a clockwise or counterclockwise direction. Since one of two possible magnetic fields is associated with each spinning electron, an additional number, the *magnetic spin quantum number*, m_s, must be assigned. It has values of $+\frac{1}{2}$ or $-\frac{1}{2}$.

Table 1.2 summarizes the symbols and values of the quantum numbers

TABLE 1.2 THE FOUR QUANTUM NUMBERS

Name	Symbol	Values	Equivalent Symbols
Principal	n	1, 2, 3, . . .	$K, L, M, N, \ldots$
Subsidiary	l	0, 1, . . . $(n - 1)$	s, p, d, f
Magnetic	m_l	$-l, \ldots, 0, \ldots, +l$	—
Magnetic spin	m_s	$+\frac{1}{2}, -\frac{1}{2}$	—

as developed in the preceding discussion.

It is now possible to picture electron distributions other than the spherical. The simplest of the nonspherical distributions is a dumbbell-shaped distribution, with the nucleus of the atom at the center point of the dumbbell. This is the distribution achieved when $l = 1$. It has been pointed out previously when $l = 1$, m_l has three permissible values: -1, 0, and $+1$. This permits three equivalent dumbbell volumes, arranged along the x, y, and z axes of the cartesian coordinates in space. Since m_s may be + or −, two electrons may enter each dumbbell volume, giving a total of six electrons in these orbitals. These orbitals are known as the p orbitals, and electrons occupying this type of orbital are known as p electrons. The s and p orbitals are shown in Fig. 1.1. Since it is possible to have $l = 1$ in all except the lowest energy level of an atom, the particular energy level to which a given p electron belongs is indicated by adding the principal quantum number: $2p$, $3p$, $4p$, etc.

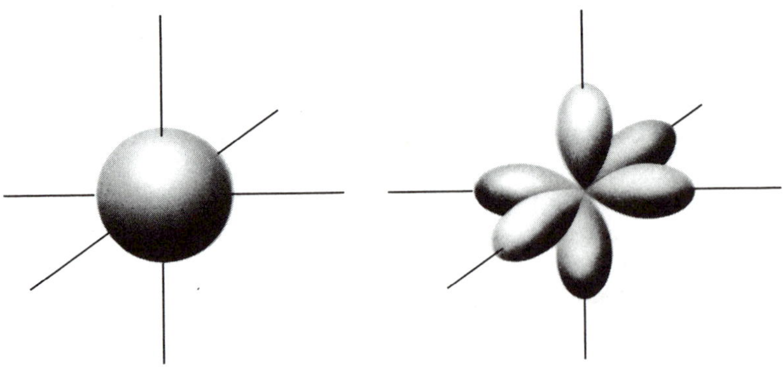

FIGURE 1.1 Boundary surfaces of s and p orbitals.

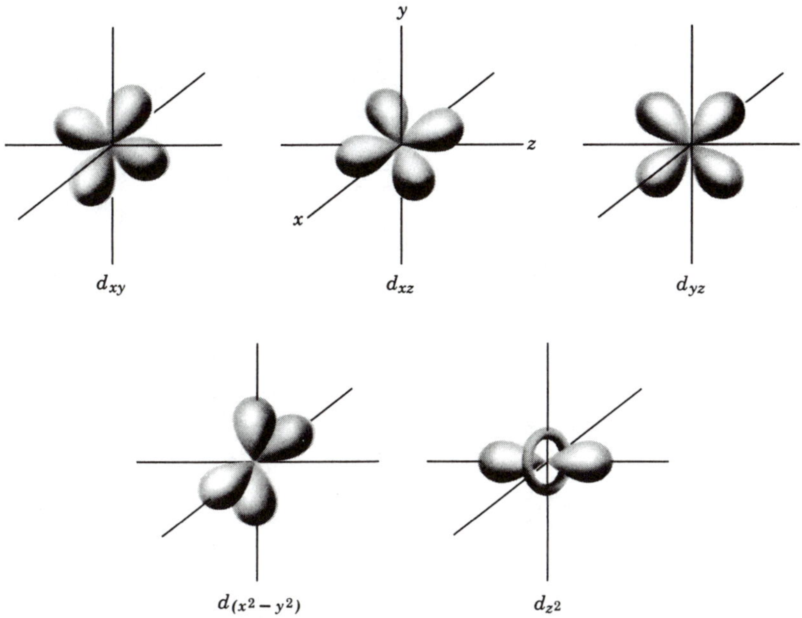

FIGURE 1.2 Boundary surfaces of the *d* orbitals. The labels are arbitrary, depending on the relationship to the cartesian coordinates. The subscripts *xy*, *xz*, and *yz* indicate the plane in which the orbital lobes are situated. The subscripts $(x^2 - y^2)$ and z^2 refer to the axes on which the axial lobes are situated.

When $l = 2$, a third arrangement of the electrons comes into being—the quadruple sausage cloud. These are the *d* orbitals, and entering electrons are known as *d* electrons. Since m_l may have values of -2, -1, 0, $+1$, $+2$, there are obviously five different sets of *d* orbitals. The five sets are shown in Fig. 1.2. Unlike the *p* orbitals, the *d* orbitals are not all equivalent in compounds. Since electrons of both + and − spin may occupy each of the five orbitals, a total of ten electrons may enter the *d* orbitals.

A third nonspherical distribution is possible when $l = 3$. Since m_l may have values -3, -2, -1, 0, $+1$, $+2$, and $+3$, and m_s may be $+\frac{1}{2}$ and $-\frac{1}{2}$, there are seven *f* orbitals, occupied by a maximum of fourteen *f* electrons.

In summary: The atomic number of an element, by giving the number of protons in the nucleus, also gives the number of electrons associated with an electrically neutral atom of the element. Each electron is identified by four quantum numbers. The principal quantum number (*n*) denotes the energy level occupied by a given electron. The subsidiary quantum

TABLE 1.3 QUANTUM NUMBERS AND ELECTRON DISTRIBUTIONS

Quantum Number	Theoretical Values Permitted	Allowed Values of Each Quantum Number for Each Orbital Type			
		s	p	d	f
n	1, 2, 3, 4, 5, . . .	1, 2, 3, 4, . . .	2, 3, 4, 5, . . .	3, 4, 5, . . . ,	4, 5, 6, . . .
l	0, 1, 2, . . . $(n-1)$	0	1	2	3
m_l	$-l, \ldots, 0, \ldots, +l$	0	−1, 0, +1	−2, −1, 0, +1, +2	−3, −2, −1, 0, +1, +2, +3
m_s	$\pm\frac{1}{2}$	±	± ± ±	± ± ± ± ±	± ± ± ± ± ± ±
Maximum electrons	$2(2l+1)$	2	6	10	14
Shape and number of orbital lobes		Spherical	Three dumbbell lobes	Five quadruple sausage lobes	(?)

number (l) determines the kinds of orbitals allowed in a given energy level. The magnetic quantum number (m_l) determines the orientation and number of suborbitals possible within each of the orbital types. Together with the latter two, the magnetic spin quantum number (m_s) determines the maximum number of electrons allowed within a given energy level. This information is summarized in Table 1.3. It is evident that the energy level with the principal quantum number 1 may have only an *s* orbital. The energy level with the major quantum number 2 may have *s* and *p* orbitals, but not *d* and *f*. That with a principal quantum number of 3 may have *s*, *p*, and *d*, but not *f*. Energy levels with major quantum numbers greater than three may have all four types of orbitals.

Before electron arrangements can be achieved two additional factors must be considered. The first, Pauli's exclusion principle, has been implied in the foregoing discussion. In effect it says that in a given atom no two electrons may have all four quantum numbers with identical values. In simpler terms, a given suborbital is fully occupied when one electron has taken over the volume of space associated with it. There is no space for a second electron with the same spin. However, an electron with an opposite spin may enter; but one quantum number, the m_s, is different.

Hund's rules for the order of filling orbitals must also be considered. In a simplified form these rules state: (1) In the ground state an electron may enter only the vacant orbital of least energy. In other words, if an orbital of lower energy is available, the entering electron may not occupy another orbital even though it is vacant. (2) Electrons must enter equivalent suborbitals singly and as far as possible with parallel spins, i.e., $m_s = +\frac{1}{2}$. Only after each of a set of suborbitals is occupied by a single electron, all of parallel spin, may a second electron, of opposite spin, enter one of the suborbitals. For example, with the three 2*p* orbitals, three electrons spinning clockwise must occupy the three possible suborbitals, one to each suborbital, before an electron spinning counterclockwise is allowed.

THE ELECTRONIC CONFIGURATION OF ATOMS

In the ground state of an atom the available electrons occupy the lowest available energy levels. In the excited state one or more electrons have been given additional quanta of energy, allowing them to occupy orbitals in higher energy levels. Only the ground state will be considered at this time. In this state, except for the inert gases, there will be at least one incompletely filled orbital in one or more of the outermost electron shells. These partially filled outermost orbitals, called *valence orbitals*, are responsible for the chemical properties of the element. With only a few

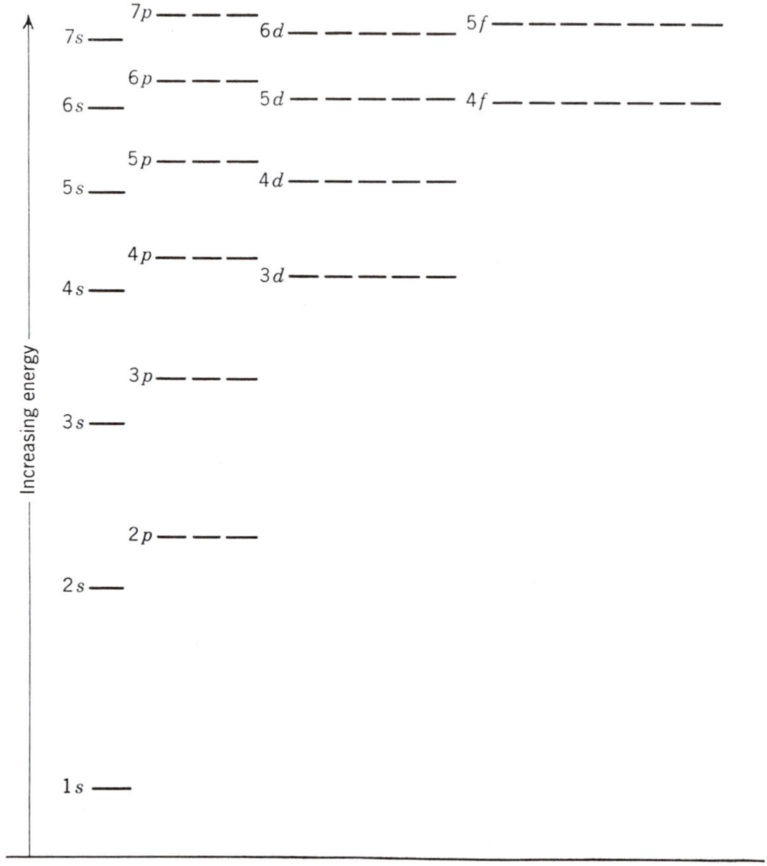

FIGURE 1.3 Basic energy levels.

exceptions the lower, fully occupied orbitals play no role in determining chemical properties. They are tightly localized around the nucleus.

The Aufbau principle. As pointed out earlier, the method by which the requisite number of electrons are fed into the orbitals overlying the nucleus of an atom is known as the Aufbau principle. It is an application of quantum theory, Hund's rules, and the Pauli exclusion principle to the elucidation of electron structures. In the process of building up the electron structures of the successive elements, each additional electron required by the increasing atomic number goes into the lowest energy level available. Except for minor cases (in the transition elements) the ground state distribution of electrons in the atom of a given element is exactly like that of the atom of the element with the preceding atomic number, with the additional electron going into the next lowest energy level. At the same time the nucleus gains enough protons and neutrons to maintain nuclear stability.

The order of accretion of electrons around the nucleus is not strictly in the order of principal quantum numbers. The seeming irregularity in filling arises from the fact that the energies of the different orbitals frequently are nearly the same as one or more orbitals in the overlying quantum groups. The approximate order of these energies in the different electron shells and subshells is shown in Fig. 1.3. However, within a principal quantum level the order of decreasing attraction of the nucleus for the electron is always *s*, *p*, *d*, *f*.

Since there are apparent discontinuities in the order of filling the electron shells, qualitative rules may be formulated as an aid in deriving electron structures. The filling of the first and second principal quantum number electron shells is straightforward. Upon complete filling of the *p* orbitals of the second energy level, the following rules apply in almost all cases. Generally an electron cannot enter a *d* orbital unless the *s* orbital of the overlying quantum group contains its two electrons. After the *s* orbital of the outermost energy level has its two electrons, electrons may enter *p* orbitals of that energy level only if the *d* orbitals of the underlying level have been completely filled. Electrons cannot enter the *f* orbitals of an energy level until one electron has entered a *d* orbital of the overlying energy level; in reality this rule is violated with several lanthanides and actinides. These rules are given pictorially in Fig. 1.4.

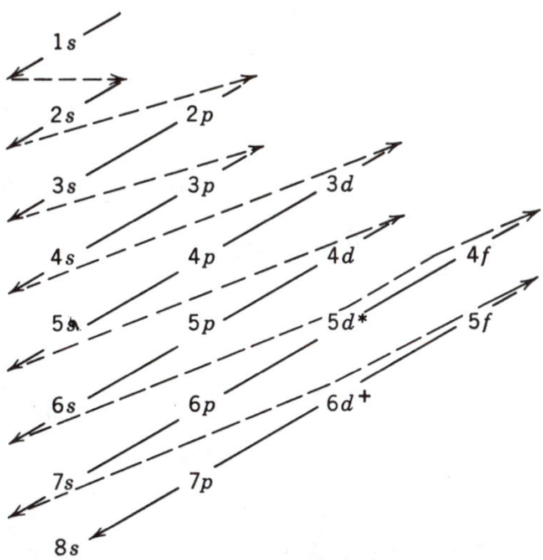

FIGURE 1.4 Order of filling of orbitals. * A single 5*d* is added before the 4*f* orbitals can be filled. + One or more 6*d* electrons must be added before the 5*f* orbitals can be filled.

TABLE 1.4 Electron Structures of Elements with Atomic Numbers 1–54

	H	$1s^1$									
	He	$1s^2$									
Alkali Metals	Li	$1s^2 2s^1$	Na	Ne Core	$3s^1$	K	A Core	$4s^1$	Rb	Kr Core	$5s^1$
Alkaline Earths	Be	$1s^2 2s^2$	Mg	”	$3s^2$	Ca	”	$4s^2$	Sr	”	$5s^2$
	B	$1s^2 2s^2 2p^1$	Al	”	$3s^2 3p^1$	Ga	”	$3d^{10} 4s^2 4p^1$	In	”	$4d^{10} 5s^2 5p^1$
	C	$1s^2 2s^2 2p^2$	Si	”	$3s^2 3p^2$	Ge	”	$3d^{10} 4s^2 4p^2$	Sn	”	$4d^{10} 5s^2 5p^2$
	N	$1s^2 2s^2 2p^3$	P	”	$3s^2 3p^3$	As	”	$3d^{10} 4s^2 4p^3$	Sb	”	$4d^{10} 5s^2 5p^3$
Chalcogens	O	$1s^2 2s^2 2p^4$	S	”	$3s^2 3p^4$	Se	”	$3d^{10} 4s^2 4p^4$	Te	”	$4d^{10} 5s^2 5p^4$
Halogens	F	$1s^2 2s^2 2p^5$	Cl	”	$3s^2 3p^5$	Br	”	$3d^{10} 4s^2 4p^5$	I	”	$4d^{10} 5s^2 5p^5$
Inert Gases	Ne	$1s^2 2s^2 2p^6$	A	”	$3s^2 3p^6$	Kr	”	$3d^{10} 4s^2 4p^6$	Xe	”	$4d^{10} 5s^2 5p^6$
			Sc	A Core	$3d^1 4s^2$	Y	Kr Core	$4d^1 5s^2$			
			Ti	”	$3d^2 4s^2$	Zr	”	$4d^2 5s^2$			
			V	”	$3d^3 4s^2$	Nb	”	$4d^4 5s^1$			
			Cr	”	$3d^5 4s^1$	Mo	”	$4d^5 5s^1$			
First-Row and Second-Row Transition Elements			Mn	”	$3d^5 4s^2$	Tc	”	$4d^6 5s^1$			
			Fe	”	$3d^6 4s^2$	Ru	”	$4d^7 5s^1$			
			Co	”	$3d^7 4s^2$	Rh	”	$4d^8 5s^1$			
			Ni	”	$3d^8 4s^2$	Pd	”	$4d^{10}$			
			Cu	”	$3d^{10} 4s^1$	Ag	”	$4d^{10} 5s^1$			
			Zn	”	$3d^{10} 4s^2$	Cd	”	$4d^{10} 5s^2$			

The methodology used for placing electrons in their proper orbitals may be demonstrated, using as an example the element of atomic number 53—iodine. The atomic number establishes the electron cloud at 53 electrons, each of which is to be placed in its proper orbital. Starting with 1*s*, the lowest orbital, two electrons are assigned to it. The proper number of electrons are assigned to each successive "lowest" energy orbital until the number of unassigned electrons will not completely fill the next orbital. The process is shown in Fig. 1.5. Compare the order for its compliance to the rules of filling given in the preceding paragraph and in Figs. 1.3 and 1.4. Three of the five 5*p* electrons will go, one each, into the three available *p* orbitals. These three electrons will have parallel spins. The remaining two electrons will then pair their spins with two of these three by entering the orbital involved with opposite spins.

The electronic structure for iodine can thus be written

$$1s^2 2s^2 2p^6 3s^2 3p^6 3d^{10} 4s^2 4p^6 4d^{10} 5s^2 5p^5$$

Note that the order is that of the principal quantum number rather than the order of filling. The superscript denotes the number of electrons in the specified orbital. If it is desirable to show the suborbitals occupied by the last five electrons, the following notation is substituted for $5p^5$: $5p_x^2 5p_y^2 5p_z^1$. In this way structures may be written for all known (and even unknown) elements. The structures of several typical elements are shown in Table 1.4. Under some circumstances it may be desirable to show the valence electron spins diagrammatically, since this information is valuable, e.g., in predicting possible hybrid bond types. (See page 75.) One method of representation is given in Fig. 1.6.

Orbital	Electrons admitted	Total number of electrons admitted	Symbolic representation
1*s*	2	2	$1s^2$
2*s*	2	4	$2s^2$
2*p*	6	10	$2p^6$
3*s*	2	12	$3s^2$
3*p*	6	18	$3p^6$
4*s*	2	20	$4s^2$
3*d*	10	30	$3d^{10}$
4*p*	6	36	$4p^6$
5*s*	2	38	$5s^2$
4*d*	10	48	$4d^{10}$
5*p*	5	53	$5p^5$

FIGURE 1.5 Order of feeding electrons into the orbitals of element of atomic number 53.

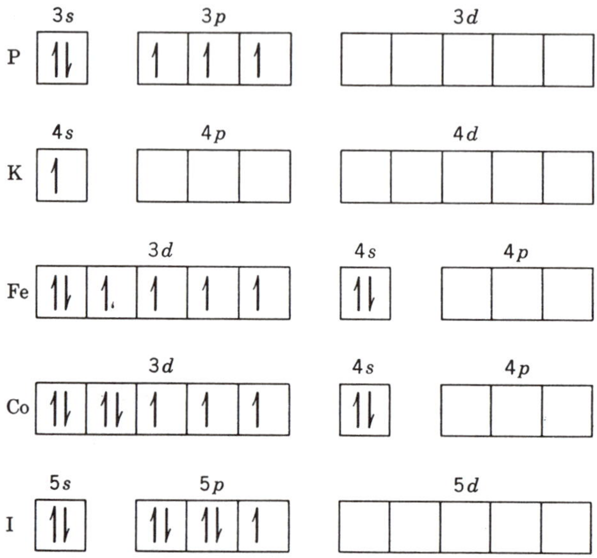

FIGURE 1.6 Method used to indicate valence electron spins.

The presentation of the Aufbau principle frequently gives the impression that the energy of a given orbital is identical in all elements and that the order of energies of the levels existing within a given atom is fixed. Both conclusions are unwarranted. The energies of the same level, even in neighboring elements, are different; i.e., the 4*s* electrons of iron and cobalt have different energies. Even within the atom of a given element the order of energies changes as the result of ionization. In calcium the 3*d* level is obviously of higher energy than the 4*s* level, since the valence electrons occupy the latter. In the next element, scandium, a 3*d* level is occupied by the one additional valence electron. However, upon ionization to the +2 oxidation state, scandium, like calcium, loses its two 4*s* electrons—not the 3*d* and one of the 4*s* electrons, as would be expected from the earlier discussion. Once an atom has achieved its ground state configuration, the energies of the levels do not necessarily remain unchanged. Electrons assume energy values that are consistent with the interactions existing between all electrons and between the electrons and the nucleus. On ionization, it is always the most loosely held electron which is lost first—not necessarily the electron predicted from the Aufbau principle. A few other examples of this change in energy with respect to ionization are:

$$\mathrm{Co}^0,\ 3d^7 4s^2 \rightarrow \mathrm{Co}^+,\ 3d^8 + e^-$$

$$\mathrm{La}^0,\ 5d^1 6s^2 \rightarrow \mathrm{La}^+,\ 5d^2 + e^-$$

Typical valence electron structures. On close examination of the electronic structures shown in Table 1.4, it is evident that the outer, or valence, shells fall into several well-defined types. The *inert gas* type has as an outer shell structure of ns^2np^6, where n represents the principal quantum number of the electron shell involved. This is known as a closed shell and is characterized by unusual stability or resistance to chemical reactions. Among the *typical elements* the *s*-valenced metals have a valence shell structure ns^x ($x = 1$ or 2); the *p*-valenced metals and the nonmetals have a valence shell structure of ns^2np^x ($x = 1, 2, \ldots, 5$). These elements are chemically reactive, with well-defined valence systems. The *transition elements* have the external structure $(n-1)s^2(n-1)p^6(n-1)d^yns^x$ ($y = 1, 2, \ldots, 10$; $x = 0, 1, 2$). These elements are metallic and are characterized by extremely variable valence. The presence of one of these elements in a compound usually results in color. Finally, the *inner transition elements*—lanthanides (rare earths) and actinides, have the shell structure $(n-2)f^z(n-1)s^2(n-1)p^6(n-1)d^yns^x$ ($z = 0, 1, 2, \ldots, 14$; $y = 0, 1$; $x = 0, 1, 2$). Since the electronic differences are deep within the electron cloud, inner transition elements are characterized by a pronounced similarity of chemical and physical properties. The similarities are so close in the rare earth series that until very recently many had been isolated only in very small, impure quantities, if at all.

PERIODIC CLASSIFICATION

Examination of the elements in the order of increasing atomic number reveals a periodic reappearance of specific types of electron configurations. As a result, it would appear that periodic classification is merely the consequence of the elucidation of atomic structures and a convenient method for grouping similar structures. This is not the case, however. The initial stages in the development of the periodic table preceded elucidation of the modern electronic structures of the elements by almost a century.[5] The periodic table was developed through careful observations of the properties of elements and compounds. In fact, periodic classification served as a basis for evaluating electronic configurations and, together with magnetic, x-ray, and spectroscopic data, ultimately supported the electronic structures elucidated.

The arrangement of the elements in the periodic table is governed by the periodic law: The physical and chemical properties of the elements are a periodic function of their atomic numbers. Actually there are several methods by which the elements may be arranged in conformity

[5] "The Prehistory of the Periodic System of Elements," Jan W. van Spronsen, *J. Chem. Educ.*, **36**, 565–567 (1959).

TABLE 1.5 Periodic Table

Orbital Filling	s: Unpaired	s: Paired	d: Unpaired Spins					d: Pairing Spins					p: Unpaired Spins			p: Pairing Spins		
Family	**IA**	**IIA**	**IIIA**	**IVA**	**VA**	**VIA**	**VIIA**	**VIII**			**IB**	**IIB**	**IIIB**	**IVB**	**VB**	**VIB**	**VIIB**	**O**
1	H	He															H	He
2	Li	Be											B	C	N	O	F	Ne
3	Na	Mg											Al	Si	P	S	Cl	A
4	K	Ca	Sc	Ti	V	Cr	Mn	Fe	Co	Ni	Cu	Zn	Ga	Ge	As	Se	Br	Kr
5	Rb	Sr	Y	Zr	Nb	Mo	Tc	Ru	Rh	Pd	Ag	Cd	In	Sn	Sb	Te	I	Xe
6	Cs	Ba	La*	Hf	Ta	W	Re	Os	Ir	Pt	Au	Hg	Tl	Pb	Bi	Po	At	Rn
7	Fr	Ra	Ac*															

Principal Quantum Number–Group Number

Nonmetals

Amphoteric elements

Transition Elements

p-valenced Metals

Alkali Metals

Alkaline Earths

Coinage Metals

Volatile Metals

Chalcogens

Halogens

Inert Gases

*Inner Transition Elements

	Unpaired Spins						**f-filling**	Pairing Spins						
Lanthanides	Ce	Pr	Nd	Pm	Sm	Eu	Gd	Tb	Dy	Ho	Er	Tm	Yb	Lu
Actinides	Th	Pa	U	Np	Pu	Am	Cm	Bk	Cf	Ei	Fm	Mv	No	Lw

with this law, including "short forms," "long forms," "three-dimensional forms," and "spiral forms." Each arrangement has some defect; none has achieved a truly unambiguous charting of the changing electronic patterns resulting with changing atomic numbers.

A modification of the so-called long form, given in Table 1.5, will be used as the basis for further discussions in the text. This form has the advantage over the commonly used short form in that the A and B groups of elements[6] are clearly separated—as they should be on the basis of their observed chemical and physical properties. In addition to the usual indication of electron structures, useful family names have been included in the table.

The usefulness of the periodic table lies in its correlation of chemical and physical data. Within a family of elements, and their corresponding compounds, a gradation of similar properties occurs. This results from the reappearance within these families of valence electron shells differing from one another by only the principal quantum number of the energy levels involved. A re-examination of Table 1.4 will bear this out. The alkaline earth family may be considered as an example. Each element has the ns^2 valence structure overlying an inert gas shell. And each of these metals is a very strong reducing agent. But differences in degree exist due to increasing atomic radius; barium, for example, is a stronger reducing agent than calcium. Because its atomic radius is larger, barium's two valence electrons are further removed from the influence of the nuclear charge and are therefore lost more easily.

It may be said in general that periodic properties, as expressed in the vertical families of the periodic table, are a function of two factors. The first is an electronic factor, which causes similarities in properties due to the common valence electron configuration of the family. Superimposed on this factor is an amplitude factor, which is responsible for the gradation of properties within the family. The amplitude factor is a function of the atomic radius of the member elements, but is also dependent to some extent on the energy level of the closed electron shell underlying the valence electrons. Thus, in the alkaline earth family, the strong reducing properties are a function of the similar valence electron structures,

[6]The student will find that confusion exists in the assignment of "A" or "B" to the respective subgroups. The usage adopted for this textbook is based on increasing atomic number. The A subgroup includes those elements of a long period in which electrons are filling the *s* and their respective underlying *d* orbitals. The B subgroup includes those elements in which two electrons are used for the final closing of the *s* and *d* suborbitals in Groups I and II and the subsequent filling of the related *p* orbitals. By means of this arrangement, the A subgroup elements match the properties of the short-period elements in Groups I and II, whereas the B subgroup elements match the properties of the short-period elements in Groups III through VII.

whereas the difference in reducing power within the series is an amplitude factor, related mainly to atomic radius. That the amplitude factor is affected by the differing closed shells is well illustrated by the differences between beryllium (with a $1s^2$ underlying shell) and magnesium (with a $1s^22s^22p^6$ underlying shell).

In the transition series, and to an even greater extent in the inner transition series, there are horizontal similarities between elements. In their lower oxidation states these elements resemble their right and left neighbors. Palladium, which is not a common element, illustrates the horizontal effect. Its properties are sufficiently similar to those of its right horizontal neighbor, silver, to make it possible to approximate its behavior from a knowledge of the behavior of silver. Horizontal similarities reach their peak in the inner transition elements; it has been difficult to obtain pure samples of some of these because of the complexities of separating their crude minerals, which are mixtures of these elements.

Another valuable relationship existing in the periodic table is that of the *bridge elements*. Bridge elements occur as pairs, consisting of the element in the first short row of the table and the element one space down and one space to the right. In general the first-row element will show many similarities to its bridge in the second row; sometimes its behavior is closer to that of its bridge element than the second member of its own family. Beryllium has chemical properties that follow its bridge, aluminum, much more closely than it follows those of magnesium, the next member of its own family. (There is usually a greater difference than expected between first long-period elements and second long-period elements.)

The following useful generalizations can be made based upon areas within the periodic table. Electronegativity[7] decreases as one moves to the left across the table and as one moves downward in the table. The opposite is of course true with respect to electropositivity. The most electronegative element is fluorine, in the extreme upper right of the table; the most electropositive known element is francium, in the extreme lower left position. Since the nonmetals are the electronegative elements, they are found to the upper and to the right side of the periodic table. A diagonal line drawn from polonium to aluminum, and from hydrogen to

[7]The concept of *electronegativity*, as forwarded by Linus Pauling, is an indication of the relative attraction of an atom for the valence electrons involved in an electron pair bond. He based his evaluations of this property on experimental bond energy data. Other methods of evaluation have also been used. Frequently, reliable experimental data for the calculations have not been available, but despite the resulting lack of exactitude, the scale has been extremely useful because of its consistent correlation with chemical behavior.

beryllium, divides the nonmetals from the metals. Those elements falling along this line are amphoteric, that is, they exhibit both metallic and nonmetallic properties.

The periodic table should be studied carefully with respect to electronic configurations and the deductions which can be drawn concerning properties of elements. It is an invaluable tool in the study of inorganic chemistry, since it provides a systemization not otherwise possible. In fact, it makes study in this field much simpler. The student, instead of learning the properties of every compound he encounters, can first study the group properties and then learn the deviations from group behavior when necessary.

TWO · ATOMIC INTERACTIONS—COMPOUND FORMATION

The atomic nature of chemical reactions has been indicated in the opening paragraphs of Chapter 1. A study of the atom with special emphasis on its electronic structure followed. The relationship between electron structure and atomic radius as well as the electrostatic interactions between differing atoms will now be considered. This will lead into a systematic development of the nature of chemical bonds. Although the initial discussion must be restricted to a pair of reacting atoms, it should not be forgotten that in *molecular* reactions *specific atoms* of the molecules are directly involved.

ELECTRON PAIR BONDS

A molecule may be considered to be two or more relatively fixed point centers of positive charge, embedded in a diffuse electron cloud—that is, the nuclei of the constituent atoms surrounded by the dense, spherical, closed electron shells and the valence electrons. This situation is achieved through chemical bonding. Chemical bonding may be defined as the process by which atoms are consolidated into aggregates whose energy content, as a result of the mutual sharing of valence electrons, is substantially lower than the sum of the energies of the separate atoms. The reduction in energy content is the result of the electrostatic interactions which take place during the time that the unreacted atoms approach one another as a preliminary to bond formation. It must suffice to say that the greater the decrease in the energy content during compound formation, the greater the stability of the resulting molecule. The inertness of the nitrogen molecule, N_2, as contrasted with the very reactive nitrogen atom results from the fact that 225 kcal. are required to restore the energy lost when $2 \times (6.02 \times 10^{23})$ atoms of nitrogen pair to form diatomic nitrogen molecules.

Bond formation process. As two atoms approach one another, three attractive-repulsive interactions come into play: There is a repulsive force between the approaching electron clouds. There is another repulsive force between the approaching nuclei. There is an attractive force between the electron cloud of one atom and the nucleus of the other atom, and vice versa. It is the balance between these electrostatic attractive-repulsive forces that determines whether or not a reaction between the approaching atoms takes place. Gravitational and magnetic effects are too small to have any significance in chemical bond formation.

If one or both of a pair of mutually approaching atoms is an inert gas, a chemical bond cannot form between them;[1] no molecule is created and no chemical reaction can take place. It must therefore be assumed that the electron structure of the inert gases, $1s^2$ for He and ns^2np^6 for the others, has special properties which militate against compound formation. These inert gas valence electron structures, known as closed shells, shield the nucleus from interaction with valence electrons of other atoms. As a result the attractive forces necessary for molecule formation cannot develop. At the same time the valence electrons of the closed shell are so firmly held by the nucleus that they cannot be attracted to other nuclei. The inert gas electron structure is achieved by reacting atoms when forming their most stable compounds. Stable compounds having other electron structures are possible, however.

Revisualize the two atoms approaching one another in space on a collision course and with sufficient kinetic energy to favor the attractive forces rather than the repulsive forces. Assume each atom in itself lacks sufficient electrons to achieve a stable inert gas shell; in other words, one or more orbitals of the valence shell are vacant or singly occupied. As the two atoms approach one another, one or more of the vacant or singly occupied valence shells may eventually overlap, and bond formation will then become possible between the atoms. The process is shown schematically in Fig. 2.1(*a–c*). Contact between bonding orbitals does not assure bond formation. If a bond does materialize, the resultant structure need not retain its original bond form.

For permanent bonding to take place between two atoms, several

[1]Inert gases are not as chemically inactive as was once believed. Very recently xenon tetrafluoride, a stable crystalline solid melting at 90°C., was prepared. Crystalline sodium perxenate has also been prepared. Evidence for existence of different xenon oxyfluorides and corresponding radon compounds has been obtained. Inert gas compounds reported earlier, e.g., hydrates, boron trifluoride addition products, are stable only at very low temperatures.

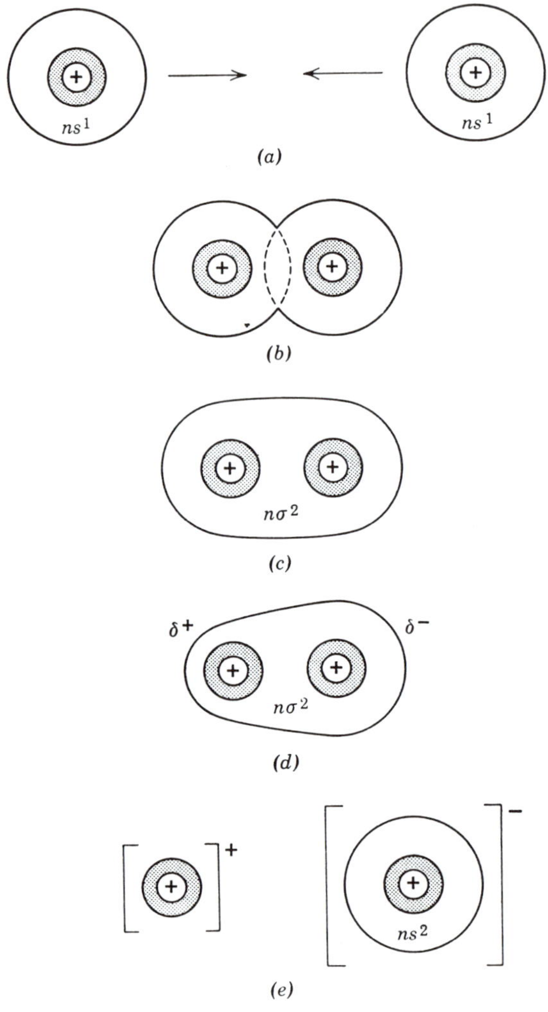

FIGURE 2.1 Electron cloud interactions: (*a*) approach, (*b*) contact and overlap, (*c*) bond formation, (*d*) polarization, (*e*) ionization. The nucleus is represented by the smallest circle with a plus sign. The underlying closed spherical shell is crosshatched.

conditions must be satisfied. Electrons must be available from the reacting atoms to achieve a stable electron pair, i.e., a pair of electrons with opposed spins occupying a single orbital. This pair must be capable of being shared between the two bonded atoms. If the reacting atoms have electronegativities of the same or nearly the same order, the resulting electron pair will be shared equally between the two atoms, and a symmetrical bonding electron cloud results.

Polarization and ion formation. If the atoms involved in an electron pair bond have somewhat different electronegativities, the electron pair will not be shared equally. As a result the density of the bonding electron cloud will shift to favor the more electronegative atomic partner. The compound becomes electrically unsymmetrical. In Fig. 2.1(*d*) the atom on the left has an excess positive nuclear charge with respect to the average electron charge density of its end of the molecule at any given instant. This end of the molecule is therefore positive with respect to the other end, which must be negative since the electron charge density here must be higher than average. The resultant molecule acts as though it has a positive and a negative pole; it is called a *dipole*. A molecule having a distorted bond of this kind is said to be *polar*, in distinction to a *nonpolar* molecule which has a symmetrical disposition of the bonding valence electrons.

In order to distinguish a polar from a nonpolar molecule and to identify the relative polarity of the atoms involved, the concept of *partial charge* has been introduced. The more electronegative combining atom will have a partial negative charge, $\delta-$; the less electronegative (or more electropositive) atom will have a partial positive charge, $\delta+$. The use of these symbols is illustrated in Fig. 2.1(*d*). The over-all charge of a polar molecule remains zero—no electrons or protons are lost during dipole formation.

The process of charge separation can take place to a greater and greater degree, resulting in greater and greater polarization of the molecule. For practical purposes the ultimate result of the increasing charge separation must be a complete shift of the bonding electron pair to the more electronegative of the bonded atoms. This final shift takes place if the electronegativities of the two atoms are extremely different. Under these circumstances a formal electron pair bond no longer can exist; the bonding electron pair is not found in the space between the atoms, but is now associated exclusively with the more electronegative of the original atoms. The atoms no longer have partial charges—rather a complete separation of charges has taken place. Ions result. The atom which has lost its share in the bonding electron pair is now positively charged, a *cation*, and the atom which has gained complete control of the valence electron pair is negatively charged, an *anion*.

The greater the difference in electronegativity, the greater the probability of ion formation. For example, potassium, of low electronegativity, and chlorine, of very high electronegativity, cannot ordinarily form permanently bonded molecules; the valence electron pair goes to the chlorine atom. At the other extreme, carbon and hydrogen have nearly equal electronegativities—they form electron pair bonds with no separation of charges. Their electronegativities do not give either the carbon or the hydrogen an advantage in electron attraction. Carbon and oxygen, on the other hand, have somewhat different electronegativities. Therefore, although they form electron-pair bonds, oxygen, due to its greater electronegativity, distorts the electron cloud in its favor. The oxygen will have a partial negative charge, and carbon consequently has a partial positive charge. A study of the table of electronegativities given in Table 2.1 should be

TABLE 2.1 ORDER OF ELECTRONEGATIVITY OF COMMON ELEMENTS

Most electropositive →	Cs, Rb, K, Na, Li, Ba,	
	Sr, Ca, Mg, Be, Al, Cd,	
	Si, B, Zn, In, Hg, Tl,	
	Pb, Sn, Bi, P, Sb, H,	
	Ge, Te, C, I, As, S,	
	Se, N, Br, Cl, O, F	← Most electronegative

useful in predicting bond behavior between the different atoms.[2]

Bond types. The student should recognize the two fundamental types of chemical bonding. The *covalent bond* is the type in which the valence electrons are shared, and gives rise to *covalent compounds.* In *ionic bonding*, the complete separation into independent ions results in an *electrostatic compound*, there is no formal bond between ions. There are, however, electrostatic interactions between the constituent ions which modify their independent behavior in solutions and which are responsible for crystal formation in the solid state.

Elementary presentation of bonding frequently results in erroneous impressions about bonding and the resulting chemical compounds. Two unfortunate misconceptions are likely to arise. One, that since the carbon compounds studied in elementary organic chemistry are almost always covalent, all organic compounds are therefore covalent—an incorrect conclusion. And, since inorganic compounds studied in general chemistry are almost all electrovalent compounds, all inorganic compounds are therefore ionic compounds—also an incorrect conclusion. Covalent and ionic compounds are found in both organic and inorganic chemistry.

Another misconception resulting from elementary presentations is the belief that the terms covalent and ionic will categorize all compounds. Nothing is further from the truth. Covalent and ionic represent the

[2]For electronegativities of all the elements see "A complete table of electronegativities," E. J. Little, Jr., and M. M. Jones, *J. Chem. Educ.*, **37,** 231–233 (1960).

ideal or extreme cases of bonding. A few compounds do approach the ideal. For example, $Na^{+}Br^{-}$ approaches the ionic ideal closely, whereas CH_4 approaches the covalent ideal. Most compounds fall between these limiting types. The salt $Na^{+}Cl^{-}$, generally considered ionic, actually has 30% covalent character. This may be interpreted to mean that at any instant thirty per cent of the bonding electrons are not formally associated with the chloride ions, as they should be in the pure ionic state, but are found in the free space between the sodium and chloride ions. The Si—O bonds of the common silicates have only 51% ionic character. Hydrogen fluoride has 57% covalent character. Custom allows classification of a given substance according to the bond type it most closely approximates.

Since the properties and chemical behavior of covalent molecules and ionic compounds differ markedly, a study must be made of the conditions for their formation. Even though the two kinds of substances must be considered separately, it should be borne in mind that most substances represent some degree of admixture of both. Ion formation will be considered first.

IONS AND THEIR PROPERTIES

Requirements for ion formation. In any consideration of why two atoms react to form stable ionic compounds, the nature both of the reacting atoms and of the resultant ions requires attention. The electron structure, electronegativity, and atomic radius must be considered for each reacting atom. Similarly, electronic structure, ionic radius, and charge must be considered for the resultant ions.

Fajan's rules relate the size of an atom to its ability to form an ion. They may be qualitatively described as follows. For a small atom the associated positive electrostatic field, having its center in the nucleus, will be strong (see Fig. 3.1). For this reason the valence electrons are tightly bound in closely localized orbitals. As a result a small atom has a high *ionization potential*, a quantitative measure of the ease with which an electron may be removed from an atom. At the same time the small atom has a high *electron affinity*, a quantitative measure of the attraction an atom has for additional electrons. Both of these quantities, particularly the latter, are directly related to a determination of electronegativities. Thus a small atom has a high electronegativity.

Because of its high electron affinity, a small atom will attract additional electrons into vacant or singly occupied valence orbitals (other conditions being favorable). Because of its high ionization potential, it cannot lose its electrons. Thus a small atom will form an anion rather than a cation.

At the other extreme are the large atoms, with their weakly bound valence electrons in rather diffuse orbitals. Their positive electrostatic field, originating in the nucleus, is weak, since field strength varies inversely with the distance from the nucleus. The valence electrons, being found in a relatively weak electrostatic field, cannot be held effectively. Therefore the ionization potential of large atoms is small. Since the positive field is weak, large atoms have little or no attraction for additional electrons. Thus their electron affinity is also small. The net result is a tendency for large atoms to lose their valence electrons; they tend to exist as cations.

The electronegativity of the reacting atoms may also be considered. It has been previously pointed out that a rather large difference in electronegativity must exist if two atoms are to form ions. From the foregoing discussion it is evident that the more electronegative member of the pair will assume the anionic role.

The nature of the valence shells of the reacting atoms and of the resultant ions determines whether or not stable ions may form. In general, a donor-acceptor relationship involving one or two electrons must develop between the two reacting atoms. This fact imposes a condition of low net charges on either of the ions involved in electrostatic compounds. Ions of high charge, regardless of sign, cannot exist. High positive and high negative oxidation numbers of elements (e.g., Mn^{4+}, N^{3-}) must not be confused with ionic valence.

Anions. Only those nonmetals with an electron structure lacking one, and in a few cases two, electrons of a completed ns^2np^6 shell can form anions. The sole exception is hydrogen, which accepts a single electron to achieve the $1s^2$ helium shell in becoming the hydride ion. Since the number of protons is less than the number of electrons, an anion always has a net negative charge. Because the positive field of the nucleus must now hold an additional electron or two, its effect on any one electron is weakened, with the result that the anionic electron cloud spreads outward and becomes more diffuse. This is reflected in the fact that the anionic radius of an element is always greater than its atomic radius.

The reason for the paucity of divalent anions and the nonexistence of true trivalent anions should be quite evident. The ns^2np^4 atom must spread its positive field very thin if it is to hold the two additional electrons acquired to achieve a ns^2np^6 shell. Only in oxygen and sulfur, the smallest elements of the ns^2np^4 structure, does the field remain sufficiently strong to hold two additional electrons. With larger atoms the greater volume through which the field must act weakens it to such an extent at any one point that a second electron cannot be attracted and held permanently to form a divalent anion.

Cations. Stable cations are formed, having several different types of electron structures. The most stable configuration is the inert gas structure ns^2np^6, the $8e^-$ shell. The expanded electron structure $ns^2np^6nd^{10}$, known as the $18e^-$ shell, also has a certain amount of stability, depending on the valence of the ion. Because of its similarity to the $1s^2$ structure of helium, the ns^2 structure is a stable cation configuration. The helium structure is actually attained by the first-row cations Li^+ and Be^{2+}. Many of the typical elements can successfully achieve the ns^2 cationic structure by partial loss of their valence electrons (e.g., Sn^{2+}, Bi^{3+}). Finally, the transition elements form stable cations, with their d electron shell partially filled. A few typical valence shell changes taking place during ion formation are shown in Table 2.2.

TABLE 2.2 Typical Valence Electron Shell Transformations During Ion Formation

Ground State Element		Ion Formed
F $(1s^22s^22p^5) + e^-$	=	F^- $(1s^22s^22p^6)$
Li $(1s^22s^1)$	=	Li^+ $(1s^2) + e^-$
Mg $(1s^22s^22p^63s^2)$	=	Mg^{2+} $(1s^22s^22p^6) + 2e^-$
Sn (Kr core $4d^{10}5s^25p^2$)	=	Sn^{2+} (Kr core $4d^{10}5s^2$) $+ 2e^-$
Zn (Ne core $3s^23p^63d^{10}4s^2$)	=	Zn^{2+} (Ne core $3s^23p^63d^{10}$) $+ 2e^-$
Mn (A core $3d^54s^2$)	=	Mn^{2+} (A core $3d^5$) $+ 2e^-$

Since electrons have been lost and the number of protons remains unchanged, a cation always has a net positive charge. When an electron has been lost, the net positive nuclear field available to hold the remaining electrons is effectively increased. Consequently the electrons are drawn closer to the nucleus, and a denser electron cloud results. For this reason the cationic radius is always less than the atomic radius of the parent element. It also follows that, since the remaining electrons are more effectively under the influence of the positive field of the cation, each successive removal of an electron becomes increasingly difficult. Thus the common cationic valences are limited to one and two. Because of the inverse relation between the strength and the distance of an electrostatic field, only a few of the very largest atoms can lose three electrons to become true trivalent cations.

Typical electron structures and their stability. The different ionic valence electron structures, together with typical examples, are shown in Table 2.3. A scaled representation of isoelectronic structures, showing the relationship between radius and charge, is shown in Fig. 2.2

Ion stability and ability to enter into additional reactions without change in oxidation number are closely related to the type of shell achieved

TABLE 2.3 Characteristic Electron Structures of Ions

Shell Type	Examples	Valence Electron Configuration	Inert Gas Shell Achieved
Zero e^-	H^+	—	—
$2e^-$ shell	Li^+, Be^{2+}, H^-	$1s^2$	He
Inert pair	In^+, Sn^{2+}, Sb^{3+}	$4d^{10}5s^2$	He
	Tl^+, Pb^{2+}, Bi^{3+}	$5d^{10}6s^2$	He
$8e^-$ shell	Na^+, Mg^{2+}, Al^{3+}	$2s^22p^6$	Ne
	K^+, Ca^{2+}, Sc^{3+}	$3s^23p^6$	A
	O^{2-}, F^-	$2s^22p^6$	Ne
	S^{2-}, Cl^-	$3s^23p^6$	A
$18e^-$ shell	Cu^+, Zn^{2+}, Ga^{3+}	$3s^23p^63d^{10}$	—
	Ag^+, Cd^{2+}, In^{3+}	$4s^24p^64d^{10}$	—
Incomplete shell	Mn^{2+}, Fe^{3+}, Fe^{2+}, Co^{2+}	$3d^5$, $3d^5$, $3d^6$, $3d^7$	—

(and ionic radius). The inert gas ($8e^-$) shell ions are most stable. These cations show little tendency to share their valence shell electrons; complexation is rare. On the other hand, many anions of this type undergo further reaction by sharing their valence shell electron pairs under certain conditions. This will be discussed at a more appropriate time.

The $18e^-$ shell ranks next in stability, but its stability is a function of the valence of the cation involved. A divalent cation of the type Zn^{2+} is as stable as an inert gas cation. But the shell tends to be unstable as a univalent cation, e.g., Cu^+. Being more diffuse than the inert gas structure, the $18e^-$ shell is more easily polarized. This accounts for the fact that these cations will readily act as centers for complexes.

The transition element cation, with its incompletely filled d orbital as the external shell, is least stable. It can undergo many different reactions. Most frequently these cations attempt to achieve closed electron shells by complexation; these complexes are, as a rule, very stable.

The electrostatic forces surrounding an ion are nondirectional. The

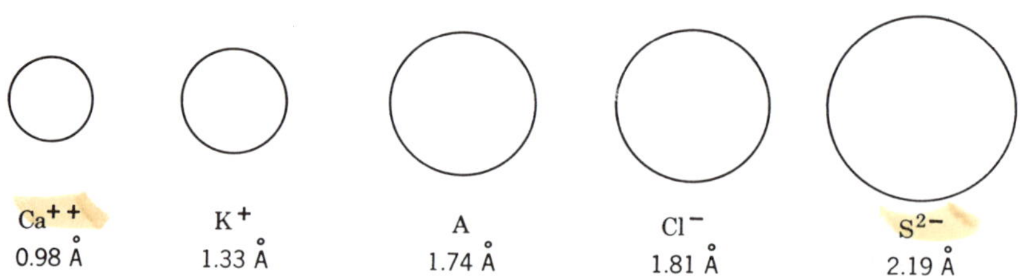

FIGURE 2.2 Relationship between size and charge for isoelectronic particles with the argon electron shell type. Radius values are given in angstrom units.

generally uniform electrostatic field, whether positive or negative, surrounds the ion in all directions in space. The field diminishes to the same degree, regardless of direction, as one moves out from the center of the ion.

In ionic compounds, the independence of the individual ions is generally maintained. There is evidence for the fleeting existence of ion-pair molecules only in the gaseous state. The independent ions exist in most solutions and in the liquid and solid state. In fact, in the solid state electrovalent compounds are closely packed arrangements of the ions, held together by their mutual electrostatic interaction.

Some misconceptions about the existence of certain highly charged cations frequently occur. Although it is usually accepted as such, iron(III) is incapable of independent existence in aqueous solution. It invariably complexes with either the solvent or another ligand present in the solution. This behavior is true of many other highly charged cations. A special form of interaction enabling highly charged cations to exist as free cations of moderate charge is illustrated by bismuth(III). It associates with an oxo group to become the bismuthyl ion $[BiO]^+$, in which form it can exist in solution. However, in the solid state discrete Bi^{+3} and O^{-2} ions exist together with a charge compensating anion such as Cl^-. Each of the ions occupies an independent position in the structure.

COVALENT BONDS

In forming covalent compounds, each atom in the molecule tends to assume a closed shell electron structure. A covalent bond results when two electrons are paired (opposed spins) and the pair is localized in the space between the two bonded atoms. Some of the factors in the formation of these bonds, and their properties will be discussed next.

σ and π bonds. Covalent bonds may be distinguished by the kind of atomic orbitals used in forming the bonds or by the type of resultant molecular orbitals. Electrons from *s* and *p* atomic orbitals most frequently form covalent bonds. As shown in Fig. 2.3(a), the simplest situation arises when a pair of *s* orbitals, each occupied by a single electron, overlap. The overlap necessary for bonding can also develop with *p* orbitals: either the head-on overlap of singly occupied *p* lobes from two atoms, as shown in Figure 2.3(b), or the overlap of an *s* and a *p* orbital. The resulting molecular orbital is known as a σ molecular orbital or a σ bond. The σ bond is the classical single bond of the organic chemist.

In addition to the head-on overlap giving a σ bond, singly occupied *p* lobes may overlap side-by-side, as shown in Fig. 2.3(c). This can only take place, however, if a σ bond already exists between the two atoms involved. This type of overlap results in the so-called π molecular orbital or π bond. A σ bond associated with a π bond is commonly termed a double bond. Indeed, if the two σ-bonded atoms each have two pairs of

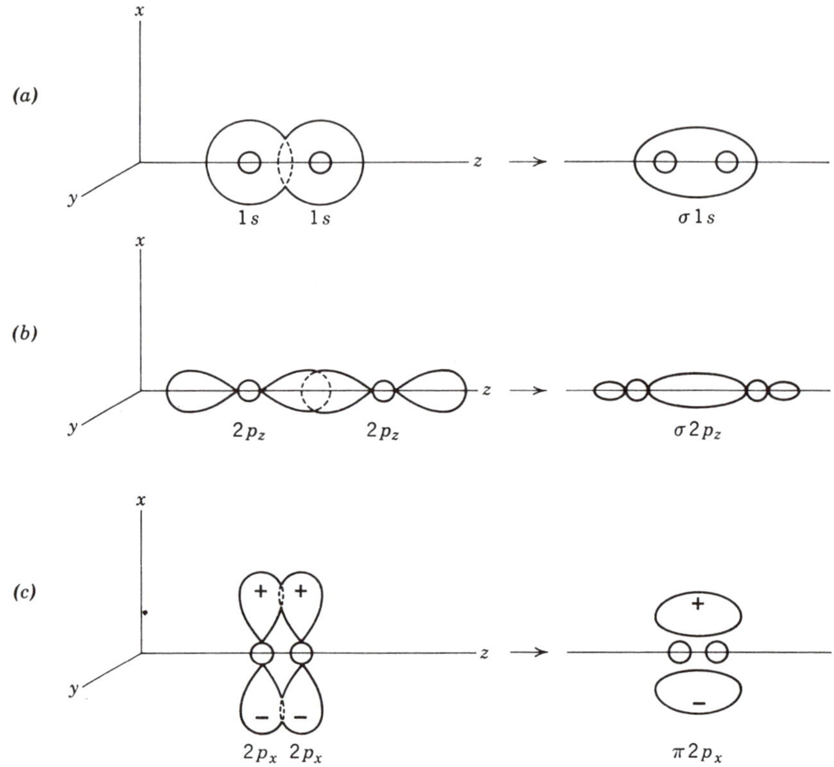

FIGURE 2.3 Formation of molecular orbitals. The representation is in cross section of the xz plane through the line of the centers of the reacting nuclei. The σ molecular bond of the π bonding illustration is not shown.

properly orientated singly occupied p lobes, they may overlap in four separate zones. Two π bonds now form, in addition to the σ bond already present. This is the classical triple bond of such substances as the nitrogen molecule, the cyanide ion, and carbon monoxide. Using molecular orbital methods, the two separate π bonds do not exist independently. Rather, the two electron clouds fuse into a single cloud of higher electron density.

Molecular bonding of the π type is restricted for the most part to elements in the first row of the periodic table. The tendency is by far strongest in carbon, located at the center of the first row. Interatomic distances shrink appreciably to achieve the overlap needed to form π bonds between atoms. Because the larger atomic radius of the heavier elements prevents sufficient overlap, these elements do not form π bonds. Three diagrammatic structures illustrating σ and π bonding in diatomic molecules are shown in Fig. 2.4.

The distance between two bonded atoms is characteristic of the atoms involved and the type of bond between them. This property of molecular bonds is known as *bond distance*, or in the case of the individual atoms involved as *covalent radius*. Bond distance is the sum of the covalent radii of the bonding partners. The covalent radius of an atom remains reasonably constant with various bonding partners, although there is a slight variation depending on spatial configuration (e.g., tetrahedral). However, π bonding, always results in shortening; the greater the multiplicity of the bond, the greater the shortening. The double-bond radius of an element is about 13% less than the normal single-bond

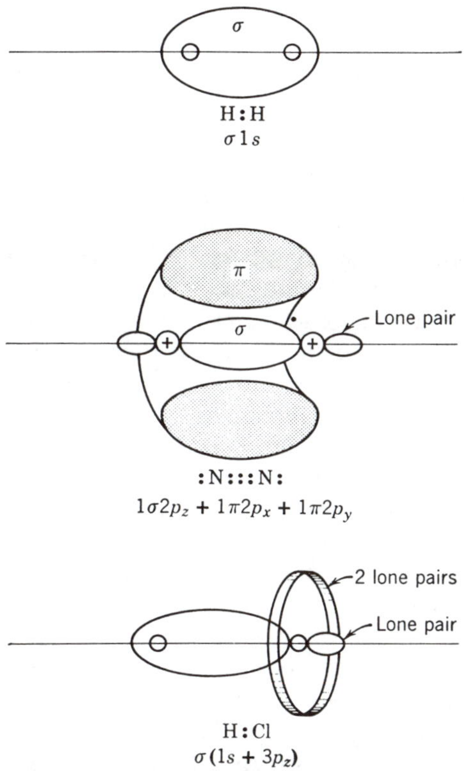

FIGURE 2.4 Schematics of molecules representing different types of molecular bonds. The π molecular orbital of nitrogen results from two pairs of equivalent $\pi 2p$ molecular orbitals. They differ by rotation through 90°, but when put together their density contributions are axially symmetrical.

radius. The triple-bond radius is about 22% less. The comparative values of the radius of the carbon atom are 0.60, 0.67, and 0.77 Å in acetylene, ethylene, and diamond, respectively. Some representative covalent radii are given in Table 2.4.

TABLE 2.4 REPRESENTATIVE COVALENT RADII

Group	Element and Radius[a]									
I	H	0.37								
III	B	0.88	Al	1.26	Ga	1.26	In	1.44		
IV	C	0.77	Si	1.17	Ge	1.22	Sn	1.41	Pb	1.46
V	N	0.74	P	1.10	As	1.21	Sb	1.38	Bi	1.51
VI	O	0.73	S	1.02	Se	1.17	Te	1.35	Po	(1.53)
VII	F	0.72	Cl	0.99	Br	1.14	I	1.33	—	—

[a]Radii are given in angstrom units.

The molecular orbitals discussed in the previous paragraphs are known as *bonding molecular orbitals*. Each bonding orbital has a so-called *antibonding* partner. The improbability of forming a helium molecule by moving two helium atoms together in such a way as to bring about overlap of their filled *s* subshells illustrates the nature of the antibonding partner. To provide space for a bonding electron pair, one electron of each helium atom must seek a higher energy level (Pauli exclusion principle). Since the energy required to promote a 1*s* electron to the 2*s* state is large, the promotion does not occur under ordinary circumstances. Rather, the repulsive forces dominate, thereby preventing an electron pair from occupying what would ordinarily be the σ bonding volume.

In addition to the electrons used in forming bonds to other atoms, many atoms have electrons already paired but not used in bond formation. The orbitals occupied by these electron pairs are known as *nonbonding* orbitals.

Hybrid bonds. On a theoretical basis many known compounds would be impossible, since bonding theory requires the formation of electron pair bonds by overlap of singly occupied atomic orbitals. Carbon has the structure $1s^2 2s^2 2p_x^1 2p_y^1$. Its 2*s* shell, being filled, is theoretically unavailable for molecular bond formation (anti-bonding electrons). Carbon should be able to form two bonds only, using each of its two singly occupied *p* orbitals. Similarly, boron with the structure $1s^2 2s^2 2p^1$ and aluminum with the valence structure $3s^2 3p^1$ should be monovalent—not trivalent—a single 2*p* or 3*p* orbital being theoretically available for bond formation in each case. Beryllium with the structure $1s^2 2s^2$ should be inert, not divalent, since it has no partially filled orbitals for bond formation.

The apparent inconsistency in orbital reaction theory may be explained by the phenomenon of *hybridization*. In many elements the paired *s* electrons of the valence shell are split, and one electron is elevated to an unoccupied *p* orbital. So-called *sp* hybrids result. Hybridization occurs most readily when the energy of the newly occupied *p* orbital is not much higher than that of the *s* orbital vacated. For the elements named above: beryllium forms two *sp* hybrids, boron and aluminum form three sp^2 hybrids, and carbon forms four sp^3 hybrids. The superscript indicates

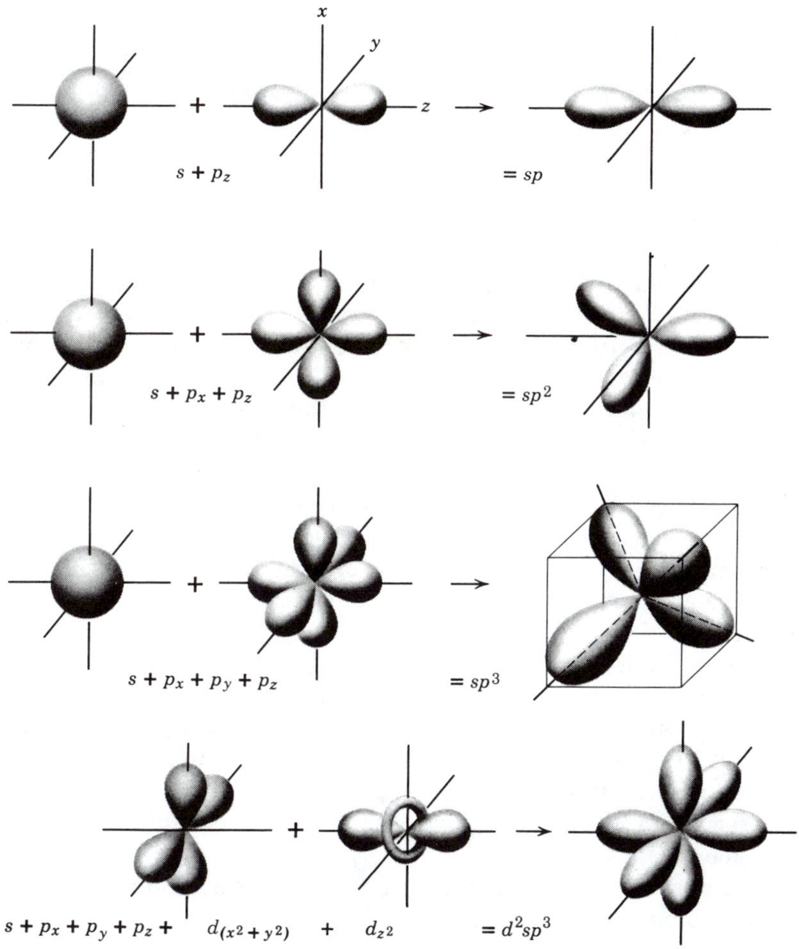

FIGURE 2.5 Formation of four hybrid bonds.

the number of the particular atomic orbital used for a given hybrid bond, e.g., one *s* and three *p* in sp^3. Except for hydrogen, helium, and the elements of the first row of the periodic table, unoccupied *d* orbitals are also available for hybridization. However, metallic properties become more predominant, especially in the long periods. The lower electron affinities and ionization potentials of these elements favor ion formation over fixed molecular electron pair bonds between atoms. The four most common hybrids are developed in Fig. 2.5.

Hybridization is important in the reactions of many atoms. Compound formation is favored by the fact that orbital overlap may be increased by hybridization. The phenomena is in accord with the fact that electrons repel one another and stay as far apart as possible, particularly if each has the same spin. It accounts for the observed molecular geometry, which is frequently different from that predictable from the disposition in space of the bonding suborbitals.

In summary, hybridization is the process by which an electron pair in the ground state of an atom is split and one of the electrons is elevated to an unoccupied higher energy orbital in order that additional electron pair bonds may be formed with other atoms.

BOND ORIENTATION AND BOND ANGLES: In order properly to complete a discussion of bond hybrids, another characteristic property of the covalent bond must be reviewed. Except for bonds involving *s* atomic orbitals, molecular bonds are directional. This fact imposes directional characteristics on covalent bonds between three or more atoms. Within a given molecule the orientation of the individual atoms with respect to one another is indicated by the *bond angle*. This is the angle between a pair of lines drawn from the nuclei of two atoms in a compound through the nucleus of a commonly bonded third atom. Some ideal bond angles are shown for various configurations in Fig. 2.6 and Table 2.5. These angles are closely approached in most appropriate compounds.

When an atom uses only pure orbitals (no hybridization), only three directional characteristics are possible. Pure *s* bonds, since they occur only singly, have no directional characteristics. Since *p* orbitals lie at right angles to one another (except for singly bonded atoms), pure *p* bonds form 90° angles. Pure *d* bonds are extremely rare. They appear to lie along the slant edges of a pentagonal pyramid, the central bonding atom at the apex.

Numerous directional possibilities exist for hybrid bond orbitals using various combinations of *s*, *p*, and *d* orbitals. The only requirement is that the orbital be allowed by the principal quantum number; it need not be occupied in the ground state of the atom. The configuration and bond angle of several typical hybrids (both compounds and ions) are given in Table 2.5 and diagrammed in Fig. 2.6.

MAXIMUM COORDINATION NUMBER: The number of possible bonds which a given atom may form is dependent upon the different kinds of atomic orbitals available, as determined by the principal quantum number, and upon the space around the atom, as determined by its size. The first-row elements are restricted to the maximum number of bonds allowed by their *s* and *p* orbitals, *d* orbitals being forbidden since $n = 2$. In the second-row elements *d* orbitals are allowed ($n = 3$) and become available for bond formation. However, the amount of space around these relatively small atoms is usually not sufficient to allow full utilization of the *d* orbitals. The maximum number of molecular bonding orbitals permitted for a given atom is known as the *maximum coordination number*. Table 2.6 summarizes this information for the elements.

A geometric factor also figures in the arrangement of a given hybrid

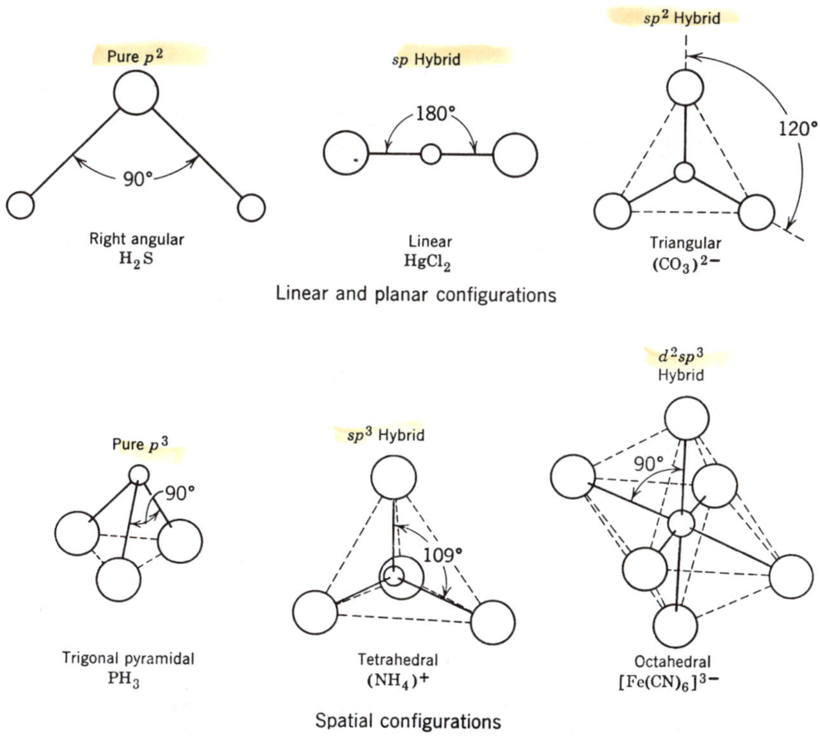

FIGURE 2.6 Directional configuration of pure and hybrid molecular bonds. Atomic sizes are relative only, and interatomic distances have been exaggerated in order to emphasize bond angles and geometric relationships.

TABLE 2.5 COVALENT BOND ORIENTATIONS

Covalence of central atom	Bond Type	Bond Angle	Spatial Configuration	Examples[a]
1	*s*(*p*)	—	linear	F_2, H_2, Li_2
2	p^2	90°	angular	$H_2\mathbf{S}$
3	p^3	90°	trigonal pyramid	$\mathbf{P}H_3$, $\mathbf{Sb}H_3$
2	*sp*(*dp*)	180°	linear	$[\mathbf{Ag}(CN_2)]^-$, $\mathbf{Hg}Cl_2$, $\mathbf{Be}Cl_2$, I_3^-, $[ICl_2]^-$
3	sp^2	120°	triangular	$\mathbf{B}Cl_3$, $[CO_3]^{2-}$, $[NO_3]^-$
4	sp^3	109°	tetrahedral	CCl_4, $[\mathbf{N}H_4]^+$, $[\mathbf{B}F_4]^-$, $[\mathbf{Be}F_4]^{2-}$, $[PO_4]^{3-}$
4	dsp^2	90°	square	$[\mathbf{Ni}(CN)_4]^{2-}$, $[ICl_4]^-$
5	dsp^3	90°, 120°	trigonal bipyramidal	$\mathbf{Fe}(CO)_5$
6	d^2sp^3	90°	octahedral	$\mathbf{S}F_6$, $[\mathbf{Si}F_6]^{2-}$

[a] The central atom of each example is shown in boldface type.

TABLE 2.6 MAXIMUM COORDINATION NUMBERS

Atomic Number	Elements	Maximum Coordination Number	Maximum Number of Shared Electron Pairs
1	H	1	1
3–9	Li–F	4	4
11–35	Na–Br	6	6
37–	Rb–	8	8

bond orbital. In general, the resultant hybrid orbitals arrange themselves to achieve maximum interorbital distances. For an atom with two *sp* hybrids a straight line provides maximum separation; for an atom with four sp^3 hybrids the tetrahedral arrangement provides maximum separation of the bonded groups. Distortion of the usual arrangements of a given bond type takes place under certain circumstances. The most common distortion is that caused by a nonbonding orbital occupied by an unshared pair. It appears that an unshared electron pair requires a greater volume of space for its orbital than a shared pair. To put it another way, the repulsive electrostatic forces of the nonbonding electron pair force the bonding pairs away from it. This is best illustrated by the sp^3 hybrid hydrides of the first-row nonmetals. The bond angles and relative configuration of the series is shown in Fig. 2.7. This series hypothetically results when successive protons of the methane molecule fall into and become a part of the nucleus of the central atom.

Coordinate covalent bonds. In covalently bonded compounds each of the two bonding atoms will generally share any remaining unpaired electrons in order to form bonds with still other partners. This process continues until each atom has shared all available unpaired electrons of its valence shell. Unlike electrostatic reactions, which require completion of a closed shell of 2, 8, or 18 electrons (except for certain transition metal cations), a covalently bound atom can achieve stability in a molecule without completely filling its valence shell; the element boron, in the boron trifluoride molecule, is an excellent example. The boron atom forms three electron pair bonds (using sp^2 hybrids) by sharing its three valence electrons, one with each of three different fluorine atoms. It thus achieves a stable valence shell consisting of only three electron pairs.[3]

[3]Molecules of atoms with incompletely filled valence shells, such as the boron atom in the boron trifluoride molecule, have interesting properties which are a function of the unused valence orbitals present. These properties will be discussed later. (See page 483.)

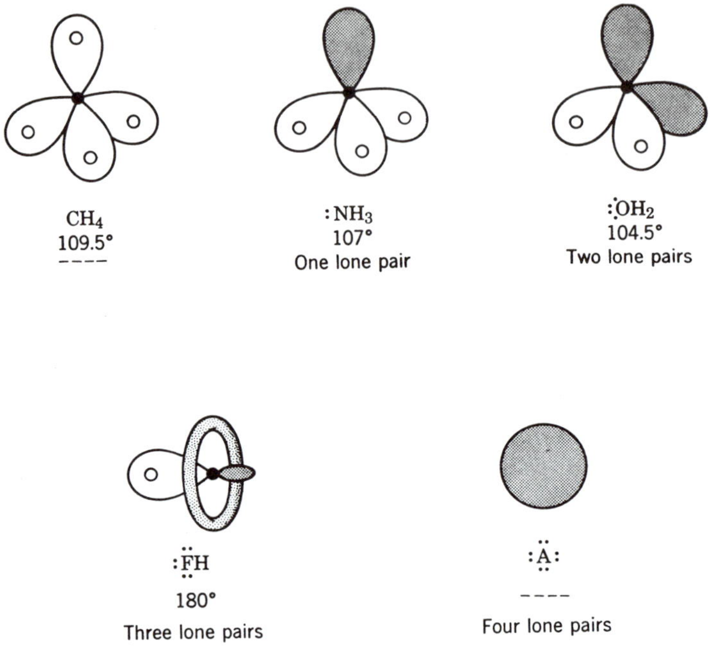

FIGURE 2.7 Distortion of normal bonding angle by nonbonding electron pairs. The illustrations use the sp^3 hybrids in an isoelectronic series. Nonbonding pairs are crosshatched. A closed valence shell tends to approach a spherical shape.

(Each fluorine atom at the same time has completed its fourth pair, giving it a stable inert gas octet.)

An atom such as boron in the compound just described seeks to complete its valence shell. Boron can accept an additional pair of electrons from a donor if proper conditions are attained, in this case by placing boron trifluoride in the presence of any source of fluoride (e.g., sodium fluoride, hydrofluoric acid). A free fluoride ion (but not an atom) has four nonbonding electron pairs in its completed valence shell; it can share at least one of these pairs. Boron, in the trifluoride, lacking one electron pair of a closed shell, can accept one of the fluoride ion electron pairs to achieve a fourth σ bond. Boron physically contributes nothing to the resultant bond—except the use of its unfilled orbital; the fluoride ion contributes both electrons. In forming the three original boron-fluorine bonds, each atom contributed an electron to the bond. Thus the distinguishing feature of this fourth σ bond is in method of formation; the bond is in other respects identical to the other three.

In order to distinguish this new donor-acceptor relationship—a covalent bond in which one atom is solely an electron pair acceptor and the other atom strictly an electron pair donor—the term *coordinate covalent bond* has been introduced. The distinction between covalent bonds and coordinate covalent bonds is in the source of the bonding electrons only. Once the bond forms, it is impossible to distinguish between the two types of bonds if both are present in a single molecule. It must be emphasized, however, that coordinate covalent bonds cannot form until all possible ordinary covalent bonds have formed.

As a result of the reaction of boron trifluoride with the fluoride ion, discussed above, the boron appears in a new structural species—the tetrafluoborate ion. (In actual usage the tetra may be dropped.) It is interesting to observe that in accepting the fourth electron pair, the triangular sp^2 configuration of the boron atom changes to the tetrahedral sp^3 configuration. Figure 2.8 summarizes the reactions described in this discussion.

Bond strength. One topic should be considered briefly: the comparative stability of covalent bonds. Bond strengths are closely related to bond energies, tables of which are found in most modern textbooks of inorganic chemistry. For simple diatomic molecules, especially in the gaseous state, bond energy is easily determined experimentally. It is simply the energy necessary to cleave the molecule into its two constituent atoms. In polyatomic molecules, however, the meaning of bond energy may be ambiguous. Since there are numerous bonds in a complex molecule, it is difficult to distinguish the individual effect of each bond on the over-all stability of the molecule. For this reason average bond

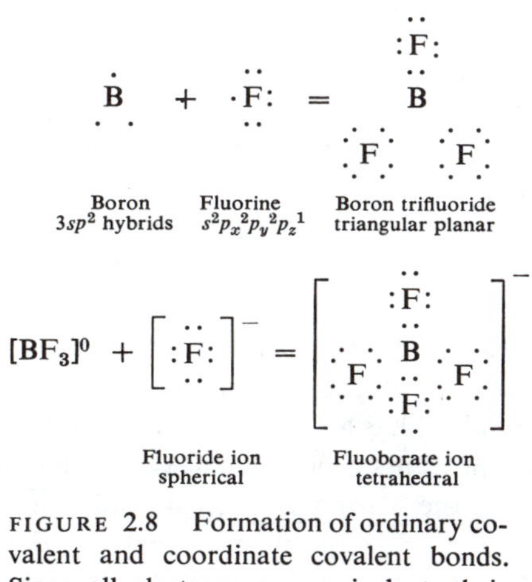

FIGURE 2.8 Formation of ordinary covalent and coordinate covalent bonds. Since all electrons are equivalent, their source has not been indicated.

energies are usually reported for polyatomic molecules. This is the energy required to break all bonds divided by the number of bonds present in the molecule. Sometimes one atom, or group of atoms, separates more easily from the molecule than others. Specific bond energies for such reactions may be reported, e.g., heat of ionization of the hydrogen ion of a monoprotic acid, heat of hydration.

Since bond energies have limited general application, it is of greater value to examine qualitatively the factors important in determining bond strengths. There are three principal factors: size of bonded atoms, kind of orbital, and polarity of adjacent atoms.

In general, size affects bond strength in two ways. First, the smaller and denser the volume occupied by the bonding pair, the stronger the resultant covalent bond. Small atoms favor dense electron clouds due to the stronger electrostatic fields from the closely neighboring nuclei of the bonded atoms. This effect is shown by comparing experimental dissociation energies of gaseous diatomic molecules of a given family of elements. In Group VB dissociation energies of N_2, P_2, As_2, Sb_2, Bi_2, are 225, 117, 92, 70, and 30 kcal./mole, respectively.

Secondly, size within a given family of elements is a function of the principal quantum number of the individual members. When unlike atoms are bonded together, the greater the difference in the principal quantum numbers of the valence shells involved, the less stable the bond. The oxygen family hydrides well illustrate this effect. Water is extremely stable, whereas hydrogen telluride is extremely unstable.

Orbital type is, of course, related directly to the subquantum numbers allowed within the valence energy level. Bond strength in general is a function of the amount of overlap possible between bonding atomic orbitals. Familiarity with orbital shapes and geometry indicates that bonds utilizing *s* orbitals will be weakest. Bonds utilizing *p* orbitals will be stronger since greater overlap is possible. Hybrids tend to give the strongest bonds.

The polarity of adjacent atoms in a molecule gives a measure of the distribution of electrons within the molecular bond volume. Therefore any factor which favors greater concentration of the bonding electrons within the bond envelop will increase bond strength, whereas factors which allow the distribution to become more diffuse will weaken or cause destruction of the bond. It has already been seen how an extreme difference in electronegativities prevents a permanent molecular bond from forming between two atoms: the strongly electronegative member of the pair achieves complete or nearly complete control of the bonding electron pair, and ions result.

Partial positive charges on each of two adjacent atoms of a molecule result in a weakening of the bond between the two atoms; the more positive the partial charges, the weaker the bond. The average electron density within the bond envelop will be below normal in this case. Hypophosphorous acid and its salts, the hypophosphites, are unstable because of the bond between a positive hydrogen (the nonacidic hydrogen) and the positive phosphorus central atom; disproportionation into phosphine and metaphosphoric acid takes place easily. Ammonia and nitrogen(III) fluoride are isoelectronic compounds; they differ in that the nitrogen atom is formally negative in ammonia and formally positive in the fluorine compound. Ammonia coordinates readily with a wide variety of electron pair-deficient substances (e.g., boron trifluoride), but the fluorine compound will not. Since electron-deficient compounds invariably have a positively charged central atom, the positive nitrogen and the positive central ion set up repulsive forces which prevent coordinate covalent bond formation between them.

Finally, compounds are unstable if the bonded atom pair has vacant bonding orbitals and exposed nonbonding electron pairs. These substances have a fleeting existence as molecules, tending to solidify into ionic crystals even at relatively high temperatures.

Delocalized bonds. Under certain circumstances increased structural stability results from delocalization of the bonding electron cloud. Modern molecular bonding theory has shown that this is the explanation of the classical concept of *resonance*. It has been observed that certain "double-bonded" compounds could be drawn with more than one classical structure, as shown in Fig. 2.9. The concomitant chemical behavior of these compounds was a stability not predicted by classical theory. The resonance concept explains this by holding that although

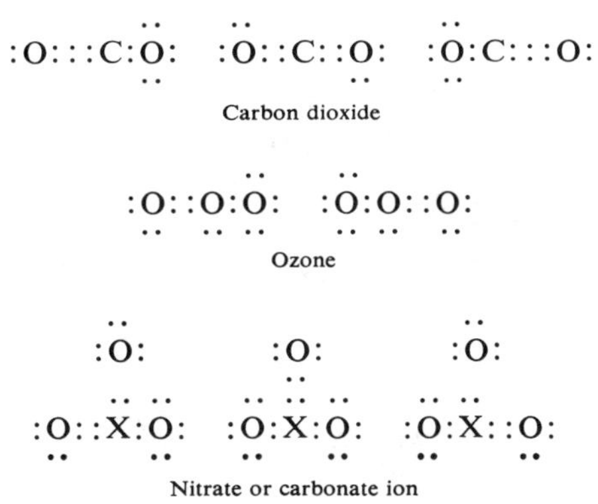

FIGURE 2.9 Classicai canonical forms of some resonance hybrids,

the *canonical forms* (the different possible written structures) do not exist as such, the actual behavior of the substance represents the average of the inherent properties of each of the possible forms, corrected for the relative contribution of each. The actual compound thus possesses properties which are an intermediate of the canonical structures, but no one all-inclusive structure can be drawn for the compound.

When applied to appropriate covalently bonded substances, resonance has been useful in accounting for the unusual stability and shorter bond lengths observed in these molecules. Stability of molecules can be determined experimentally by measuring heat of reaction. Classical thermochemistry allows the heat of reaction to be broken down to give the contributions of each of the bonds involved; from the average bond energies thus obtained the heat of formation of most common substances can be accurately predicted. With resonance hybrids, however, the agreement between calculated and experimentally observed energy values is poor; the heat given off is less than calculated. Since a stable molecule gives off less heat on decomposition than an unstable molecule of the same type, the order of divergence noted predicts greater stability for resonance hybrids—the experimentally observed fact. The difference between the calculated and experimental heats of reaction is known as stabilization energy or *resonance energy*.

Although the classical concept of resonance has been valuable for qualitative predictions, uncritical use together with an inadequate understanding of resonance forms has led to misinterpretations. The assignment of physical reality to the forms, e.g., "the molecule exists in this resonance form fifty per cent of the time," exemplifies complete disregard for the fundamental tenets of the resonance concept. For this reason modern interpretation based on the electronic structure of this type of molecule provides a more fundamental characterization.

Using the electronic interpretation the basic structure of the molecule is still determined by the σ-bonded framework. One fundamental difference, a delocalization of the "double bonds," is imposed. The restrictive pairing of their electrons between neighboring atoms is no longer allowed. Rather, the π-bond electrons are accommodated in molecular orbitals formed by overlaps which extend over two, three, or more neighbors, or even over the entire molecule. The resulting streamer type of structures are more difficult to draw than the classical resonance forms. Thus a descriptive representation of the modern interpretation is dependent upon the artistic abilities of the worker. A few of these modern structures are shown in Fig. 2.10. Borazole has been included to give an inorganic example of the annular ring π bonds found in benzene and other unsaturated cyclic organic compounds.

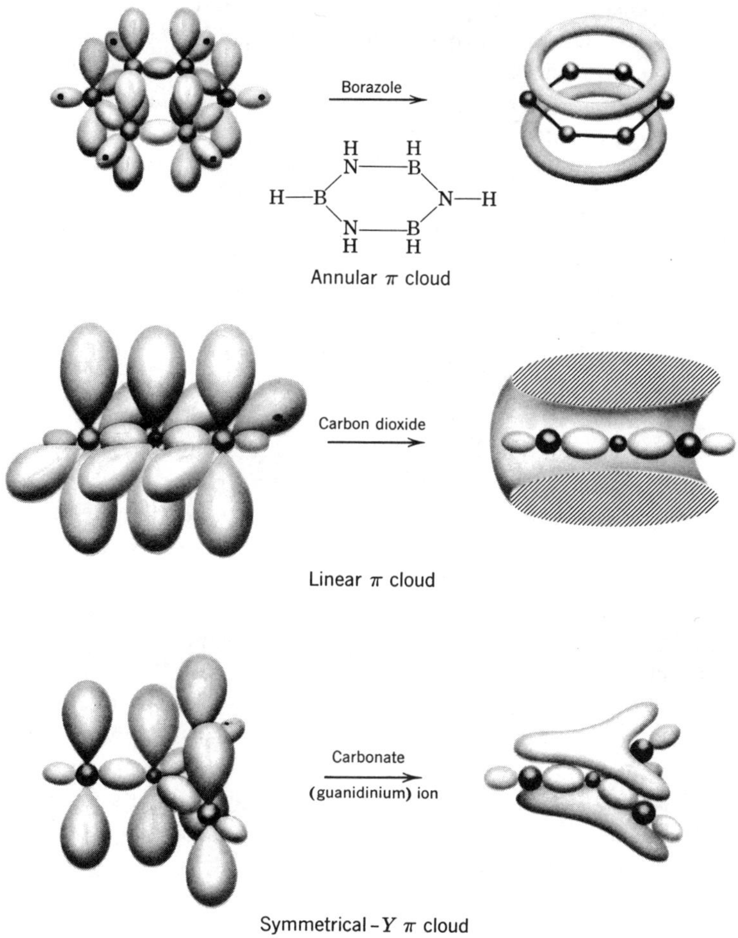

FIGURE 2.10 Nonlocalized molecular bonds.

The density of the π-bond cloud is not necessarily uniform over the entire molecule. There may be substantial differences in the π bonding in different regions of the bond. This is reflected in the extent to which an underlying π bond is shortened and strengthened. Consequently, a local difference in the π-bond cloud will be detected chemically by a preferred attack site in these molecules, at the π bond which has been least shortened. If the cloud is uniform, there is no preferred site of chemical attack.

Experimental verification of the π-bond shortening that accompanies molecular π bonds has been obtained from x-ray evidence. Not only have the classical cases been fully verified, but the existence of this type of electron cloud structure has turned up in unexpected places. The sulfate ion structure with electron pair orbitals can be constructed entirely without π bonding. However, x-ray techniques indicate a shortening of the sulfur-oxygen bonds from the expected 1.78 Å to 1.51 Å. This may be interpreted as indicating the presence of a certain amount of π bond, which must arise by the utilization of vacant *d* orbitals of the sulfur atom and the filled nonbonding *p* orbitals of the oxygen atoms. This interpretation helps to account for the unusual stability of this ion.

One type of bonding between atoms has purposely been omitted from the preceeding discussion—the *metallic bond.* It exists between atoms of the true metals, including Groups IA, IIA, IB, the transition elements, and the inner transition elements. The metallic bond may be best described as a complete nonlocalization of all valence electrons. Pauling has described the metallic bond as a supermolecular resonance system.

In the solid state a true metal is built up by "closest packing" of the atoms minus their valence electrons. This may be best described by representing the atoms as spheres, all of one size, and packing them in such a way that there is a minimum of empty space. The requisite number of valence electrons are then fed into the empty spaces between the spherical atoms. The electrons move freely throughout the structure, no one atom having any directed claim to certain electrons. Completely nonlocalized bonds result. (This is not to be confused with the valence shells of ions in which the electrons are localized, although the resultant electrostatic field is not.) The complete freedom of movement of the valence electrons within the metallic crystal easily accounts for all of the properties associated with the metallic elements; electrical conductivity, heat conductivity, malleability, etc.

In this chapter the reasons why atoms react and their manner of reaction have been examined. There is no single complete answer to this problem. However, a close re-examination of the thread of thought developed should show two underlying factors—electronic structure and atomic size.

THREE · MOLECULES—THEIR INTERACTIONS AND PROPERTIES

Molecular and ionic particles, while chemically stable, usually cannot exist in complete independence of one another (except possibly the permanent gases under low pressures). Since all particles[1] are built up from a positively charged nucleus and the negatively charged electron cloud, all particles will be surrounded by electrostatic fields. The electrostatic field interactions are responsible for the physical properties of a given substance under given conditions and, further, determine the magnitude of such properties. Solution formation and solution properties are also a function of these interactions. In addition, chemical reactions (e.g., complex formation) are made to occur between substances which would otherwise be chemically stable.

The gradual understanding of electrostatic field interactions has resulted in a gradual evolution of chemical theory. The Arrhenius theory of ionization, when applied to aqueous solutions of ionic species, failed because it could not account for the strong electrostatic interactions existing between the cations and anions. In order to provide an acceptable explanation the Debye-Hückel theory of electrolytes was introduced. In this chapter the origin and effects of electrostatic fields will be studied.

ELECTROSTATIC FIELDS

A zero net charge is a requirement imposed on any chemical system. Systems with a net positive or net negative charge do not exist as such. This rule also generally applies to the individual particles making up a

[1]In the following discussion the term particle will be used to indicate any one of the ultimate free units making up the chemical environment: ions, molecules, or free atoms.

chemical system. The molecule or the atom even in excited states, must have a zero net charge. If one particle assumes a net charge, as in an ionic system, another particle (or particles) must assume a net compensating charge of opposite sign so that the system's over-all charge remains zero. Ion formation is illustrated by the equation:

$$Fe + I_2 = Fe^{2+} + 2I^-$$

Note that in this system the net charge $(2^+) + 2(-)$ is zero.

Because of its zero net charge a molecule or free atom should have no electrostatic field associated with it; but if one does exist, it is at best a weak field. On the other hand, an ion will always be the center of an electrostatic field. The strongest interactions between particles involve at least one ion.

Ionic fields. The field surrounding an ion is uniform in all directions, the only restriction being the inverse rule governing the diminution of field strength with distance from its source. The strength of an ionic field is a function of several factors: ionic radius, ionic charge, and valence shell type. For a given net charge the larger the ion, the weaker the net electrostatic field acting on approaching particles. The greater the distance between an ion and an approaching particle, the smaller the effect of the ion's field on the approaching particle. On the other hand, the greater the net charge on an ion the stronger its field. Thus for a given series of ions of a given ionic radius and similar valence shells, the field of a divalent ion must be much stronger at a given distance from the ion than that of a univalent ion. Similarly, a trivalent ion has a much stronger field than a divalent ion. The relationship between field strength and cationic size and charge is illustrated diagrammatically in Fig. 3.1.

The effect due to the valence electron structure is more difficult to visualize. The shielding effects of the various valence shells differ. An $8e^-$ inert gas outer shell more effectively shields out the positive electrostatic field arising from the nucleus than does an $18e^-$ shell. The incompletely filled valence shell of a transition element cation provides the least shielding. The negative field of the inert gas electron cloud more effectively couples with the positive field of the nucleus, leaving a relatively small net electrostatic field surrounding the particle as compared to the stronger field resulting from the less effective coupling of an unclosed shell. The effect of valence shell type on electrostatic field can be well illustrated by studying the stability of the hydrates of a series of cations. In hydrate formation coupling takes place between the electrostatic field of the ion and the strong dipole field of the water molecule. Only the smallest members of the alkali metal cation series, with their inert

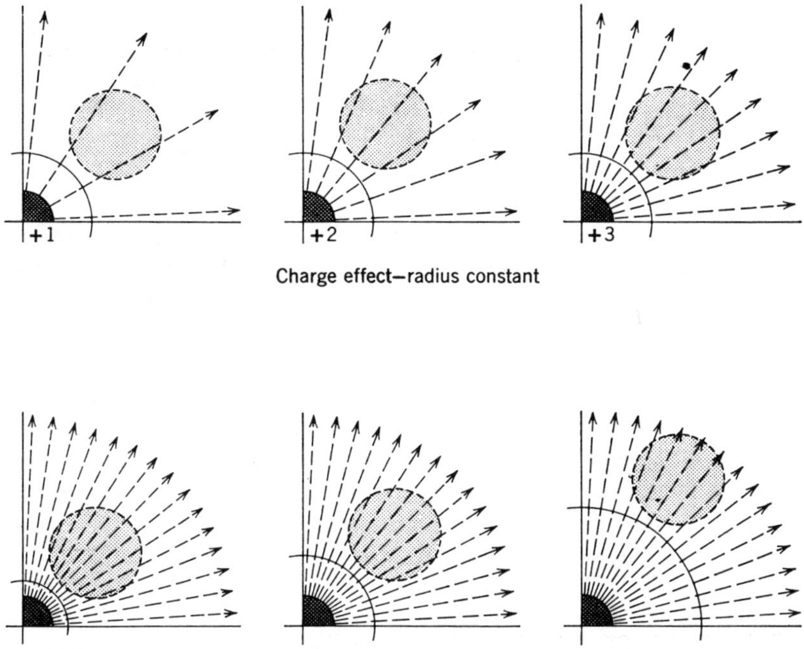

FIGURE 3.1 Cationic electrostatic field strength. One quadrant of the positive field is shown by the dotted rays originating in the nucleus (small shaded circle). The electron shell boundary is represented by the larger concentric circle. The number of rays drawn is proportional to the cationic charge. Comparative field strengths are represented by the number of rays intersecting an approaching particle (the crosshatched circle).

gas valence shells, can form stable hydrates; cesium ion does not hydrate. On the other hand, cations with $18e^-$ shells form stable hydrates, and transition element cations with their open shells form numerous stable hydrates. Obviously, the effectiveness with which the nuclear field is screened by the valence shell will affect the final net strength of the electrostatic field of a given series of ions of different shell type.

Dipoles. Despite the net electrical neutrality required of atoms and molecules, these particles may have distorted electron cloud structures which result in the appearance of electrostatic fields around these particles. These distorted electron structures are due to a tendency of electrons to drift toward one part of a molecule, leaving the rest of the molecule with a net electron deficiency; the over-all net charge remains zero. These so-called dipole structures may arise because of structural features within the molecule itself (see page 29) or from forces arising outside of the particle. The process by which dipoles arise is known as *polarization*. The strength of a dipole is measured by its *dipole moment*.

The existence of a dipole depends upon the arrangement and comparative electronegativity of the atoms comprising a given molecule. If a symmetrical arrangement of atoms exists within the molecule, ordinarily no dipole can be expected. On the other hand, an unsymmetrical arrangement will result in the appearance of a dipole. A comparison of carbon dioxide and carbon tetrachloride with carbon monoxide and chloroform illustrates the relationship existing between symmetry and the tendency for electron cloud deformation. Carbon dioxide and carbon tetrachloride have no dipoles even though strongly electronegative oxygen and chlorine atoms, respectively are present. Carbon dioxide is a symmetrical linear molecule with carbon occupying the center position, the oxygen atoms to each side effectively cancelling one another's polarizing effects on the electron cloud. The strongly electronegative chlorine atoms are tetrahedrally arranged around carbon in carbon tetrachloride—a symmetrical arrangement. On the other hand, the unsymmetrical molecules carbon monoxide and chloroform, with the same electronegative groups, do have dipoles. Dipoles of this type, which exist because of conditions within the molecule, are known as *permanent dipoles*. The effect of the distorted electron clouds is to surround these molecules with a polar electrostatic field. The field resulting from a permanent dipole will ordinarily be of lesser strength than a field arising from ionic charge.

Dipoles due to external electrostatic fields may appear on particles which show no polarization under ordinary conditions. External fields react with the electron cloud of a symmetrical molecule or an atom and distort it, attracting electrons to or driving electrons from the part of the affected particle closest to the field. Under these conditions an *induced dipole* arises. An induced dipole exists only as long as the particle remains under the influence of the field causing the distortion. Ions, especially those which do not have an inert gas valence shell, are affected by external fields; their electron clouds become deformed. Diagrammatic representation of two permanent dipoles and the formation of an induced dipole on a spherical electron cloud is shown in Fig. 3.2. The deforming electrostatic field may arise from the plates of a charged condenser, as shown in the diagram, or it may be a field associated with an ion, a permanent dipole, or even another induced dipole.

It can thus be seen that when molecules, ions, or free atoms are brought into close proximity of one another, interreactions due to the

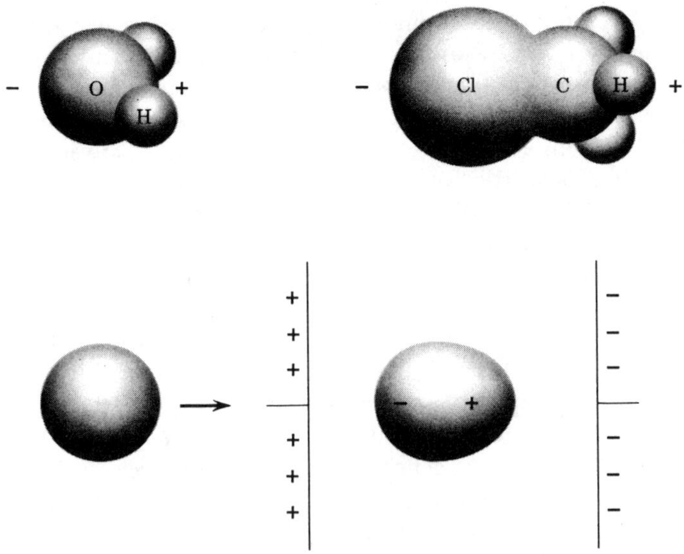

FIGURE 3.2 Permanent and induced dipoles.

electrostatic fields associated with particles will generally arise. The extent of the interactions will be a function of the strength of the electrostatic fields associated with the reacting particles. This will be true regardless of the state of the system—solid, liquid, gaseous, in solution, or as colloids—in which the particles are found. In order to give an over-all picture, each of the possible different interparticle electrostatic interactions will be discussed and typical examples given.

TYPES OF PARTICLE INTERACTIONS

Ion-ion interactions. The strongest of the interparticle interactions is the ion-ion. The interaction between positive and negative fields surrounding the cations and anions of a given salt are responsible for holding the ions within their sphere of influence in a crystal structure. Since the field surrounding an ion is spherical, a given ion will pack as many ions of opposite charge as can be grouped around it, the arrangement being modified only to the extent necessary to provide the required over-all neutrality.

In solution the same interactions take place between the ions involved. However, due to the freedom of movement and the effect of the solvent the interactions between the ions are limited. The degree of interaction of the ions is related to the properties of the solvent, especially its dielectric constant, and is directly proportional to their concentration. (Interactions between the ions and the solvent will not be considered at this point.) In dilute solutions the ions can move further apart on the average and thus will spend more of their time in the weaker portions of a neighboring ion's electrostatic field. If dilution is great enough the ions may separate sufficiently from one another to act as independent particles. It is largely because of this electrostatic interaction between ion fields that colligative properties of solutions of electrolytes fail to show theoretical values, but upon dilution tend to approach the theoretical norm. Thus sodium chloride should theoretically cause a freezing point depression of two times the molal freezing point constant in aqueous solution. However, owing to electrostatic interactions, two ions per molecule are not effectively available, and the freezing point is less than the expected factor of two times the molal freezing point constant. This factor is important in the preparation of isotonic solutions.

The effect of charge magnitude on ionic interactions in solution is especially interesting. It is generally assumed that all electrolytes exist as ions when dissolved in ionizing solvents. This ideal is approached rather easily by monovalent ions, especially in dilute solution. However, ionization can be severely restricted in similar solutions involving polyvalent ions. As ionic valence is increased, the strength of the electrostatic field surrounding the ion increases markedly and ionic interactions increase correspondingly. In solution the electrostatic fields of a divalent cation and a divalent anion interact so strongly that very little existence as independent ions is possible even in dilute solutions. Rather, these solutions act as though closely associated ion-pairs are present. However, these associated ion-pairs must not be called molecules; covalent bonds do not form. Ionic salts such as calcium sulfate, copper(II) sulfate, and tin(II) sulfate illustrate this behavior in aqueous solution. Except at extreme dilutions, various experimental techniques to determine the degree of ionization of salts (e.g., depression of the freezing point and conductivity) fail to show the presence of significant numbers of ions of these salts. The electrostatic fields due to the double charges existing on these ions are so intense that independent behavior of the ions becomes transitory.

Trivalent cations ordinarily cannot exist independently in solution in ionizing solvents. Either they react with the solvent, if it is polar, or cation-anion combinations of various kinds result. An aqueous solution of iron(III) chloride may serve as an example of the latter. Depending on the concentration of the iron compound and the amount of chloride ion available from other compounds present (HCl or NaCl), the following species may form in aqueous solution. (See top of next page.)

Because of the strong electropositive field surrounding the iron(III) ion, increasing chloride concentration favors increasing formation of the

$$Fe^{3+} + 6H_2O = [Fe(H_2O)_6]^{3+}$$
$$[Fe(H_2O)_6]^{3+} + Cl^- \rightarrow [FeCl(H_2O)_5]^{2+} + H_2O$$
$$[FeCl(H_2O)_5]^{2+} + Cl^- \rightarrow [FeCl_2(H_2O)_4]^+ + H_2O$$
$$[FeCl_2(H_2O)_4]^+ + Cl^- \rightarrow [FeCl_3(H_2O)_3]^0 + H_2O$$
$$[FeCl_3(H_2O)_3]^0 + Cl^- \rightarrow [FeCl_4(H_2O)_2]^- + H_2O$$

higher chloro complexes of the iron(III). The chloride ion more effectively gives the cation a share in its electrons, its field interactions being stronger than those of the water molecule dipole. Thus water will be successively displaced from the coordination sphere by the more strongly interacting chlorides if they are available.

Ion-permanent dipole interactions. Ion-permanent dipole interactions such as the iron(III) hydrates cited in the preceding discussion are very common. The bonds formed are relatively strong. The hydrates of cations, especially those having inert gas shells, are usually of this type. Sodium ion exists in a highly hydrated state in aqueous solution. Its formula should be written $Na(H_2O)_x^+$. The strong dipole-cation field interaction is responsible for the hygroscopic property of many sodium salts. The silver ammonia complex cation is another example of this type of interaction.

There is one ion-dipole interaction which occurs with such frequency that it has been given a distinctive name, the *hydrogen bond.* The hydrogen bond results when a hydrogen, already bonded to an electronegative atom of a compound, forms a second bond with the lone pair of some other strongly electronegative group, e.g., fluoride. The bond was originally interpreted by N.V. Sidgwick as a coordinate link, with hydrogen assuming a coordination number of two. However, hydrogen has only one stable bonding orbital, 1*s*, and must therefore be limited to a covalency of one. The second bond must be due to electrostatic interaction between the hydrogen atom and the strongly electronegative second bonding partner.

Hydrogen bonding is limited to the most electronegative elements—F, O, N, and, sometimes, Cl. Evidence indicates that with a few exceptions (e.g., FHF^-) the bonding hydrogen is closer to one of its bonding partners than to the other. The bond strength of the $H \cdots F$ bond is about 10% of the σ bond and has a bond length about twice as long. The $H \cdots O$ bond is about three fourths, and the $H \cdots N$ about one half as strong as the $H \cdots F$. The bond energies lie in the range 3–10 kcal./mole; except for covalent and ion-ion electrostatic interactions, hydrogen bonds are stronger than any other bonds forming between molecules.

Hydrogen bonding may be intramolecular or intermolecular. The *o*-hydroxy aromatic aldehydes and *o*-nitrophenols are typical examples of the former. As a result the properties of these molecules are markedly different from those of their *m*- and *p*-isomers. Intermolecular hydrogen bonding is widespread, ranging from simple types such as the dimeric acetic acid molecule and many anion hydrates to hydrogen-bonded macromolecular structures such as ice and the dihydrogen phosphates. The first-row hydrides—H_2F_2, H_2O, and H_3N—owe their unusual properties to the presence of hydrogen bonds; the effect of hydrogen bonding on the boiling points of these substances in comparison to the other members of the families is shown in Table 3.1. Water with two possible

TABLE 3.1 THE EFFECT OF HYDROGEN BONDING ON BOILING POINTS

Molecule[a]	Boiling Point, °C.	Molecule	Boiling Point, °C.
CH_4	−164	(2) H_2O	+100
SiH_4	−112	H_2S	−61
GeH_4	−88	H_2Se	−41
SnH_4	−52	H_2Te	−2
(1) NH_3	−33	(1) H_2F_2	+20
PH_3	−88	HCl	−85
AsH_3	−55	HBr	−67
SbH_3	−18		
BiH_3	+22	HI	−36

[a] When hydrogen bonds are present in a given substance, the number of such bonds formed per molecule is indicated parenthetically before its formula.

hydrogen bonds per molecule (2H and 2 unshared electron pairs per molecule) shows the greatest divergence in boiling point. Fluorine, which forms the strongest hydrogen bonds, is restricted to one bond per molecule since only one hydrogen is available per molecule.

Hydrogen bonds are of interest in many biological substances. The string-like molecules of cellulose are bound to one another by hydrogen bonds to form cellulose fibers. Polypeptide chains are connected into sheets by =N— H · · · O— bonds. Helical molecular chains of nucleic acid are held together by hydrogen bonds between purine and pyrimidine rings.

A group of substances, selected to show the variety of structures resulting from hydrogen bonding, are shown in Fig. 3.3.

Dipole-dipole interactions. Bonds due strictly to dipole interactions between molecules give rise to the final group of bondings. Four different

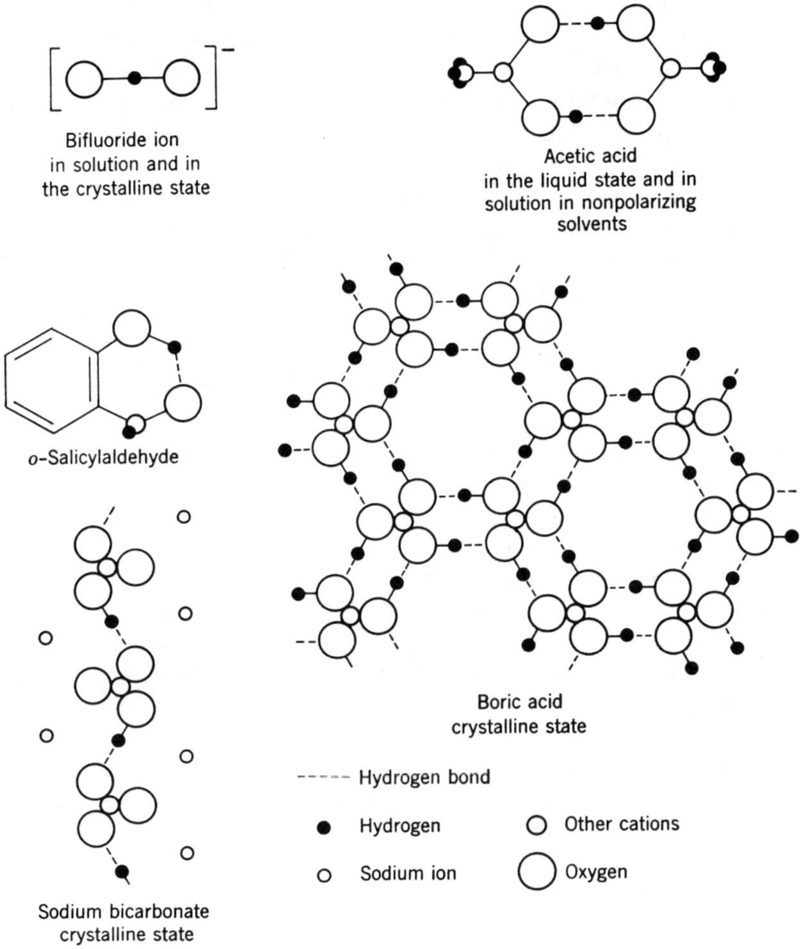

FIGURE 3.3 Structures involving hydrogen bonding.

interactions are possible: permanent dipole-permanent dipole, ion-induced dipole, permanent dipole-induced dipole, and induced dipole-induced dipole. The two arrangements resulting from dipole interactions are shown in Fig. 3.4.

The interaction between two permanent dipoles gives rise to a relatively strong association between the affected molecules. Due to its very strong permanent dipole, water exists as a dimer in steam at 100°C. Even at the relatively high kinetic energies of this temperature the association is maintained.

Dipoles may be induced on otherwise symmetrical molecules by ions or the permanent dipole of another molecule. The triiodide ion is the result of a dipole induced on the iodine molecule by the iodide ion, followed by association of the two interacting particles. Many unexpected crystalline hydrates are due to the reaction of the strong permanent dipole of water with symmetrical molecules or free atoms to give induced dipoles on these particles. These induced dipoles then react with the water to give hydrates. Examples of this interaction type are crystalline hydrates of various saturated gaseous hydrocarbons, e.g., $C_5H_{12}(H_2O)_x$, and inert gases, e.g., $He(H_2O)_x$. In the latter case the spherical electron cloud of the helium atom is sufficiently polarized by the water dipole to form the crystalline hydrate. The 17 aquo hydrate of butane is the basis of a newly devised method for the preparation of potable water from saline water.

The weakest of all electrostatic interactions are those of the induced dipole-induced dipole type. When two otherwise symmetrical molecules or single atoms move sufficiently close to one another, the nucleus of one begins to interact with the electron cloud of the other and vice versa. This distorts the electron cloud of each of the reacting particles sufficiently to induce a dipole. The forces giving rise to such attractions are called van der Waals forces. A wide variety of interparticle associations are due to these interactions. Many organic substances solidify in this way—e.g., naphthalene, hexachlorethane, ethane. The inert gases liquefy and ultimately solidify due to formation of induced dipoles. The gas is compressed and cooled until the individual atoms are close enough to one another to allow the nucleus-electron cloud interactions necessary to

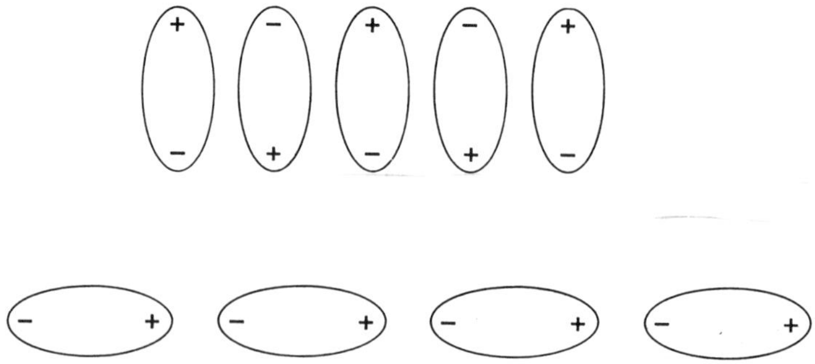

FIGURE 3.4 The two possible arrangements of molecules resulting from dipole interactions.

produce dipoles. After this stage is reached liquefication and solidification proceeds easily.

PHYSICAL PROPERTIES AND INTERPARTICLE BONDING

Physical properties frequently give a good indication of the bonding forces producing a given crystalline solid. The melting point of a substance is directly related to its molecular weight and the type of forces causing interparticle bonding in the crystalline solid. Although the contribution of molecular weight to the melting point of a substance is not of prime interest at this time, its contribution must be considered since it acts simultaneously with the bonding forces in determining a substance's melting point. In general for a series of corresponding compounds in a given family of elements the melting point increases with increasing molecular weight. This is illustrated by the melting points of silane and germane, which are 88° and 108°K, respectively. The rule as generally applied may be stated: Other factors being equal, the melting or boiling point of a substance of greater molecular weight will be higher than that of a substance of lesser molecular weight. This rule should be kept in mind during the discussion which follows. It will usually explain apparent inconsistencies.

Completely covalently bonded substances have the highest melting and boiling points. The bonding is directed in space, and the bonding electrons are restricted to the bond volume. This means that the bond forces are effectively concentrated, and the bond strength is the highest of any type. Actually a crystal whose binding forces are from strictly covalent bonds is a single continuous molecule. There is no discontinuity in the interatomic bonding at any point in the crystal. Because of the high energies needed to break a covalent bond, melting points of these substances are of the order of 1300°K, or higher. Some covalently bonded substances, e.g., the diamond, have no melting point by ordinary standards. Other examples of crystalline materials having complete covalent bonding are aluminum oxide, silicon carbide, and quartz.

Electrostatically bonded crystals have binding energies which are ordinarily less than those of the covalently bonded crystals. Most true electrolytes fall into this group and generally have melting points ranging from 600–1500°K. The electrostatic forces surrounding these ions are strong, but since they are dissipated through a large volume of space due to their spherical distribution, rather than being concentrated into bonding regions as in the covalent bond, they are weaker than the binding forces existing in the covalent bond. Therefore less energy is required to free an ion from its neighbors in a crystal, and a lower melting point results.

Molecular crystals bonded by hydrogen bonds are found in the next melting point range. Crystals whose particles are held together by permanent dipoles are found in the same general range. These substances are hard and brittle and, except for the simple hydrides, (e.g., H_2F_2), melt in the range of about 300–700°K.

Crystalline substances having the lowest melting points are those bonded by van der Waals forces (induced dipole-induced dipole). For the most part these substances are liquids or permanent gases at room temperature and pressure.

Many common compounds have more than one of these bonding forces responsible for maintainence of their crystal forms. Of the three directions in space within a crystalline substance, bonding in one or two directions may be covalent while the bonding in the remaining directions may be due to ionic electrostatic forces, dipole forces, or van der Waal's forces. Talc and graphite are such substances. They consist of slabs, one molecular layer thick, of covalently bonded atoms. These slabs or layers are held together by van der Waals forces. This accounts for their characteristic slippery feel when rubbed between the fingers; instead of the particles rolling over one another and giving a gritty feel, the layers slide over one another easily. The simple pressure of the fingers overcomes the van der Waals forces holding the layers together. The micas also are layer compounds of a similar general arrangement, but the covalent layer structures are cemented together by the electrostatic fields of alkali metal cations located between the layers. Simple finger pressure no longer separates the layers. But because the electrostatic bonding between the layers is weaker than the covalent bonding within the silicate layer, the micas cleave readily into layers, as evidenced by the ease with which layers may be peeled from one another. The different mechanical properties of these substances well illustrate the difference in strength of the various binding forces. Other common properties of substances are frequently related to the type of force holding the system together. Hardness and viscosity are determined in large measure by the forces discussed.

Table 3.2 summarizes the foregoing discussion by giving some examples of crystalline substances, their bonding forces, and their melting points.

A source of confusion frequently arises because organic molecules are covalently bound intramolecularly but usually dipole bound intermolecularly. Since relatively low melting points are generally associated with organic compounds, the erroneous conclusion is drawn that covalent bonding is easily broken and therefore weak! It must be remembered that the covalent bonds within the organic molecule are not ordinarily broken on melting. It is the various electrostatic fields existing between the individual molecules which are overcome when the crystal melts.

TABLE 3.2 Relative Strength of Crystal Binding Related to Melting Point

Substance	Type of Bonding[a]	Melting point, °K, 1 atm.
Al_2O_3	covalent	2323
SiO_2	covalent	1983
NaCl	ionic	1081
LiH	ionic	953
Pentaerythritol	hydrogen	535
KH_2PO_4	hydrogen	529
H_3BO_3	hydrogen	458 (dec.)
$KHCO_3$	hydrogen	~400 (dec.)
p-Salicylaldehyde	hydrogen	390
glucose	hydrogen	385
water	hydrogen	273
Propane · $17H_2O$	dipole (p. + i.)	320 (4.5 atm.)
$Xe \cdot 6H_2O$	dipole (p. + i.)	273 (dec.)
$A \cdot 6H_2O$	dipole (p. + i.)	248 (dec.)
$HgCl_2$	dipole (i)	550
$\gamma C_6H_6Cl_6$	dipole (i)	385
P_4	dipole (i)	317
o-Salicylaldehyde	dipole (i)	266
$SnCl_4$	dipole (i)	243
CO_2	dipole (i)	217
CH_4	dipole (i)	91
A	dipole (i)	84
N_2	dipole (i)	68

[a] p. = permanent, i. = induced.

COMPLEX COMPOUNDS

The most comprehensive illustration of the application of intermolecular and interionic forces is found in the modern theory of "complex" substances. Simple ions and simple and more complex molecules do not necessarily exhaust their capacity for chemical and physical reaction immediately. The capacity for the formation of coordinate covalent bonds is usually not exhausted until all available unoccupied orbitals have been occupied by previously nonbonding electron pairs. The permanent and induced electrostatic fields associated with molecules and ions can cause further interaction between particles in order to more fully satisfy their combining tendencies.

Complex substances are much more widespread than usually realized. Many substances are complexes but not treated as such. Sulfate ion, nitrate ion, and the tetraquacopper(II) cation are complex ions, but because of their stability (and the traditional way of regarding them before the advent of modern complex theory) are usually treated as simple ions. Many common molecules may similarly be treated as complexes—e.g., methane, tetracarbonylnickel(0). Many crystalline substances are complex macromolecules. For example, calcium carbonate has a complex coordination structure extending in all directions throughout each crystal. Each calcium ion is in octahedral coordination with six oxygens, each from a different carbonate group. Each carbon is coordinated triangularly with three oxygens, each of which is coordinated with a different calcium ion. Of course, many substances—e.g., tetraamminecopper(II) ion, dicyanatoargentate(I) ion, calcium citrate—are correctly recognized as complexes.

A review of the elementary terminology of modern complex theory is in order at this time. The *central unit* (ion or neutral atom) is the particle about which the *ligands* are arranged. A ligand is, therefore, any group which occupies a position in the coordination sphere of the central unit. The maximum coordination number of the central unit determines the limit of the number of simple ligands which may coordinate with it. The relationship existing between the principal quantum number and the maximum coordination number of the central unit has been discussed (page 43). Maximum coordination number need not be achieved for stable complex formation.

The central unit and complex stability. The qualitative factors affecting complexation should also be understood before a discussion of theory is undertaken. The central unit may be electrically neutral (an atom) or positively charged (a cation). It cannot carry a net negative charge—anions never act as central units in a complex. The stability of a complex is directly related to the magnitude of the formal charge of the central unit; the tetraoxochromate(VI) ion has a greater inherent stability than the tetrachlorochromate(III) ion. Cationic radius also affects the stability of a series of complexes with the same ligands. Since the strength of the positive field of a cation is inversely proportional to the distance from its center, larger cations of a given charge will form weaker complexes than the smaller cations; a first-row transition element cation forms more stable complexes than third-row members of its family, other factors being equal.

The cation's valence shell structure is important in determining the stability of its complexes. In general the more easily a valence shell is polarized by an approaching ligand, the more stable the resulting complex.

Since the *d*-shell is a mushy, diffuse highly polarizable electron cloud, ligands can penetrate deeply into its outer portions. On the other hand, the inert gas shell with its tightly packed spherical electron cloud is highly resistant to deformation by outside fields. The $18e^-$ shell is intermediate to these two shells in all respects. Thus the order of complex stability as a function of the valence shell is: transition element shell, most stable; $18e^-$ shell, intermediate stability; inert gas shell, minimum stability if at all capable of forming complexes. An additional factor dependent on shell type is the availability of unoccupied low-energy orbitals. Transition elements with unoccupied *d* orbitals form especially stable complexes.

The role of the properties of the central unit on the stability of a complex may be summarized by restating the extremes. At the most stable end of the scale is found the transition element cation, of high formal charge, small ionic radius, and one or more unoccupied *d* orbitals. At the other end of the scale are found the inert gas shell cations, of low formal charge and large ionic radius. The former has a prototype in titanium in the titanates. The latter reaches its epitome in cesium which is incapable of acting as the central unit for a complex, not even a hydrate.

The central atom of a complex structure can serve as a center of symmetry just as the carbon atom does in its compounds. In fact, the multiplicity of arrangements theoretically possible outnumber those of the carbon atom, restricted as it is to the tetrahedral structure. Mirror image pairs occur frequently. These asymmetrical forms are optical isomers. *Cis* and *trans* isomerism can exist in tetrahedral, square planar, and octahedral structures. The octahedral arrangement, with its six positions as variables, offers by far the largest number of possibilities.

The ligand and stability of complexes. In order to be effective as a ligand a substance must be or have within its structure one or more highly electronegative atoms capable of acting as electron pair donors. Thus all ligands are or include nonmetals. The most effective in this respect are P, C, S, Cl, N, O, and F. A ligand may be an anion or it may be a neutral molecule. The halides and halogenoids, the best known anionic ligands, are in order of coordination strength $CN^- > OCN^- > SCN^- > F^- > Cl^- > Br^- > I^-$. Other anions, including complex anions, are equally capable of acting as ligands. The hydroxyl ion falls between chloride and fluoride in this scale. The ability of a complex anion to act as a ligand falls off rapidly with increasing charge of a very small central unit and the compactness or tightness of its "ligand" electron cloud. Thus nitrate (N^{+5}) and especially perchlorate (Cl^{+7}) are considered to be incapable of acting as ligands and are for that reason used as the anion of choice in solutions intended for study of free uncomplexed cations. Some of the neutral molecule ligands in order of coordination strength are $NH_3 > RNH_2 > R_2NH > R_3N > H_2O > ROH > R_2O > R_3P > R_2S$ (where R represents an alkyl or aryl group).

The size of the ligand plays a role in two ways. Coordination stability falls off with increasing electron cloud size, as can be seen in the halogen series. Frequently, maximum coordination cannot be achieved when large anions act as ligands. Silicon tetrafluoride will coordinate readily to form stable six coordinate fluosilicate ion, $[SiF_6]^{2-}$. On the other hand, silicon tetrachloride is unable to achieve further coordination with chloride ions. The relatively small Si^{4+} is incapable of achieving its maximum coordination number of six in this latter case even though the chloride ion will share its electrons more readily than the fluoride ion.

The polarizability of the ligand, as well as its ability to polarize the valence shell of the central ion, is also important. In general, the greater the polarizability of a ligand, the more effective it is in forming stable complexes. This fact is demonstrated by the complexation behavior of water and ammonia. The permanent dipole of water is stronger than that of ammonia. In aqueous media hydrate formation should be favored rather than ammoniate formation. This is not, however, the entire story. The water molecule is relatively nonpolarizable, whereas the ammonia molecule is highly polarizable. In the presence of a strong electrostatic field an induced dipole appears on the ammonia molecule in addition to its permanent dipole. Under these conditions the sum of the permanent dipole and the induced dipole must be considered. For the ammonia molecule this sum exceeds that of the water molecule. As a result, even in aqueous media, ammonia will complex preferentially with cations with strong electrostatic fields, e.g., with the transition metal cations. Tetraamminecopper(II) is stable in the presence of ammonia, even in aqueous solution. On the other hand, lithium ion can induce little if any polarization in ligands; lithium ion hydrate is stable in the presence of ammonia. The stronger permanent dipole of the water molecule determines the course of this latter reaction.

When relatively intricate molecules or ions serve as ligands, several other factors must be considered. A ligand containing two electronegative atoms sufficiently separated so that each atom may interact with the central ion is said to be *bidentate*. If the ligand can form three bonds with the central ion, it is a *tridentate* ligand. A ligand with two or more donor atoms must have two additional properties for cyclic complexes to form readily. First, the number of atoms must be such that counting the central atom, a five-, six-, or seven-member ring may form. Four- and eight-member rings may form under certain conditions but are unstable. Second, the molecular chain must be flexible in order to bend back on itself and thereby form two bonds with the central atom. Ethylenediamine is a flexible bidentate ligand, easily forming five-member rings with a central ion. Thiocyanate also has the two necessary electronegative atoms, one at each end of the ion. But it fails as a bidentate ligand because its chain of atoms is linear, incapable of flexing to form a ring, and even if it could flex, an unstable four-member ring would result.

When complexation involves such ring closure, the complex is known as a *chelate*. The process of closing the ring is known as *chelation*. Ring closure in complexes results in increased stability for the complex. With a bidentate ligand the cyclic form is stable with respect to the linear form. Nickel usually appears with a coordination number of four; however, with ethylenediamine the unusual coordination number of six is stabilized, nickel chelating with three of the amine molecules. Figure 3.5 shows some chelates. Bidentate ligands may be involved in *cis-trans* isomerism. The two coordinating positions must always be *cis*, however; the strain would be too great for the *trans* configuration to form.

General factors in stability of complexes. Additional factors not directly related to the structure and properties of the central atom and ligand are important in complex formation. The concentration of both the central atom and the ligand is of great importance. The effect of concentration is expressed by the *instability constant*, K_{inst}. In general terms the equilibrium between a complex and its constituent ions is expressed by the equation

$$[MX_x]^{+(m-xn)} \rightleftharpoons M^{m+} + xX^{n-}$$

and the instability constant is

$$K_{inst} = \frac{[c_{M^{m+}}] \cdot [c_{X^{n-}}]^x}{[c_{MX_x}{}^{+(m-nx)}]}$$

The extent to which a given complex decomposes is directly proportional to the concentration of the free central particle present, and to the xth power of the concentration of the free ligand, but it is inversely proportional to the concentration of the actual complex present in the equilibrium mixture. Since the concentration of the complex is in the denominator, its concentration is increased by increasing the concentration of either of its component substances, especially its ligand since its concentration appears as a power. From the structure of the constant it can be seen that for stable complexes the instability constant will be a very small

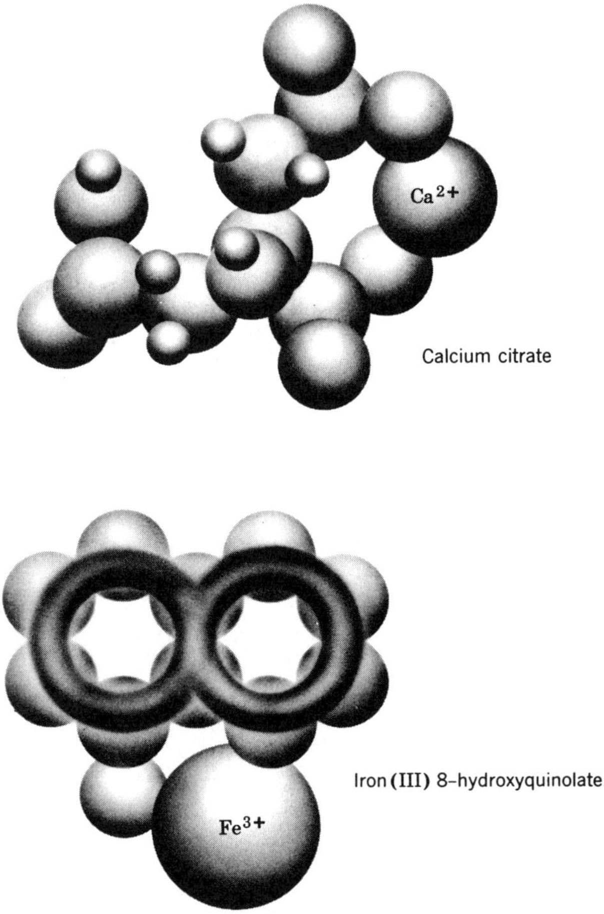

FIGURE 3.5 Chelates.

number. The instability constants of a few common complexes are given in Table 3.3.

When a given central ion and a given ligand are capable of forming more than one complex, concentration of the ligand in relation to the given concentration of the central atom is important in another way. As an example iron(III) can form two complex ions with fluoride ion, $[FeF_4]^-$ and $[FeF_6]^{3-}$. Increasing concentration of fluoride, with iron(III) concentration constant, will favor the formation of more and more of the higher complex. However, the higher complex cannot form (except to the small extent allowed by the simultaneous solution of the instability

TABLE 3.3 INSTABILITY CONSTANTS OF SOME COMPLEXES

Complex[a]	K_{inst}	pK_{inst}
$[Fe(C_2O_4)]^+$	4×10^{-10}	9.4
$[Fe(C_2O_4)_2]^-$	1.6×10^{-7}	6.8
$[Ca(EDTA)]^{2-}$	2×10^{-11}	10.7
$[Fe(EDTA)]^{2-}$	5×10^{-15}	14.3
$[Fe(EDTA)]^-$	8×10^{-26}	25.1
$[Ni(NH_3)_6]^{2+}$	4×10^{-9}	8.4
$[Ca(Ci)]^-$	6.75×10^{-4}	3.17
$[Ca(HCi)]^0$	8.1×10^{-4}	3.1
$[Ca(H_2Ci)]^+$	8.0×10^{-2}	1.9
$[Cu(Ci)]^-$	6.2×10^{-15}	14.2

[a] EDTA represents the anion of ethylenediaminetetraacetic acid. Ci represents the anion of citric acid.

constant expressions for the two complexes) until each iron(III) has achieved a coordination number of four. It is only then that fluoride will be available for the conversion of the tetrahedral four-coordinated complex into the octahedral six-coordinated form. Or if an iron(III) compound such as the sulfate is added to a solution of $[FeF_6]^{3-}$, the hexafluoro complex ions will give up enough fluorides to allow the newly introduced simple ions to achieve the $[FeF_4]^-$ structure.

Temperature and pressure are also important external factors. This is especially true for gaseous ligands, e.g., ammonia. Increasing the temperature or reducing the partial pressure of ammonia over the solution results in a loss of gaseous ammonia. As a consequence the concentration of the ligand decreases, and the usual concentration effects come about. Some complex radicals decompose in an attempt to maintain the proper equilibrium concentration of free ligand, which in turn is lost to increase the partial pressure over the solution. A vicious cycle is set up which is completed only by the total decomposition of the complex, unless temperature is lowered or the partial pressure of the gaseous ligand over the solution is increased.

COMPLEXATION THEORY

Historical. The recognition of the existence of substances known today as complexes dates back to the early nineteenth century. However, theory and chemical knowledge had not developed sufficiently for an understanding of the structure of these substances. Formulae were written in terms of reacting ratios, as is demonstrated by the old formula for the potassium salt of the tetraiodomercurate(II) ion—$2KI \cdot HgI_2$.

The first attempt to bring order to the field was by A. Werner. He introduced the concept of residual or secondary valences in addition to the well-known normal or primary valence. He also postulated that residual valences were directed in space. Early in this century N. V. Sidgwick, building on ideas first advanced by G. N. Lewis, put Werner's views into terms of the coordinate covalent bond. It was thought that each central atom would accept electron pairs until an inert gas structure was achieved. An element's effective atomic number (E.A.N.) was considered to equal that of the next inert gas. With continuing work, however, too many exceptions to this rule were found.

In the early thirties L. Pauling introduced three new ideas. He explained stereochemical arrangements by the concept of orbital hybridization. He suggested the possibility of π bonding between certain ligands and the central atom. Finally he was able to show that the magnetic properties of complexes could be used to elucidate the valence electron structure of the central atom.

From these beginnings three main theories gradually evolved: the valence bond, the molecular orbital, and the electrostatic. (Note the similarity to those developed for an explanation of atomic reactions.) Each of the three can account reasonably well for certain of the various characteristics of complex substances. But unanswered problems remained: orbital-spectral correlation, higher than calculated magnetic moments, four-coordination achievement by square planar or tetrahedral arrangements, unsatisfactory energy balance calculations.

Ligand field theory. A significant breakthrough occurred in the early 1950's when L. E. Orgel and others recognized the value of the crystal field theory[2] when combined with the electrostatic theory. After further refinement of the combined theories by application of certain aspects of molecular orbital theory, ligand field theory was developed. The discussion which follows will be restricted to a qualitative analysis of the salient features of this theory. Quantitative aspects such as calculation of binding energies and spectral correlation to electron slitting are beyond the scope of this text book.

[2] Originally proposed by Hans Bethe for the theoretical elucidation of the structure of crystalline materials. A crystal is considered an orderly array of electrostatically interacting particles. Physicists had successfully used the theory for the explanation of spectroscopic and magnetic properties of crystals before the chemist applied it to complex substances, using similar reasoning.

As a first approximation the ligand field theory considers a complex as an orderly arrangement of electrostatically interacting particles about a central charge. The essence of the theory is that the five *d* orbitals, while equal in energy in the *gaseous* metal ion, become differentiated in the presence of the electrostatic field due to the ligands surrounding it. Those orbitals of the central ion lying toward the ligands are raised in energy with respect to those lying away from the ligands—a process known as *splitting*. By using the resultant low-energy orbitals for the electrons, the system is stabilized with respect to a random filling of the *d* orbitals.

Three sets of the quadruple *d* orbitals, the d_{xy}, d_{yz}, and d_{xz}, lie in the three planes of the Cartesian axes (see Fig. 1.2). The lobes lie along a 45° line passing through the origin. Together the lobes of these three sets of orbitals, collectively given the symbol d_ϵ, point toward the midpoints of the slant edges of a regular octahedron.

The four $d_{x^2-y^2}$ lobes lie along the x and y axes. Two lobes of the d_{z^2} orbital lie along the z axes, and the remaining two lobes are compressed into a doughnut shape lying in the xy plane. Essentially the six axial lobes of these orbitals, collectively denoted as d_γ, point toward the corners of a regular octahedron.

Certain electrostatic interactions are possible between the valence electrons of a central atom and the field of an approaching ligand. A repulsive action is set up between the valence electrons of the central atom. This repulsion is stronger between two paired electrons in the same orbital than between two unpaired parallel spin electrons in adjacent orbitals. A repulsive force also exists between the valence electrons of the central atom and the electrons of the approaching ligand. Ligands with a strong field (e.g., cyanide ion) cause the electrons of the central ion to concentrate as electron pairs in nonbonding d_ϵ orbitals. The interaction of an unpaired valence electron of the central atom with a strong electrostatic field is sufficient to overcome the natural resistance to pairing electrons of opposed spin in the same orbital (Hund's rule). On the other hand, if a ligand with a weak field (e.g., water, fluoride ion) approaches the valence electrons of the central atom, the electrons tend to spread out in singly occupied orbitals, and the weak field will even allow unpairing of existing pairs. Considering water and fluoride ion as weak field ligands may at first glance seem inconsistent with the known strong dipole of water and the high electronegativity of the small fluoride ion. However, both of these groups, in common with all weak field ligands, lack one essential property: they are relatively nonpolarizable. The dense, tightly held valence shells of the fluoride ion and the oxygen atom of the water molecule (for practical purposes two protons imbedded within the electron cloud of the oxygen) are among the least polarizable substances. For strong ligand-central atom interactions, the central atom must be able to bring about some polarization of the approaching ligand.

Consider, for descriptive purposes, the formation of a hypothetical complex by allowing six ligands to approach a central ion in octahedral configuration. The lobes of the two axial orbitals will lie in a direct line with the approaching ligands (Fig. 3.6). The lobes of the d_ϵ orbitals will be directed so that each lobe lies midway between the neighboring approaching ligands. Owing to its direct orientation toward the electrostatic fields of the approaching ligands, an electron occupying one of the axial orbitals will have a higher potential energy than an electron in one of the d_ϵ orbitals.

If each of the ten electrons which may theoretically occupy the d orbitals of the central unit is successively fed into the ten available orbitals, the dependence of electron configuration upon ligand field strength becomes apparent. Since an electron seeks the orbital of lowest energy, the first three electrons will enter the d_{xy}, d_{yz}, and d_{xz} orbitals, respectively. Irrespective of ligand these will be the orbitals of lowest energy for the three initial electrons with an octahedral arrangement. However, the orbital which a fourth electron would enter depends upon the ligand. With a weak-field ligand the energy required to pair the fourth electron with an electron already present in a d_ϵ orbital will be greater than that of an unpaired electron in the high-energy axial (d_γ) orbitals; the electron will enter an axial orbital with a spin parallel to the existing spins. If, on

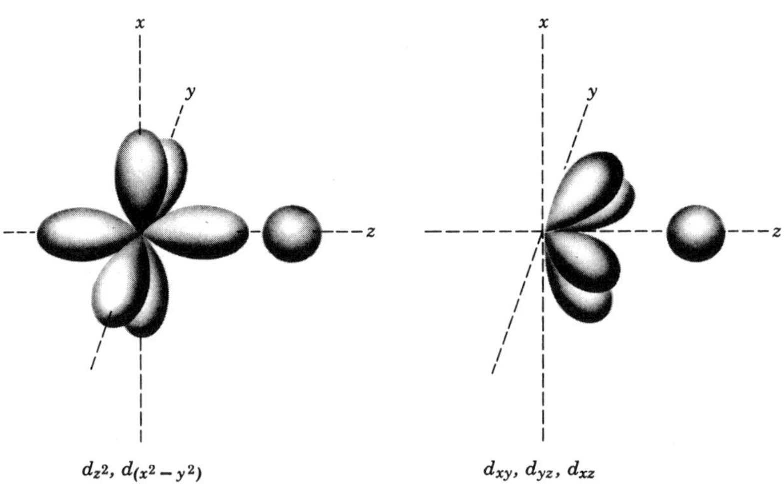

FIGURE 3.6 Alignment of d_γ and d_ϵ orbitals with respect to an approaching ligand. For simplicity only two lobes of each the d_{xz} and d_{yz} orbitals are shown. The doughnut lobe of the d_{z^2} orbital is omitted for the same reason.

the other hand, a strong-field ligand is present, the energy to pair the fourth electron in an occupied d_ϵ orbital will be less than that necessary to force an electron into one of the axial orbitals. Here spin pairing will take place. Similarly, the orbital occupied by the fifth, sixth, seventh, and eighth electron, respectively, depends upon the field strength of the approaching ligand. The final two electrons would again be independent of the ligand, irrespective of the previous history. In any case the d_ϵ orbitals must be full and vacancies can exist only in the axial orbitals.

It is thus apparent that for only the d^4, d^5, d^6, d^7, and d^8 ion shells is there a choice of the orbital configuration for the entering electron—the choice being a function of the ligand field. In the d^1, d^2, and d^3 ions the d_ϵ orbitals each receive one electron, and in the d^9 and d^{10} each axial orbital receives its final pairing electron, the ligand field having no effect on the electron configuration in these five cases. To reiterate, in octahedral configuration it is only in the d^4 through d^8 structures that the electron configuration and spin depend upon the field strength of the ligand. Orbital splitting for the octahedral arrangement, together with the configurations discussed in the next paragraph, are summarized in Fig. 3.7.

Similar reasoning can be developed for the splitting of the d orbitals required for the other common geometric arrangements of ligands. Square planar arrangements are essentially the octahedral arrangement with no ligands on the z axis. This is achieved if the four ligands on the xy plane are allowed to move in closer to the central ion; the two ligands on the z

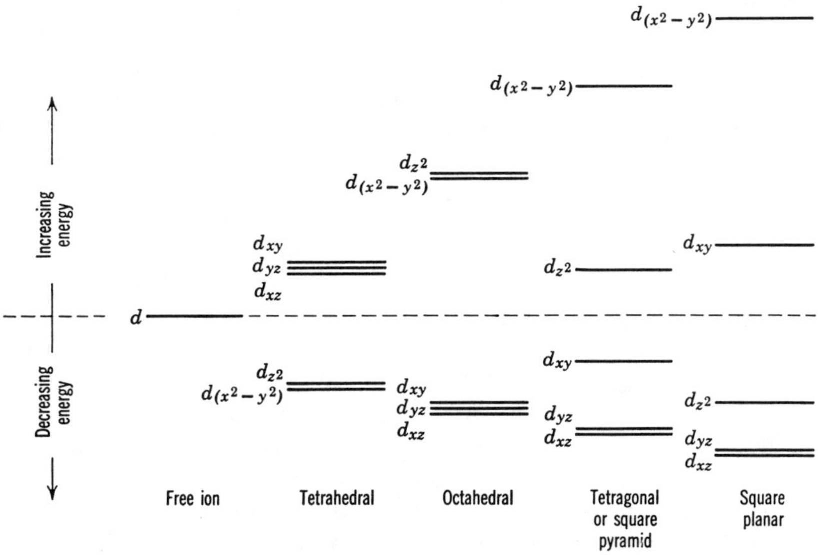

FIGURE 3.7 Splitting of d orbitals related to various regular structures.

axis must now move out. In practice this occurs most readily with ions (or atoms) having the d^8 and d^9 structures, although d^7 can also give the configuration. These electron structures are found in Ni(II), Cu(II), and the Au(III) ions, ions which commonly show the square planar arrangement. With structures d^6 and lower the octahedral arrangement is preferred.

The tetrahedral arrangement forms with nontransition elements and with transition elements and ions with the structures d^0, d^5, and d^{10}. In general each orbital involved must be equally filled to achieve this arrangement. Examples are $[SO_4]^{2-}$, $[Zn(NH_3)_4]^{2+}$, $[HgI_4]^{2-}$, $[AlCl_4]^-$, $[FeCl_4]^-$, $[FeO_4]^{2-}$, $[CoCl_4]^{2-}$, $[MnO_4]^-$, $[CrO_4]^{2-}$.

It is evident that a knowledge of the ligand field strength is essential for an accurate evaluation of the interactions in complex substances. The order for some of the more common ligands is $I^- < Br^- < SCN^- < Cl^- < OH^- < RCO_2^- < F^- < H_2O < C_2O_4^{2-} <$ glycine < pyridine < NH_3 < ethylenediamine < NO_2^- < 1–10 phenanthroline < CN^-.

A few classical examples of application of the ligand field theory will serve as further clarification. Consider the element iron, its cations, and their complexes. Iron itself has the ground state structure $3s^23p^63d_{xy}{}^23d_{yz}{}^1 3d_{xz}{}^13d_{z^2}{}^13d_{(x^2-y^2)}{}^14s^2$. In iron(III) the $4s^2$ and one of the paired $3d$ electrons is lost, leading to the $3s^23p^63d_{xy}{}^13d_{yz}{}^13d_{xz}{}^13d_{(x^2-y^2)}{}^13d_{z^2}{}^1$ structure. If six water molecules or six fluoride ions now move up in octahedral array, the d electrons remain unpaired, since the repulsive effects of the water molecule's fields are not sufficient to provide the additional energy required to pair the two electrons occupying the axial orbitals with the existing electron in say the d_{yz} and d_{xy} orbitals. The electron configuration of the iron in the complex remains that of the simple iron(III) ion. On the other hand, the electrostatic interaction of the approaching strong-field cyanide ions with the electron field of the axial lobes causes these lobes to assume a high-energy character. It now requires less energy to pair electrons in the d_{xy} and d_{yz} orbitals than to maintain them in the unpaired state in the axial orbitals. The electron configuration for the cyanide complex becomes $3s^23p^63d_{xy}{}^23d_{yz}{}^23d_{xz}{}^1$. These structures have been confirmed experimentally in several ways. The most spectacular proof is found in the *paramagnetic* properties of the substances discussed. The strength of the magnetic field of the aquo and fluorine complexes of iron(III) is concomitant with the value calculated for that which should be present with five unpaired electrons. Similarly, theory and experiment both show one unpaired electron in the cyanide complex. Spectral studies and oxidation behavior patterns further confirm these studies.

The salient features of the ligand field theory of complexes have been summarized in Table 3.4. The complex ions have been selected to show

TABLE 3.4 COMPLEX FORMATION AS RELATED TO LIGAND TYPE AND ELECTRON STRUCTURE

Complex	d Electrons	Structure: Hybrid	Structure: Geometry	Electron Configuration[a]: d_{xy} type	axial	s	p	$d_{xy}d_{yz}$	Spin Properties[b]: Unpaired e^-	Exp.	Calc.
$[AlF_6]^{3-}$	d^0	sp^3d^2	Octahedral			××	×× ×× ××	×× ××	0	0	0
$[AlCl_4]^-$	d^0	sp^3	Tetrahedral			××	×× ×× ××		0	0	0
$[Fe(H_2O)_6]^{2+}$	d^6	sp^3d^2	Octahedral	•• • •	• •	××	×× ×× ××	×× ××	4	5.2	4.9
$[Fe(H_2O)_6]^{3+}$	d^5	sp^3d^2	Octahedral	• • •	• •	××	×× ×× ××	×× ××	5	5.9	5.9
$[FeCl_4]^-$	d^5	sp^3	Tetrahedral	• • •	• •	××	×× ×× ××		5		5.9
$[Fe(CN)_6]^{4-}$	d^6	d^2sp^3	Octahedral	•• •• ••	×× ××	××	×× ×× ××		0	0	0
$[Fe(CN)_6]^{3-}$	d^5	d^2sp^3	Octahedral	•• •• •	×× ××	××	×× ×× ××		1	2.0	1.7
$[Ni(NH_3)_4]^{2+}$	d^8	sp^3	Tetrahedral	•• •• ••	• •	××	×× ×× ××		2	2.6	2.8
$[Ni(CN)_4]^{2-}$	d^8	dsp^2	Square planar	•• •• ••	•• ××	××	×× ××		0	0	0
$[Co(CN)_6]^{3-}$	d^6	d^2sp^3	Octahedral	•• •• ••	×× ××	××	×× ×× ××		0	0	0
$[Cu(NH_3)_4]^{2+}$	d^9	dsp^2	Square planar	•• •• ••	•• ××	××	×× ×× •		1	1.9	1.7
$[Zn(NH_3)_4]^{2+}$	d^{10}	sp^3	Tetrahedral	•• •• ••	•• ••	××	×× ×× ××		0	0	0

[a] Nonbonding electrons = •; bonding electrons = ×.
[b] Experimental and calculated magnetic properties are in Bohr magnetons.

both strong- and weak-field ligands so that their effect on electron configuration of the central cation can be evaluated for the different *d* electron structures. The relationship between the different nonbonding electron structures and magnetic properties and geometric arrangement is also brought out. Both the d^2sp^3 octahedral hybrid, known as the *inner-orbital* complex, and the sp^3d^2 hybrid, known as the *outer-orbital* complex, are shown. In general the inner-orbital type is more stable and shows greater covalent character than the outer-orbital type.

Unusual complexes. A few interesting and special aspects of complex structures remain. The following discussion is necessarily brief and intended only as an introduction.

In the years since L. Pauling first suggested their existence π bonds of several types have been fully substantiated. Carbon monoxide and other ligands with first-row donor atoms can form π bonds by overlap between a filled *d* orbital and a vacant *p* orbital of the ligand. In carbon monoxide the σ bond is formed by overlap of the unshared electron pair of carbon with an unoccupied *d* orbital of the metal. The electrons of one of the π bonds between carbon and oxygen are allowed to go to the oxygen, and the resulting pair of vacant *p* lobes on carbon can now overlap with a pair of filled d_{xy} lobes of the metal to give the π bonding. The stability of $Ni(CO)_4$ and the nonexistence of the isoelectronic $Zn(CO)_4^{2+}$ are due to π bonding in the nickel compound. Zinc has no vacant *d* orbital for the stabilizing π bonds. Cyanides and cyanide-containing ligands undergo similar reactions.

Another type of π bond requires an acceptor atom from the second or later period on the ligand. Sulfur, phosphorus, and arsenic serve especially well in this regard. Here a vacant *d* orbital is required on the acceptor atom; it overlaps with a filled orbital of the central atom to give a π bond. Phosphorus trifluoride, the diarsines, and diphosphanes exemplify ligands of this type.

A most interesting recent discovery involves proof of the fact that π-bonded ligands (e.g., C_2H_4) can coordinate in such a way that their π electron cloud overlaps a vacant bonding orbital of the central ion to serve as the bond. The compound $K[PtCl_3C_2H_4]$ is such a structure, the ethylene C=C group being perpendicular to the plane of the resulting square planar complex. "Sandwich" compounds have also received much attention recently. Here the annular ring π electrons of cyclopentadienyl derivatives have been used for bonding purposes. The metal ion is found sandwiched between the annular π clouds of two parallel cyclic structures. These compounds conceivably could find application in pharmaceutical chemistry. Some members of the ferrocene family have been shown to be excellent screening agents for ultraviolet light in a

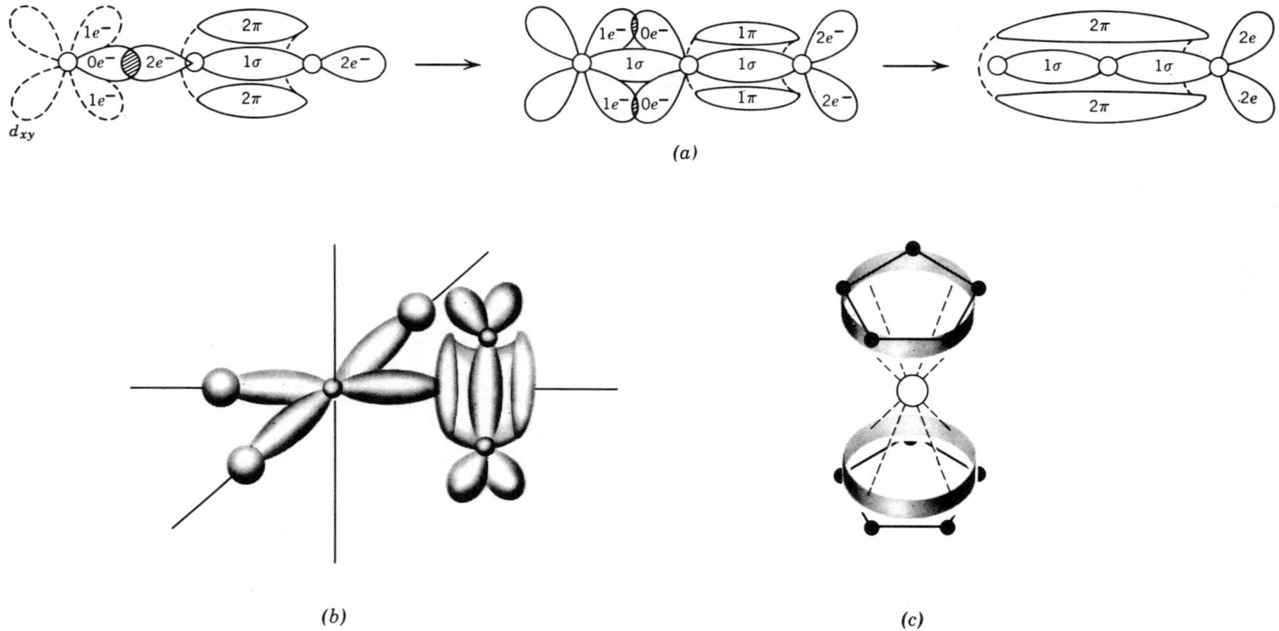

FIGURE 3.8 Examples of π bonding in complex substances: (*a*) bond formation mechanism between Ni and CO in $Ni(CO)_4$; (*b*) $[PtCl_3(C_2H_4)]^-$; (*c*) chromium dicyclopentadiene—annular π clouds shown as circular bands.

nonpharmaceutical application. Ferrocene itself, $Fe(C_5H_5)_2$, might serve as an organic carrier of iron in high concentrations. Figure 3.8 illustrates a few examples of π bonding.

FOUR · CLASSIFICATION OF CHEMICAL REACTIONS

In the preceding chapters chemical reactivity has been considered from the atomistic or structural point of view. The transfer and sharing of electrons between atoms have been studied in relation to atomic structure and electrostatic properties. The field forces associated with particulate species and their interactions with those of neighboring particles have been investigated.

Although such fundamental interpretation of the physical and chemical properties of particles and their reactions is essential for an understanding of the nature of reactions, a practical classification of reactions is equally useful. Such a classification can be based on the type of reactants involved and certain characteristic phenomena always accompanying their interaction. On this basis almost all inorganic chemical reactions classify into one of three categories: oxidation-reduction, the reaction of an oxidizing agent with a reducing agent; neutralization, the reaction of an acid with a base; and complexation, the reaction of a substance capable of acting as a central unit with one or more ligands. In addition it is frequently useful to include physical reactions, e.g., precipitations, and hydrogen bond formation as reaction types.

Closer examination reveals a relationship between this practical classification of reactions and certain of the atomic phenomena which have been earlier discussed. Oxidation-reductions involve electron transfers between two different particles. Such reaction may involve the transfer of an electron to a simple monoatomic particle, as in the reduction

$$\tfrac{1}{2}Cl_2 + e^- = Cl^-$$

or it may involve relatively complex species, as in the oxidation

$$MnO_4^{2-} = MnO_4^- + e^-$$

Acid-base reactions, using Brønsted definitions, may be considered as proton transfers.

$$HCl + H_2O = H_3O^+ + Cl^-$$

Complexation may take place either by reaction between a suitable central unit and ligand molecules or ions, or by a rearrangement or substitution of ligands of an existing complex.

$$[Cu(H_2O)_4]^{2+} + 4NH_3 = [Cu(NH_3)_4]^{2+} + 4H_2O$$

Physical reactions of interest ordinarily involve the appearance of a heterogeneous system, usually by separation of insoluble substances from solution.

$$Ag^+ + Cl^- = AgCl\downarrow$$

$$H_2CO_3 = H_2O + CO_2\uparrow$$

As with complexation, physical reactions involve interaction of the electrostatic field forces of the particles participating. Table 4.1 summarizes this classification of reactions. An observed reaction may involve two or more different reaction types simultaneously or consecutively, as, for example,

TABLE 4.1 CLASSIFICATION OF CHEMICAL REACTIONS

Reaction Type	Reactants	Mechanism
Oxidation-reduction	Oxidizing agent + reducing agent	Electron transfer (complete or sharing)
Neutralization	Acid + base	Proton transfer
Hydrogen bonding	Molecules having strongly electronegative groups	Proton sharing
Complexation	Central unit + ligand(s)	Interaction of particle field forces
Physical	—	Interaction of particle field forces

$$SO_3^{2-} + 2H_3O^+ = H_2SO_3 + 2H_2O$$

followed by

$$H_2SO_3 = H_2O + SO_2\uparrow$$

Complexation and hydrogen bonding have already been considered in Chapter 3. Oxidation-reduction reactions and acid-base reactions will now be considered. Solubility-insolubility relationships will be treated in Chapter 8.

OXIDATION-REDUCTION REACTIONS

Two different numbers, descriptive of the state of reacted atoms, have been used thus far: valence and coordination number. They have evolved over the years with the growth of chemical knowledge and its systemization. As used in this textbook, valence refers to the charge of an ionic particle, and coordination number refers to the number of atoms which may be bonded to the central atom of a molecular grouping or complex ion. Another number, the *oxidation number* (oxidative valence, oxidation state, formal charge), has evolved at the same time, and is at times confused with valence. Oxidation number will be defined as the difference between the number of valence electrons *formally* assigned to a reacted atom and the number of valence electrons that atom would normally have in the unreacted ground state. The sign of the oxidation number is positive if the reacted atom has fewer valence electrons than when unreacted, and negative if it has more electrons than when unreacted. For simple monoatomic ions valence and oxidation number must be identical.

Evaluation of oxidation numbers. Certain formal rules must be established and followed in the evaluation of the oxidation number of an atom when present in a molecule or complex ion. It is generally assumed, irrespective of bond type, that the central atom of a molecule or complex is acting as an electron donor and that the atoms bonded to it are acting as electron acceptors. The oxidation state of coordinated oxo- groups is considered minus two at all times and that of coordinated hydrogen is considered plus one.[1] The determination of the oxidation number of the central atom thus becomes a counting process, i.e., the counting of the number of electrons formally lost to the coordinated atoms or ligands, adjusted, when necessary, for the residual charge in the case of ionic groups. Figure 4.1 illustrates the methodology involved in determination of oxidation numbers. The method used is independent of the actual route of synthesis of the compound, being concerned only with the number of electrons gained or lost irrespective of previous history. The arbitrary nature of the reasoning and methodology often leads to unusual oxidation states, e.g., fractions instead of whole numbers, or if incorrectly viewed as

[1]There is an exception in each case. Peroxo-oxygens each have a minus one oxidation number. The hydride ion also has a minus one state. In this state hydrogen is acting as the first member of the halogen family. It can assume this character only in the presence of the most electropositive elements.

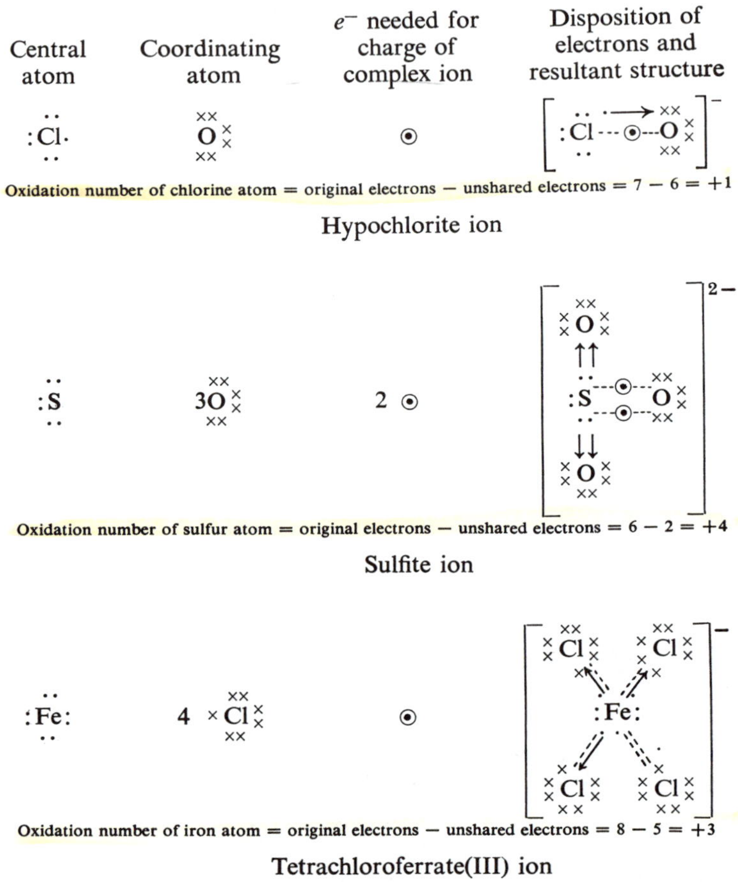

FIGURE 4.1 Determination of oxidation numbers. The formal identification of electrons as to source by using different symbols is a bookkeeping procedure only. All electrons are equivalent irrespective of source or disposition.

valences, some very unusual valence states. The oxidation number of a free element is obviously zero.

Uses of oxidation numbers. The arbitrary concept of oxidation numbers proves useful in several ways. The conventional concept of oxidation-reduction is defined in terms of change in oxidation number. Upon oxidation an element's oxidation number becomes more positive; on reduction it becomes more negative. (-1 is more positive than -3 and $+3$ is more negative than $+5$.) In analysis a knowledge of the change in oxidation number is necessary for calculation of normalities of oxidizing or reducing standard solutions. The equivalent weight is the molecular weight divided by the change in oxidation number taking place during the titration.

Oxidation numbers, however, have no predictive values as to the course of an oxidation-reduction or, as commonly abbreviated, a *redox reaction.* Thus the chemist must turn to other information in order to determine the course of these reactions. Redox reactions involving only ions or atoms present no particular problem in this regard, since previously developed methods based on particle radius, particle charge, and valence shell type are useful in these cases. With polyatomic particles, however, the interactions between the associated atoms changes the availability of the valence electrons for redox reactions. As a result the redox properties are different from those of the unassociated simple ions or atoms. The theoretical prediction of the course of redox reactions involving complex molecules or ions must await not only adequate solution of the equations for the interactions taking place within the complex particle, but also the experimental evaluation of many parameters required in these calculations.

Theory of redox reactions. A good reducing agent is essentially a substance that has valence electrons which are freely available for the reduction of a substance in its oxidized state. Similarly, a good oxidizing agent is a substance which readily accepts (or has a strong affinity for) electrons from some other substance. The process of transferring the electron from the reducing agent to the oxidizing agent is known as oxidation-reduction. In general terms a redox reaction may be represented as follows:

$$R = R^+ + e^- \quad \text{Reducing agent is oxidized} \tag{1}$$

$$O + e^- = O^- \quad \text{Oxidizing agent is reduced} \tag{2}$$

$$R + O = R^+ + O^- \quad \text{Over-all redox reaction} \tag{3}$$

The electron always ends up with that atom which has the strongest affinity for it, the strongest oxidizing agent. It is evident that upon completion of the reaction, theoretically if not practically, the original oxidizing agent has been transformed into a reducing agent and the reducing agent has been transformed into a possible oxidizing agent.

The electrochemist has developed the best method for the measurement of the tendency for electron transfer between reducing and oxidizing agents. In the course of his efforts to convert chemical energy directly into electrical energy, rather than into the kinetic energy (heat), he found that spontaneous redox reactions developed positive potentials (voltages) when set up in the form of stylized electrochemical cells. Nonspontaneous reactions, on the other hand, show negative potentials when set up conventionally. At the same time he found that the larger the numerical value of these potentials, the more rapid and complete the reaction if positive, and the more inert if negative.

These facts can be expressed in the Nernst equation:

$$\mathscr{E} = \mathscr{E}^0 - \frac{0.059}{n} \log Q_a$$

where $\mathscr{E}$ is the potential, in volts, of an electrochemical cell set up in the conventional manner. $\mathscr{E}^0$ is the standard potential, which is defined as the potential of the above cell when all reactants and products are at unit activity. The constant term, 0.059, represents the combination of various constants. It is valid only at 25°C. The symbol n is the number of electrons transferred for the reaction as written. Since equations frequently have several forms, the numerical coefficients of which must always be multiples, n will depend upon the equation selected. This does not lead to different values of $\mathscr{E}$, depending on the equation selected, since compensating changes appear in Q_a. Q_a is the activity quotient. It is set up using conventions identical to those used in setting up an equilibrium constant. It differs from the equilibrium constant in that the concentrations used in its evaluation are actual experimental values rather than equilibrium concentrations.[2] If the selected concentrations are equilibrium concentrations, then $Q_a \equiv K$. The same equation must be used for setting up Q_a as was used to determine n. Using Equation 3,

$$Q_a = \frac{[R^+][O^-]}{[R][O]}$$

In order to use the Nernst equation, $\mathscr{E}^0$ must be known. Its values are not available for over-all reactions but may be evaluated from *standard electrode potentials*. As the equation for a redox reaction may be separated into two half-reactions—the reduction and the oxidation, respectively—the

[2]For a strict theoretical application of equilibrium principles, activities must be used. Activity may be defined qualitatively as the "effective concentration." In solutions, because of the electrostatic interactions between the ions present and between the ions and the solvent molecules, concentrations do not give a true measure of the reacting capacity of a substance. Except for very dilute solutions, reacting capacity is usually less than the stoichiometric concentration. Since activities tend to approach molar concentrations in dilute solution, molar concentrations are usually used in calculations, as in this textbook. The true relationship between activities and molar concentrations is expressed by

$$a = \gamma c$$

where γ is the activity coefficient, a is the activity, and c is the concentration. Activity coefficients are determined experimentally from freezing point and electrode potential data.

electrochemical cell consists of two *half-cells*: one half-cell in which the oxidation is taking place and one in which the reduction is taking place. The standard potentials $\mathscr{E}^0_{1/2}$ for each of the half-cells of an electrochemical cell can be determined experimentally, and extensive tables are available.[3] Some standard electrode potentials of interest in inorganic pharmaceutical chemistry are found in Table 4.2. Just as the over-all redox equation may be obtained by summing the two half-reactions, the over-all $\mathscr{E}^0$ for an electrochemical cell may be obtained by algebraically summing $\mathscr{E}^0_{1/2}$ values for each half-cell:

$$\mathscr{E}^0_{\text{over-all}} = \mathscr{E}^0_{1/2,\,\text{oxidation}} + \mathscr{E}^0_{1/2,\,\text{reduction}}$$

Care must be exercised in selecting the $\mathscr{E}^0_{1/2}$ values to be substituted. If the half-reaction of the cell is exactly that given in the table, the $\mathscr{E}^0_{1/2}$ in the table is used as given. But if the half-reaction, although identical in the substances involved, has the products and reactants in a reverse order, the sign of the $\mathscr{E}^0$ value given in the table must be reversed before summing.

Applications in pharmaceutical chemistry. Inorganic pharmaceutical chemistry has many preparations and procedures which involve redox reactions. A better appreciation of the application of redox potentials to calculations will be achieved by their application to some of these cases. The preparation of iron(II) iodide from the elements is a simple example. The half-reactions for the reaction may be written:[4]

$$\begin{aligned} Fe &= Fe^{2+} + 2e^- \quad \text{(oxidation)} \\ I_2 + 2e^- &= 2I^- \quad \text{(reduction)} \end{aligned}$$

These half-reactions may be summed algebraically to give the over-all reaction

$$I_2 + Fe = Fe^{2+} + 2I^-$$

The evaluation of the $\mathscr{E}^0_{\text{over-all}}$ for this reaction is

$$\mathscr{E}^0_{\text{over-all}} = \mathscr{E}^0_{1/2,\text{oxidation}} + \mathscr{E}^0_{1/2,\text{reduction}}$$

Examining the half-reactions, it is seen that the oxidation is exactly as

[3]Unfortunately two systems, the American and the European, are in use for recording standard potentials. The obvious difference is a reversal of the sign of the potential. This textbook uses the American convention. Students with a good background in elementary theory will find an interesting review of this problem by T. S. Licht and A. J. de Bethune, *J. Chem. Educ.*, **34,** 433–440 (1957).

[4]A requirement which must always be observed in balancing redox equations is that the number of electrons given off in the oxidation half-reaction must equal the number of electrons required for the reduction reaction. This is another application of the requirement that over-all electrical neutrality must be maintained in chemical systems.

TABLE 4.2 STANDARD POTENTIALS

Half-cell Reaction	$\mathscr{E}^0_{1/2}$ (volts)
$Li \rightleftharpoons Li^+ + e^-$	3.02
$Na \rightleftharpoons Na^+ + e^-$	2.71
$Al \rightleftharpoons Al^{3+} + 3e^-$	1.67
$3OH^- + H_2PO_2^- \rightarrow HPO_3^{2-} + 2H_2O + 2e^-$	1.60
$2OH^- + SO_3^{2-} \rightleftharpoons SO_4^{2-} + H_2O + 2e^-$	0.93
$3H_2O + H[H_2PO_2] \rightarrow H_3PO_3 + 2H_3O^+ + 2e^-$	0.59
$S^{2-} \rightleftharpoons S + 2e^-$	0.51
$Fe \rightleftharpoons Fe^{2+} + 2e^-$	0.44
$Fe \rightleftharpoons Fe^{3+} + 3e^-$	0.036
$H_2 + 2H_2O \rightarrow 2H_3O^+ + 2e^-$	0.00
$H_2S + 2H_2O \rightarrow S^0 + 2H_3O^+ + 2e^-$	−0.14
$Sn^{2+} \rightarrow Sn^{4+} + 2e^-$	−0.15
$Bi + 3H_2O + Cl^- \rightarrow BiOCl + 2H_3O^+ + 3e^-$	−0.16
$H_2SO_3 + 5H_2O \rightarrow SO_4^{2-} + 4H_3O^+ + 2e^-$	−0.17
$2S_2O_3^{2-} \rightleftharpoons S_4O_6^{2-} + 2e^-$	−0.17
$Cl^- + Ag \rightarrow AgCl + e^-$	−0.22
$2Cl^- + 2Hg \rightarrow Hg_2Cl_2 + 2e^-$	−0.27
$Bi + 3H_2O \rightarrow BiO^+ + 2H_3O^+ + 3e^-$	−0.32
$4OH^- \rightarrow O_2 + 2H_2O + 4e^-$	−0.40
$2I^- \rightarrow I_2 + 2e^-$	−0.54
$3I^- \rightarrow I_3^- + 2e^-$	−0.54
$4H_2O + HAsO_2 \rightleftharpoons H_3AsO_4 + 2H_3O^+ + 2e^-$	−0.56
$ClO^- + 2OH^- \rightleftharpoons ClO_2^- + H_2O + 2e^-$	−0.66
$H_2O_2 + 2H_2O \rightleftharpoons O_2 + 2H_3O^+ + 2e^-$	−0.68
$Fe^{2+} \rightarrow Fe^{3+} + e^-$	−0.77
$2Hg \rightarrow Hg_2^{2+} + 2e^-$	−0.79
$Ag \rightarrow Ag^+ + e^-$	−0.80
$Hg \rightarrow Hg^{2+} + 2e^-$	−0.85
$Hg_2^{2+} \rightarrow 2Hg^{2+} + 2e^-$	−0.92
$HNO_2 + 4H_2O \rightarrow NO_3^- + 3H_3O^+ + 2e^-$	−0.94
$2OH^- + Cl^- \rightarrow ClO^- + H_2O + 2e^-$	−0.94
$2Br^- \rightarrow Br_2(aq) + 2e^-$	−1.09
$2H_2O + Cl^- \rightarrow HClO + H_3O^+ + 2e^-$	−1.49

given in the table of standard potentials, whereas the reduction is the reverse of that in the table. This means that the $\mathscr{E}^0_{1/2,\text{oxidation}}$ will be used as given in the table, but $\mathscr{E}^0_{1/2,\text{reduction}}$ requires the reversal of the sign given in the table.

$$\mathscr{E}^0_{\text{over-all}} = (+0.44) + (+0.56)$$
$$= +1.00 \text{ volts}$$

The Nernst equation,

$$\mathscr{E} = \mathscr{E}^0 - \frac{0.059}{n}\log Q_a$$

on substitution becomes

$$\mathscr{E} = (+1.00) - \frac{0.059}{2}\log\frac{[Fe^{2+}][I^-]^2}{[Fe][I_2]}$$

The concentration of reaction product, iron(II) iodide, is initially negligibly small; during the course of the reaction its concentration rises to a predetermined value.

By convention, reactants (or products) in the solid state rather than in solution are considered to have unit activity. Consequently the numerical value of the activity quotient, Q_a, must be <1 on substitution of appropriate concentrations. Since the log of a number less than one is negative, it follows that $\mathscr{E} > +1.0$ volt at all times during the course of the reaction. The actual numerical values involved at various stages of the reaction are given in Table 4.3. Since the electrode potential of the theoretical cell for this synthesis is positive, the reaction is spontaneous.

TABLE 4.3 EFFECT OF CONCENTRATION OF REACTION PRODUCT ON $\mathscr{E}$

Fraction Reacted	Moles Fe^{2+}	Moles I^-	Q_a	$-(0.059/2)\log Q_a$	$\mathscr{E}$
0.25	0.05	0.10	5×10^{-4}	+0.10	+1.10
0.50	0.10	0.20	4×10^{-3}	+0.07	+1.07
0.75	0.15	0.30	1.4×10^{-2}	+0.05	+1.05
1.00	0.20	0.40	3.2×10^{-2}	+0.04	+1.04

In practice the reaction stops at the iron(II) stage. The iron(III) state should be possible, at least theoretically, but application of electrode potential theory explains why it is not. Assuming the reaction is possible,

$$2Fe^{2+} = 2Fe^{3+} + 2e^- \quad \text{(oxidation)}$$
$$I_2 + 2e^- = 2I^- \quad \text{(reduction)}$$
$$2Fe^{2+} + I_2 = 2Fe^{3+} + 2I^- \quad \text{(total reaction)}$$

Applying the Nernst equation,

$$\mathscr{E}^0_{\text{over-all}} = (-0.77) + (+0.54) = -0.23$$

$$\mathscr{E} = -0.23 - \frac{0.059}{2}\log\frac{[Fe^{3+}]^2[I^-]^2}{[Fe^{2+}]^2[I_2]}$$

$\mathscr{E}^0_{\text{over-all}}$, being negative, already indicates a nonspontaneous reaction. However, the concentration effect must be considered since through the activity quotient, Q_a, it can change the sign of the over-all $\mathcal{E}$ for the reaction. Assuming the same initial quantities of reactants as in the immediately preceding illustration, the limiting concentration of iodide ion will again be 0.4*M*. Iodine will again have unit activity during the course of the reaction. It is the Fe^{3+}/Fe^{2+} ratio which will be significant in determining the final value of Q_a. The effect of changing this ratio is shown in Table 4.4. It is evident from these calculations that iron(III)

TABLE 4.4 EFFECT OF RELATIVE CONCENTRATIONS ON $\mathcal{E}$ VALUE

(Fe^{3+}/Fe^{2+}) Ratio	$[I^-]^2$	Q_a	$-(0.059/2) \log Q_a$	$\mathcal{E}$
1/2	1.6×10^{-3}	4×10^{-4}	+0.10	−0.13
1/4	1.6×10^{-3}	1×10^{-4}	+0.12	−0.11
1/10	1.6×10^{-3}	1.6×10^{-5}	+0.15	−0.08
1/100	1.6×10^{-3}	1.6×10^{-7}	+0.21	−0.02
1/1000	1.6×10^{-3}	1.6×10^{-9}	+0.27	+0.04

can form to only a very small extent from iron(II) during this reaction. Only when ratios approach 1 iron(III) ion per 1000 iron(II) ions does the $\mathcal{E}$ for this reaction become positive. Thus for practical purposes the oxidation of iron(II) to iron(III) by iodine may be considered negligible, and the reaction of elemental iron and elemental iodine produces iron(II) iodide, not iron(III) iodide.

From a theoretical point of view, the study of electrode potentials shows that the electron affinity of the iodine atom is sufficiently great to cause the iron atom to surrender two of its valence electrons. However, the iodine atom's electron affinity is not sufficient to cause iron to surrender a third electron. In fact, methods which theoretically should yield iron(III) iodide yield instead a mixture of iron(II) iodide plus elemental iodine, the iron(III) iodide undergoing autooxidation as quickly as it forms.

Another interesting example of the use of electrode potentials to explain the efficacy of certain procedures exists in the use of sodium thiosulfate as an antoxidant for potassium iodide solution. The reactions taking place are

$$2S_2O_3^{2-} = S_4O_6^{2-} + 2e^- \quad \text{(oxidation)}$$

$$I_2 + 2e^- = 2I^- \quad \text{(reduction)}$$

$$2S_2O_3^{2-} + I_2 = S_4O_6^{2-} + 2I^- \quad \text{(total reaction)}$$

$\mathscr{E}^0_{\text{over-all}}$ may be calculated

$$\begin{aligned}\mathscr{E}^0_{\text{over-all}} &= (-0.17) + (+0.54)\\ &= +0.37\end{aligned}$$

The Nernst equation for this reaction becomes

$$\mathscr{E} = (+0.37) - \frac{0.059}{2}\log\frac{[S_4O_6{}^{2-}][I^-]^2}{[S_2O_3{}^{2-}]^2[I_2]}$$

Using the *N.F. XI* preparation as a basis for calculations, the concentration of iodide is approximately $4M$ and the thiosulfate approximately $0.003M$. The concentrations of the tetrathionate ion and the elemental iodine must each be negligibly small in the freshly prepared solution and may be neglected; thus

$$\begin{aligned}&= (+0.37) - 0.03\log\frac{[4]^2\cancel{[S_4O_6{}^{2-}]}}{[0.003]^2\cancel{[I_2]}}\\ &= (+0.37) - 0.03(-5.77)\\ &= +0.20\end{aligned}$$

Since the potential is positive, thiosulfate must afford protection to this preparation by maintaining the iodine in the form of the iodide ion. It is obvious on examination of the form of Q_a that if the solution is excessively exposed to air, the conversion of thiosulfate to tetrathionate will result in an increase in the numerical value of this factor to the extent that it cannot be neglected. As a result the value of $\mathscr{E}$ becomes less positive. Eventually the protective action will be lost, and elemental iodine will appear.

The entire subject of electrochemical cells and the meaning of their potentials is of necessity a matter of convention. All conventions must be rigidly adhered to if meaningful results are to be obtained. A thorough discussion of the conventions used in the preceding simple applications of electrochemistry to pharmaceutical chemistry will be found in elementary physical chemistry textbooks.

In addition to their analytical applications, electrode potentials are able to predict the course of redox reactions. This arises from the sound experimental basis for the values of the constant terms which must be used in the calculations, and the relative simplicity of these calculations. At this point it is the only method which is capable of predicting, with reasonable accuracy, the tendency for electron transfer between a donor and acceptor atom, regardless of type of resultant bonding. Although the theoretical discussion of electron transfers given in this chapter is essential to a fundamental understanding of chemical reactions involving electron transfers, it is incapable of reasonably simple and accurate quantitative calculations. The available experimental data required for these calculations is inadequate both in quantity and quality, and the equations involved are too complex for practical calculations at this stage. As more data and better solutions of the fundamental equations are achieved, the predictive value of this more fundamental approach should be enhanced.

Mechanisms of redox reactions. The mechanism of redox reactions has been the subject of great interest during the past few decades. The question of the means by which electrons are transferred from the oxidizing agent to the reducing agent is important since electrons are incapable of independent existence in ordinary chemical situations—in solution or even in gaseous media except under very high vacuum. For this reason it appears essential that contact, probably through a temporary bond, must be achieved as a first step in redox reactions. Two ways probably exist for achieving this contact. The first is the formation of a temporary bond of the Lewis acid-base type (page 94). The second is the ability of an electron to penetrate the electron cloud of the reducing agent upon collision.

In the first type of redox mechanism the formation of an intermediate compound which includes both oxidant and reductant is essential. During the existence of the intermediate the shift of the molecular electron cloud resulting in the redox reaction takes place. This is followed by a dissociation of the intermediate. The intermediates of some redox reactions have been identified, and a few are given in Table 4.5.

Although several cases of electron transfer without an intermediate compound have been identified, a clear explanation has not evolved.

TABLE 4.5 REDOX MECHANISMS

Oxidizing Agent	Reducing Agent	Intermediate	Resultant Substances
Fe^{3+}	I^-	FeI_2^+, $FeI(OH)^+$	Fe^{2+}, I_2
Fe^{3+}	Sn^{2+}	$[Fe:Cl:SnCl_3]^+$	Fe^{2+}, Sn^{4+}
NO_2^-	HSO_3^-	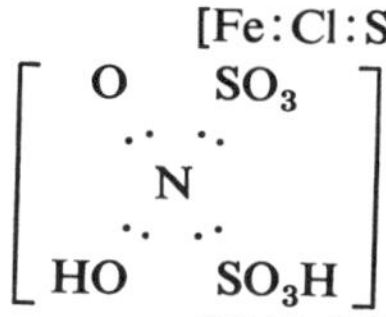 $[(HO)ONSO_3]^-$	N_2, SO_4^{2-}
BrO_3^-	HSO_3^-	$[O_2BrOSO_3H]^{2-}$	Br_2, SO_4^{2-}

Apparently this type of transfer involves ions which are weakly solvated. Collisions can occur directly between the ions involved, the electron being able to approach very closely the electron cloud boundary of the ion to be reduced. The electron possibly then makes its way from one particle to another easily. This type of redox is illustrated in solutions having both ferrocyanide and ferricyanide ions present. The extra valence electron of the iron(II) is actively exchanged with iron(III) atoms. The exchange is direct, since it has been shown that the cyanide ligand units are tightly held and cannot be removed during the electron transfer. This type of redox has been explained as "quantum mechanical tunneling." A more thorough discussion of this subject will be found in modern textbooks of inorganic chemistry.

ACID-BASE REACTIONS

Theories of acids and bases. The oldest acid-base theory is that due to Arrhenius, traceable to his theory of electrolytic dissociation, first proposed in 1884. The definitions which grew from this theory are familiar to all students of elementary chemistry: An acid is a substance which can give up hydrogen ions. A base is a substance which can give up hydroxyl ions. Neutralization is the reaction of an acid with a base to give a salt and water. Hydrolysis, the converse of neutralization, requires a separate definition (page 110).

The limitations of the Arrhenius concept become evident when the following reactions are considered:[5]

$$Na^+(OH)^- + H^+Cl^- = Na^+Cl^- + H_2O \quad \text{(water solution)}$$

$$OAc^- + H_2O = HOAc + OH^- \quad \text{(water solution)}$$

$$Na^+(NH_2)^- + (NH_4)^+Cl^- = Na^+Cl^- + 2NH_3 \quad \text{(liquid ammonia solution)}$$

$$H_2SO_4 + HNO_3 = (H_3SO_4)^+ + NO_3^- \quad \text{(hydrogen sulfate solution)}$$

$$C_5H_5N + AlCl_3 = C_5H_5NAlCl_3 \quad \text{(chlorobenzene solution)}$$

Each of these equations represents a neutralization, yet only the first qualifies as such by the Arrhenius theory. In the light of modern knowledge, it will be seen that the Arrhenius system of acids and bases is a special case of the more sophisticated acid-base systems.

As early as 1905, E. C. Franklin realized that water is not unique as a

[5]The form HOAc will be used to represent acetic acid throughout this textbook. The acetate ion will be written OAc^-.

medium for acid-base reactions (nor as an ionizing solvent). He recognized the ammonia system of acids and bases, a recognition which led to the gradual evolution of the general theory of solvent systems of acids and bases. Accordingly, the basis for a given acid-base system is the autoionization of the parent solvent to give rise to the acid and base ions. A compilation of solvents and their respective acid and base ions is given in Table 4.6. The extent of autoionization of these solvents is very small.

TABLE 4.6 ACID-BASE ION SYSTEMS BASED ON THE FRANKLIN THEORY OF SOLVENT SYSTEMS

Solvent	Autoionization Reactions	Unsolvated Cation
H_2O	$H_2O + H_2O = H_3O^+ + OH^-$	H^+
NH_3	$NH_3 + NH_3 = NH_4^+ + NH_2^-$	H^+
HOAc	$HOAc + HOAc = H_2OAc^+ + OAc^-$	H^+
SO_2	$SO_2 + 2SO_2 = [SO{\cdot}SO_2]^{2+} + SO_3^{2-}$	SO^{2+}
$COCl_2$	$COCl_2 + COCl_2 = [COCl{\cdot}COCl_2]^+ + Cl^-$	$COCl^+$

Water has an ion product, K_w, of 1.008×10^{-14} at 25°C.; ammonia has an ion product of approximately 10^{-22} at its boiling temperature, −33°C. Some neutralizations involving different solvent systems are

$$Na^+(OH)^- + H_3O^+Cl^- = Na^+Cl^- + 2H_2O \quad \text{(water)}$$
$$Na^+(NH_2)^- + NH_4^+Cl^- = Na^+Cl^- + 2NH_3 \quad \text{(liquid ammonia)}$$
$$(SO{\cdot}SO_2)^{2+}Cl^-_2 + Cs^+_2SO_3^{2-} = 2Cs^+Cl^- + 3SO_2 \quad \text{(liquid sulfur dioxide)}$$

The Franklin theory of acids and bases has direct significance in pharmaceutical chemistry because it can account for certain observed properties of the several nitrogen hydride substituent groups. The acidic nature of the ammonium ion and the amine salt cations is explained, as is the basic nature of the amine group. The amine group of the ammonia system is the analog of the hydroxyl group of the water system, and it is thus apparent why the acid amides have an acidic character. Examination of $—CONH_2$ and $—SO_3NH_2$ shows that they are the analogs of the $—COOH$ and $—SO_3OH$, respectively.

In 1923, J. N. Brønsted and Lowry independently proposed the proton donor theory of acids. Since I. M. Lowry subsequently did little further work on the theory, it is frequently identified as the Brønsted theory of acids and bases. Accordingly, an acid is defined as a substance which can give up a proton, i.e., a proton donor. A base is defined as a substance which can accept a proton, i.e., a proton acceptor. In accepting a proton, the base becomes an acid; in losing its proton, the acid becomes a base.

A *conjugate pair* is an acid and base related to one another by the presence or absence of a proton. It should be stated emphatically that the process of proton transfer implied in defining a conjugate pair is not neutralization.

Acids may be cations, neutral molecules, or anions. Bases are usually neutral molecules or anions; very few cationic bases are known. Table 4.7 gives a comprehensive series of conjugate pairs. On examination of

TABLE 4.7 CONJUGATE PAIRS OF ACIDS AND BASES

Acid	Base	Acid	Base
HCl	Cl^-	H_3PO_4	$H_2PO_4^-$
$HOAc$	OAc^-	$H_2PO_4^-$	HPO_4^{2-}
H_3O^+	H_2O	HPO_4^{2-}	PO_4^{3-}
H_2O	OH^-	$[Al(H_2O)_6]^{3+}$	$[Al(H_2O)_5(OH)]^{2+}$
NH_4^+	NH_3	$[Al(H_2O)_5(OH)]^{2+}$	$[Al(H_2O)_4(OH)_2]^+$
NH_3	NH_2^-	$H_3BO_3 \cdot H_2O$	$[B(OH)_4]^-$
H_2CO_3	HCO_3^-	$C_2H_5NH_3^+$	$C_2H_5NH_2$
HCO_3^-	CO_3^{2-}	$C_6H_5SO_3NH_2$	$(C_6H_5SO_3NH)^-$

this list it will be seen that some substances enter into two different conjugate pairs. Water serves to illustrate: It is the acid and base, respectively, of two different conjugate pairs. Oxonium ion and hydroxyl ion do not constitute a conjugate pair; conjugate pairs may differ in constitution by a single proton only. A strong acid loses its proton readily, whereas a weak acid holds its proton tenaciously. It follows that the conjugate base of a strong acid is a weak base, and the conjugate base of a weak acid is a strong base.

Neutralization is the transfer of the proton from the acid of one conjugate pair to the base of a second conjugate pair. In neutralization the proton seeks the strongest base, or, conversely, the strongest acid loses its proton to the strongest base present. A general equation for neutralization may be written

$$HB_1 + B_2 = HB_2 + B_1$$

where HB_1 and B_1 represent one conjugate pair and HB_2 and B_2 the other. Some typical neutralizations follow:

$$HCl + H_2O = H_3O^+ + Cl^- \quad (4)$$

$$H_2O + OAc^- = HOAc + OH^- \quad (5)$$

$$NH_4^+ + NH_2^- = NH_3 + NH_3 \quad (6)$$

$$HSO_4^- + OH^- = H_2O + SO_4^{2-} \quad (7)$$

$$[Al(H_2O)_6]^{3+} + H_2O = H_3O^+ + [Al(H_2O)_5(OH)]^{2+} \quad (8)$$

$$H_2PO_4^- + PO_4^{3-} = HPO_4^{2-} + HPO_4^{2-} \quad (9)$$

$$H_3O^+ + HCO_3^- = H_2CO_3 + H_2O \quad (10)$$

Some useful facets of the theory are illustrated. Equation 4 explains why hydrogen chloride is not hydrochloric acid until "ionization" has taken place. Equations 5 and 8 suggest the fact that the type of reaction commonly known as hydrolysis is actually a neutralization.

Neutralization, in the Brønsted sense, is simply the application of the ideas developed when discussing electrostatic properties and bonding. The proton will find its way to the group (base) having the strongest negative electrostatic field. Consider the solution of hydrogen chloride in water. The water molecule has a covalent radius of 0.67 Å, and the chloride ion has an ionic radius of 1.81 Å. Since water is a polar molecule, the oxygen carries a net negative charge on the side away from the hydrogens. Considering its small radius, the net negative electrostatic field of the water dipole at bonding distances is stronger than the disperse spherically distributed field of large radius from the chloride ion's single negative charge. Consequently, water is a stronger base than chloride ion, and the proton of hydrogen chloride is transferred to the water molecule in aqueous media.

The Brønsted theory has proven of special value in the treatment of aqueous acid-base reactions. It recognizes the dual role of water. Since inorganic medicinals usually involve water or are based on aqueous systems, the Brønsted theory is especially well adapted for explanation of acid-base phenomena in this field. Hydrolysis and the use of salts as acids and bases are explained with special clarity. For these reasons the Brønsted theory will be used as a basis for the discussion of acid-base phenomena in this textbook.

One of the most comprehensive acid-base theories is that due to G. N. Lewis. Acids are defined as ions or molecules whose structures contain an atom with at least one vacant suborbital in its valence shell. The following are examples of Lewis acids: the proton, which lacks its $1s^2$ electron pair; aluminum chloride, which lacks one pair of electrons of a completed $2s^22p^6$ octet; tin(IV) chloride, which lacks two electron pairs of a $5s^25p^65d^4$ structure. Bases are defined as ions or molecules whose structures include an electronegative atom having a nonbonding electron pair available for sharing. Substances such as hydroxyl or chloride ions and ammonia or water molecules have nonbonding electron pairs and are therefore Lewis bases. Neutralization is defined as the formation of a coordinate covalent bond between an acid and a base. Some diagrammatic equations showing typical Lewis neutralizations are given in Fig. 4.2.

From a theoretical point of view the Lewis interpretation lends a unifying approach to neutralization processes. By its use many otherwise seemingly unrelated reactions are clarified and brought together. However,

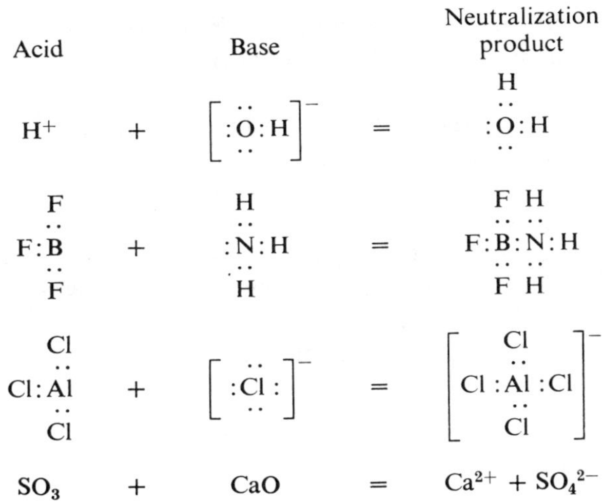

FIGURE 4.2 Lewis neutralizations. The three non-bonding electron pairs of each of the covalently bonded fluoride and chloride ions are not shown.

from a practical point of view its broad scope is not always an advantage. If accepted, every reaction involving coordinate covalent bond formation becomes a neutralization. For the present at least the more restrictive Brønsted theory seems of greater utility for interpretation of acid-base phenomena in inorganic pharmaceutical chemistry.

The Usanovich acid-base theory is even more comprehensive than the Lewis approach. It includes all possible examples of acid-base behavior. Oxidation-reduction reactions are considered an aspect of neutralization phenomena. Like the Lewis theory, it suffers from being too general. For practical purposes it would seem more logical to keep the classical division of reactions as acid-base reactions, coordination reactions, and oxidation-reduction reactions rather than bringing all together in one generalized theory. Those interested will find further details on acid and base theories in advanced inorganic chemistry textbooks.

Strengths of acids and bases. Although from a descriptive point of view it is of value to recognize that the strength of an acid is measured by the availability of its proton, a quantitative basis of evaluation is required. Formerly *degree of dissociation*, α, was used to express this parameter, which is the fraction of acid or base existing in the ionic form. Its value is limited, however, because its numerical value is not only a function of temperature and solvent, but also varies with concentration. In addition the effect of common ions and pH cannot be evaluated by its use.

A more useful measure of acid (and base) strength is the *ionization constant*, which is based on equilibrium considerations. Although dependent on temperature and the solvent used, an ionization constant has the advantage that within its limits it is independent of the concentration of the acid or base, pH, and common ion effects. Ionization constants may be derived for any slightly ionized substance, e.g., water, mercury(II) chloride. However, in practice their application has largely been limited to acids and bases.

Consider the general equation for a Brønsted neutralization:

$$HB_1 + B_2 = HB_2 + B_1$$

Its equilibrium constant is written

$$K = \frac{[HB_2][B_1]}{[HB_1][B_2]}$$

In an aqueous system B_2 becomes water and the constant may be written

$$K = \frac{[\mathrm{H_3O^+}][B_1]}{[HB_1][\mathrm{H_2O}]}$$

In an aqueous solution the concentration of water is very large compared to the concentration of the other substances present. Any change in its concentration caused by a shift in concentration of the other constituents of the solution must be negligible when compared to the total water present. For this reason $[H_2O]$ may be considered a constant. The usual mathematical practice of combining all constants as one new constant gives

$$K_a = K[\mathrm{H_2O}] = \frac{[\mathrm{H_3O^+}][B_1]}{[HB_1]}$$

where K_a is known as the acid ionization constant.

In a similar manner the base ionization constant, K_b, may be derived:

$$B_1 + \mathrm{H_2O} = \mathrm{OH^-} + HB_1$$

$$K_b = K[\mathrm{H_2O}] = \frac{[\mathrm{OH^-}][HB_1]}{[B_1]}$$

Although it is not obvious from direct examination, K_a and K_b of a given conjugate pair are related through the parent solvent. The ionization constants may be rewritten to give, respectively,

$$\frac{K_b}{[\mathrm{OH}^-]} = \frac{[HB_1]}{[B_1]}$$

$$\frac{K_a}{[\mathrm{H_3O^+}]} = \frac{[B_1]}{[HB_1]}$$

Equating and rearranging,

$$\frac{K_a}{[\mathrm{H_3O^+}]} = \frac{[\mathrm{OH}^-]}{K_b}$$

$$K_a \cdot K_b = [\mathrm{H_3O^+}][\mathrm{OH}^-]$$

or

$$K_a \cdot K_b = K_w$$

The relationship may be illustrated using the NH_4^+, NH_3 conjugate pair.

$$\mathrm{NH_4^+ + H_2O = NH_3 + H_3O^+}$$

$$K_a = \frac{[\mathrm{NH_3}][\mathrm{H_3O^+}]}{[\mathrm{NH_4^+}]} = 5.7 \times 10^{-10}$$

$$\mathrm{NH_3 + H_2O = NH_4^+ + OH^-}$$

$$K_b = \frac{[\mathrm{NH_4^+}][\mathrm{OH^-}]}{[\mathrm{NH_3}]} = 1.8 \times 10^{-5}$$

$$K_a \cdot K_b = [5.7 \times 10^{-10}][1.8 \times 10^{-5}] = 1.03 \times 10^{-14} \sim K_w$$

The use of both K_a and K_b in discussions may easily result in some confusion. For this reason K_a will ordinarily be used in this textbook. If required, K_b can be derived from K_a from their relationship to K_w.

Since oxonium and hydroxyl ion concentrations are important in the discussion of the acid-base strength of a solution, the essential details of pH must also be reviewed. pH is a defined term, originated by S.P.L. Sorenson in 1909 to simplify the writing of the very small oxonium ion concentrations encountered in many biological fluids. The basic definition is

$$\mathrm{pH} = -\log a_{\mathrm{H_3O^+}}$$

where a is activity. In practical calculations molar concentration is usually substituted for activity, giving Sorenson's original definition,

$$\mathrm{pH} = -\log [\mathrm{H_3O^+}]$$

A similarly defined term exists for hydroxyl ion:

$$\mathrm{pOH} = -\log [\mathrm{OH^-}]$$

(A pH meter measures the activity of the oxonium ion, not its molar concentration.)

The relationship existing between pH and pOH in aqueous media may be shown using K_w.

$$K_w = [H_3O^+][OH^-] = 1 \times 10^{-14}$$

Taking the log of both sides of the final equality gives

$$\log [H_3O^+] + \log [OH^-] = -14$$

which on multiplying by (-1) becomes

$$-\log [H_3O^+] - \log [OH^-] = 14$$

On the basis of the definitions for pH and pOH previously given, this becomes

$$\text{pH} + \text{pOH} = 14$$

However, since K_w is temperature dependent, the constant 14, relating pH to pOH, is valid only at 25°C. High concentrations of solute will also change its value.

The application of the pH concept is frequently rather haphazard and often in complete disregard of the underlying facts of its definition and water dependence. Being based on K_w, pH measurements are without meaning in nonaqueous systems or even in mixed solvent systems in which one member is water. For this reason the determination or assignment of a pH, as, for example, to nonaqueous base ointments or hydroalcoholic solutions, is meaningless. pH and pOH measurements are also temperature dependent. This fact is frequently ignored despite introduction of relatively large errors, e.g., pH + pOH $\sim$ 12 at 100°C. Measurements at temperatures other than 25°C. are without meaning unless compensated for the change in K_w. (Most pH meters have an electronic temperature compensation feature.)

Although they have no sound theoretical basis, pH measurements in nonaqueous systems can be valuable for control purposes. Thus if the finish point of a nonaqueous base ointment gives a "reading" of 5.0 pH units, this point may be used as the finish point for subsequent batches provided all measurements are made at the same temperature. However it must be remembered that this is a reproducible number only; it is no the actual pH of the ointment.

Another frequently unwarranted practice is the assignment of numerica values to measured or calculated pH or pOH, indicating an accurac impossible with present pH scales. Even under the most ideal conditioi accuracies beyond ±0.03–0.05 pH units are impossible of achievemen

TABLE 4.8 IONIZATION CONSTANTS OF SELECTED ACIDS AT 25°C.

Acid	K_a	pK_a
Acetic	1.8×10^{-5}	4.72
(Aliphatic alcohols)	$\sim 10^{-16}$	16.
Ammonium ion	5.6×10^{-10}	9.26
Arsenious	6×10^{-10}	9.22
Arsenic	5×10^{-3}	2.30
Benzoic	6.3×10^{-5}	4.20
Boric	5.8×10^{-10}	9.24
Carbonic (1st proton)[a]	4.5×10^{-7}	6.4
Carbonic (2nd proton)	5.6×10^{-11}	10.2
Citric (1st proton)	8.7×10^{-4}	3.06
Citric (2nd proton)	1.8×10^{-5}	4.74
Citric (3rd proton)	4×10^{-6}	5.40
Ethane	$\sim 10^{-34}$	34.
Gluconic	3.6×10^{-4}	3.56
Hexaquaaluminum(III) ion	$1.3^{+} \times 10^{-5}$	4.85
Hexaquachromium(III) ion	1.5×10^{-4}	3.82
Hexaquacalcium(II) ion	$\sim 10^{-13}$	13.
Hexaquairon(II) ion	5×10^{-9}	8.3
Hexaquairon(III) ion	6.3×10^{-3}	2.22
Hexaquazinc(II) ion	2×10^{-10}	9.7
Hydrochloric	$\sim 10^{+7}$	−7
Hydrocyanic	7.2×10^{-10}	9.14
Hydrofluoric	7.2×10^{-4}	3.14
Hydrogen peroxide	2.4×10^{-12}	11.6
Oxonium ion	55.5	−1.74
Hydrosulfuric (1st proton)	5.7×10^{-8}	7.24
Hydrosulfuric (2nd proton)	1.2×10^{-15}	14.9
Hydroxyl ion	$\sim 10^{-36}$	36.
Hypochlorous	3.6×10^{-8}	7.44
Hypophosphorous	1×10^{-2}	2.0
Lactic	1.4×10^{-4}	3.86
Monoethanolammonium cation	3.0×10^{-10}	9.5
Nitrous	4.5×10^{-4}	3.34
Oxalic (1st proton)	6.5×10^{-2}	1.19
Oxalic (2nd proton)	6.0×10^{-5}	4.22
Phenol	1.3×10^{-10}	9.6
Phosphoric (1st proton)	5.0×10^{-2}	1.3
Phosphoric (2nd proton)	5.9×10^{-8}	7.2
Phosphoric (3rd proton)	4.8×10^{-13}	12.3
Phosphoric-Meta	5.8×10^{-10}	9.24
Silicic (H_4SiO_4) (1st proton)	2.0×10^{-10}	9.7
Sulfuric (1st proton)	~ 100	−2.0
Sulfuric (2nd proton)	1×10^{-2}	2.0
Sulfurous (1st proton)	1.7×10^{-2}	1.76
Sulfurous (2nd proton)	6.2×10^{-8}	7.20
Tartaric (1st proton)	1.0×10^{-4}	3.0
Tartaric (2nd proton)	3.2×10^{-5}	4.5
Water	1.1×10^{-16}	16.

[a] The sum of the dissolved CO_2 plus carbonic acid molecules is used in this calculation. If dissolved carbon dioxide is subtracted, as it should be, the value of K_a becomes 2.5×10^{-4} and pK_a becomes 3.68. The K_a for the second ionization is unaffected.

Certain electrochemical and thermodynamic assumptions necessary to establish the pH scale introduce an error of this magnitude.

Electronic art is far beyond the accuracy of the pH scale. pH meters are available with readout reproducible to 0.001 pH units. In the light of the foregoing discussion these three-decimal readings are spurious. Only the first significant figure is accurately measured, the second place is an estimate, and the third-place digit has no significance. An instrument of this type can, however, be used to duplicate accurately solutions of identical, albeit not exactly known, pH.

By use of pH measurements and analytically determined concentrations, it is possible to determine K_a and K_b experimentally. Since the values of these constants are very small numbers for weak and very weak acids and bases, pK_a and pK_b have been introduced. The defining relationships are

$$pK_a = -\log K_a$$

and

$$pK_b = -\log K_b$$

As with the pH calculation when dealing with oxonium ion, the use of pK_a and pK_b conveniently eliminates the need for very small numbers.

pK_a values give a convenient means of comparing the strength of different acids and bases. A range of ionization constants and corresponding pK values for selected acids in inorganic pharmaceutical chemistry are found in Table 4.8.

STRONG ACIDS: Although the relative strength of most acids is clearly indicated, a problem exists with respect to the strongest acids and the strongest bases. No acid stronger than oxonium can be accurately evaluated because of such an acid's rapid and almost quantitative conversion to oxonium ion in aqueous media.

$$HClO_4 + H_2O = H_3O^+ + ClO_4^-$$

Similarly, bases stronger than hydroxyl ion are immediately converted to hydroxyl ion by reaction with water.

$$PO_4^{3-} + H_2O = OH^- + HPO_4^{2-}$$

Accordingly, all strong acids (e.g., nitric, perchloric) appear to have approximately the same acid strength in water. The solvent is said to have a *leveling effect* on the acid strength. This also holds true for the strongest bases.

That the acid strengths of such strong acids are not the same, however, is indicated in concentrated solutions and by their differing catalytic power. Attempts at evaluation in other solvents, especially aprotic solvents with little leveling affect, have established a reasonable scale of pK_a values for strong acids. The transfer of these results to water solutions has not been resolved. However, it can be postulated on the basis of these different experimental methods that in aqueous solution the very strong acids arrange in the order of strongest first: perchloric, hydriodic, hydrobromic, hydrochloric, nitric, sulfuric (1st proton).

Types of acids. Examination of Table 4.8 reveals three general types of acids on the basis of structure: hydroacids (e.g., H_2S and HCl); solvated ions (e.g., hexaquairon(III) and hexaquaaluminum cation); and hydroxy acids (e.g., C_6H_5COOH and H_2CO_3). The structural features and ionization mechanisms of each type will now be considered.

HYDROACIDS: The acid strength of the simple hydrides (e.g., HCl, H_2S, H_2O) is determined by two distinct factors. First, within a given family of elements acid strength increases with increasing atomic weight (ionic radius). There is a weakening of the negative electrostatic field of the larger anions due to increased volume, and thus the proton is held less firmly. The second factor is a horizontal relationship within a given period of the periodic table. As atomic weight increases, acid strength increases. This is a function of increasing electronegativity. Size effects are negligible since successive members of a period have almost identical radii. The first-row hydrides—NH_3, H_2O, and H_2F_2—illustrate the electronegativity effect on acid strength very clearly. They have pK_a values of 23, 16, and 4, respectively. The hydrides of the halogen family illustrate the size effect; H_2F_2 and HCl have pK_a values of 3.1 and -7, respectively.

The protonated cations, of which the ammonium and oxonium ions and the amine salt cations are most important, may be considered as hydroacids.

HYDRATED CATIONS AS ACIDS: The hydrated cations have not been recognized as acids in the field of inorganic pharmaceutical chemistry despite their importance in several pharmaceutical and medicinal preparations. The general equation for their ionization in water solution may be written

$$[M(H_2O)_x]^{n+} + H_2O = [M(H_2O)_{(x-1)}(OH)]^{(n-1)+} + H_3O^+$$

The acid strength determinants of aqua cations are exactly those which determine the stability of complex ions: charge, valence shell type, and ionic radius of the central ion. (Ligand effects are nearly constant since water initially is the only ligand present.) The higher the positive charge and the smaller the radius, the valence electron shell being constant, the greater the polarizing effect of the central cation on the coordinated water molecules. The net result is an additional displacement of the oxygen-hydrogen bonding electrons at the expense of the hydrogen atom. This effect is enhanced if the central atom has an unclosed valence electron shell. If the polarization is sufficiently strong, the bond between the oxygen and one of the protons of a complexed water molecule will be so weakened that the proton can be attracted to a stronger base, e.g., a solvent water molecule.

The aqua cations of aluminum, magnesium, and sodium illustrate these structural factors as determinants of acid strength. Charge decreases from aluminum through magnesium to sodium, whereas ionic radius increases in the same order. Aluminum ion, with its high charge and small radius, has a strong positive electrostatic field which can react strongly with the coordinated water molecules. The electron cloud of the coordinated water molecules are so distorted that the bond to one of the protons is too weak to prevent its loss to a stronger base. The second proton of the former water molecule, now a ligand hydroxyl group, is strongly bonded to the oxygen. The formal negative charge of this radical strengthens the bond to the remaining proton. However, a proton may be lost from a second complexed water molecule, but not as easily as the first. Compared to aluminum, magnesium, with its smaller positive charge and larger ionic radius, has a weaker electrostatic field and as a result cannot polarize its coordinated water molecules as effectively. It can do so only under most favorable conditions, i.e., in the presence of very strong base. The weak polarizing effect of the sodium ion is no longer sufficient to allow ionization of a proton.

HYDROXY ACIDS: This final group is by far the largest and most diverse of the three types of acids. Several general factors enter into the determination of their acid strength. In addition certain subgroups have special features which are operative in strength determination.

The number of nonprotonated oxygens on the central atom of a hydroxy acid has a direct bearing on acid strength, irrespective of the central atom involved. If there are no nonprotonated oxygens on the molecule, the acid is very weak with pK_a values usually in the range 9–10. Boric acid, $B(OH)_3$, phenol, C_6H_5OH, and hypochlorous acid, HOCl, are good examples. If the molecule has one nonprotonated oxygen atom, the pK_a will be about 2; phosphoric acid, $OP(OH)_3$, and sulfurous acid, $OS(OH)_2$, are good examples. Acids with two or three nonprotonated oxygens are strong acids, e.g., sulfuric, $O_2S(OH)_2$, and perchloric, $O_3Cl(OH)$. For polyprotic acids, these rules apply only to the neutral acid molecule; the charged ionic acids resulting from their partial neutralization do not follow these rules.

The central atom of a hydroxy acid affects its acid strength through its electronegativity, atomic volume, and through its formal charge. The greater its electronegativity, the greater the attraction of the central atom

TABLE 4.9 EFFECT OF CHARGE ON ACID STRENGTH

Acid	Charge of Central Ion	Number of Nonprotonated Oxygens	pK_a (1st proton)
H_4SiO_4	+4	0	10
H_3PO_4	+5	1	1.3
H_2SO_4	+6	2	−2
$HClO_4$	+7	3	∼−10
$HClO_3$	+5	2	−1
$HClO_2$	+3	1	2.0
$HClO$	+1	0	7.2

for the bonding electrons of the acid molecule. If the central atom is also very small and of high charge, the electrons are strongly attracted to it, the oxygen-hydrogen bonds are weakened, and a strong acid results. These effects are illustrated in Table 4.9.

Hydroxy compounds are distinctive in that they may act as acids or bases, depending on the bond at which ionization takes place. Ionization modes are illustrated in Fig. 4.3. If the central atom has a low charge and an electronegativity of less than 1.7, basic ionization predominates, as represented by reaction (a) of Fig. 4.3. If, however, the central atom has a high charge and an electronegativity of greater than 1.7, acidic properties predominate, as shown in (b). The more these variables shift in value toward either extreme, the stronger the basic or acidic properties, respectively. The effect of electronegativity on the mode of ionization is illustrated by the compounds sodium hydroxide and "chlorine hydroxide"

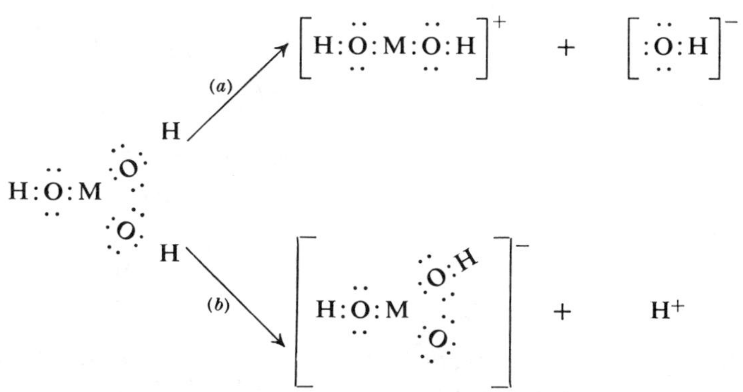

FIGURE 4.3 Ionization modes of hydroxy compounds: (*a*) basic ionization predominates; (*b*) acidic ionization predominates.

(hypochlorous acid, HOCl). Charge effects are formally identical, being +1 for both sodium and chlorine in these compounds. Sodium hydroxide is a strong base; electropositive sodium is incapable of polarizing the hydroxyl groups sufficiently to cause a weakening of the oxygen-hydrogen bond. But hypochlorous acid is a weak acid, and the chlorine atom is incapable of independent existence as a cation. Because of its strong electronegativity, it polarizes the hydroxyl group. Thus the tendency for the bonding electrons to shift from the σ bond between hydrogen and oxygen results in weakening this bond sufficiently to allow acid ionization.

Central atoms of intermediate charge and with electronegativities at or near 1.7 give rise to amphoteric hydroxy compounds. The chromium series of hydroxides serve as good examples of charge effect on the acid-base behavior of hydroxy compounds. Chromium(II) hydroxide shows only basic properties. Chromium(III) hydroxide is amphoteric; it gives rise to both cationic chromium(III) and the dioxochromate(III) anion (chromite). Chromium(VI) hydroxide is acidic but does not exist as such. An electrically neutral, highly hydroxylated complex disproportionates; that is, it splits out one or more molecules of water to form residual molecules having a minimum number of hydroxyl groups. Chromic acid, $O_2Cr(OH)_2$ or H_2CrO_4, always gives rise to salts in which the chromium(VI) is acting in its acidic capacity. Table 4.10 illustrates the chromium series of compounds.

TABLE 4.10 EFFECT OF A CENTRAL ION'S FORMAL CHARGE ON ACID-BASE PROPERTIES

	Neutral Hydroxylated Complex	Ionization Mode
Chromium(II) (basic)	$Cr(OH)_2$	$Cr(OH)_2 = Cr^{2+} + 2(OH)^-$
Chromium(III) (amphoteric)	$Cr(OH)_3$	$Cr(OH)_3 = Cr^{3+} + 3(OH)^-$, or $\begin{cases} Cr(OH)_3 = HCrO_2 + H_2O \\ HCrO_2 = (H^+) + CrO_2^- \end{cases}$
Chromium(VI) (acidic)	$Cr(OH)_6$	$\begin{cases} Cr(OH)_6 = H_2CrO_4 + 2H_2O \\ H_2CrO_4 = 2(H^+) + CrO_4^{2-} \end{cases}$

During a neutralization reaction there is a decreasing ease of ionization of the successive protons of polyhydroxy acids. L. Pauling estimates that the K_a values for these successive ionization steps should have the relative ratios of $1:10^{-5}:10^{-10}$. It is easy to understand this effect if it is remembered that for each proton lost there remains with the anion one electron in excess of the total number of electrons normally associated with the atoms constituting the molecule. Although the resultant negative charge is shown formally as residing on the oxygen atom of the former hydroxyl group, actually there is a partial shift of the electron cloud from that oxygen to all other bonds in the molecule. As a result the net electron density of the σ bonds between all remaining hydroxy hydrogen and oxygen atoms tends to increase, and the bonds are strengthened. Concomitantly the acid strength is weakened; a second ionization is difficult, and the removal of a third proton (in triprotic acids) can be effected only by extremely strong bases. Phosphoric acid, H_3PO_4, and its two anion acids, $H_2PO_4^-$ and HPO_4^{2-}, illustrate the effect of succeeding proton removals on the magnitude of pK_a (see Table 4.8).

A pK_a of about 5 is normal for unsubstituted aliphatic and aromatic monocarboxy acids. Substitution of an electronegative element for one of the hydrogen atoms on the carbon atom adjacent to the carboxy group causes an increase in acid strength. This *inductive effect* results from the electron-attracting power of the introduced electronegative group, e.g., Cl, NH_2, another COOH. The σ electron cloud between the hydrogen and oxygen of the carboxyl group is deformed and weakened; the proton ionizes more readily. "Di" and "tri" substitution on this carbon increases the effect. The chlorine-substituted acetic acids, shown in Table 4.11, illustrate how successive substitution increases acid strength. Although introduction of an electronegative group on the carbon atom some distance from the carboxyl group has an effect on acid strength, it is less pronounced. The α- and β-chlorine-substituted propionic acids are illustrative; the pK_a of the β-compound is but slightly greater than that of propionic acid itself, as shown in Table 4.11. The added methyl group

TABLE 4.11 THE EFFECT OF ELECTRONEGATIVE SUBSTITUENTS ON ACID STRENGTH

Acid		pK_a
Acetic acid	CH_3COOH	4.76
Monochloracetic acid	$ClCH_2COOH$	2.86
Dichloracetic acid	$Cl_2CHCOOH$	1.30
Trichloracetic acid	Cl_3CCOOH	0.89
Propionic acid	CH_3CH_2COOH	4.87
α-Chlorpropionic acid	$CH_3CHClCOOH$	2.80
β-Chlorpropionic acid	CH_2ClCH_2COOH	4.08

which converts acetic acid to propionic acid has little or no effect on the pK_a.

Types of bases. Brønsted bases are usually anions (e.g., OH^-, OAc^-) or neutral molecules (e.g., H_2O). There are a few cations which are capable of acting as bases, e.g., $[Al(H_2O)_4(OH)_2]^+$.

ANIONS: The hydroxyl ion is probably the one base which finds widest application in neutralizations. Historically, hydroxide and base were synonymous. In order for a hydroxyl compound to act as a base the cation associated with the hydroxyl ion must be at least moderately electropositive, have a low charge, and a closed valence electron shell. Consideration of the restrictions imposed by these conditions indicates that the strong hydroxide bases, both in the crystalline state and in solution, must be ionic compounds, e.g., NaOH, $Ba(OH)_2$, TlOH. If the cation involved has a high charge, and particularly if it is not of the inert gas type, it is likely that the mutual polarization of the cation and hydroxyl groups involved will lead to increasing covalent character of the bond between them. In these cases the "hydroxide" is likely to act as a hydroxy acid, e.g., $Cr(OH)_3$ or $Zn(OH)_2$.

The anions arising from monohydrogen molecular and ionic acids are Brønsted bases, e.g., Cl^-, SO_4^{2-}, PO_4^{3-}. The anions of polyhydroxy acids with one or more intact hydroxyl groups may act as acid or bases, e.g., HSO_4^-, HCO_3^-, $H_2PO_4^-$. This latter type of anion may lose its proton(s) to a stronger base or may accept protons from a stronger acid. The following equations represent the neutralization mechanism for anionic bases; the acid is not specified.

$$OH^- + (H^+) = H_2O$$
$$PO_4^{3-} + (H^+) = HPO_4^{2-}$$
$$HSO_4^- + (H^+) = H_2SO_4$$

NEUTRAL MOLECULES: The two neutral molecules most commonly acting as bases are water and ammonia. Other neutral bases exist (e.g., hydrogen sulfate, organic amines), but their ability to act as bases in neutralizations is usually not the property of direct interest. The following equations represer' the neutralization mechanism for neutral bases:

$$H_2O + (H^+) = H_3O^+$$
$$NH_3 + (H^+) = NH_4^+$$
$$RNH_2 + (H^+) = RNH_3^+$$
$$H_2SO_4 + (H^+) = H_3SO_4^+$$

Examination of these bases shows that their structure characteristically includes a highly electronegative atom having one or more nonbonding electron pairs. One of the nonbonding pairs is available to the proton for σ bond formation.

Salts as acids and bases. The acid-base reactions of salts in aqueous media must be discussed for two reasons. First, the use of salts as acids and bases or as a source of oxonium or hydroxyl ion may be more convenient than using the corresponding free acid or base. Subsidiary buffer effects, a better control of the reaction, and a less critical need for exact attainment of equivalence point are typical advantages. Second, even a limited reaction between water and one or more of the ions arising from a salt will upset the pH of a given solution. This may become critical when working with pH-dependent pharmaceutical systems. The theoretical and general implications are to be considered in the following paragraphs; the practical applications will be considered in Chapter 7.

In considering a salt as a possible acid or base both the cation and anion must be studied. If the cation is protonated or hydrated or if it hydrates on solution, it will most likely act as an acid. Only anhydrous cations, or those hydrated cations which have little or no polarizing action on the coordinated water molecules, may be disregarded as acids. The alkali metal cations are typical of those cations having very limited or no acid properties in aqueous solution. Only the very weak anion bases (e.g. perchlorate, nitrate) may generally be disregarded. It is well to remember that the very strong acids have very weak conjugate anion bases.

The reaction or lack of reaction is a function of the relative strengths of the acids and bases of all possible conjugate pairs present in a given solution. Neutralization proceeds to apparent completion only when the K_a values of the acids of the competing conjugate pairs are quite different. For instance, a strong acid (oxonium ion, pK_a, -1.74) will react with a strong base (hydroxyl ion, pK_a, 16) with a reaction efficiency approaching 100%; for example,

$$H_3O^+ + OH^- = H_2O + H_2O$$
$$HB_1 + B_2 = B_1 + HB_2$$

The reaction between a weak acid and the base of another weak acid will be incomplete; an equilibrium will be established. A strong acid and the base of another strong acid will behave similarly. The reaction between ammonium chloride and sodium borate is illustrative. (Sodium and chloride ions may be neglected for reasons previously stated. The structure of the borate ion is discussed on page 485.)

$$NH_4^+ + [B(OH)_4]^- \rightleftharpoons NH_3 + B(OH)_3 + H_2O$$
$$2\,H_2O \rightleftharpoons H_3O^+ + OH^-$$

The extent of reaction is readily calculated from a knowledge of the ionization constants of the two acids involved and the fact that at equilibrium the oxonium ion concentration must be the same for both acids. The ionization constants for ammonium ion and boric acid are

$$K_a' = \frac{[NH_3][H_3O^+]}{[NH_4^+]} \tag{11}$$

$$K_a'' = \frac{[B(OH)_4^-][H_3O^+]}{[B(OH)_3]} \tag{12}$$

Equations 11 and 12 are each solved for $[H_3O^+]$ and equated to give, on subsequent rearrangement of terms,

$$\frac{K_a'}{K_a''} = \frac{[B(OH)_3][NH_3]}{[B(OH)_4^-][NH_4^+]} \tag{13}$$

In order to establish the concentration relationships between the species of ions present *electroneutrality,* EN, and *mass balance,* MB, are used.

EN: $[NH_4^+] + [H_3O^+] = [B(OH)_4^-] + [OH^-]$
since $[H_3O^+] < [NH_4^+]$ and $[OH^-] < [B(OH)_4^-]$
EN: $[NH_4^+] = [B(OH)_4^-]$
MB: $B_{total} = [B(OH)_3] + [B(OH)_4^-]$
$N_{total} = [NH_4^+] + [NH_3]$
$B_{total} = N_{total}$
$[B(OH)_3] + [B(OH)_4^-] = [NH_4^+] + [NH_3]$
$[B(OH)_3] + [B(OH)_4^-] = [B(OH_4)^-] + [NH_3]$
$[B(OH)_3] = [NH_3] = x$

assuming equal volumes of 0.2M solutions of ammonium chloride and sodium borate are mixed and substituting the appropriate values into (13)

$$\frac{5.6 \times 10^{-10}}{5.8 \times 10^{-10}} = 0.97 = \frac{[x][x]}{[0.1 - x]^2}$$

Extracting the square root of both sides of the equation and solving for x, the value of $0.049M$ is obtained for the concentration of both the boric acid and ammonia molecules. Thus the reaction proceeds almost halfway to completion at equilibrium, a fact to be expected considering the almost equal ionization constants of the two acids involved.

The pH of this solution may now be calculated, based on either the ammonium ion or the borate ion.

$$K_a' = \frac{[NH_3][H_3O^+]}{[NH_4^+]} \qquad K_b'' = \frac{[B(OH)_3][OH^-]}{[B(OH)_4^-]}$$

$$5.6 \times 10^{-10} = \frac{(0.049)x}{(0.1 - 0.049)} \qquad K_b'' = \frac{K_w}{K_a} = \frac{1 \times 10^{-14}}{5.8 \times 10^{-10}}$$

$$x = [H_3O^+] = 5.6 \times 10^{-10} \qquad 1.7 \times 10^{-5} = \frac{[0.049]x}{[0.1 - 0.049]}$$

$$pH = 9.24$$

$$pOH = 14 - 9.24 \qquad x = [OH^-] = 1.8 \times 10^{-5}$$

$$pOH = 4.76 \qquad pOH = 4.75$$

Of particular interest to the pharmaceutical chemist are those salt reactions in which water plays a role. These are the reactions commonly termed hydrolyses.[6] The water molecule may function as an acid or a base, depending on the salt involved; for example,

$$\underset{B_1}{(Na^+)OAc^-} + \underset{HB_2}{H_2O} = \underset{HB_1}{HOAc} + \underset{B_2}{OH^-} + (Na^+)$$

$$\underset{HB_1}{[Al(H_2O)_6]^{3+}} + \underset{B_2}{H_2O} = \underset{B_1}{[Al(H_2O)_5(OH)]^{2+}} + \underset{HB_2}{H_3O^+}$$

The method of calculating the extent of reaction between either or both of a salt's ions and water depends on the acid-base properties of the ions. For convenience salts are classified on the basis of the comparative acid-base strength of the cation and anion involved. The four resulting classes of salts, together with other pertinent data, are given in Table 4.12.

Solutions of salts in class 1 are always neutral, or nearly so. Neither the cation nor anion can react with water to a sufficient extent to cause an upset of the normal equal concentrations of the oxonium and hydroxyl ion of water. These salts result from the neutralization of a strong acid with a strong base, e.g., sodium chloride.

Solutions of salts in class 2 are alkaline in reaction. The degree of alkalinity is directly proportional to the strength of the base; sodium

[6] Although the proton donor theory does not require a special definition for hydrolysis, the use of the term is so widespread that it should be considered briefly. Hydrolysis is a special case of solvolysis—the reaction between a dissolved solute and its solvent. Examination of the usual definition, the reaction between a salt and water to give the corresponding acid and base, indicates a neutralization with an ion of the salt acting as one member of the first conjugate pair, and water acting as one member of the second conjugate pair.

TABLE 4.12 CLASSIFICATION OF SALTS FOR CALCULATION OF EXTENT OF HYDROLYSIS

Class	Cation		Anion		Constant for Calculation of Extent of reaction
	Classification[a]	Example	Classification	Example	
1	Weak acid	$Na(H_2O)_x^+$	Weak base	Cl^-	—
2	Weak acid	$Na(H_2O)_x^+$	Strong base	OAc^-	K_b
3	Strong acid	NH_4^+	Weak base	Cl^-	K_a
4	Strong acid	NH_4^+	Strong base	OAc^-	K_aK_b/K_w

[a] The terms "weak" and "strong" are relative.

acetate solutions are mildly alkaline, whereas tertiary sodium phosphate solutions are strongly basic. Since the cation of these salts may be disregarded, calculations to determine the extent of reaction with water and the pH are based on the characteristics of the anion. Using a $0.1M$ solution of sodium acetate, the calculations proceed as follows:

$$OAc^- + H_2O = HOAc + OH^-$$
$$B_1 + HB_2 = HB_1 + B_2$$

The calculation is based on the K_b of acetate ion,

$$K_b = \frac{K_w}{K_a} = \frac{[HOAc][OH^-]}{[OAc^-]}$$

EN: $[Na^+] + [H_3O^+] = [OAc^-] + [OH^-]$

since $[Na^+] > [H_3O^+]$

$[Na^+] = [OAc^-] + [OH^-] = 0.1$

$[OAc^-] = 0.1 - [OH^-]$

MB: $[OAc]_{total} = [OAc^-] + [HOAc] = 0.1$

$[0.1 - [OH^-]] + [HOAc] = 0.1$

$[OH^-] = [HOAc] = x$

Appropriate values may now be substituted into K_b:

$$\frac{1 \times 10^{-14}}{1.8 \times 10^{-5}} = 5.55 \times 10^{-10} = \frac{[x][x]}{[0.1 - x]}$$

The quadratic equation may be used to solve for hydroxyl ion concentration ($7.4 \times 10^{-6}M$). However, since the acetate ion is not a particularly strong base, it is assumed that the extent of reaction is such that $[0.1 - x] \sim 0.1$. Using this approximation, the solution becomes

$$5.55 \times 10^{-11} = x^2$$

$$[OH^-] = 0.74 \times 10^{-5} = 7.4 \times 10^{-6}M$$

$$pOH = 5.13$$

$$pH = 14 - 5.13 = 8.87$$

Class 3 of the salts has an anion which for practical purposes does not react with water. Ammonium chloride is a typical example:

$$NH_4^+ + H_2O = NH_3 + H_3O^+$$

It is evident that calculations will be based on the K_a of the acid.

$$K_a = \frac{[NH_3][H_3O^+]}{[NH_4^+]}$$

$$\text{EN:}\quad [NH_4^+] + [H_3O^+] = [Cl^-] + [OH^-]$$

$$\text{since}\quad [Cl^-] \gg [OH^-]$$

$$[NH_4^+] + [H_3O^+] = [Cl^-] = 0.1$$

$$[NH_4^+] = 0.1 - [H_3O^+]$$

$$\text{MB:}\quad N_{total} = [NH_4^+] + [NH_3] = 0.1$$

$$[0.1 - [H_3O^+]] + [NH_3] = 0.1$$

$$[NH_3] = [H_3O^+] = x$$

Assuming a $0.01M$ solution of ammonium chloride as a basis for calculations,

$$5.6 \times 10^{-10} = \frac{[x][x]}{[0.01 - x]}$$

Since the ammonium ion is not a strong acid, it is assumed that $[0.01 - x] \sim 0.01$. Using this approximation, the solution becomes

$$x^2 = 5.6 \times 10^{-12}$$

$$[H_3O^+] = 2.37^- \times 10^{-5}$$

$$pH = 5.6$$

Class 4 is more complicated since both the cation and anion of these salts are capable of reacting with water in different acid-base reactions. The general equations for the two simultaneously occurring reactions are

$$HB_1 + H_2O = B_1 + H_3O^+$$

$$B_2 + H_2O = HB_2 + OH^-$$

Summing these equations, an over-all ionization constant may be written

$$HB_1 + B_2 + 2H_2O = (B_1 + H_3O^+) + (HB_2 + OH^-)$$

$$K_i = \frac{[B_1][H_3O^+]}{[HB_1]} \cdot \frac{[HB_2][OH^-]}{[B_2]} \tag{14}$$

The constants of the respective salts are readily recognized:

$$K_i = K_a^{HB_1} \cdot K_b^{B_2}$$

The expression for the ionization constant, Equation 14, may now be rewritten

$$K_a^{HB_1} \cdot K_b^{B_2} = \frac{[B_1][HB_2]}{[HB_1][B_2]} \cdot [H_3O^+][OH^-] \tag{15}$$

The latter two terms are recognized as K_w, giving

$$\frac{K_a^{HB_1} \cdot K_b^{B_2}}{K_w} = \frac{[B_1][HB_2]}{[HB_1][B_2]} \tag{16}$$

This general equation may now be used to determine the extent of reaction and resultant pH for any salt in class 4. Its application may be illustrated using a 0.01*M* ammonium acetate solution:

$$[NH_4^+] + [OAc^-] \rightleftharpoons [NH_3] + [HOAc]$$
$$2H_2O \rightleftharpoons [H_3O^+] + [OH^-]$$

$$\frac{K_a^{NH_4^+} K_b^{OAc^-}}{K_w} = \frac{[NH_3][HOAc]}{[NH_4^+][OAc^-]} \tag{17}$$

$$\begin{aligned}
\text{EN:}\quad & [NH_4^+] + [H_3O^+] = [OAc^-] + [OH^-] \\
\text{assume}\quad & [NH_4^+] > [H_3O^+] \text{ and } [OAc^-] > [OH^-] \\
\text{then}\quad & [NH_4^+] = [OAc^-] \\
\text{MB:}\quad & N_{total} = [NH_4^+] + [NH_3] = 0.01 \\
& OAc_{total} = [OAc^-] + [HOAc] = 0.01 \\
\text{thus}\quad & [NH_4^+] + [NH_3] = [OAc^-] + [HOAc] \\
\text{subst.}\quad & [OAc^-] + [NH_3] + [OAc^-] + [HOAc] \\
& [NH_3] = [HOAc] = x
\end{aligned}$$

Substituting appropriate values into (17)

$$\frac{(5.6 \times 10^{-10})\left(\dfrac{1 \times 10^{-14}}{1.8 \times 10^{-5}}\right)}{(1 \times 10^{-14})} = \frac{x \cdot x}{(0.01)(0.01)}$$

$$(3.1 \times 10^{-5})(10^{-4}) = x^2$$
$$x = [NH_3] = [HOAc] = 5.6 \times 10^{-5}$$

The pH of the solution may now be calculated, on the basis of either ion.

$$K_a^{NH_4^+} = \frac{[NH_3][H_3O^+]}{[NH_4^+]} \qquad K_b^{OAc^-} = \frac{[HOAc][OH^-]}{[OAc^-]}$$

$$5.6 \times 10^{-10} = \frac{[5.6 \times 10^{-5}][H_3O^+]}{[0.01]} \qquad \frac{1 \times 10^{-14}}{1.8 \times 10^{-5}} = \frac{[5.6 \times 10^{-5}][OH^-]}{[0.01]}$$

$$[H_3O^+] = 1 \times 10^{-7} \qquad [OH^-] = 1.0 \times 10^{-7}$$
$$pH = 7 \qquad pH = 7$$

Additional examples, applied to appropriate pharmaceutical cases, will be discussed later.

Anhydrides. Acid and basic anhydrides must be discussed briefly since they may serve as acids or bases. Anhydrides may be defined as substances which react with their parent solvent to give either acids or bases. The oxides are the acid-base anhydrides of the water system; metallic oxides react with the water to give bases and nonmetallic oxides give rise to acids. Equations for typical reactions, written in the Brønsted sense are

$$CaO + H_2O = Ca^{2+} + 2OH^-$$
$$CO_2 + 2H_2O = H_3O^+ + HCO_3^-$$

Depending on the relative equivalents reacted, these two solutions will undergo neutralization on mixing:

$$Ca^{2+}(OH^-)_2 + H_3O^+ + HCO_3^- = CaCO_3\downarrow + 3H_2O$$

or

$$Ca^{2+}(OH^-)_2 + 2H_3O^+ + 2HCO_3^- = Ca^{2+}(HCO_3^-)_2 + 4H_2O$$

An acid anhydride and basic anhydride do not require an aqueous medium for reaction, but may react directly:

$$CaO + CO_2 = CaCO_3$$

It is probable that traces of water are essential for this reaction to proceed. The Brønsted theory is incapable of explaining this reaction, since the theory requires a proton transfer as a concomitant of neutralization. General acid-base theories, such as the Lewis or the Usanovich, are needed for a satisfactory explanation. Some anhydrides of interest in inorganic pharmaceutical chemistry are shown in Table 4.13.

TABLE 4.13 SELECTED ACID AND BASIC ANHYDRIDES IN PHARMACEUTICAL CHEMISTRY

Oxide	Acid	Base
CO_2	H_2CO_3	—
SO_2	H_2SO_3	—
As_2O_3	H_3AsO_3	—
P_2O_5	H_3PO_4	—
CrO_3	H_2CrO_4	—
K_2O	—	KOH
CaO	—	$Ca(OH)_2$
BaO	—	$Ba(OH)_2$

FIVE · INORGANIC NOMENCLATURE

The average student, pursuing his introductory studies in chemistry, may have memorized the names of the compounds encountered as a matter of course. But perceiving the rules of nomenclature, as well as some of the problems in unambiguously naming inorganic compounds, was probably not achieved. If chemical theory is to be successfully applied to the problems of pharmaceutical chemistry, a review of inorganic nomenclature is important at this time. Immediate recognition by the student of any compound under discussion is required. But a good knowledge of nomenclature goes beyond the student stage; it is an essential tool for the practitioner in the field.

Systematic nomenclature only will be considered at this time. Common or trivial names will be introduced as appropriate occasions arise. Fortunately, the use of Latin names is rapidly disappearing from pharmacy; treatment similar to that established in the compendia will be followed. Only in cases where it is impossible to recognize the English equivalent directly will Latin names be introduced. For example, it is obvious that sodii chloridum is the Latin name for sodium chloride or at least a sodium-chlorine compound, whereas hydrargyrum gives no clue to its identity, mercury.

Two problems must be considered in the naming of a given compound. The first is the order of citation of the constituents of that compound. The second is the unambiguous indication of stoichiometry, i.e., the oxidation state of each of the elements involved. Stoichiometry becomes important whenever an element may exist in more than one oxidation state. Although presentation will center about the currently recommended rules for indication of stoichiometry, the still widespread use of some older equivalent forms necessitates their inclusion in the discussion.

ORDER OF CITATION

Order of citation is generally well defined. In English and German the name of the most electropositive constituent is placed first in the compound's name.[1] In salts and salt-like compounds the cation must be the most electropositive element and is usually easily recognized, e.g., sodium chloride, aluminum sulfate. In covalent compounds the order is sometimes not as obvious, and a table of electronegativities must be consulted. The IUPAC recommends the following order for the nonmetals in compounds: B, Si, C, Sb, As, P, N, H, Te, Se, S, At, I, Br, Cl, O, F. Examples are lithium hydride, hydrogen telluride, and boron trichloride.

For ternary and more complicated compounds, the rules regarding order of citation are equally straightforward. The cationic elements of the compound are named first in alphabetical order. The anionic species are then named in the same fashion and both sets are combined to yield the overall name. The only exception to this procedure is for acidic hydrogen which is named last among the cations. If a radical has a commonly accepted trivial name, e.g., nitrate, ammonium, sulfate, thiocyanate, etc., it is treated as though it is an elementary particle. Thus, the salt $NaNH_4HPO_4$ is properly named ammonium sodium hydrogen phosphate.

Basic salts which contain oxide or hydroxide anions are named in the same manner despite the fact that the oxide or hydroxide associates with the cation when the salt ionizes. For example, BiOCl, which yields the ions BiO^+ and Cl^- in solution is named bismuth chloride oxide. In a similar manner, HgOHCl is mercury(II) chloride hydroxide.

STOICHIOMETRY INDICATION

Indication of stoichiometry presents no problems if each of the elements (or radicals) involved has a single oxidation state. Aluminum oxide is obviously Al_2O_3. However, the names carbon oxide and iron sulfate are not definitive, since carbon and iron each have two oxidation states. Thus carbon oxide fails to indicate whether CO or CO_2 is intended; the iron sulfate does not distinguish between $FeSO_4$ and $Fe_2(SO_4)_3$. A number of methods have been devised to address this problem. The three most frequently encountered are described here.

ous and ic system. Cations with variable oxidation states may be identified as to actual oxidation state in a given compound by several

[1] In the Romance languages this order of citation is reversed.

methods. The classical scheme, first proposed by L. B. Guyton de Morveau, C. L. Berthołet, A. F. de Fourcroy, and A. L. Lavoisier, assumes a limit of two oxidation states for the cation. The suffix *ous* is used to indicate the lower and *ic* the higher oxidation state. By appending the proper suffix to the root of the name of the element or radical involved, a compound can be uniquely identified, *providing the element has only two oxidation states.* On this basis the iron compounds cited in the preceding paragraph become ferrous sulfate and ferric sulfate, respectively. If more than two oxidation states exist, this classical system breaks down. MnO, Mn_2O_3, and MnO_2 obviously cannot be uniquely named by this system. For this reason the method should be avoided. It is included here only because of its use in compendia and the older textbooks and literature.

Stock method. The most frequently used method today is that originally devised by A. Stock. The electrochemical valence of an element is indicated by means of Roman numerals, placed in parentheses without a hypen immediately after the name of the element to which it refers. Thus the three manganese oxides become manganese(II) oxide, manganese(III) oxide, and manganese(IV) oxide. They are read manganese two oxide, etc.

Ewens-Bassett System. This is a relatively new procedure which was first introduced in 1970. When the compound to be named is known to produce clearly identifiable ions in solution, it is permissible to use Arabic numbers in parentheses to indicate the charge on the ion. Therefore, $FeCl_2$ is named iron (2+) chloride and $FeCl_3$ is iron(3+) chloride. But, MnO_2 is *not* manganese (4+) oxide because it does not give clearly identifiable ions in solution. It may be named either manganese dioxide or manganese(IV) oxide.

In some cases neither the Stock nor the Ewens-Bassett methods are satisfactory or desirable. Here Greek numerical prefixes may be used (Table 5.1). They should precede without a hypen the name of the constituent to which they refer. The prefix mono is usually omitted; by

TABLE 5.1 GREEK NUMERICAL PREFIXES[a]

$\frac{1}{2}$	hemi (semi)	6	hexa
1	mono	7	hepta
$1\frac{1}{2}$	——(sesqui)	8	octa
2	di (bi)	9	ennea
3	tri (ter)	10	deca
4	tetra (quadri)	11	hendeca
5	penta	12	dodeca

[a] Commonly used Latin prefixes are shown parenthetically.

convention it is understood that lack of a prefix indicates the mono compound. Illustrations of prefix usage are carbon dioxide, dinitrogen tetraoxide, nitrogen dioxide, tetramethyl ammonium chloride. This procedure is most frequently used in naming binary compounds of nonmetals.

Classical oxo and oxo substituted anions. Complex oxygenated anions are named by using the root of the name of the central atom together with the appropriate suffix and, if necessary, the appropriate prefix. For acids the suffix *ic* indicates an oxidation number for the central atom just higher than that of the *ous* state. For salts the suffixes are *ate* and *ite* for the higher and lower state, respectively. If still higher or still lower oxidation states exist, prefixes are used. The prefix *hypo,* used with *ous* or *ite* indicates a lower oxidation state than that of the central atom in the *ous* compound. Similarly the prefix *per,* used with *ic* or *ate,* indicates a higher oxidation state than that existing for the *ic* compound. The proper use of these prefixes and suffixes, using chlorine as the central atom, is demonstrated in Table 5.2. Since the names of the parent acids vary somewhat from the names of their salts, acid nomenclature has been included. It should be noted that the valence of the complex ion itself is constant, even though the oxidation number of the central ion changes in passing through the series.

If one or more of the oxygens in a complex named by the preceding system is replaced by another element, the change is indicated by using as a prefix the root of the name of the element or group substituted. Examples are KCNO, potassium cyanate, and KCNS, potassium *thio*-cyanate; Na_3AsO_4, sodium arsenate, and Na_3AsS_4, sodium *tetrathio*-arsenate; K_2HgI_4, *tetraiodo*mercurate; Na_2SiF_6, sodium *hexafluoro*silicate. If substitution is incomplete, Greek prefixes should be added; e.g., $Na_2S_2O_3$ should be sodium monothiosulfate, although in practice the "mono" is not used. The group substituting for the oxygen need not have a minus two charge. The usual rules as to coordination number and the summing of charges apply in determining the over-all net charge of the complex anion thus substituted.

Some of the more common elements substituting for oxygen in the complex anions are sulfide, fluoride, chloride, cyanide, and peroxide. The prefixes used are *thio, fluoro, chloro, cyano, and peroxo* respectively. Although thio is the recommended prefix for sulfur, *sulfo* is frequently used. Thus while thiocyanate and sulfocyanate name the same substance, thiocyanate is preferred. Many older fluorine-substituted compounds are named using the abbreviated prefix *fluo*, e.g., sodium fluoborate.

For peroxide substitution *per* was formerly used. However, this leads to confusion with the per acids and salts previously discussed. To dis-

TABLE 5.2 NOMENCLATURE OF COMPLEX OXYGEN ANIONS

Relative Oxidation State of the Central Atom	Acids		Salts		Illustrations Using Anions with a Chlorine Central Atom		
	Prefix	Suffix	Prefix	Suffix	Oxidation Number	Acids	Salts
Negative	hydro	ic	—	ide	−1	hydrochloric acid (hydrogen chloride)	chlorides, Cl^-
Least positive	hypo	ous	hypo	ite	+1	hypochlorous acid	hypochlorites, ClO^-
	—	ous	—	ite	+3	chlorous acid	chlorites, ClO_2^-
	—	ic	—	ate	+5	chloric acid	chlorates, ClO_3^-
Most positive	per	ic	per	ate	+7	perchloric acid	perchlorates, ClO_4^-

tinguish the substitution of a true peroxide group (O_2^{2-}) for oxygen in compounds, the prefix *peroxo* should always be used. Figure 5.1 illustrates the differences.

An anion may include two or more central ions. Greek numerical prefixes are used to indicate the number of central ions within the complex in these cases, e.g., potassium dichromate, $K_2Cr_2O_7$, disodium trisilicate, $Na_2Si_3O_7$, and sodium tetraborate, $Na_2B_4O_7$.

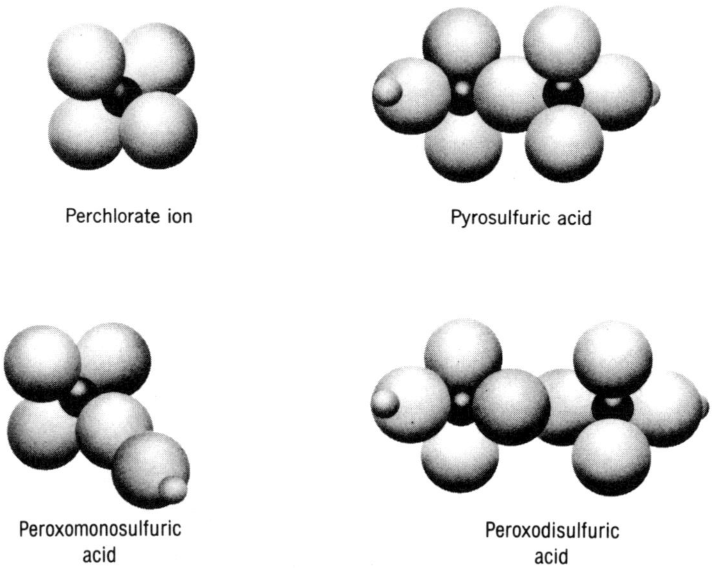

FIGURE 5.1 Structures and nomenclature to differentiate between peroxo and per compounds

Polyprotic acids and their salts. For polyprotic acids all possible acid salts must be unambiguously identified and distinguished from the normal salt. For the acid salts it is recommended that the word *hydrogen,* together with the proper Greek numerical prefix, be used before the name of the anion. The sodium salts of sulfuric acid are thus sodium *hydrogen* sulfate, the acid salt; sodium sulfate, the normal salt. The sodium salts of phosphoric acid are sodium ***dihydrogen*** phosphate and disodium ***monohydrogen*** phosphate, the two acid salts of this acid; ***trisodium*** phosphate, the normal salt. In practice the prefixes ordinarily used for acid salts with sodium are dropped to give sodium dihydrogen phosphate and sodium *mono*hydrogen phosphate, respectively. A knowledge of the number of hydrogen ions present in the salt unambiguously fixes the number of sodium ions.

In the older literature acid salts of diprotic acids were indicated by the Latin prefix *bi* with the anion name, e.g., sodium *bi*sulfate. Still another method of indicating acid salts has been the insertion of "acid" between the cation and anion names, e.g., sodium *acid* sulfate. These usages are to be discouraged, although the student should be able to recognize them.

For triprotic acids several other systems of nomenclature have been used in the past. One system names the cation and anion in the usual order and prefixes this with the term *basic*. The term basic is in turn prefixed with the Greek numerical prefix indicative of the number of equivalents of base necessary to react with a mole of the acid to give the salt in question. The sodium phosphates, in the previous order, become *monobasic* sodium phosphate, *dibasic* sodium phosphate, and *tribasic* sodium phosphate.

Another system of nomenclature for the triprotic acids employs the words *primary*, *secondary*, and *tertiary*. In the previous order the sodium phosphates become *primary* sodium phosphate, *secondary* sodium phosphate, and *tertiary* sodium phosphate. These nomenclature systems are summarized in Table 5.3, illustrated by the sodium salts of phosphoric

TABLE 5.3 NOMENCLATURE OF THE PHOSPHATES

Formula	Preferred Name	Other Commonly Used Names		Name in Compendia[a]
NaH_2PO_4	sodium dihydrogen phosphate	monobasic **sodium phosphate**[b]	primary sodium phosphate	sodium biphosphate
Na_2HPO_4	sodium monohydrogen phosphate	dibasic **sodium phosphate**[b]	secondary sodium phosphate	sodium phosphate
Na_3PO_4	sodium phosphate	tribasic sodium phosphate	tertiary sodium phosphate	

[a]Used through USP XX.

[b]Became official title beginning with Supplement 3 of USP XX.

acid. Although no longer the preferred usage, each of the foregoing will still be found in use.

Another problem in naming acids and their salts arises from the fact that hydration of a parent acid anhydride can theoretically be stopped at various intermediate stages to give acids of different degrees of hydroxylation. In order to distinguish between these acids the prefixes *meta*, *pyro*, and *ortho* have been adopted. Degree of hydroxylation proceeds from meta, the least, through pyro to ortho, the most highly hydroxylated form of the acid. The phosphoric acids serve as an excellent illustration. Phosphorus pentoxide (correctly, tetraphosphorus decaoxide) hydrates in stages to give varying degrees of hydroxylation, as shown in Fig. 5.2. By custom, omission of the prefix implies the ortho acid and its salts. Modern usage substitutes the prefix *di* for *pyro*. This follows from the fact that the pyro acids are actually the first member of a series of polymerized acids known as poly acids. With phosphorus, silicon, boron, etc. complex systems of poly acids exist. The second member of the polyphosphate series of acids, triphosphoric acid, is illustrated in Fig. 5.2.

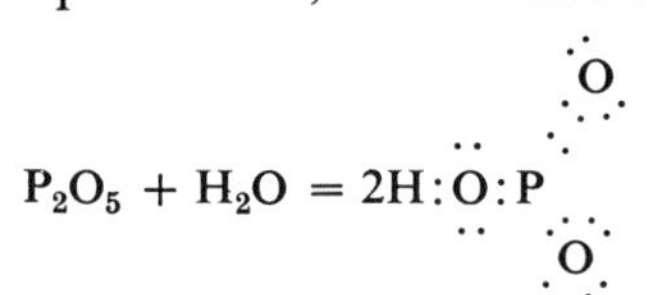

Metaphosphoric acid

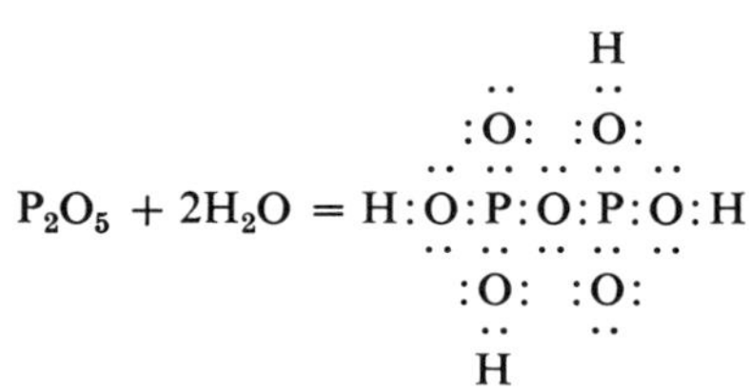

Pyrophosphoric acid or diphosphoric acid

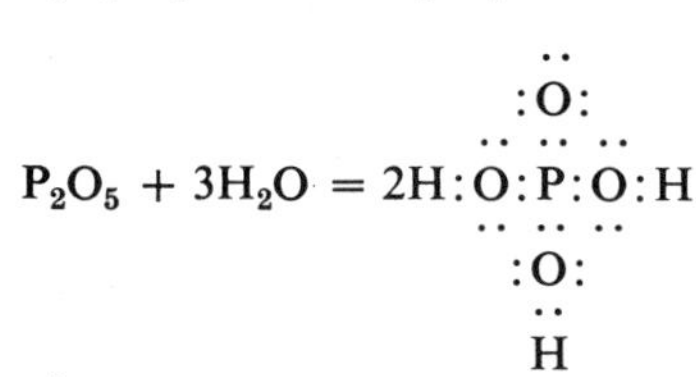

Orthophosphoric acid, commonly phosphoric acid

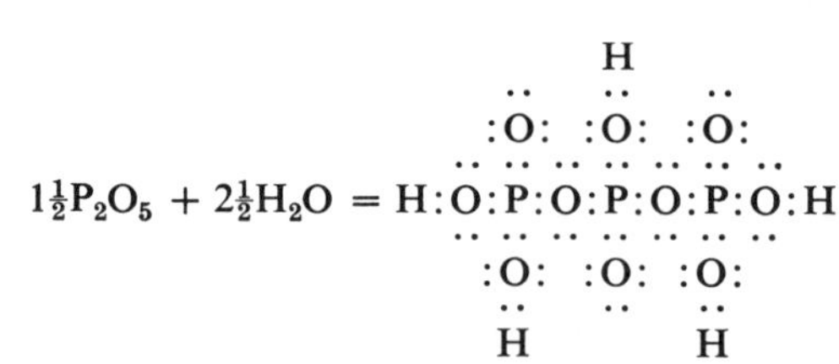

Triphosphoric acid

FIGURE 5.2 Nomenclature of the common phosphoric acids.

The method of nomenclature for those substances known as addition compounds should be mentioned. This group includes such substances as the hydrates, alcoholates, etherates and ammoniates. These are named by writing the correct names of the component substances separated by a hyphen and indicating their proportions by using Arabic numbers separated by a slash and enclosed in parentheses. This may be illustrated by $BF_3 \cdot NH_3$, boron trifluoride-ammonia (1/1).

A distinct system of nomenclature has been devised for naming substances which involve coordination compounds or complex ions. If the compound consists of a cation and anion, either or both of which may be coordination complexes, the usual order of naming cation first is used. The following rules apply to the complex portion of the compound or to the entire compound if it is a neutral complex.

The ligand or ligands are named first, their number being indicated by the appropriate Greek prefixes. Vowels between the prefix and the anion name are not elided; thus when four NH_3 groups are the ligands, the proper name is tetraammine, not tetrammine. Where two or more different ligands are involved, the order of citation is strictly alphabetical ignoring any Greek numerical prefixes. Therefore, the combination of ligands $Cl_2(NH_3)_4$ is named tetraammine dichloro. With the exception of NH_3(ammine) and H_2O(aqua), the names of neutral ligands are unchanged. The endings of anionic ligands are modified to end in *o*, e.g., hydroxo, cyano, chloro, etc. Some of the more common ligands are given in Table 5.4.

The name of a given complex ion or neutral complex always concludes with citation of the metal acting as the central unit. In neutral or cationic complexes, the name of the central ion is cited unchanged. In anionic complexes, the suffix *-ate* is added to the root name of the central ion. For certain central atoms, Latin root names are used, e.g., *plumb-* for lead, *aur-* for gold, *argent-* for silver. The oxidation state of the central atom is indicated by a Roman numeral in parentheses after its name (Stock system). As an alternate the Ewens-Bassett system may be used to indicate the charge on the *entire* ion.

If a ligand is unduly complicated or if the use of *di-*, *tri-*, or *tetra-* would be ambiguous, the prefixes *bis, tris,* or *tetrakis* (indicating 2, 3, or 4, respectively), are used. When one of these special prefixes is used, the name of the ligand is enclosed in parentheses.

Complexes also exist in which more than one atom of the central ion is present, the atoms being joined by a bridging ligand such as peroxo. In such polynuclear complexes the name of the bridging ligand is prefixed by the Greek letter μ. The rules of the nomenclature of complexes are illustrated by the compounds in Table 5.5.

TABLE 5.4 PREFERRED NOMENCLATURE FOR COMMON LIGANDS

Ligand	Name in Complexes[a]
F^-	fluoro
Cl^-	chloro
Br^-	bromo
I^-	iodo
O^{2-}	oxo
OH^-	hydroxo
O_2^{2-}	peroxo
S^{2-}	thio
S_2^{2-}	disulfido
$-NO^-$	nitrosyl
$-NO_2^-$	nitro (nitrito)
$-CN^-$	cyano
$-CNS^-$	thiocyanato
$-NCS^-$	isothiocyanato
SO_3^{2-}	sulfito
SO_4^{2-}	sulfato
$S_2O_3^{2-}$	thiosulfato
$C_2O_4^{2-}$	oxalato
H_2O	aqua
NH_3	ammine
CO	carbonyl
$H_2NCH_2CH_2NH_2$	ethylenediamine (abbreviated en)

[a] Equivalent forms are given in parentheses.

This necessarily brief survey of inorganic nomenclature should in no way be considered the complete picture. There are many aspects of nomenclature, both for ordinary compounds and coordination complexes, which are beyond the scope of this textbook. However, mastery of the items which have been covered will provide the student with a sufficient background in systematic nomenclature for recognition of almost all inorganic compounds he is apt to meet in the field of inorganic pharmaceutical chemistry. The reader interested in the presently recognized complete rules of inorganic nomenclature is referred to the 1970 Report of the Commission on Nomenclature of Inorganic Compounds, IUPAC.[2]

[2]Nomenclature of Inorganic Chemistry, Definitive Rules 1970, 2nd Ed., Butterworths, London 1971. Reprinted in *Pure and Applied Chemistry, 28,* 1 (1971).

TABLE 5.5 Modern Nomenclature for Some Complexes

	Formula	Name[a]
Cationic:	$[Fe(H_2O)_6]^{2+}$	hexaaqua iron(II) ion hexaaqua iron(2+) ion
	$[CoCl_2(NH_3)_4]^+$	tetraamminedichlorocobalt(III) ion tetraamminedichlorocobalt(3+) ion
	$[Cr(en)_3]^{+++}$	tris(ethylenediamine) chromium(III) ion tris(ethylenediamine)chromium(3+) ion
	$[(NH_3)_5CoO_2Co(NH_3)_5]^{4+}$	decaammine-μ-peroxodicobalt(III) ion decaammine-μ-peroxodicobalt(4+) ion
Neutral:	$[Ni(CO)_4]$	tetracarbonylnickel(0)[b]
Anionic:	$K_4[Fe(CN)_6]$	potassium hexacyanoferrate(II) potassium hexacyanoferrate(4−)
	$Na_2[Fe(CN)_5(NO)]\cdot 2H_2O$[c]	sodium pentacyanonitrosylferrate(III) —water (1/2) sodium pentacyanonitrosylferrate(2−) —water (1/2)
	$K_2Na[Co(NO_2)_6]$	dipotassium monosodium hexanitrocobaltate(III) dipotassium monosodium hexanitrocobaltate(2−)

[a]Both Stock and Ewens-Bassett names are given for each formula.

[b]Ewens-Bassett does not apply.

[c]Sodium nitroprusside.

PART II · INORGANIC CHEMISTRY IN THE PRACTICE OF PHARMACY

In Chapters 6 through 11, substances important to the pharmaceutical chemist but usually without direct medicinal action will be studied. This study is integrated with those chemical and physical operations, involving inorganic reagents, essential in the preparation of medicinals and their dosage forms. The pharmacist uses several terms to identify these aspects of pharmaceutical chemistry: Secundum artem, adjunct, adjuvant, vehicle, carrier, and pharmaceutical necessity. The authors suggest an all-inclusive term, pharmacy aid, *to cover both the substances used and the preparative aspects.*

SIX · WATER—THE SOLVENT OF CHOICE

Water is an interesting, extremely unusual liquid. Its properties do not follow those of the other hydrides, either those found in its own group or those of neighboring groups of the periodic table. Together with ammonia and hydrogen fluoride—the first member of Group VB and Group VIIB, respectively—water is distinguished from other covalent hydrides by the strong hydrogen bonds which exist between neighboring molecules, in both the liquid and solid states. Although oxygen is second only to fluorine in the strength of the hydrogen bonds it forms, hydrogen bonding reaches its peak in water because of the two protons per molecule available for bonding. Ammonia and hydrogen fluoride are limited to one hydrogen bond per molecule; the former has only one open coordination site, and the latter has only one available proton per molecule.

Because of the multiplicity and strength of the hydrogen bonds in water, many of its physical properties are of much greater magnitude than would normally be expected. Boiling point, melting point, heat of vaporization, heat of fusion, specific heat, surface tension, viscosity, dielectric constant—all have higher values than expected for a covalent hydride. Because of the greater bond length of the hydrogen bond, ice has an extremely open structure. Although this structure persists to a certain extent in the liquid state, the first effect on melting is a partial collapse of the open structure. Thus results one of the most striking properties of water: the density of ice is less than that of liquid water between 0 and 4°C. This phenomenon is an inversion of the normal density relationship existing between a given solid and its molten form. As a result, instead of expanding like other liquids, water contracts when heated from 0 to 4°C.

The unusual properties of water are important to the biological as well as the chemical world. It is interesting to speculate on our physical environment if it were based on a normal liquid rather than water. Because of its high specific heat and density inversion, water is responsible for a much more constant and milder climate. The crust of ice which forms on bodies of water in temperate zones during the winter prevents them from freezing into a solid mass. This is due not only to the insulating effect of the ice layer, but also to the fact that no fresh water is exposed to freezing (if the density of water were normal, any ice formed earlier in the season would sink to the bottom, thereby exposing more water to freezing). In addition, the high specific heat of water makes possible the storing of large quantities of heat in the oceans in sunny climates, heat which is given up to the colder air in the temperate zones.

Man's settlement of the torrid zones would have been more difficult were it not for the high heat of vaporization of water. This property is important in cooling through the process of perspiration. Each gram of water evaporating from the skin carries away with it 540 calories. A more normal liquid can carry off only about 50 calories for each gram evaporated under similar conditions.

CHEMICAL PROPERTIES

Under most ordinary conditions water is relatively stable chemically. This is evident from a consideration of the two equilibria existing in pure water. The first, a disproportionation reaction,

$$2H_2O \leftrightharpoons 2H_2 + O_2$$

is for practical purposes nonexistent at ordinary temperatures. Even at 2000°K. the decomposition of water into its elements is less than one per cent. The second, the ionic dissociation reaction,

$$2H_2O \leftrightharpoons H_3O^+ + OH^-$$

is equally insignificant. For pure water, $K_w = 1.008 \times 10^{-14}$ at 25°C. This dissociation constant varies directly with temperature; increasing dissociation gives a K_w approaching 10^{-12} at 100°C.

Despite and because of the relative stability of water, it is a versatile substance. Its uses depend on its ability to act as a solvent, as a ligand, as an oxidizing or a reducing agent, and as an acid or a base

Redox. The redox properties of water are generally of minor importance in pharmaceutical systems. Water will act as an oxidizing agent toward any element able to displace one or both of its protons as hydrogen gas. This can be done by any metal above hydrogen in the activity series. However, at room temperature, this reaction is negligible for most metals (e.g., iron, aluminum, zinc). Active elements (e.g., lithium, sodium, calcium) capable of reacting with water at room temperature are not generally encountered as metals in pharmaceutical practice, except in synthetic work. The following examples are illustrative:

$$2Na + 2H_2O = 2NaOH + H_2\uparrow \quad \text{(room temperature)}$$
$$2Fe + 3H_2O\text{(steam)} = Fe_2O_3 + 3H_2\uparrow \quad \text{(100°C. or higher)}$$
$$Hg + H_2O = \text{no reaction}$$

Water is also capable of acting as a reducing agent. However, the probability of encountering a reaction of this type in pharmaceutical practice is even less than the probability of encountering water as an oxidizing agent. There is only one element, fluorine, which is strong enough to bring out the latent reducing power of water:

$$2H_2O + 2F_2 = 2H_2F_2 + O_2\uparrow$$

Acid-base. Pharmaceutical reactions in which the water molecule as such is decomposed are usually the result of water's ionizing ability, yielding the hydronium and hydroxyl ions,

$$2H_2O = H_3O^+ + OH^-$$

These ions then take part in typical acid-base reactions. The acid-base properties of water have been discussed in Chapter 4. Hydrolysis has also been studied in Chapter 4.

Hydration. Water takes part in many reactions in which the identity, if not the properties, of the water molecule is preserved. The most typical is the hydration reaction. Because of its strong permanent dipole, water frequently assumes a role as a ligand in complex substances. Almost all cations form one or more hydrates. Only the very largest inert gas cations are unable to form aquo complexes, but it is possible that in aqueous solution even they have transitory existence as loosely bonded hydrates. Divalent cations, due to their stronger electrostatic field, are more highly hydrated than the monovalent cations of approximately the same size. Lithium, potassium, cesium, and magnesium illustrate these facts; the number of water molecules in their aquo complexes are 4, 2, 0, and 10, respectively. In addition, water may be present in complexes to a more limited extent, sharing the coordination positions of the central atom with other ligands.

Solvent properties. To the pharmaceutical chemist and pharmacist the solvent property of water is probably of greatest immediate importance. For most liquid preparations it is the solvent of choice. Since it is the predominant constituent of the body (60–70%), problems of toxicity, metabolism, excretion, etc., are generally nonexistent. Resort to a non-aqueous system is made only after all possible aqueous systems, solutions, suspensions, or mixed aquo-solvents have been tried and found unsatisfactory. The water-ethanol system is frequently a useful substitute solvent for substances slightly soluble in water, but it has a disadvantage: alcohol accentuates the saline taste of bromides and other salts.

Water acts as a solvent for an unusually wide range of substances. Its solvent action is coupled to one or more properties of water: its high dielectric constant, the small size of the molecule, the strong permanent dipole of the molecule, and the availability of its protons for hydrogen bonding.

The water molecule is one of the smallest known. This small size enables it to penetrate into the free space within a crystal. As a result its dissolving action involves an attack from all directions, literally, "boring from within." Larger solvent molecules, not being able to penetrate into the crystal's free space, can attack only on the surface. As a result the bonds within the crystal may be sufficiently strong to prevent solution by large solvent molecules.

The strong permanent dipole of water is important in the dissolution of many substances. When the water molecule enters a crystal, favorable electrostatic interactions enable it to form hydrates with one or more of the constituent units of the solute. Solubility in water is thus enhanced by hydration, since the resulting hydrate, with its water shell, is more similar to the solvent water. The old chemical saw, "like dissolves like" applies. In some cases the ability of water to form hydrogen bonds with the solute particles aids its dissolving action.

Heat is released as hydration energy during the solution process. In order for a substance to dissolve, sufficient energy must be supplied from some source to overcome the bonding energy of the crystal, i.e., the heat of crystallization or crystal energy. This energy must be available when a crystal melts, or goes into solution. The hydration process may supply more energy than necessary to break the crystal bonds, as evidenced by a warming effect as solution is effected.

The high dielectric constant of water favors the dissolution of ionic substances. Water molecules can effectively insulate a cation from the electrostatic field of a neighboring anion, and vice versa. Water entering an ionic crystal reduces the attractive electrostatic forces between cations and anions. Consequently it is easier for the surface ions to separate from the crystal and enter into solution. After solution, water continues this role by reducing the extent of interaction between ions.

Catalytic properties. Water is essential, frequently in traces only, for many well-known reactions. If thoroughly dried, hydrogen and oxygen do not react, but in the presence of traces of moisture will react violently. In other cases the mechanism of reaction is different when water is present than when it is absent. Water is generally considered to have a catalytic action on oxidations.

NATURAL WATERS

Before considering the preparative methods for the different waters used in pharmaceutical operations, it is necessary to study the raw water from which they are obtained. Natural water, or raw water, may be runoff water collected in reservoirs, well water, or fresh water from rivers or lakes. A knowledge of the substances present in the raw water is necessary so that procedures may be devised for the removal of those which might have a deleterious effect. These impurities may be suspended solids or dissolved substances.

Impurities present. Depending on the source, natural waters have varying amounts and kinds of suspended solids. The soil particles, plant and animal debris, and various organisms can be readily removed by settling and filtering.

Dissolved substances may be classified in several ways. The categories dissolved gases and dissolved solids will be used in this text. As the result of exposure to the atmosphere, water will usually be saturated or nearly saturated with nitrogen, oxygen, and carbon dioxide. The quantity of each gas dissolved will be a function of its partial pressure, as governed by Henry's law, and will be independent of the other gases present. Some waters will have other dissolved gases, mainly hydrogen sulfide and sulfur dioxide which are generally found in natural water from springs and wells. They originate in the strata through which underground water percolates.

Dissolved solids vary widely. Man-made dissolved impurities and pollutants include sewage, industrial wastes (both organic and inorganic), and pesticides, herbicides, and fertilizers used in agriculture. The latter especially may enter the water supply in run-off water or by seepage through the ground into aquifers. Organic metabolic and decomposition products from plants and animals living in the water-shed also are present.

Inorganic dissolved solids range widely in concentration and somewhat in variety. Generally the concentration of a given dissolved substance is a function of the water source. Runoff water, gathered in the huge water sheds of cities such as New York, is low in dissolved inorganic salts. On the other hand, water obtained from wells and springs will usually be high in dissolved inorganic materials, the quantity depending upon the length of time the water percolated through the underlying rock and soil strata. The particular dissolved solids present will be a function of the type of soil and rock through which the water percolated. For example, carbon dioxide-rich water will always be high in calcium hydrogen carbonate after passing through limestone.

Hardness. For convenience of classification inorganic dissolved solids are divided on the basis of the heat liability of their anions. *Temporary hardness* is defined as bicarbonate hardness. Since bicarbonates are unstable at the temperature of boiling water, temporary hardness is also defined as "hardness removed by boiling." Relatively soluble bicarbonates are converted to insoluble carbonates on boiling.

$$2HCO_3^- = H_2O + CO_2\uparrow + CO_3^{2-}$$

Since the most common cations present are calcium, magnesium, and iron(III), it is the carbonates of these cations which precipitate.

$$Ca^{2+} + CO_3^{2-} = CaCO_3\downarrow$$

Permanent hardness is due to salts with anions other than the bicarbonate ion. Sulfate and chloride are most common. Their salts are stable to boiling.

Man is responsible for the contamination of natural water with various inorganic compounds not ordinarily present. Industrial wastes, as, for example, copper salts or cyanides, may make their way into raw water. Radioactive fallout is today a matter of concern, and accidental contamination by radioactive wastes is always a possibility.

Special natural waters. Many "mineral waters" are commercially available. All are natural waters containing dissolved salts ordinarily not present. Sulfur waters contain dissolved hydrogen sulfide, lithia waters have dissolved lithium salts, and siliceous waters have traces of soluble alkali silicates. Carbonated waters, both natural and manufactured (soda water or seltzer, depending upon the regional usage of the term), are water charged with carbon dioxide under pressure. Upon release of the pressure these waters effervesce. Natural carbonated waters usually have high calcium and magnesium levels.

Some natural waters contain one or more dissolved salts which may have an effect on body functions. Purgative waters have relatively high concentrations of magnesium and sulfate ions—ions which have a cathartic effect (see cathartics, page 577). Waters with a very high total salt content usually have a constipating effect, because such a large volume of the body's water is needed to flush the assimilated salts out through the kidneys. Fluorosis is a harmless condition of the teeth occurring in regions whose water is high in fluoride ion content (page 366). Nitrates are normally present in surface waters in small amounts but may occur in ground waters in larger amounts. Although harmless themselves, nitrates are reduced to nitrites in the intestinal tract. Nitrites are readily absorbed and combine with the hemoglobin, producing the condition known as methemoglobinema. Nitrate concentrations are generally too low to cause this condition in adults and children, but infants are more susceptible; 80 p.p.m. nitrate have proven toxic.

POTABLE WATER

Potable water is natural water rendered safe and suitable for human use. It is the water supplied by a municipal water plant. Facetiously it is known under such names as "branch water," and "aqua tapus."

Preparation. The methods used and the extent of treatment are determined by several factors, all of which depend upon the nature and quantity of undesirable constituents in the available natural water. Taste, odor, color, pH, total solids, and pathogenic organisms must be considered. The treatment selected will consist of one or more unit operations in the following order: coagulation, settling, filtration, chlorination, and taste and odor improvement. Fluoridation is also performed in many communities. Figure 6.1 is a block diagram of a complete water treatment operation.

The degree and extent to which each unit operation is carried out will vary from community to community. When the water is obtained from a large, well-protected water shed, minimum treatment, usually centering around assurance of bacterial purity, is required. On the other hand, polluted water from a lake or stream that also serves as a disposal system for industrial waste, sewage, and runoff water will require a very thorough

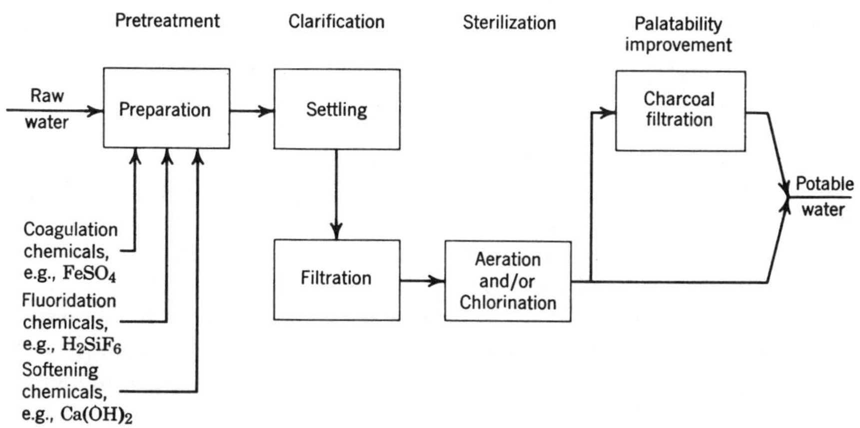

FIGURE 6.1 Unit operations in preparation of potable water.

and efficient treatment in order to assure its users a safe, if not always palatable, water supply.

Some natural waters require little or no clarification; the majority, however, do. Settling is accomplished by allowing the water to flow into huge basins (settling tanks) under conditions such that inflow at one end and outflow from the other do not cause any appreciable movement of the water while in the basin. Since the suspending action of flow is absent, large particles settle and remain at the bottom of the basin. Periodic removal of these settlings is easily achieved on draining and flushing the basins. Many waters have suspended particles too fine to be removed by simple settling. These waters are filtered through graded beds of sand and gravel. If the suspended material is exceptionally fine or colloidal, it is necessary to add coagulating materials. Ferrous or aluminum sulfate are used for this purpose. The iron(II) is first oxidized by air to form iron(III),

$$2Fe(H_2O)_6^{2+} + \tfrac{1}{2}O_2 + H_2O = 2Fe(H_2O)_6^{3+} + 2OH^-$$

Both iron(III) and aluminum salts hydrolyze in very dilute solutions to give gelatinous precipitates of their respective hydrated hydroxides,

$$Fe(H_2O)_6^{3+} + 3H_2O = Fe(OH)_3(H_2O)_3\downarrow + 3H_3O^+$$

These precipitates coagulate the last traces of fine solids and are themselves caught in the pores of the filter beds, improving the efficiency of filtration.

Dissolved solids are ordinarily not removed; they range from 25 to 1000 p.p.m. If a raw water is exceptionally *hard*, lime or ammonia may be added during the pretreatment stage. Base neutralizes the soluble bicarbonate ions, precipitating calcium, magnesium, and iron(III) ions as carbonates to the extent allowed by the bicarbonate originally present,

$$HCO_3^- + OH^- = CO_3^{2-} + H_2O$$
$$Ca^{2+} + CO_3^{2-} = CaCO_3\downarrow$$

The precipitated normal carbonate is removed during settling and filtration. Only bicarbonate hardness will be reduced by base treatment. Other chemicals such as the metaphosphates are available to reduce hardness more completely. However, further reduction of hardness is uneconomical when applied to all water passing through a water plant.

Pathogenic and other microorganisms are controlled by oxidative procedures. Aeration will destroy anerobic bacteria. With badly polluted waters the more drastic action of chlorine or its compounds is required. Calcium hypochlorite, in powder or tablet form or sodium hypochlorite solution are both excellent oxidizing germicides and are frequently employed instead of chlorine. Molecular hypochlorous acid has been found to be more effective than chlorine or hypochlorite ion, both of which are converted to hypochlorous acid by adjustment of the water under treatment to an acid pH. Acidic media repress the ionization of the hypochlorous acid formed from the chlorine, and convert the strongly basic hypochlorite ion to hypochlorous acid,

$$Cl_2 + 2H_2O = HOCl + H_3O^+ + Cl^-$$
$$ClO^- + H_3O^+ = HOCl + H_2O$$

A more thorough review of the chemistry involved will be found in the discussion of the germicidal action of these substances (page 449).

When chlorine is used, an equivalent of hydronium ion appears for every mole of chlorine added. If very high chlorination rates are necessary, an excessive acidity results; ammonia is then added to increase the pH to a more acceptable level.

$$NH_3 + H_3O^+ = NH_4^+ + H_2O$$

Although for practical purposes this should complete the treatment process, frequently a taste improvement procedure is necessary, especially when the raw water used is highly contaminated with algae and protazoal growths, and/or municipal and industrial wastes. Aeration will reduce odors and tastes to some degree. But for excessively offensive tastes and odors filtration through charcoal beds is necessary. This filtration is usually applied to only a limited fraction of the total water processed.

Under rare circumstances water can be excessively colored. Most commonly industrial wastes or higher than usual quantities of decayed plant matter are responsible. Adsorption on charcoal or chlorination will remove or destroy the colors due to organic dyes, etc. The chlorine levels required are however higher than those ordinarily necessary for sterility control; the resultant water will usually have a strong taste and odor.

Fluoridation is also necessary in communities with fluoride-deficient water sources. It is recognized by the dental and medical professions as being the one most effective means for providing this essential element. Sodium fluorosilicate, fluorosilicic acid, and, less extensively, ammonium fluorosilicate are used for this purpose. Fluoride ion becomes available on dissociation of the complex fluorosilicate ion in a very dilute solution.

$$H_3O^+ + HSiF_6^- = H_2SiF_6$$
$$H_2SiF_6 = H_2F_2 + SiF_4$$
$$SiF_4 + 4H_2O = 2H_2F_2 + H_4SiO_4\downarrow$$
$$H_2F_2 + 2H_2O = 2H_3O^+ + 2F^-$$

In concentrated solutions these dissociation reactions would take place only to a negligible degree.

A fluoride ion level of 0.8–1.3 p.p.m. is recommended; 1 p.p.m. is commonly used. The level of natural fluoride present in the raw water must be considered in determining the quantity of fluosilicate necessary to bring the fluoride ion concentration up to the desired 1 p.p.m. level. The upper limit quoted must not be exceeded.

Standards for potable water. Potable water must comply with the standards issued by the Environmental Protection Agency, EPA. In 1980 water quality standards were issued by the EPA covering its 64 priority toxic pollutants. Water to be used for preparation of Purified Water must meet EPA standards.

Waste Water. With the present interest in environmental cleanup the treatment of waste water, both municipal and industrial, becomes important. The processing of waste water is basically similar to that for the preparation of potable water. Industrial operations frequently use organic flocculants as aids. The EPA's requirements in federal funding of advanced treatment of municipal waste result in water having less than 30mg/L suspended solids and material that consumes oxygen in water by biochemical action.

Water-deficient areas. Many areas of the world are faced with insufficient water supplies. The average per capita consumption of water in modern industrial communities is very high; in the United States it has reached 2000 tons per year. Based on present rates of water consumption, water shortages in the future are predicted. Thus active research has been initiated to develop methods capable of desalting water from the oceans and brackish inland waters. Many new ideas have been advanced for separation of solids from sea water. Efficient distillation, ion exchange, propane hydrate crystallization, electrolysis methods, ice crystal formation are all being actively investigated. Potable water recovery plants are in operation in some waterless areas, e.g., Saudi Arabia, where inexpensive fuel is available.

SOFTENED WATER

The dissolved solids remaining in potable water are objectionable for many reasons. Depending on end use, the water is subjected to suitable treatments designed to change the nature of or remove the unwanted salts to the extent necessary.

Preparation. The oldest and simplest of these operations is *water softening*. Softening is accomplished by neutralization-precipitation, chelation, or ion exchange procedures. Lime treatment for removal of temporary hardness has been discussed. Any basic material capable of neutralizing the bicarbonate will accomplish the same effect; in practice only lime or ammonia treatments are economically feasible on a large scale.

For small-scale operations, in which only the water necessary for a certain operation need be treated, chemicals are available for the complete removal of objectionable cations. Sodium carbonate, either as washing soda or soda ash, sodium silicates, or borax are satisfactory. Hydrolysis of each of these salts imparts an alkaline reaction to the water,

$$CO_3^{2-} + H_2O = OH^- + HCO_3^-$$
$$B_4O_7^{2-} + 7H_2O = 2OH^- + 4H_3BO_3$$
$$SiO_4^{4-} + 4H_2O = 4OH^- + H_4SiO_4\downarrow$$

Magnesium and iron(III) precipitate either as the insoluble hydroxides or as complex basic salts of the acid anion.

$$Mg(H_2O)_6^{2+} + 2OH^- = Mg(OH)_2\downarrow + 6H_2O$$

Calcium is precipitated as the carbonate when sodium carbonate is used; with silicate and borax it precipitates as an insoluble salt.

Chemical removal of total hardness does not reduce the salt content of the treated water. Soluble sodium salts of the different anions originally present in the water remain in solution, one equivalent of soluble salt for every equivalent of insoluble compound precipitated. Chemically softened water has an alkaline reaction because of the hydrolysis of the slight excess of softening chemical necessary.

Softened water obtained by cation exchange techniques using zeolites, permutites, or greensands is identical to that obtained by chemical softening. It is alkaline in reaction, and an equivalent of sodium ion is substituted for each equivalent of di- and trivalent cation removed. Since the cation removed remains in the exchanger bed, no precipitate forms.

The chemistry of these cation exchange processes is similar to that of the cation exchange taking place during the demineralization process to be discussed shortly (page 143). It differs in the release of sodium rather than oxonium ions and in the nature of the macroanions present in the exchanger. The structure and properties of the zeolites and related substances will be discussed (page 296).

The use of chelating agents provides a final method for preparing soft waters. Sodium polyphosphates (e.g., calgon) are most commonly used for this purpose, although substances such as tetrasodium ethylenediaminetetraacetate may be used. No precipitate forms when a chelating agent is used. The cations present are rendered inactive by the chelation process; the chelates formed are water soluble. Since the hardness property is associated with the free cation, hardness has been effectively overcome even though a physical removal has not been effected. Again, equivalent quantities of sodium ion are released into the water. Partial hydrolysis of the polyphosphates imparts an alkaline reaction to the softened water.

Properties. Regardless of the method used softened water has an alkaline pH. The degree of alkalinity depends upon the original hardness, very hard waters becoming highly alkaline as a result of treatment. If the water is intended for washing and cleansing purposes this presents no problem since soaps and most detergents are more efficient in alkaline media. However, such relatively alkaline water may be unsatisfactory for general home use. Some individuals have a skin sensitivity to alkali; for others the alkaline taste is objectionable.

Softening may be mandatory for waters with high salt contents, especially if conditions of use encourage the precipitation of salts. Only the greatest offenders in this respect are removed—by substituting the less objectionable sodium ion. Total salt content of the water remains unchanged. Although a complete removal of electrolyte (such as that achieved by the demineralization process) would seem to be preferable, cost differentials make this impractical where softening is adequate for the end use intended.

By far the greatest application of water softening is to water intended for cleansing purposes. In this application it is a soap-saving device. The calcium or magnesium forms an insoluble precipitate with the soap.

$$Ca^{2+} + 2C_{17}H_{35}COO^- = Ca(C_{18}H_{35}O_2)_2 \downarrow$$

To the extent that the soap is precipitated from solution (one equivalent of soap per equivalent of calcium) it is of no value for cleansing purposes and thus represents an economic loss. The advent of modern detergents has lessened the need for water softening in this application.

PURIFIED WATER

When the dissolved solids remaining in potable water are objectionable they are removed to give "Purified Water." This may be accomplished by distillation, ion-exchange, reverse osmosis, etc.

Distilled water. Distillation depends on the fact that almost all dissolved solids are nonvolatile and can therefore be separated by boiling off the water as steam and re-condensing it. But dissolved gases and volatile solids can be removed only with special stills and by special procedures. Distilled water, because of the temperature used, is and can be maintained in a sterile condition with proper care.

The preparation of distilled water of high purity requires careful techniques, special stills and condensers, and specialized chemical processing. Because of the solvent action of water for glass, ordinary glass stills and receivers cannot be used; condensers of block tin are usually substituted. Special glasses are now available, however, which are relatively unattacked by water. In order to eliminate as much volatile material as possible stills are operated under conditions which prevent the complete condensation of the steam, the theory being that the uncondensed steam sweeps the included gases from the condenser undissolved. At the same time the first portions of the distillate, up to one fifth with small still volumes, are discarded. Because of the washing action of the steam bubbles in the liquid within the still pot, the first distillate is always richer in volatiles. A concentration effect exists within the still pot. A newly formed bubble is poorer in volatile impurity than the liquid; the volatile impurity therefore tends to pass into the steam bubble and is swept out of the still during the initial period of boiling.

The last fractions are never gathered. For best results one third of the original water is not volatilized. As distillation proceeds, the still pot water becomes progressively richer in nonvolatile substances and bumping and entrainment become progressively more pronounced. Bumping and entrainment are responsible for carrying along submicroparticles of the nonvolatile substances.

Dissolved or suspended carbonaceous material is removed by chemical techniques. To prepare an especially pure water, a two-stage distillation is used. In the first stage ordinary distilled water is redistilled over acid permanganate. Enough solid permanganate is added to give the water a distinct dark purple color. A nonvolatile acid must be used; a few drops per liter of glacial phosphoric acid is the usual choice. The second stage is a redistillation from alkaline permanganate. Barium hydroxide is the most satisfactory base.

The chemistry involved is twofold. The permanganate oxidizes the organic material to carbon dioxide and water. The fact that permanganate is its own indicator is of special value. A rapid discharge of its color indicates high levels of reducing materials. Distillation must be stopped before complete decolorization even though large quantities of water remain in the still pot. During the acid distillation the carbon dioxide

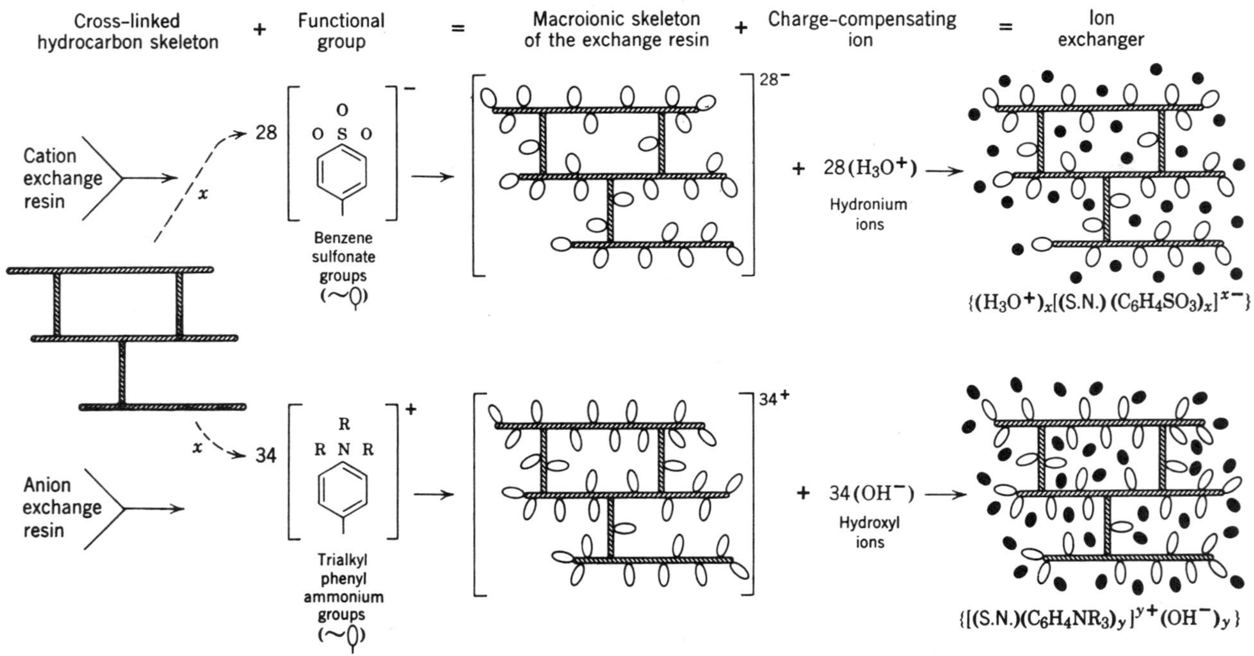

FIGURE 6.2 The structure of demineralization resins. There are no formal bonds between the hydronium or the hydroxyl ions and their associated macroionic resin skeletons. The electrostatic forces of the respective resins hold an equivalent number of free ions within their structural confines. The term (S.N.) represents the structural network of the respective resins. The numbers 28 and 34 are arbitrary.

escapes freely with the steam as formed, contaminating the distillate to some degree. The barium hydroxide very effectively prevents the escape of carbon dioxide during the second distillation; the carbon dioxide is neutralized and highly insoluble barium carbonate precipitates. Like the alkali carbonates, it is heat stable under the conditions existing in the still pot. Alkali hydroxides are unsatisfactory because the carbonate formed remains in solution and tends to increase frothing and spraying during the later stages of distillation. The distillate from the second stage must be protected from the atmosphere to prevent resaturation with carbon dioxide.

For water of intermediate purity only the alkaline distillation is employed if carbon dioxide is objectionable; otherwise a single acid distillation is preferable. For extreme purity a third distillation into a block tin condenser, in a stream of carbon dioxide-free air, is needed.

To determine the purity of water its conductivity is measured. Conductivity is proportional to the concentration of ionizing substances present in solution. Meters reading directly in p.p.m. rather than in the usual electrical units are available. Special distilled waters have specific conductivities in the range of 0.5–1.0 μmho at 20°C.

Demineralization.[1] This process depends upon ion exchange resins. Two different resins are necessary for complete removal of salts: a cation exchanger and an anion exchanger. These resins are arranged in alternate beds or mixed beds, the water being passed through the bed at room temperature.

CATION REMOVAL: Cation exchangers are essentially macroanionic crystals associated with equivalent hydronium ions. These hydronium ions are held in holes and channels within the anionic skeleton. The resin particle is a linear cross-linked polymer; negative charges are introduced by integral sulfonate, carboxylate, or phenolate groups at numerous points in the polymer's structure. A schematic representation is shown in Fig. 6.2.

The action of a cation exchanger depends upon the relative strength of the interactions existing between the negative field of the macroanion and the positive field of the hydronium ion as compared to the strength of interaction possible between the macroanion and the metallic cations present in the water being demineralized. The usual rules for strength of interactions developed in Chapters 2 and 3 apply. The strong fields of the divalent and trivalent cations, which are the predominant cationic impurities in water, interact more strongly with the anionic field than the

[1] Demineralized water was first given official recognition in the fourth supplement to *U.S.P. XIV.*

Cation exchange reaction

$$\underset{\text{Cation exchange resin}}{\{(H_3O^+)_x[(S.N.)(C_6H_4SO_3)_x]^{x-}\}} + \underset{\text{Ions in the untreated water}}{\{mCa^{2+} + nMg^{2+} + p(SO_4)^{2-} + r(CO_3)^{2-} + sCl^-\}}$$

$$= \underset{\text{Partially discharged cation exchange resin}}{\{(Ca^{2+})_m(Mg^{2+})_n(H_3O^+)_{(x-2m-2n)}[(S.N.)(C_6H_4SO_3)_x]^{x-}\}}$$

$$+ \underset{\text{Partially treated water}}{\{2(m + n)H_3O^+ + p(SO_4)^{2-} + r(CO_3)^{2-} + sCl^-\}}$$

Anion exchange reaction

$$\underset{\text{Anion exchange resin}}{\{[(S.N.)(C_6H_4NR_3)_y]^{y+}(OH^-)_y\}} + \underset{\text{Ions in the partially treated water}}{\{2(m + n)H_3O^+ + p(SO_4)^{2-} + r(CO_3)^{2-} + sCl^-\}}$$

$$= \underset{\text{Partially discharged anion exchange resin}}{\{[(S.N.)(C_6H_4NR_3)_y]^{y+}(OH^-)_{(y-2p-2r-s)}(SO_4{}^{2-})_p(CO_3{}^{2-})_r(Cl^-)_s\}}$$

$$+ \underset{\text{Treated water}}{\{2(m + n)H_3O^+ + (2p + 2r + s)OH^-\}}, \text{ with immediate neutralization}$$

$$2(m + n)H_3O^+ + (2p + 2r + s)OH^- = 4(m + n)H_2O$$

FIGURE 6.3 Reactions during demineralization of water. The bicarbonate hardness of untreated water has been removed by addition of lime; it is assumed that only the indicated cations and anions are present. (S.N.) represents the structural network of the respective resins. R may be one, two, or three different alkyl groups. All subscripts and coefficients are arbitrary, but the following conditions hold: $(m + n) = (p + r + \frac{1}{2}s) < x$ or y; x and y need not be equal.

monovalent proton. The usual effects of ionic radius and valence shell structure are noted. Having a stronger field than the large cations, small cations will be attracted more strongly and preferentially taken up by the resin. There are no formal bonds between the macroanion resin structure and the hydronium or metal cations. The forces at play are exactly identical to those found in an ionic crystal—undirected electrostatic field forces.

As water passes over and through the resin, the cations remain in the resin. To preserve the necessary electrical neutrality the resin particles release an equivalent quantity of hydronium ion to the water. The anions present in the water are unaffected. Note that the net result of passage over a cation exchanger is to convert a dissolved salt to its corresponding acid. Figure 6.3 summarizes this action.

ANION REMOVAL: The anion exchanger has essentially the same resin skeletal structure. However, the reactive groups are usually quaternary ammonium groups which effectively convert each resin particle into a macrocation. To preserve electrical neutrality a requisite number of free hydroxyl ions are included in the interstitial spaces and in the surface layer of the resin particle. The structure of the anion exchange resin is shown diagrammatically in Fig. 6.2.

The mechanism of reaction for the anion exchanger is dependent on the relative strength of the electrostatic fields and polarizability of the hydroxyl ion on the one hand, and the dissolved anionic impurities of the water on the other. The relatively nonpolarized, univalent, hydroxyl ion reacts weakly with the positive field of the resin. As a result hydroxyl ions are easily displaced by the more polarizable and frequently higher-charge anions present in the water. An equivalent quantity of hydroxyl ions are discharged into the water to maintain the over-all electrical neutrality. Figure 6.3 summarizes this action. Cations are unaffected by passage through the anion exchanger bed. As with the cation exchangers there is no formal bond between the anion exchanger macrocation and the anions present; typical nondirected electrostatic forces exist.

Another class of anion exchangers, the so-called weak, substitute a tertiary amine group for the quaternary ammonium group. These weak exchangers are incapable of holding weak acids (e.g., carbonic), but are effective for all strong or moderately strong acids.

CHEMISTRY OF DEMINERALIZATION: The net effect of passing water containing electrolytes over a cation exchanger is the release of hydronium ions equivalent to the cations taken up. In the anion exchanger an equivalent quantity of hydroxyl ions is exchanged for the anions sorbed. Since the hydronium and hydroxyl ions must also be quantitatively equivalent, the net result of the demineralization process is the substitution of one equivalent of water for each equivalent of dissolved salt originally present. The over-all reactions are summarized in Fig. 6.3. These reactions may be simplified, using calcium sulfate as the illustrating material,

$$Ca^{2+} + [(H_3O^+)_2R_c^{2-}] = 2H_3O^+ + [Ca^{2+}R_c^{2-}]$$

$$SO_4^{2-} + [R_a^{2+}(OH)_2^-] = 2OH^- + [R_a^{2+}SO_4^{2-}]$$

$$2H_3O^+ + 2OH^- = 4H_2O$$

Since demineralized water has not been subjected to heating, its sterility is not assured. However, with cleanliness and care the demineralizer beds and storage tanks may be maintained without growth of organisms. In fact, a table-top demineralizer with a 15-gal. glass receiver used by the students in the author's general physical chemistry laboratory required cleaning because of microorganism contamination only three times over a ten-year period.

REGENERATION OF RESIN: The length of time an exchanger bed operates effectively depends on its exchange capacity, that is, the number of equivalents of ions it can exchange. A regeneration procedure is initiated before exhaustion of exchange capacity. The bed is isolated from service and treated with a strong acid to regenerate the cation exchange bed and with a strong base to regenerate the anion exchange bed. The procedure is a mass effect. The concentration of the respective regenerating ion (e.g., hydronium in the cation exchanger), is so high that at equilibrium it effectively replaces the mineral cations present on the exchanger. After equilibrium is achieved, the beds are back-washed until the excess acid (or base) is removed. The beds may then again be placed in service.

Comparisons. The effectiveness of distillation versus demineralization in the removal of various impurities from water is compared in Table 6.1.

TABLE 6.1 COMPARISON OF THE EFFECTIVENESS OF DISTILLATION AND DEMINERALIZATION PROCEDURES

	Distillation	Demineralization
Strong Electrolytes	Good	Good (one stage can equal several distillations)
Weak Electrolytes	Poor	Good (even silicates, carbonates, borates, and phenolates can be removed)
Nonelectrolytes	Poor, depends on volatility	Ineffectual
Sterility	Sterile (pyrogen-free after repeated distillation)	Possible, but requires care

Neither demineralization nor a single distillation removes pyrogen from the treated water. Purified water is therefore unsuitable for preparation of parenterals.

Demineralizers in pharmacy. Because of the convenience and availability of small demineralization units, many retail pharmacists are in a position to prepare their own purified water. The capacity of these small units must be related to the solids content of the potable water available. Flow rates of 5–20 gal./hr. can be maintained. The units are equipped with an electronic indicator of the quality of the effluent, thereby making it no problem to determine cartridge replacement time. Long-term costs of these units are generally lower than the cost of distilled water purchased from outside sources. The pharmaceutical industry has found it economically feasible to use large scale demineralizers for many years.

In general the long-range economics of demineralizers depends upon the content of the dissolved solids in the available potable water. With salt contents of roughly 500 p.p.m. and higher the costs of chemicals and labor for more frequent regeneration of fixed beds, or the cost of replacement cartridges for small portable units, become excessive in comparison to distillation procedures.

SPECIAL PURPOSE WATERS

In addition to Purified Water, *U.S.P. XXI* recognizes five other waters with designated uses. Three are recognized for parenteral use. Water for Injection is water purified by distillation or reverse osmosis. It is pyrogen free and is intended as a solvent in parenterals to be sterilized after preparation. If the parenteral is to be prepared under aseptic conditions and not sterilized in the final container, the Water for Injection must first be rendered sterile and then protected from bacterial contamination. Sterile Water for Injection is Water for Injection suitably sterilized and packaged in single dose units. It is used as the solvent for sterile parenteral products which must be distributed dry because of the limited stability of their solutions. Bacteriostatic Water for Injection is Sterile Water for Injection to which one or more antibacterials have been added. It is distributed in units from which multiple withdrawals may be made. The two other waters are Sterile Water for Irrigation and Sterile Water for Inhalation. The former is Sterile Water for Injection but may be dispensed in containers with volumes greater than one liter. The latter is water prepared by distillation or reverse osmosis and sterilized. It is intended for preparation of inhalations, e.g., Sodium Chloride Inhalation, and for use in humidifiers. Table 6.2 summarizes some of the standards for the five waters discussed.

Pyrogens are the metabolic by-products of living organisms or the decomposition products of dead microorganisms. Various techniques have been devised to aid in the production of pyrogen-free water. Oxidizing agents such as hydrogen peroxide or alkaline permanganate (page 141) may be added to the water before the final distillation. The carbonaceous material is converted to carbon dioxide and water.[2] Another aid is the use of pyrogen and bacterial retentive filters. Steam passes through these filters before being condensed, thereby effecting removal of the undesirable particulate materials. The retention of bacteria and pyrogen, however, proceed by different mechanisms. Bacteria are removed by true filtration action. The cells are trapped in the pores of the filter; continued use

---

[2]*USP XXI*, p. 1124.

TABLE 6.2 Standards for Compendial Waters

	PH	Total Solids Limit	Bacterial Purity Test	Pyrogen Test	Added Substances
Purified Water	5.0-7.0	0.001%	---[a]	---	none
Water for Injection	5.0-7.0	0.001%	---	meets	none
Sterile Water for Injection	5.0-7.0	30mL[b] or less-0.004% 30-100mL-0.003% 100mL-1L-0.002%	meets	meets	none
Bacteriostatic Water for Injection	4.5-7.0	---	meets	meets	one or more antibacterials allowed
Sterile Water for Irrigation	5.0-7.0	0.001%	meets	meets	none
Sterile Water for Inhalation	4.5-7.5	---	meets	---	none

[a]Meets FDA standards for potable water.

[b]Allowed limits depend on container size.

actually improves the filter action since pore size decreases with accumulation of bacterial debris. On the other hand, pyrogens are removed by adsorption on the reactive surface of the filter. As adsorption takes place, the available free surface area decreases progressively, and pyrogen retention falls off correspondingly. The limited capacity of a pyrogen retentive filter restricts its use to a rather short period of time.

Sterile Water for Injection has higher allowed total solids than the other official waters. Being distributed in small volumes, there is an increased opportunity for reaction between the water and the glass because of the relatively large glass surface area exposed per unit volume. This water-glass reaction is accelerated during the period of elevated temperatures required for sterilization. (See page 303).

SELECTION OF SUITABLE WATER

Since the pharmaceutical chemist and pharmacist have several different waters available for their use, suitability and economics frequently must be considered in choosing a water suitable for a given purpose. Economics is simple; the less treatment water receives the lower its cost per unit volume. Thus potable water has the lowest net cost; softened water is next. Purified Water follows, unless the water used in its preparation had an extremely high solids content. Water suitable for parenterals requires the most extensive treatment and extreme care in handling; it has the highest unit cost.

Selection is no problem with solutions intended for parenterals or irrigations; there is no alternative for sterile, pyrogen free water. Thus, depending on the drug involved and the mode of handling, Water for Injection, Sterile Water for Injection, or Bacteriostatic Water for Injection must be used for injections and Sterile Water for Irrigation for irrigations. In inhalations Sterile Water for Inhalations is required.

Purified Water is used for the preparation of all dosages requiring water except parenterals, irrigations, and inhalations. It is also required for the preparation of reagents and test solutions.

Potable water (no longer cited officially), is the raw material for the preparation of all official waters of the U.S.P. While it may be used in the preparation of drug substances covered in the compendia it may not be used in the preparation of dosage forms, reagents, and test solutions.

While potable water and the official Purified Water, *U.S.P. XXI,* meet the same standards for bacterial purity (those required for potable water by the EPA), because of wide variations in properties and composition (pH, dissolved gases, total solids) it is not possible to specify potable water instead of purified water in pharmaceutical practice.

The pH of potable water varies widely, depending on the raw water available and the exact treatment procedures used. Unless it has been adjusted by the use of ammonia, the pH of chlorinated water will be in the acid range: the more chlorine used the lower the pH. Even without chlorination the continual exposure of water to air in reservoirs, settling basins, and especially during aeration leads to saturation with carbon dioxide, which will impart a distinctly acid reaction to the water. Softened waters have an alkaline pH, the actual alkalinity being directly related to the hardness of the original water and the extent of treatment.

The pH of water must be considered when dissolving pH-sensitive substances. The natural dyes of fluid extracts and the synthetic dyes in many syrups are *indicators*; therefore the color of these dyes is a function of the pH of the medium in which they are dissolved. Solubilities of many substances are pH dependent; precipitates ranging from slight turbidities to actual residues may appear. Finally the stability of many substances to oxidation is pH dependent.

Residues of chlorine or hypochlorite are usually present in potable water. Since these substances are strong oxidizing agents, easily reducible materials dissolved in water in which they are present will react to the extent of their residual concentration. Ordinary distillation procedures effectively remove chlorine only with careful techniques. Dissolved chlorine gas will pass over to the condenser with the first portions of the vapor. Addition of an excess of thiosulfate to water before distillation will convert both impurities to nonvolatile salts.

$$Cl_2 + 2S_2O_3^{2-} = S_4O_6^{2-} + 2Cl^-$$
$$ClO^- + 2S_2O_3^{2-} + 2H_3O^+ = S_4O_6^{2-} + Cl^- + 3H_2O$$

Demineralization and reverse osmosis units will remove hypochlorite, and chlorine too, since it exists in the equilibrium.

$$Cl_2 + 2H_2O = 2H_3O^+ + Cl^- + ClO^-$$

Dissolved solids vary widely in potable waters, ranging as high as 0.1%. If dissolved solids are high, the possibility of precipitations involving either the cations or anions present (or both) must be considered. In water with low dissolved solids turbidities can result, but with high dissolved solids actual residues may separate. This precipitation is especially likely when using water of high dissolved solids in alkaline media or mixed aquo-organic solvents, since both tend to decrease the solubility of many electrolytes. A few illustrations involving possible solubility problems follow.

It was provident to use purified water in the preparation of Compound Sodium Borate Solution, *N.F. XI*; calcium and magnesium borates are insoluble and would precipitate with hard waters (softened waters would not present this problem). In preparing Aromatic Ammonia Spirit, *U.S.P. XXI,* insoluble carbonates of the divalent (and trivalent if present) cations could form if water having even an intermediate solids content is used. With lower solids clouding would take place.

$$HCO_3^- + NH_3 = NH_4^+ + CO_3^{2-}$$
$$M^{2+} + CO_3^{2-} = MCO_3 \downarrow$$

Preparation of Diluted Alcohol with water having high dissolved solids leads to separation of insoluble calcium and magnesium salts. Even at the 0.02% level the solution is slightly turbid, but no precipitate separates in mixing.

In the preparation of some solutions, the presence of a high solids level in the water would present no problem: Aluminum Subacetate Topical Solution requires calcium carbonate in its preparation; the final step is filtration. Sulfurated Lime Topical Solution, *U.S.P. XXI,* uses calcium hydroxide for its preparation.

EMERGENCY TREATMENT OF WATER

Sterilization techniques. Conditions sometimes arise when sources of potable water are suspect or no safe water is available. Under these conditions emergency procedures must be set up to assure water free of pathogenic organisms. Selection of the proper reagent requires a knowledge of the effectiveness of the germicidal agent and assurance that it and its residues are nontoxic. Since germicidal effectiveness is closely related to concentration, a knowledge of effective concentrations must be at hand. Time of exposure to the reagent is also important since some organisms are more resistant than others. Some common chemicals useful for emergency water sterilization and the particulars for their use are given in Table 6.3.

If no suitable chemical reagent is available, the water may be boiled.

TABLE 6.3 EMERGENCY STERILIZATION OF WATER

Reagent	Effective Concentration	Contact Time for Effective Sterilization
Strong Iodine Tincture (*U.S.P. XXI*)	2 drops/liter	30 minutes
Iodine Tincture (*U.S.P. XXI*)	7 drops/liter	30 minutes
Halazone[a] (*U.S.P. XXI*)	tablet (4 mg.)/liter	30 minutes

[a]Halazone = *p*-(Dichlorosulfamoyl)benzoic acid.

A minimum period of twenty minutes of active ebullution should be effective, but at higher altitudes the time should be extended since the reduced atmospheric pressure lowers the boiling point of water.

Radioactive fallout and removal techniques. Today man lives with an ever-present threat of radioactive contamination of water supplies. This can arise accidentally from radioactive wastes and the by-products of various procedures using radioactive substances, or it can be the result of radioactive fallout from atomic and hydrogen bomb explosions.

The extent of radioactive contamination will be a function of the water source of a municipal supply and the time lapsed between contamination and intended use. Where watersheds are some distance from the city, dilution, natural decay, and adsorption by soil particles will probably reduce levels sufficiently for consumption on an emergency basis—except for a limited time immediately following contamination. On the other hand, water obtained from a nearby river or lake may require the lapse of several days before decay and adsorption reduce the contamination sufficiently for safe emergency use. Water obtained from underground sources should generally be free of contamination. The great length of time required for percolation of the water from the surface, coupled with the natural adsorption of the radioactive substances by the soil through which the water percolates (especially clays), should result in nonradioactive water.

Emergency treatment of supplies follows the same general procedures used for ordinary water treatment. If the municipal water plant is intact, normal settling, coagulation, and filtration will remove up to 90% of the radioactive contaminants—all suspended insoluble material and varying amount of dissolved compounds. Chemical water softening techniques (except chelation) can effectively reduce dissolved solids, especially strontium-90. Treatment with phosphate must be carefully controlled with respect to pH and calcium-phosphorous ratios if effective precipitation is to occur. Clay suspensions will effectively remove cesium-137.

Most effective for removal of dissolved solids (after a preliminary filtration) are the techniques used to produce purified water: namely, ion-exchange and distillation. Since radioactive fallout is largely cationic, only a cation exchanger is required, but a mixed bed will be slightly more effective. Either of these two techniques will successfully remove better than 99% of dissolved radioactive solids.[3]

MISCELLANEOUS TOPICS

Home pools. Home swimming pools have opened another area of water treatment and control. Algae control requires use of copper sulfate or quaternary ammonium salts. Soda ash or alum is used for pH control. Hypochlorites (calcium hypochlorite tablets or powder and sodium hypochlorite solutions) are used to maintain sterility. If not properly controlled, chlorine compounds have an undesirable bleaching action on materials contacted. To circumvent this, nontoxic iodine-based germicides have been introduced. These iodine compounds are usually in a complex form. The chemistry and action of most of these compounds are discussed in Chapter 16 (page 457).

Heavy water. The element hydrogen has three isotopes. It is the only element whose isotopes have been named. Some pertinent information concerning these isotopes is given in Table 6.4. Since ordinary hydrogen is almost entirely H-1, its properties are essentially those of that isotope.

---

[3] The radioactivity hazards of exposed foods, if due to relatively short-lived isotopes, can be largely eliminated by storage for ten times the half-life of the particular isotope. Iodine-131 (half-life, 8.14 days) appeared in the milk from dairy herds in several areas near the atom bomb test sites in Nevada. By removing this milk from the fluid milk market and using it for the production of storable products (butter and cheese), the hazard involved in its consumption was eliminated.

TABLE 6.4 THE HYDROGEN ISOTOPES

Isotope	Symbol	Atomic Weight	Atomic Number	Per cent in Natural
Protium	H	1.00812	1	99.9844
Deuterium	D	2.014708	1	0.0156
Tritium	T	3.01707	1	—

Heavy water is deuterium oxide—D_2O. Its properties are very similar to ordinary water, the differences being due to the greater mass of its molecule. This is reflected in a higher specific gravity, boiling point, melting point, etc. Its surface tension and dielectric constant are lower, making D_2O a poorer electrolyte solvent. It undergoes all the chemical reactions of ordinary water.

Deuterium oxide and deuterium-labeled compounds have been useful in studying reaction mechanisms by isotope tracer techniques. The difference in physical properties due to the presence of deuterium in a compound is sufficient to be detected by such standard methods as density and mass spectrograph measurements.

Drying agents. Although water has been called the universal solvent, at the opposite extreme it may also be called a universal impurity. When vapor density is used as a basis for determination of such properties as surface tension, boiling point, and molecular weight, the experimental values are frequently a function of the rigor of the method used in drying the sample. The melting point of sulfur has been raised from 112.50°C. to 117.50°C. by complete dehydration. The boiling points of bromine and mercury are much higher if these substances are thoroughly dried for long periods of time over phosphorus pentoxide. When minute quantities of water are added to these especially dried substances, the physical property being measured returns to its generally accepted value.

The relative efficiency of drying agents is of importance in pharmaceutical work. Many of the more commonly used dehydrating agents—calcium chloride, copper(II) sulfate—are relatively inefficient in this respect. Table 6.5 lists a range of drying agents. The specific drying agent is selected according to whether there will be direct contact between the drying agent and the material being dried or whether the drying will take place through a common inert atmosphere. Direct contact will give more rapid drying, but the drying reagent may react chemically with the substance being dried. The rate of drying is dependent on the exposed surface area of the drying agent.

TABLE 6.5 EFFICIENCY OF DRYING AGENTS[a]

Copper(II) sulfate	1.4
Zeolites[b]	4.0–1.0
Zinc chloride	0.8
Calcium chloride (granular)	0.15
Calcium oxide	0.2
Sodium hydroxide	0.16
Calcium sulfate	0.004
Hydrogen sulfate	0.003
Alumina	0.003
Magnesium perchlorate	5×10^{-4}
Phosphorus pentoxide	2×10^{-5}
Sodium metal[c]	1×10^{-23}
Calcium hydride[c]	1×10^{-23}
Cold trap at −194°C	1×10^{-23}

[a] The ratings are in milligrams of water per liter of air (25°C.) at equilibrium.

[b] The efficiency of the zeolites is a function of the cation present. The order of efficiency is Li^+, Na^+, K^+. The synthetic zeolites (molecular sieves) used as drying agents for air give the upper limit; as a drying agent for organic solvents the lower limit may be attained.

[c] It should be remembered that hydrogen gas is evolved in these cases. Adequate ventilation is a necessary caution if large quantities of moisture are to be removed.

Note to the Student

At this point a series of study and self-appraisal aids is being initiated. Similar sections will close all succeeding chapters. Although except for chapter outlines such guidance material has not appeared in the theoretical section, the subject matter covered in the first five chapters has been integrated into appropriate study aids in this and the remaining section of the textbook. The make-up of these student aids is as follows.

Each chapter closes with a chapter outline showing the physical organization of the material discussed and at the same time setting out the major topics covered. This outline is followed by a list of technical terms or words introduced in the chapter and words or terms which must be understood for an appreciation of the material covered in the chapter. Any terms appearing in the chapter outline will not be repeated, e.g., potable water appears in the present chapter outline and is therefore omitted from the list. A section designed to appraise the student's proficiency in the nomenclature of most compounds appearing in the chapter follows.

A unit consisting of questions designed for self-evaluation of the degree of mastery of the theoretical and practical aspects of the subject matter appears next in the arrangement. Since officially recognized compounds and preparations are ordinarily not identified in the text material an opportunity is provided at this time for consultation of the compendia for student identification of these substances. Since an ability to write equations for reactions is one of the best tests of mastery of chemical concepts and properties of compounds, a closing section provides an over-all review based on equation writing.

IMPORTANT TERMS. Define or identify in a single concise sentence if possible.

Aeration, chelation, chlorination, coagulation, filtration, fluoridation, settling, solvation, ion-exchange, reverse osmosis, bacterial filter, carbonaceous, catalyst, dehydration, density, detergent, dielectric constant, dipole, heat of vaporization, hydride, hydrogen activity series, hydrogen bond, K_w, methemoglobin, micromicrocurie, nonelectrolyte, permanent hardness, pyrogen, pyrogen filter, purgative water, radioactive fallout, solute, solvent, specific heat, strontium-90, temporary hardness, turbidity, zeolite.

NOMENCLATURE OF INTEREST. Identify each of the following by older chemical names and, when important, by trivial names.

Potassium tetraoxomanganate(VII), sodium hexafluorosilicate(IV), monoxochlorate(I) ion, calcium oxide, hydrogen peroxide, iron(III) bicarbonate.

SELF-APPRAISAL QUESTIONS

1. Distinguish between proton, hydrogen ion, hydrogen atom, hydrogen molecule, hydronium ion, hydride ion, hydroxyl ion, and the water molecule. Show the valence electron structures and bonding when necessary.

2. Describe the water molecule as to its structure and physical properties (i.e., polarizability, bond angles).

3. Hydrogen sulfide and hydrogen selenide, the second and third hydrides of the oxygen family, are fixed gases, but hydrogen oxide, the first member of the series, is a liquid. The reverse of these physical states is normal; hydrogen oxide should be gaseous and hydrogen selenide

possibly a liquid or solid. Explain the fundamental differences of these molecules which are responsible.

4. Distinguish between raw, potable, softened, purified, and parenteral waters. Use an analytical basis.

5. Give the usual stages in the preparation of potable water and the purpose of each.

6. Explain why the following properties of potable water must vary widely from community to community: pH, total solids, taste and odor, and salts present. Why is the presence of excessive detergent a problem?

7. How is the character of a given softened water dependent on the character of the potable water from which it is prepared? Why is purified water independent of the character of the potable water from which it is prepared?

8. Identify the ions commonly responsible for each, permanent and temporary hardness. Why are iron(II) salts absent?

9. What is meant by "solids by evaporation"?

10. How does the action of a zeolite as used in water softening procedures differ from the action of a cation exchange resin as used in demineralization? How are their actions similar?

11. Name the two overriding considerations in preparation, storage, and use of waters intended for parenteral uses. What "caution" must always be observed when preparing solutions to be administered via injection?

12. Identify each of the waters officially recognized in the current compendia. Compare the allowed limits of various properties, etc.

13. The following are available: potable water, pH, 4.5, dissolved solids, 300 p.p.m. (largely as calcium bicarbonate and chloride); demineralized water, pH, 6.5, dissolved solids <10 p.p.m.; softened water, pH, 9.0, dissolved solids, 400 p.p.m. (largely sodium chloride and sulfate). Which of these waters would be suitable for the following preparations: $0.005N$ silver nitrate solution, 50% ethanol, 0.24% calcium sulfate solution, 1% sodium palmitate solution, $0.1N$ sodium nitrite. (Assume room temperature.)

14. Explain why water obtained from a well is unlikely to be contaminated by radioactive fallout.

15. How can dissolved radioactive salts be removed from water?

PRACTICE IN EQUATION WRITING. Write balanced equations describing each of the following situations. Use only those structural species actually involved in the reaction.

16. How ammonia reduces acidity of water.

17. How alkaline permanganate (in barium hydroxide solution) destroys organic compounds in water. (Use HCOOH as the formula of the organic compounds.)

18. Show why dissolved strontium-90 will be precipitated by addition of lime to contaminated water.

19. How does heat destroy temporary hardness?

20. Give the reaction taking place when chlorine is dissolved in water.

21. Write the equations for the reactions taking place when sodium hydride, quicklime, and copper(II) sulfate, respectively, act as dehydrating agents.

22. Using simplified equations, show how dissolved calcium sulfate is affected on passing through a water softener; through a demineralizer.

23. Using simplified equations, show why an exhausted anion exchange resin is regenerated by sodium hydroxide treatment.

24. Give the reaction taking place when tetrasodium ethylenediaminetetraacetate is used to treat hard water. Assume iron(III) chloride is the only salt species present.

25. Show why water having a high dissolved solids content would be unsatisfactory for dilution of a barium hydroxide solution.

26. Show the reaction taking place when aluminum sulfate is used as a coagulant.

SEVEN · ACIDS, BASES, AND BUFFERS IN PHARMACEUTICAL PRACTICE

Acid-base phenomena serve two distinct roles in pharmaceutical chemistry. The first, the subject of the present chapter, involves acid-base reactions employed by the pharmacist. These include neutralization reactions in analytical and preparative procedures, the use of conjugate pairs as buffers, and the specialized acid-base systems of effervescent mixtures. The second role, which is the subject of Chapter 15, involves the use of acids and bases as therapeutic agents in the control and adjustment of the pH of the gastrointestinal tract, body fluids, and urine.

NEUTRALIZATION PROCEDURES IN THE PRACTICE OF PHARMACY

Neutralization procedures are used for the preparation of suitable salts, the conversion of salts into more suitable forms, and in analytical procedures involving acid-base titrations.

Criteria in the selection of a suitable acid or base. When a neutralization procedure is decided upon, theoretical and practical factors must be considered. Acid and base strength of the members of the reacting conjugate pairs must be compared. Generally the acid-base properties of the solvent must also be considered. The choice of the neutralizing reagent will be conditioned by the extent of reaction required. If complete conversion is necessary, a large ΔpK_a is required; that is, the acid formed during the neutralization process must have a large pK_a (small ionization constant) compared to that of the original acid.

$$\underset{\text{Reagent}}{HB_1} + \underset{\text{Reactant}}{B_2^-} \rightarrow \underset{\text{Products}}{B_1^- + HB_2} \qquad \text{where } pK_{a,1} \ll pK_{a,2}$$

On the other hand, if only partial neutralization is required, the pK_a

values of the acids involved may be nearly equal.

$$HB_1 + B_2^- \rightleftharpoons B_1^- + HB_2 \qquad \text{where } pK_{a,1} \sim pK_{a,2}$$

The effect of excess reagent must also be considered, e.g., its effect on the pH of the finished solution.

The reaction by-products may present problems. For this reason it is advantageous to have water or carbonic acid as the surviving acid of a neutralization procedure. Water becomes part of the solvent already present; carbonic acid, especially at elevated temperatures, escapes to the atmosphere as carbon dioxide. The carbon dioxide can be completely expelled by boiling the solution briefly, providing no heat-liable substances are present.

The pharmaceutical chemist must also be aware of problems in handling, use, and storage not only for his own and his associates' safety, but also for the safety of the patient. The compendia recognize hazardous acids and bases by the inclusion of *cautions* which describe the nature of the hazard and/or the precaution to be taken. The most common sources of storage difficulties are inadequate closures or exposure to heat. Inadequate closures lead to vaporization and loss of gaseous compounds and to exposure to moisture, oxygen, and carbon dioxide. Oxygen will oxidize acids or bases with reducing properties, e.g., hydriodic acid. As the anhydride of carbonic acid, carbon dioxide can react with bases (e.g., lime water, soda lime, sodium hydroxide) having a conjugate acid with a larger pK_a than carbonic acid. Moisture may be absorbed on exposure to air by hygroscopic substances, e.g., sodium hydroxide. (If a substance can absorb enough water from the air to become liquid, it is said to be deliquescent.) Dehydrating agents such as concentrated sulfuric acid and quicklime must also be protected from air by tight closures. Acids and bases existing as crystalline hydrates (e.g., sodium carbonate, citric acid) and having higher vapor pressures than the relative humidity of the atmosphere must be protected by tight closures. These efflorescent substances tend to lose varying portions of their water of hydration to the atmosphere. Since hydrates lose water of hydration at higher temperatures, they must be protected from heat. (For a general discussion of these factors and their effect on stoichiometry see page 237.)

Because of possible incompatabilities the pharmaceutical chemist must be aware of all other chemical properties of a given acid or base. The effect of other ions or molecules on the solubility behavior of an acid or base ion must be considered. Some acids and bases are capable of acting as redox reagents. Some molecular acids and many of the anion bases are capable of acting as ligands; resultant complexation frequently will change the anticipated behavior of a preparation.

TABLE 7.1 PROPERTIES AND NOMENCLATURE OF ACIDS AND BASES

Name		Per cent Concentration	pK_a[a]	Cautions	Other Properties
Chemical	Common				
Hydrochloric	Muriatic	35–37	—	—	High volatility of HCl
Nitric	Aqua fortis	67–71	—	—	Strong oxidizing agent
Phosphoric	(ortho)	85–88	1.3	—	Most phosphates are insoluble at pH >7
Sulfuric	Oil of vitriol	94–98	(−2)	Always add to water on dilution	—
Sodium dihydrogen phosphate	Primary sodium phosphate	($1H_2O$)	7.2	—	—
Acetic	—	36–37	4.72	—	—
Citric	(Monohydrate or anhydrous)	—	3.06	—	Chelating agent, dilute solution ferments
Lactic (DL)	—	85–90	3.86	—	Hygroscopic
Tartaric (DL)	(Monohydrate)	—	3.0	—	Chelating agent
Ammonia solution	Spirit of Hartshorn	27–30	9.28	Caustic, irritating vapors	High volatility of NH_3
Calcium oxide	Quicklime	(95%)	—	Caustic	Absorbs H_2O and CO_2 strongly
Calcium hydroxide	Slaked lime	(95%)	—	Caustic	Absorbs CO_2 strongly
Potassium bicarbonate	—	(99%)	10.2	—	—
Potassium carbonate	Pearl ash (anhydrous)	($1\frac{1}{2}H_2O$)(99%)	(3.8)	—	—
Potassium hydroxide	Caustic potash	(>85%)	—	Caustic	Absorbs CO_2 strongly
Sodium bicarbonate	Baking soda	(99%)	10.2	—	—
Sodium carbonate	Soda ash	anhydrous	(3.8)	—	—
	—	($1H_2O$)(>99.5%)	—	—	(Slight effervescence in dry air)
	Washing or sal soda	($10H_2O$)(99%)	—	—	Effervescent
Sodium hydroxide	Caustic soda	(>95%)	—	Caustic	Absorbs H_2O snd CO_2

[a] Meaningless values for very strong acids and bases are not given. First ionization only is given for polyprotic acids. Values are 25°C.

Acids and bases of pharmaceutical interest. A compilation of the compounds most frequently used for pharmaceutical neutralization procedures is found in Table 7.1. Acids (or bases) whose pharmaceutical uses are other than for neutralizations, such as boric, hydriodic, hypophosphorous acids, have not been included in the table.

Since few of the most useful acids and bases exist in a stable crystalline state suitable as a primary standard, most must be used in the form of standard solutions. The normality of these solutions must be known to an accuracy commensurate with the use for which they are intended. Concentration data for the preparation of standard acid solutions have been compiled in Table 7.2. The data for the concentrated acids refer to the most commonly available form; more concentrated and "fuming" acids are also available. Similar data for the common bases have been compiled in Table 7.3.

The preparation of smaller volumes of standard solutions having concentrations within a prescribed range is adequately covered in textbooks of quantitative analysis.[1] However, the preparation of larger volumes of standard solutions, (e.g., 10 liters of sulfuric acid) of exactly specified concentrations (e.g., 0.1000*N*) is customarily neglected. A suitable procedure for preparation of the solution given in the preceding statement follows: The desired volume of purified water (or carbon dioxide-free water) is measured into a carboy of sufficient capacity.[2] A large graduated cylinder (two-liter) is sufficiently accurate for measuring the water. The actual volume of water used must be adjusted for three factors. First, a known overage must be taken to allow for withdrawal of several samples for standardization procedures. Second, the volume to be taken must be adjusted for the volume of concentrated sulfuric acid to be added. Third, an intentional error must be introduced; for ease of manipulation it is desirable to have an initial working concentration higher than that required, e.g., 0.105*N*. The latter may be accomplished by using a volume of water purposely low, or a volume of concentrated acid slightly higher than calculated.

The resulting $0.1^{+}N$ solution is brought to exact normality by a stepwise volume adjustment, based on successive determinations of its titer. Illustrative calculations for this discussion may be followed in Fig. 7.1. The actual normality of the prepared acid is determined by titration of a sample with a standard base. This information is used to calculate the

---

[1] D.A. Skoog and D.M. West, *Fundamentals of Analytical Chemistry,* 4th Ed., Saunders, New York, 1982, pp. 6-14.

[2] The carboy should have a volume somewhat larger than needed. This facilitates mixing of the solution by simply rolling the partly filled carboy back and forth on a desk top.

TABLE 7.2 CONCENTRATION DATA FOR AQUEOUS ACIDS

Acid	Designation	Concentration			Molecular Weight	Specific Gravity	Ml. for 1 L. of 1*M* Solution
		% (w/w)	Gm./L.	Molarity			
HCl	Conc.	36.5	424	11.6	36.5	1.18	86.2
	—	10.0	105	2.88	—	1.05	347.
H_2SiF_6	Conc.	34.0	448	2.65	144.1	1.32	377.
HNO_3	Conc.	67.0	938	14.9	63.02	1.40	67.1
H_3PO_4	Glacial	85.0	1436	14.7	98.0	1.69	68.0
	—	10.0	105	1.07	—	1.05	—
H_2SO_4	Conc.	96.0	1762	18.0	98.1	1.84	55.6
	—	10.0	107	1.08	—	1.07	—
HOAc	Glacial	99.6	1045	17.4	60.05	1.05	57.5
	—	36.0	376	6.26	—	1.045	160.
	—	5.0	50.3	0.84	—	1.006	—
$H_3(C_6H_5O_7)$	—	66.0	863	4.1	210.2	1.31	243.9
	—	10.0	104	0.49	—	1.04	—
$H(C_3H_5O_3)$	DL	87.0	1044	11.6	90.1	1.20	86.3

TABLE 7.3 CONCENTRATION DATA FOR AQUEOUS BASES

Base	Designation	Concentration			Molecular Weight	Specific Gravity	Ml. for 1 L. of 1*M* Solution
		% (w/w)	Gm./L.	Molarity			
NH_3	Conc.	28.0	251.	14.8	17.03	0.898	67.5
	—	10.0	95.8	5.6	—	0.958	179.
$Ca(OH)_2$	Saturated	0.17	1.65	0.022	74.10	1.00	—
KOH	—	50.0	757.	13.5	56.11	1.51	74.1
	—	10.0	109.	1.95	—	1.09	—
Na_2CO_3	—	10.0	110.	1.04	106.0	1.10	—
NaOH	—	50.0	763.	19.1	40.0	1.53	52.4
	—	10.0	111.	2.78	—	1.11	—

Problem: Preparation of 10.0 liters of exactly 0.1000*N* sulfuric acid. To allow for samples withdrawn for analysis (titration with standard NaOH), 10.5 liters of approximate solution are prepared.

Calculated concentrated acid required, 29.2 ml.		
Actual volume concentrated acid used (to give overage)		30 ml.
Volume of water used (accurate within 50 ml.)		10470 ml.
Working Volume		10500 ml.
Removed for rinsing pipette and 50-ml. sample for analysis		80 ml.
Remaining Volume		10420 ml.
Normality based on titration—0.1029		
Theoretical volume (if diluted to exact desired normality)	10720 ml.	
Less actual volume	10420 ml.	
Water required for exact dilution	300 ml.	
Actual volume of water added		280 ml.
New Working Volume		10700 ml.
Removed for rinsing pipette and three 50-ml. samples for analysis		200 ml.
New Remaining Volume		10500 ml.
Normality based on titration—0.1004		

FIGURE 7.1 Sample calculations for preparation of solutions of exact normality. The final computations for normality is within the accuracy range of the desired normality.

hypothetical volume of the sulfuric acid solution, if diluted to exactly the desired normality. However, since the volume of the prepared acid is not known to exact volumetric standards, it is not provident to increase the volume to the calculated volume at this time. The volume of water used for dilution should be somewhat less than the difference between the remaining volume of acid and its theoretical volume; for example, if this difference is 300 ml., only 280 ml. of the water should be added. The diluted acid is thoroughly mixed, and a new titration with standard base is performed. The new normality and the new theoretical volume for exactly 0.1000*N* acid are now calculated. If necessary another dilution with a slightly less than theoretical volume of water is made, and the titer is redetermined. This method of approximation, approaching the desired normality from the high side, is repeated until the exact normality is reached. Three stages of titration will generally suffice.

Bases of choice. The three bases selected most frequently in pharmaceutical practice are the hydroxyl ion, bicarbonate ion, and carbonate ion, usually in the form of their sodium salts. The potassium compounds are rarely used. The sodium ion is preferred because of its physiological compatability, the solubility it imparts to its compounds, and the almost complete absence of acid-base properties as the hydrated cation. A discussion of the pertinent properties of each of the three sodium compounds follows.

SODIUM HYDROXIDE: Sodium hydroxide (sodium hydrate, caustic soda) is commercially available in the form of lumps, sticks, pellets, and flakes. Like all alkali bases (page 207), it is heat stable and very soluble in water. It generates considerable heat on dissolving in water. The compound is extremely deliquescent; on exposure to air it will absorb moisture and become liquid. Being a strong base, it will strongly absorb carbon dioxide from the air. As a result the solid and all solutions except the concentrated are contaminated to some degree with sodium carbonate. Therefore sodium hydroxide must always be stored in tightly sealed containers.

Carbonate-free sodium hydroxide may be prepared by one of two procedures. The more common method depends upon the insolubility of sodium carbonate in a concentrated (50%) solution of sodium hydroxide. The clear supernatant liquid from such a solution is added to carbon dioxide-free water for preparation of carbonate-free solutions of this base. Sodium carbonate is also insoluble in ethanol; when sodium hydroxide is dissolved in ethanol, the accompanying carbonate settles out.

Since there can be no accurate knowledge of the amount of absorbed moisture and sodium carbonate present with a given sample of sodium hydroxide, weighing techniques are unsuitable for measurement of stoichiometric quantities. Accurate measurement of the compound is possible only by use of standard solutions, preferably carbonate free.

Because of the presence of carbonate the titer of sodium hydroxide solutions depends upon the indicator used to determine the end point of standardization. If the base is titrated with a standard acid using phenolphthalein as the indicator (pH $\pm$ 8.5), the reactions are

$$OH^- + H_3O^+ = 2H_2O$$
$$CO_3^{2-} + H_3O^+ = HCO_3^- + H_2O$$

However, if methyl orange is used as the indicator (pH $\pm$ 5), an additional reaction takes place:

$$HCO_3^- + H_3O^+ = H_2CO_3 + H_2O$$

Extreme care must be exercised when neutralization with sodium hydroxide is a procedural step in the preparation of a pharmaceutical solution. An excess of this reagent is intolerable, regardless of mode of use—topical, oral, or via injection. A solution having excess sodium hydroxide will range from irritating to corrosive, depending on the amount of excess base. As a very strong base, sodium hydroxide has the advantage of being able to neutralize almost all of the acids commonly used in pharmacy. Even neutralization of a weak acid will proceed satisfactorily.

SODIUM BICARBONATE: Sodium bicarbonate (baking soda) is at the other extreme as a base, both as to properties and handling. Being a weak base, the bicarbonate ion has limited value for neutralization of acids. Only strong and moderately strong acids will react. Weak acids, e.g., boric, cannot be neutralized.

Sodium bicarbonate presents no serious storage or handling problems. It is not sensitive to moisture or carbon dioxide. It is obtainable in sufficient purity to make possible direct weighing of a required sample. Should an overage be accidentally used, the resultant solution cannot be more alkaline than a solution of bicarbonate.

Bicarbonates are heat liable. For this reason heat sterilization can be carried out only under special conditions when a bicarbonate is present in an injectable (page 247). In addition, sodium bicarbonate must not be stored in areas exposed to excessive heat (e.g., shelves over a steam line). Nor can solutions of this compound be boiled. The following equations illustrate the reactions on thermal decomposition:

$$2HCO_3^- = CO_3^{2-} + H_2CO_3$$
$$H_2CO_3 = H_2O + CO_2\uparrow$$

Due to the proton of the anion, sodium bicarbonate is also capable of acting as a weak acid.

SODIUM CARBONATE: Sodium carbonate is, in most respects, the intermediate of sodium hydroxide and sodium bicarbonate. It exists as an anhydrous salt or as one of three known hydrates. The decahydrate (washing soda, sal soda) crystallizes from aqueous solution at temperatures below 32°C. This salt is efflorescent, changing to the monohydrate on exposure to dry air. Its use in pharmaceutical chemistry is limited.[3] A heptahydrate also exists, crystallizing from aqueous solutions in the range 32–35.4°C. This hydrate has no commercial or practical value.

The anhydrous salt (soda ash) crystallizes from aqueous media above 109°C., or it is formed by heating any of the hydrates. (Although sodium

[3] There are economic reasons for the limited industrial use of this compound. By weight it is 63% water. Thus the equivalent available per unit is less than half that available in the monohydrate. Shipping costs also mediate against shipment of large quantities of water.

bicarbonate is extremely heat liable, sodium carbonate is heat stable except at extremely high temperatures.) It absorbs moisture if exposed to the atmosphere, forming the monohydrate.

The monohydrate crystallizes from aqueous media between 35.4–109°C. It also forms if the anhydrous salt is wet with a stoichiometric quantity of water. Because of its comparative stability to moisture and carbon dioxide, the monohydrate is usually selected for use in the pharmaceutical and chemical industries. It will, however, slowly absorb moisture and carbon dioxide:

$$CO_3^{2-} + H_2O + CO_2 = 2HCO_3^-$$

This reaction is responsible for the caking of sodium carbonate.

The carbonate ion is a moderately strong base. Its aqueous solutions are alkaline (pH $\sim$ 10) due to the reaction

$$CO_3^{2-} + H_2O = HCO_3^- + OH^-$$

which proceeds to about 3% in a $0.1N$ solution. Carbon dioxide is lost very slowly on boiling, probably due to the reaction

$$HCO_3^- + H_2O = H_2CO_3 + OH^-$$
$$H_2CO_3 = H_2O + CO_2 \uparrow$$

Carbonate ion is a useful reagent for the neutralization of a wide range of acids, including most common organic acids. In fact, the carbonate ion is a sufficiently strong base to cause many phenols to go into solution as their phenoxide ions.[4]

Because of its stability and purity sodium carbonate monohydrate, like sodium bicarbonate, may be measured by direct weighing. But like sodium hydroxide, its strong basic properties require that an excess be avoided.

The properties of the three sodium bases are shown for comparison purposes in Table 7.4.

Acids of choice. HYDROCHLORIC ACID: Because of the physiological inertness of the chloride ion, hydrochloric acid[5] is usually the acid of choice for pharmaceutical neutralizations. All except the very weakest bases can be neutralized. The disadvantages of this acid are analogous to those encountered in the use of sodium hydroxide for neutralizations. Excesses must be avoided, and standard solutions must be prepared. In

---

[4] *J. Chem. Educ.*, **35,** 568 (1958).

[5] Muriatic acid is a less pure industrial form of hydrochloric acid. It usually has a pronounced yellow color.

TABLE 7.4 A COMPARISON OF THE PERTINENT PROPERTIES OF SODIUM COMPOUNDS USED AS BASES

Acid-Base Information	Sodium Hydroxide	Sodium Carbonate	Sodium Bicarbonate
Base	OH^-	CO_3^{2-}	HCO_3^-
Conjugate acid	H_2O	HCO_3^-	H_2CO_3
pK_a	16.	10.2	6.4
Approximate pH of a solution of the sodium salt	13–14	10–11	8.0
Comparative strength as a base	Very strong	Moderate	Weak
Storage problems of crystalline form:			
Moisture	Deliquescent	Effervesces in dry air; hygroscopic in moist atmospheres	—
CO_2	Strongly absorbent (forms Na_2CO_3)	Absorber in moist atmosphere (forms $NaHCO_3$)	—
Measurement of equivalents	Accurately measured volume of standard solutions only	Weighing	Weighing
Acids neutralized	All common	Most common except extremely weak	Strong and moderately strong acids only ($pK_a > 6$)

addition, the volatility of the dissolved hydrogen chloride molecules causes a slow decrease in titer, especially if stored in improperly stoppered bottles. Hydrobromic and hydriodic acids are stronger acids than hydrochloric acid. But because of their instability in oxidizing media, they have limited pharmaceutical application.

SULFURIC ACID: Sulfuric acid[6] is suitable for both analytical and synthetic neutralization procedures. However, for injectables and substances intended for oral ingestion, the accompanying sulfate ion if in high concentration may be objectionable. Phosphoric acid is sometimes used in neutralizations when the resultant solutions or salts are to be used as medicinals. But being a weaker acid, its neutralizing action is unsatisfactory for some of the weaker bases.

SODIUM DIHYDROGEN PHOSPHATE: Sodium dihydrogen phosphate is a suitable crystalline acid for neutralization of strong bases only. It is sufficiently stable to allow preparation of solutions by weight methods.

[6] Sulfuric acid prepared by the contact process will generally be more satisfactory for pharmaceutical uses. Because the impurities present in the sulfur (e.g., arsenic, selenium) will poison the catalyst and quickly reduce its efficiency, a pure grade of sulfur must be used. Consequently, contact acid should have lower levels of these trace impurities than sulfuric acid produced by the chamber process. Lead levels should also be lower for the contact acid. Almost all new installations are contact plants.

It has also the additional advantage of being a sufficiently weak acid to allow slight overages without introduction of excessively acid pH conditions.

ORGANIC ACIDS: Several organic acids (e.g., acetic, citric, lactic, and tartaric) also are of value in preparation of pharmaceuticals. Potassium hydrogen phthalate is an excellent primary acid standard, but has no direct pharmaceutical applications.

Survey of specific neutralization applications. Acids and bases are generally used in pharmaceutical practice as neutralizers. A few examples are sodium bicarbonate in Ascorbic Acid Injection; sodium hydroxide in Solution of Nitromersol; acetic acid in the preparation of Aluminum Subacetate Solution; hydrochloric acid in Lidocaine Hydrochloride Injection.

Acids and bases may be used to improve the stability of preparations by providing optimum pH conditions or by repressing reactions. Diluted ammonia water represses the hydrolysis of ammonia carbonate in Aromatic Ammonia Spirit and thereby lessens the loss of the dissolved gases. This effectively extends the useful life of the preparation. Citric acid is included in Ephedrine Sulfate Syrup to provide a stabilizing acid medium. Zinc insulin is stabilized by adding sufficient 0.1*N* hydrochloric acid to give a pH of 2.5–3.5. Pure Glycyrrhiza Extract preparation requires the addition of ammonia to the percolate before boiling to reduce the volume.

Extraction procedures may be made more efficient by including an acid in the solvents used for the recovery of drugs with acid functional groups. The *N.F. XI* recommends the inclusion of 0.05% hydrochloric acid or 0.2% tartaric acid in the water used for the extraction of crude drugs and percolates. Ergot Fluidextract uses a menstruum consisting of 2 volumes of hydrochloric acid and 98 volumes of diluted alcohol.

BUFFERS AND THEIR ROLE IN THE PRACTICE OF PHARMACY

Definitions. A buffer system may be defined on the basis of its active constituents—an aqueous solution of a Brønsted acid and its conjugate base, or on the basis of its properties—a solution which has a capacity for neutralization of added acids or bases with a minimum shift in pH. The resistance to shift of pH is a relative property, dependent on such factors as the concentration of each constituent, the ratio of constituent concentrations, the pK_a of the buffer acid, and the pK_a of an added acid or the pK_b of an added base. Each of these factors will be discussed in turn, but a qualitative description of buffer action will first be given for background.

Buffer mechanisms. A conjugate pair, acting as a buffer system, may act in neutralization reactions either as a proton donor or a proton acceptor. This may be represented symbolically:

$$B_1H \rightarrow B_1^- + H^+ \qquad (1)$$

Buffer acid as a proton donor

$$B_1^- + H^+ \rightarrow HB_1 \qquad (2)$$

Buffer base as a proton acceptor

The actual pH of a buffer solution is determined by the extent of reaction between the buffer conjugate pair and water, usually written,

$$B_1H + H_2O \rightleftharpoons H_3O^+ + B_1^- \qquad (3)$$

and

$$B_1^- + H_2O \rightleftharpoons OH^- + HB_1 \qquad (4)$$

with

$$[H_3O^+][OH^-] = 10^{-14}$$

If the allowed oxonium ion concentration is large, a buffer of low pH results; if it is small, a buffer of high pH results.

If an acid HB_2 with a K_a larger than that of the buffer acid HB_1 is added, the following net reaction must take place:

$$HB_2 + B_1^- \rightarrow HB_1 + B_2^- \qquad (5)$$

rather than

$$HB_2 + H_2O \rightarrow H_3O^+ + B_2^- \qquad (6)$$

For this reason the oxonium ion concentration, within limits, remains relatively constant; the pH of the solution is undisturbed.

If, however, an acid with a K_a smaller than that of the buffer acid is added to the buffer solution, for practical purposes no reaction can take place. Since HB_2 is a weaker acid than the buffer acid, the reaction

$$HB_2 + B_1^- \leftarrow HB_1 + B_2^- \qquad (7)$$

cannot proceed in the forward direction.

If a base B_2^- stronger than the buffer base B_1^- is added to the buffer solution, the following net reaction must take place:

$$HB_1 + B_2^- \rightarrow B_1^- + HB_2 \qquad (8)$$

rather than

$$H_3O^+ + B_2^- \rightarrow H_2O + HB_2 \qquad (9)$$

Again, the net result is an almost constant oxonium ion concentration. Within limits no significant shift in pH takes place. If a weaker base is added, it will be unable to attract protons from the protonated form of stronger buffer base.

The question may be raised as to the involvement of the oxonium ion, hydroxyl ion, or water molecules in buffer action. It is entirely possible that they are involved, however; as quickly as one is formed or consumed, the fundamental underlying equilibrium of Equation 3 will undergo a compensating shift. The net effect on addition of a foreign acid or base to the buffer must be that described by Equations 5 and 8, respectively. This may be illustrated by writing the alternate neutralization reactions:

$$HB_2 + \mathrm{H_2O} \rightarrow \mathrm{H_3O^+} + B_2^- \qquad (10)$$

$$\mathrm{H_3O^+} + B_1^- \rightarrow HB_1 + \mathrm{H_2O} \qquad (11)$$

On adding Equations 10 and 11,

$$HB_2 + \cancel{\mathrm{H_2O}} + \cancel{\mathrm{H_3O^+}} + B_1^- \rightarrow \cancel{\mathrm{H_3O^+}} + B_2^- + HB_1 + \cancel{\mathrm{H_2O}} \qquad (12)$$

On elimination of similar quantities, the net result is that expressed by Equation 5.

Effective buffer range. The selection of a buffer, effective within a certain pH range, is essentially the selection of a conjugate pair whose acid has the proper pK_a. Fortunately, the selection usually does not involve any difficulties because of the relationship which exists between the pH of the buffer system, the pK_a of the acid, and the concentrations of the constituent acid and base. This relationship may be derived by rearrangement of the expression for K_a, based on Equation 3.

$$K_a = \frac{[\mathrm{H_3O^+}][B_1^-]}{[HB_1]} \qquad (13)$$

Equation 13 may be rewritten in terms of H_3O^+:

$$[\mathrm{H_3O^+}] = \frac{K_a[HB_1]}{[B_1^-]} \qquad (14)$$

Equation 14 is now written in its log form:

$$\log [\mathrm{H_3O^+}] = \log \left\{\frac{K_a[HB_1]}{[B_1^-]}\right\} \qquad (15)$$

Separation of terms gives

$$\log [\mathrm{H_3O^+}] = \log K_a + \log \frac{[HB_1]}{[B_1^-]} \qquad (16)$$

Both sides of the equation are multiplied by -1.

$$-\log [\mathrm{H_3O^+}] = -\log K_a - \log \frac{[HB_1]}{[B_1^-]} = -\log K_a + \log \frac{[B_1^-]}{[HB_1]} \qquad (17)$$

Terms equivalent to pH and pK_a are obvious. The equation may be rewritten

$$\boxed{pH = pK_a + \log \frac{[B_1^-]}{[HB_1]}} \tag{18}$$

It has been shown that the pH of an aqueous buffer system must be equal to the sum of the pK_a of its acid and the log of the ratio of the molar concentration of the base to the molar concentration of the acid. If the concentration of the base and acid are equal, the last term becomes zero and the pH of the buffer is identical to the pK_a of the acid used.[7] It is evident that the selection of a buffer for a given pH range reduces to the selection of a conjugate pair whose pK_a is approximately the center pH of the desired range.

Effective buffer capacity. Having selected a suitable conjugate pair on the basis of the pK_a of the acid involved, the actual concentrations of acid and base to be used must be considered. Examination of Equation 18 reveals two aspects: the ratio of acid concentration to base concentration, and the actual concentration of each. The former determines the relative resistance to change in pH on addition of a small amount of foreign acid or base; the latter determines the total useful buffering capacity of the system. Each of the effects may be illustrated with a common buffer system.

The acid, $H_2PO_4^-$, of the common phosphate buffer has a pK_a of 7.21. The pH of the system is expressed as

$$pH = 7.21 + \log \frac{[HPO_4^{2-}]}{[H_2PO_4^-]}$$

If a range of $[HPO_4^{2-}]/[H_2PO_4^-]$ ratios is selected, the pH of the respective buffer solutions may be calculated. Knowing the pH of the resulting buffer solution, the effect of adding a specified quantity of foreign acid may be calculated since the net effect of such an addition is to change the [base]/[acid] ratio—here the $[HPO_4^{2-}]/[H_2PO_4^-]$ ratio. Using HB_2 to represent the foreign acid, the reaction may be represented as

$$HPO_4^{2-} + HB_2 = H_2PO_4^- + B_2^-$$

Similarly, the addition of a foreign base B_2^- will act to change the $[HPO_4^{2-}]/[H_2PO_4^-]$ ratio:

$$H_2PO_4^- + B_2^- = HPO_4^{2-} + HB_2$$

---

[7] In quantitative analysis this relationship gives the pH at half neutralization in acid-base titrations.

Thus, the pH of a buffer will change upon addition of a foreign acid or base because of a change in the ratio of buffer acid to buffer base.

The change in pH produced by the addition of one milliequivalent (me.) of a foreign acid or base has been calculated for various $[HPO_4^{2-}]/[H_2PO_4^-]$ ratios. The results are summarized in Table 7.5. The indicated values of [base] and [acid] are considered to represent the milliequivalents of base and acid present. Since the respective calculated pH and ΔpH are intended for comparison purposes only, it has been necessary to carry the calculation to the thousandths place. In actual practice differences in the third place, and to a certain extent in the second place, are meaningless (page 99).

Examination of the data shown in Table 7.5 reveals that the addition of one milliequivalent of a foreign acid or base has the least effect on the pH of the buffer system when the [base]/[acid] ratio is unity, ΔpH = 0.009. As the [base]/[acid] ratio is increased or decreased, the addition of one milliequivalent of foreign acid or base produces an ever-increasing change

TABLE 7.5 [BASE]/[ACID] RATIOS AND CALCULATED BUFFERING EFFICIENCY

	+ 1 me. Foreign Acid	Initial Conditions	+ 1 me. Foreign Base
[Base]/[Acid]	9/191	10/190	11/189
pH	5.881	5.929	5.973
ΔpH	0.048	—	0.044
[Base]/[Acid]	49/151	50/150	51/149
pH	6.719	6.731	6.742
ΔpH	0.012	—	0.012
[Base]/[Acid]	99/101	100/100	101/99
pH	7.199	7.208	7.217
ΔpH	0.009	—	0.009
[Base]/[Acid]	149/51	150/50	151/49
pH	7.674	7.685	7.697
ΔpH	0.012	—	0.012
[Base]/[Acid]	189/11	190/10	191/9
pH	8.443	8.487	8.535
ΔpH	0.044	—	0.048
[Base]/[Acid]	999/1001	1000/1000	1001/999
pH	7.207	7.208	7.209
ΔpH	0.0008	—	0.0008

in the pH of the system. For the 50/150 and 150/50 ratios, ΔpH is 0.012; for the extreme ratios 10/190 and 190/10 a change of about 0.05 pH units is produced. It is evident that as the [base]/[acid] ratio departs more and more from unity, the ability of the buffer system to resist a change in pH decreases. It follows that the poorest buffering is exhibited by a buffer solution having a significant predominance of either its acid or its conjugate base. These facts are graphically presented in Fig. 7.2.

It is also necessary to consider the actual concentration of the buffer components. This effect can be shown by again considering the phosphate buffer system and comparing the effect of addition of foreign acid or base to a buffer having 100 me. of each component as opposed to one having 1000 me. of each component. Referring to Table 7.5, it is seen that for the 100/100 ratio the addition of 1 me. of either acid or base results in a pH shift of 0.009 units, whereas for the 1000/1000 ratio buffer system the pH shift is only 0.0008 pH units per equivalent of added foreign acid or base. Obviously the higher the individual concentrations, the greater the buffering capacity of the system.

Selection of a suitable buffer. CHEMICAL ASPECTS: From the chemist's point of view the selection of a suitable buffer system for a given situation requires consideration of four factors.

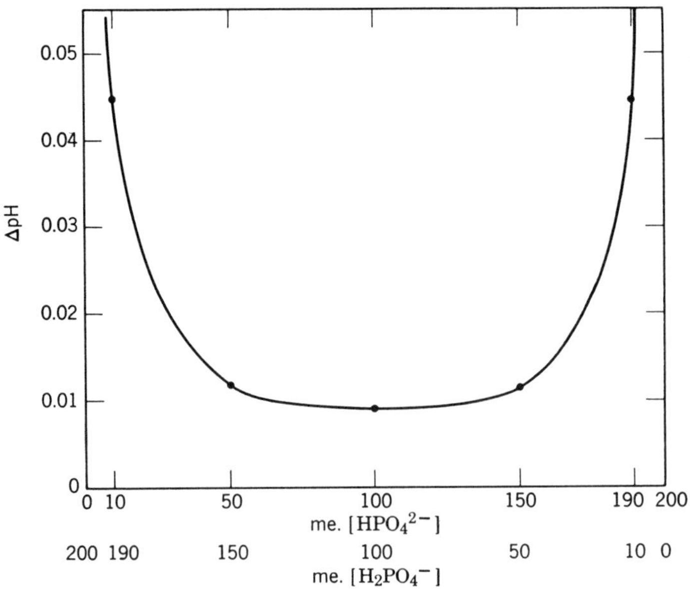

FIGURE 7.2 Effect of [base]/[acid] ratio on buffering effectiveness. Calculated pH shift produced by addition of 1 me. of a strong acid or a strong base to various $[HPO_4^{2-}]/[H_2PO_4^-]$ ratios.

1. Selection of a conjugate pair to give the proper pH range. This is done on the basis of the pK_a of the acid; it should approximate the numerical value of the desired center pH of the buffer.

2. Selection of the proper [base]/[acid] ratio for the buffer solution. This ratio will be unity, or as close to unity as practical, since it has been shown that the best resistance to change in pH exists at a 1:1 ratio.

3. Selection of the actual concentration of each buffer compound (on the basis of the ratio selected). The higher the actual concentration selected, the better the buffering capacity of the system for a given [base]/[acid] ratio. Practical considerations (e.g., solubility) frequently limit the usable range of concentrations.

4. Chemical stability of the buffer system must also be considered since it frequently tempers the usefulness of a given conjugate pair as a buffer and the concentrations of base and acid allowed. In general only solubility, complexation, and extraneous acid-base reactions need be considered; most common buffers are stable to the redox conditions encountered in the practice of pharmacy.

Alkaline buffers must be protected from the air since dissolved carbon dioxide is sufficiently strong as an acid to cause a shift to more acid pH values. Strongly alkaline buffers cannot be stored for any length of time in glass containers; high concentrations of hydroxyl ion react with the silicate tetrahedra of the glass structure. If a volatile acid or base (e.g., HCl, NH_3, CO_2) is involved, tight closures must be used. Otherwise pH shifts will take place due to loss of the volatile component of the buffer conjugate pair.

The possibility of the formation of insoluble compounds must be kept in mind. This is especially important to the pharmaceutical chemist since germicidal and antiseptic solutions of heavy metal ions are frequently buffered. Many of these (e.g., zinc) form insoluble salts with common buffer ions, particularly under alkaline conditions. Similarly, many buffer substances (e.g., citrate) may reduce effective metal ion concentrations by complexation or chelation.

The four chemical considerations in the selection of a suitable buffer pair may be demonstrated by a discussion of some common buffers. The identity of the conjugate pair involved is obvious in most buffers, e.g., the acetic acid and acetate ion of the acetate buffer. The hydrated sodium ion present in the acetate buffer is too weak an acid to affect the buffer system. The conjugate pairs of the citric acid-dihydrogen citrate system, the hydrogen chloride-chloride ion system, and the phosphate system previously cited are equally obvious.

The acid-base conjugate pair of other published buffer systems is not always as obvious. The 6.0–8.0 buffer system of the Clark and Lubs buffer mixtures,[8] illustrates one of these cases; it also illustrates a common preparative technique. Although the finished buffer solution is a dihydrogen-monohydrogen phosphate buffer, only dihydrogen phosphate is used in its preparation. The required monohydrogen phosphate is produced *in situ* by the reaction.

$$OH^- + H_2PO_4^- = HPO_4^{2-} + H_2O$$

This is accomplished by mixing predetermined volumes of standard sodium hydroxide and sodium dihydrogen phosphate. The proportions used give the [base]/[acid] ratios required for the desired pH. In a similar manner phthalic acid is produced *in situ* by the addition of the proper number of equivalents of hydrochloric acid to the solution of potassium hydrogen phthalate to give buffers in the pH range 2.2–3.2.

The conjugate pairs of some buffers are even less obvious. McIlvaine's buffer system,[9] a so-called wide-range buffer, uses varying aliquots of 0.1*M* citric acid and 0.2*M* sodium monohydrogen phosphate solutions to cover the pH range 2.2–8.0. Depending upon the relative volumes of the two solutions mixed, five different conjugate pairs are possible within its buffer range. The acids of these pairs and their pK_a values are citric acid, 3.1; dihydrogen citrate ion, 4.7; monohydrogen citrate ion, 5.4; phosphoric acid, 2.1; dihydrogen phosphate ion, 7.2. At the lowest pH range citric acid and its dihydrogen anion constitute the principal conjugate pair. Being the strongest base of the system, and because of favorable concentration gradients, the added monohydrogen phosphate reacts with a portion of the citric acid to produce the required base ion:

$$HPO_4^{2-} + H_3Ci = H_2PO_4^- + H_2Ci^-$$

As larger proportions of the monohydrogen phosphate solution are used to prepare the buffers in more alkaline pH ranges, conjugate pairs based on acids with successively larger pK_a values become the effective buffers. At the pH of 8.0 the buffer pair would be essentially the dihydrogen-monohydrogen phosphate conjugate pair. Concentration gradients are such that the small amount of citric acid added for practical purposes contributes its entire supply of protons to the monohydrogen phosphate ion to give the necessary dihydrogen phosphate.

PHYSIOLOGICAL ASPECTS: The pharmaceutical chemist must consider additional problems generally of no concern to the chemist *per se*.

[8] R.G. Bates, *Determination of pH*, 2nd Ed., Wiley, New York, 1973, p. 460.

[9] *Ibid.*, p. 132.

Some of the buffers used in chemistry have limited use in pharmacy because of their toxicity. Borate buffers belong to this category. Their toxicity precludes employment in any preparation intended for internal use. They do find limited use in ophthalmic preparations, gargles, etc. The phthalate buffer presumably has toxic effects rendering it unsuitable for use in pharmaceuticals.

In addition, possible secondary effects of the buffer compounds must be considered, especially in injectables. Substances readily metabolized via normal pathways (e.g., citrate) or normally present in the body fluids (e.g., phosphate) are the basis of the popular buffers because they introduce no complications unless used in large excess. In other words, the possible poisonous nature of the buffer, its side effects, and ultimate disposition if ingested or injected must be considered in selecting a buffer for a pharmaceutical product.

Many buffer systems are good growth media for molds and microorganisms; most systems using organic acids are in this category. Some inorganic buffers e.g., phosphate are good mineral media. Since acidic metabolic products are produced by the organisms, the pH of a contaminated buffer slowly becomes more acid. Contamination by microorganisms is evidenced by clouding of a clear solution. Microorganism contamination is no problem in acid buffers below a pH of about 3 since strong acid solutions are germicidal. Similarly, an alkaline pH greater than 10 presents no problems with microorganisms because of the dissolving action of strong base on protein.

Roles of buffers in pharmacy. One of the most important uses of buffers in pharmaceutical preparations is to assure stable pH conditions for the medicinally active compounds present in a medicinal. There are many aspects of this role. Solubility of compounds is frequently controlled by providing a medium of suitable pH. Many inorganic salts (e.g., iron(III) cation, phosphates, borates) are soluble in acid media, but precipitate in alkaline media. Similarly, many insoluble organic substances with acidic functional groups are soluble in alkaline media. Amines are soluble in acidic media, but almost all are insoluble in alkaline media. (The role of pH in the solubility behavior of compounds will be discussed in Chapter 8.)

The color of many natural dyes, present in fluid extracts, etc., are pH dependent. The familiar red color of cherry and raspberry syrups depends upon an acid pH. Both become pale yellow to nearly colorless in alkaline media. The colors of many synthetic medicinals are similarly pH dependent, e.g., phenolsulfonphthalein, phenolphthalein.

The stability of certain compounds to redox conditions will be a function of pH if water or one of its ions is a reactant or product. Epinephrine is a classic example. It is rapidly oxidized (by dissolved oxygen) to adrenochrome and further oxidation products in alkaline media. For this reason its solutions are buffered as near as possible to their most stable pH range—2.5–3.0. (Epinephrine Injection must have a pH within the range 2.5-5.0 to meet *U.S.P. XXI* standards). Similarly, ascorbic acid and penicillin are unstable in an alkaline pH.

Another series of compounds is structurally unstable within certain pH ranges, usually due to autooxidations or disproportionations giving rise to insoluble solids or gases. Sodium thiosulfate and sodium polysulfide preparations must be maintained under alkaline conditions to prevent separation of sulfur. Nitrites become brown in acid media owing to formation of colored nitrogen oxides. The acid and normal salts of carbonates and sulfites cannot be maintained at acid pH conditions owing to separation of gaseous carbon dioxide and sulfur dioxide, respectively.

It is impossible to go beyond this general exposition of those compounds dependent on maintenance of proper pH. Inorganic substances requiring buffering or other pH control will be covered fully in more appropriate sections. A survey of organic substances showing pH incompatabilities is outside the scope of this book.

A consideration of patient comfort is a second reason for use of buffers in pharmaceutical preparations. Injectables and preparations for oral or external use are irritating if their pH differs greatly from that normal for the particular tissues involved. The more extreme the departure from normal pH, the greater the discomfort to the patient. An extremely acid or alkaline pH must be avoided because of tissue damage.

Ideally the pH of a preparation will be as near as possible to that of the body fluids at the site of administration. However, many drugs are unstable at physiological pH, and compromises may be necessary. Here solutions are buffered at a pH as close as possible to physiological, but at which the drug is still reasonably stable. The blood has a good buffering capacity, and body secretions such as the lachrymal fluid have a limited buffer capacity; this tends to overcome at least partially irritation from administration of drugs buffered at nonphysiological pH for stability reasons.

Buffers may also be necessary, to provide optimum pH conditions for the activity of a medicinal. The inclusion of sodium dihydrogen phosphate with methenamine is the classical example (Methenamine and Monobasic Sodium Phosphate Tablets, *U.S.P. XXI*). Although usually not considered from this point of view, the adjustment of the pH of sodium hypochlorite to lower values increases the germicidal effectiveness of the preparation (page 450).

In control work it is frequently necessary to prepare buffers of known pH for use as standards. Calibration of pH meters and preparation of indicator color standards require accurately known buffer solutions. If the solutions for preparation of the various analytical buffers are not available, standard solutions of acids and bases may be substituted when extreme accuracy is not required. The pH of some standard solutions of acids and bases is given in Table 7.6. For the mildly acid regions either a

TABLE 7.6 pH OF STANDARD ACIDS AND BASES

Molarity	HOAc	HCl	NH_3	NaOH
1.0	2.4	0.1	11.8	14.1
0.1	2.9	1.1	11.3	13.1
0.01	3.4	2.0	10.8	12.1
0.001	3.9	3.0	10.3	11.1
0.0001	—	4.0	—	—

saturated solution of potassium hydrogen tartrate (25°C.) or a 0.05*M* solution of potassium hydrogen phthalate are useful. The pH of the solutions are 3.57 and 4.01, respectively, at 25°C. The latter solutions are especially useful since their temperature coefficients of hydrogen ion activity are unusually small.

Common buffers of pharmacy. Except for the borate buffers, the buffers of interest in pharmaceutical chemistry are those based on physiological acids and bases. The phosphate buffer is especially valuable. It is compatible physiologically, since except for the carbonic acid-bicarbonate system it is the most important buffer system of the blood and body fluids. Of equal importance, its best buffering action is within the physiological range—the pK_a of the dihydrogen phosphate ion is 7.2. The sodium (or, to a lesser extent, the potassium) salts usually present no solubility problems in preparation of the buffer.

The phosphate buffer system has two primary disadvantages. The first centers about the insolubility of many cations at neutral and alkaline pH. This limits its use in collyria or gargles which employ zinc salts, if alkaline conditions are required. If ingested regularly via oral preparations, problems can arise due to the precipitating action of phosphate on calcium, iron, and magnesium in the intestinal tract. The phosphate buffer readily supports mold growth at room temperatures. Storage in a refrigerator will control mold growth for reasonable periods. Growths can also be controlled chemically, depending upon ultimate use, by inclusion of Zephiran Chloride ®, 0.002%, or phenylmercuric nitrate, 0.05%.

Phosphate buffer solutions within the pH range 5.9—8.1 may be prepared with varying proportions of the dihydrogen and monohydrogen salts. Preparative information is given in Table 7.7. If more acid buffer ranges are required, phosphoric acid may be used with dihydrogen phosphate.

The borate buffer systems find wide use despite several inherent disadvantages. First, the poisonous nature of boric acid and the borates precludes their use in injectables and should do so for other applications. Second, the pK_a of boric acid (9.24) is such that its best buffering capacity is in the mid-alkaline pH ranges, rather than at physiological ranges. However, since it is used so frequently in solutions containing heavy metal ions (e.g., silver, zinc), these alkaline pH ranges cannot be used. In acid pH ranges the buffering capacity of the system is poor, especially

TABLE 7.7 Modified Phosphate Buffer System (Sorenson)

pH Desired	Acid Solution,[a] ml.	Base Solution,[b] ml.	Gm. NaCl for Isotonicity
5.9	90	10	0.52
6.2	80	20	0.51
6.5	70	30	0.50
6.6	60	40	0.49
6.8	50	50	0.48
7.0	40	60	0.46
7.2	30	70	0.45
7.4	20	80	0.44
7.7	10	90	0.43
8.0	5	95	0.42

[a] Acid solution: sodium dihydrogen phosphate (monohydrate) 9.20 Gm./L. purified water; (*M*/15).
[b] Base solution: sodium monohydrogen phosphate (heptahydrate 17.86 Gm./L. purified water; (*M*/15).

for acid. The buffer is sensitive to carbon dioxide. Since borates are bacteriostats, the usual problem of organism growth is minimized; but mold growth can take place when the solutions are kept for long periods at room temperature.

The buffer systems based on borate are those of Feldman, pH 7–8.2; Gifford, pH 6–7.8; and Atkins and Pantin, 7.6–11.0.[10] However, the adjustment for isotonicity is poor in these solutions. A modification has been prepared which is slightly hypotonic for all combinations.[11] Table

[10]Martin and Cook, *Remington's Practice of Pharmacy*, 12th ed., Mack Publishing Co., Easton, Pa., 1961, pp. 359–362.
[11]Arthur Puccini, M.S. thesis, submitted to the Graduate School, Rutgers—The State University, New Brunswick, New Jersey.

7.8 gives the pertinent information for its preparation. Buffers prepared using these solutions are exceptionally stable, showing only a slight alkaline shift owing to carbon dioxide loss. Some growth of microorganisms appears, with only a very slight pH shift, after one month of storage at room temperatures. No mold growth appeared when stored

TABLE 7.8 REFORMULATED ISOTONIC BORATE BUFFER SOLUTIONS

pH	Acid Solution,[a] ml.	Base Solution,[b] ml.
6.0	200	0.6
6.2	200	1.0
6.4	200	1.6
6.6	200	2.5
6.7	200	3.3
6.8	200	4.1
6.9	200	5.2
7.0	200	6.4
7.1	200	7.8
7.2	200	10.0
7.3	200	12.2
7.4	200	14.6
7.5	200	17.9
7.6	200	21.0
7.7	200	25.3
7.8	200	30.6
7.9	200	35.4
8.0	200	40.8

[a] Acid solution: boric acid 12.00 Gm. and potassium chloride 2.857 Gm./1000 ml. purified water.

[b] Base solution: sodium carbonate (anhydrous) 13.24 Gm./1000 ml. purified water.

for three months under refrigeration. The freezing-point depression for all solutions falls within the experimental range of 0.51–0.53°C. These solutions resist pH change exceptionally well on dilution.

The chemistry of the usual borate buffer preparations may be confusing since only boric acid is generally present in the preparative solutions. However, the required borate ion is produced on mixing, since the carbonate ion is a somewhat stronger base than the borate ion. (A small amount of buffering capacity results from the bicarbonate-carbonate buffer pair produced on reaction.) The equations involved may be written

$$H_3BO_3 + 2H_2O = H_3O^+ + [B(OH)_4]^-$$
$$H_3O^+ + CO_3^{2-} = HCO_3^- + H_2O$$

The tartrate, citrate, and acetate buffers are three other commonly used buffer systems. They have the advantage of being physiologically compatible, since they are normal products of carbohydrate metabolism. The pK_a of each of the acids is such that the best buffer action is in the weak acidic ranges. However, they are prone to mold growth. Preparation of acetate buffers has an additional problem in the variability of the pH of the sodium acetate solutions used in its preparation. Freshly prepared solutions of this salt have been found to have pH values ranging from 5.0 to 9.0, depending on the age and condition of the sodium acetate.

Acetate is a good buffer for silver-containing ophthalmic preparations since silver acetate is soluble. Equal volumes of 0.06*M* sodium acetate (fresh salt) and 0.0166*M* acetic acid give a pH of 6.1.

Various buffer solutions are available to serve as pH references in control work.[12]

Adjustment of pH. Before leaving the subject of buffers, it should be pointed out that certain operations termed buffering in pharmaceutical practice may not be strictly interpreted as buffering. The operations referred to would be better described as "adjusting pH." The operation consists of the addition of an acid or base in sufficient quantities to overcome a particular acid-base problem by shifting the pH of the solution to a more satisfactory range.

Such buffering operations usually can be recognized by the difficulty in identification of the buffer conjugate pair. In many of these solutions, a potential buffering capacity is created for reasons similar to those developed during the discussion of the Clark and Lubs buffers; i.e., as a result of partial neutralization an acidic medicinal and its base (or vice versa) may constitute a buffer pair.

A common example of this usage is the addition of sodium bicarbonate as a "buffer" to sodium hypochlorite solution in the preparation of Diluted Sodium Hypochlorite Solution (*N.F.XIV*). Sodium hypochlorite solution itself is strongly alkaline owing to hydrolysis of the hypochlorite ion:

$$ClO^- + H_2O = HClO + OH^-$$

In addition, manufacturing procedures allow a small excess of sodium

[12] *U.S.P. XXI*, p. 1258, 1420.

hydroxide or sodium carbonate to remain unreacted in order to improve the stability of the finished preparation.

If it is assumed that the pH of the solution is controlled by the hypochlorous acid-hypochlorite ion conjugate pair, and that the pH of the solution is approximately 11, the ratio of base to acid can be calculated in the usual way:

$$\mathrm{pH} = \mathrm{p}K_a + \log \frac{[\text{Base}]}{[\text{Acid}]}$$

$$11 = 7.46 + \log \frac{[\text{ClO}^-]}{[\text{HClO}]}$$

$$\log \frac{[\text{ClO}^-]}{[\text{HClO}]} = 3.54$$

$$\frac{[\text{ClO}^-]}{[\text{HClO}]} = 3.6 \times 10^3$$

Thus in the raw sodium hypochlorite solution 1 acid molecule is present for every 3600 hypochlorite ions.

On "buffering" with sodium bicarbonate, the pH of the diluted hypochlorite solution is reduced to approximately 8. This is due to the fact that protons are available from bicarbonate ion, first for the reaction

$$HCO_3^- + OH^- = H_2O + CO_3^{2-}$$

and also to some extent for the reaction

$$HCO_3^- + ClO^- \rightleftharpoons HClO + CO_3^{2-}$$

At first glance the second reaction appears impossible, since bicarbonate ion is a weaker acid ($\mathrm{p}K_a = 10.25$) than hypochlorous acid ($\mathrm{p}K_a = 7.46$). However, this reaction is aided by a mass effect; the concentration of hypochlorite ion is in very great excess. The new ratio of the hypochlorite conjugate pair in the finished Dakin's solution may be calculated:

$$8 = 7.46 + \log \frac{[\text{ClO}^-]}{[\text{HClO}]}$$

$$\log \frac{[\text{ClO}^-]}{[\text{HClO}]} = 0.54$$

$$\frac{[\text{ClO}^-]}{[\text{HClO}]} = 3.5$$

Although the hypochlorite ion still predominates, the hypochlorous acid present has been greatly increased by treatment with bicarbonate. By virtue of its much higher concentration the hyperchlorous acid-hypochlorite system is the dominant buffer. However, the bicarbonate-carbonate system is also present, albeit in a subsidiary role. Since the solution is at pH 8, it must adjust its [base]/[acid] ratio to the condition of the predominant hypochlorite-hypochlorous acid system.

$$8 = 10.25 + \log \frac{[CO_3^{2-}]}{[HCO_3^-]}$$

$$\log \frac{[CO_3^{2-}]}{[HCO_3^-]} = -2.25$$

$$\frac{[CO_3^{2-}]}{[HCO_3^-]} = 5.6 \times 10^{-3}$$

It is obvious that the added bicarbonate still exists predominantly as such, as expected from its pK_a. However, it has contributed sufficient protons through neutralization reactions to reduce the pH to satisfactory levels. The solution is no longer necrotic to tissues.

EFFERVESCENT MIXTURES

A special application of acid-base technology is found in effervesent mixtures. These preparations serve as media for the administration of a drug. On solution in water the resultant reaction mixture is essentially a saturated solution of carbon dioxide. Effervescent mixtures have several advantages. The evolution of the carbon dioxide provides an efficient stirring action, insuring thorough solution of the administered drug. Absorption in the stomach may be hastened, and the carminative action of the carbon dioxide reduces flatulence. Palatability is improved because of the taste-masking properties of the solution: the pleasant, slightly acid taste and prickly sensation mask the unpleasant taste of salty drugs.

Two different types of compounds are the essential components of an effervescent mixture: a base as the source of carbon dioxide and an acid as the source of protons. Sodium bicarbonate is the compound of choice as the carbon dioxide source, although potassium bicarbonate may be used. Sodium bicarbonate releases a greater volume of carbon dioxide per gram than any other physiologically compatible soluble carbonate. It is also reasonably soluble, giving a clear solution even when incompletely reacted. The particle size of the bicarbonate crystals is important in controlling reaction rate: if too coarse, the evolution of carbon dioxide is too slow; if too fine, the carbon dioxide evolution will be too violent. A sodium bicarbonate suitable for these preparations has particle size distributions such that 73% passes a 65-mesh screen, 33.5% passes through an 80-mesh screen, and 7.5% passes a 100-mesh screen.

The proton source must be an acid of sufficient strength to release its protons to the bicarbonate ion for formation of carbonic acid (pK_a, 6.4). At the same time it must not be so strong that the slight excess normally used imparts an excessively acidic taste to the preparation. For ease of handling, storage, etc., the acid must be a solid. Usually an organic acid—tartaric, citric, or sodium dihydrogen phosphate—is used.[14] Irrespective of the proton source the essential reactions of all effervescent mixture formulations are

$$H_3O^+ + HCO_3^- = H_2CO_3$$
$$H_2CO_3 = H_2O + CO_2\uparrow$$

Several variables must be controlled in these formulations. The rate of reaction is important. Reaction must take place in a reasonably short time, but must be slow enough to prevent the reaction mixture from going "wild" on the addition of water. This may be accomplished by control of particle size and by the inclusion of a normal salt of the acid used as a diluent. However, the diluent may be the normal salt of other acids. These diluents must have suitable taste qualities. A buffering action is also present. Once reaction is initiated, this buffer action controls the number of hydronium ions available for reaction with the bicarbonate ion at any given instant.

Although it is possible to dispense effervescent mixtures as two separate powders, to be mixed at the time of use, e.g., Compound Effervescent Powders (Seidlitz Powders), *N.F. XII,* it is much more convenient to have all components combined in a single powder or tablet. This requires control of moisture content at levels below those which will initiate the chemical reaction while in storage. In this form the preparation must also be protected from the moisture in the air. This can be accomplished by sealing a single dose in a moisture-proof package.

At present effervescent mixtures find widest application in various proprietary formulations. Most commonly they are used as vehicles for laxatives, antacids, and pain relievers. Powdered "soft drinks" are effervescent mixtures with suitable flavoring and sweetening agents.

STUDY AND SELF-APPRAISAL AIDS

IMPORTANT TERMS. Define or identify in a single concise sentence if possible.

---

[14] Baking powders, which are chemically analogous to effervescent mixtures, also make use of aluminum sulfate as the acid. This compound is unsatisfactory for effervescent mixtures because of the precipitation of insoluble aluminum hydroxide on reaction.

Buffer, buffer capacity, conjugate pair, constant boiling hydrochloric acid, deliquescent, effervescent, efflorescent, hygroscopic, ionization constant, neutralization, pH, pK_a, pK_b, wide-range buffer.

NOMENCLATURE OF INTEREST. Identify each of the following.

Aqua fortis, baking soda, boracic acid, borax, calcium hydrate, calx, caustic potash, caustic soda, cream of tartar, fuming sulfuric acid, glacial phosphoric acid, muriatic acid, oil of vitriol, pearl ash, potash, Rochelle Salt, Seidlitz Powders, sal soda, soda ash, monobasic sodium phosphate (*USP XXI),* sodium hydrate, dibasic sodium phosphate (*USP XXI*), soluble tartar, Spirit of Hartshorn.

SELF-APPRAISAL QUESTIONS

1. Give the electron structures of sulfuric, nitric, hydrochloric, acetic, and boric acids. Arrange these acids in order of acid strength. Use them to illustrate the various structural factors such as central atom charge which are determinants in acid strength. Does the relationship between the number of unprotonated oxygens and acid strength hold in each case?

2. Distinguish between a strong base and a strong acid in terms of Brønsted definitions.

3. Distinguish between ammonia and ammonium ion. Explain on structural and theoretical bases why the former is a very weak acid but the latter is only a moderately weak acid.

4. Why is ammonia solution unsuitable as a base for most neutralization procedures?

5. Why is it necessary *always* to add concentrated sulfuric acid to the diluent water? Explain your answer on a strictly theoretical basis.

6. Hydriodic acid is a very strong acid but is not used for neutralization procedures. Why?

7. Both sodium hydroxide and concentrated sulfuric acid absorb moisture strongly from the atmosphere if improperly stoppered. The former is said to be deliquescent, whereas the latter is merely described as a dehydrating agent. Is there any reason for this distinction?

8. Compare the heat liability of sodium hydroxide, sodium bicarbonate, and sodium carbonate.

9. Sodium hydroxide, sodium carbonate, and sodium bicarbonate are used for their ability to act as bases. Explain on a theoretical or structural basis why their K_b values are quite different. Why is the sodium ion disregarded in consideration of base strength? How would the corresponding potassium salts compare as to base strength?

10. Compare the pros and cons for the use of sodium hydroxide, sodium carbonate, and sodium bicarbonate as neutralizing agents in pharmaceutical preparations.

11. Distinguish between lime, quicklime, slaked lime, milk of lime, and lime water.

12. Calcium hydroxide is often erroneously classified as a weak base. Explain.

13. For a two-year period a pharmacist maintained his lime water supply by adding purified water to his stock bottle whenever its level had dropped about one half. Since there were plenty of solids remaining in the bottle, he did not add calcium hydroxide at any time. Criticize his procedure.

14. List those acids and bases which are official in the current compendia. Identify those which have cautions.

15. What physiological factors must be considered in the selection of an acid or base, or a buffer, intended to be present in a finished pharmaceutical product?

16. Calculate the pH of 0.1*M* and 0.001*M* solutions of acetic acid and hydrochloric acid. How do these values compare with those given in Table 7.6? Explain.

17. You wish to neutralize completely an acid which has the structure RCHClCOOH (R is an alkyl group of 1–4 carbons). Which of the following compounds would be suitable: sodium monohydrogen citrate, sodium dihydrogen phosphate, sodium carbonate, sodium bicarbonate, potassium hydroxide, tertiary sodium phosphate, tetramethyl ammonium hydroxide?

18. Suppose the previous question were reworded to read "neutralize partially" instead of "neutralize completely." How would the list of suitable bases change?

19. You wish to neutralize completely an ammonia solution. Which of the following should be suitable as acids for this purpose: hydrochloric acid, potassium bicarbonate, citric acid, lactic acid, sodium dihydrogen phosphate, sodium bisulfate, boric acid, ammonium chloride?

20. How many ml. of 0.2000*N* sodium hydroxide would be required to neutralize exactly 1.0 Gm. nicotinic acid during preparation of 100 ml. of Nicotinic Acid Injection? (40.65 ml)

21. The pH of Niacin Injection is given as 4-6 in *U.S.P. XXI.*[15] What per cent of the original niacin is still in the acid form after neutralization

[15]*U.S.P. XXI,* p. 731.

if the pH of the finished solution is 5.17? The pK_a of niacin is 4.87. (33.4%)

22. If the pharmacist used 41.0 ml. of 0.2000M NaOH solution in the neutralization of problem 20, what is the calculated pH of the resulting solution? Is this solution safe as an injection? Assume an initial 1 g. niacin and a final volume of 100ml. (10.8)

23. Using the information given in problems 20–22, calculate the number of ml. of the 0.2000N sodium hydroxide required to achieve a pH of 5.17.

24. How many grams of $Na_2CO_3 \cdot H_2O$ are needed to dissolve 1.00 Gm. nicotinic acid to give a final volume of 100.0 ml. of solution having a pH of 5.17? (Assume that the carbonate ion is a sufficiently strong base to cause 100% neutralization of nicotinic acid.)

25. On a theoretical basis, would sodium bicarbonate be a suitable base for dissolving nicotinic acid in preparation of Nicotinic Acid Injection?

26. A buffer is to be effective in the pH range of 4.0–7.0. Which of the following conjugate pairs would be suitable: lactic acid and lactate ion; monohydrogen oxalate and oxalate ions; phosphoric acid and dihydrogen phosphate ion; monohydrogen sulfate and sulfate ions; monohydrogen tartrate and tartrate ions; benzoic acid and benzoate ion; carbonic acid and bicarbonate ion; monohydrogen citrate and citrate ions?

27. Referring to your answer to problem 26, which of these pairs would be generally unsatisfactory for buffering medicinal preparations, particularly injections? Why?

28. Show by suitable equations how the acetate buffer (equimolar in both constituents) will respond to the addition of one drop, respectively, of a 0.1M solution of the following: phosphoric acid, tetraethylammonium hydroxide, ammonia water, boric acid. Estimate which substance should theoretically cause the greatest shift in pH. Why?

29. Why was the condition "addition of one drop" imposed in problem 28? What advantage is gained by use of equimolar concentrations of the buffer pair?

30. Theoretically the pH of a buffer having equal equivalents of its acid and base should be identical to the pK_a of the acid. Referring to Table 7.7, it is seen that the pH of the phosphate buffer with equal equivalents of the acid and base is actually given as 6.8. Explain.

31. All alkaline buffers, especially those moderately and strongly alkaline, have two storage problems. Identify these problems. Explain.

32. Identify three buffers currently in use in pharmaceutical preparations. Identify the conjugate pair serving as the buffer pair. What is the effective buffer range of each? What are the advantages and disadvantages of each?

33. Compound Effervescent Powders are described in *N.F. XII*[15]. Give the reason(s) for each of the following: (*a*) Why are standard sieve sizes specified for each of the constituents? (*b*) What is the role of each constituent? (*c*) Why the blue papers and the white papers? (*d*) Why is the potassium sodium tartrate mixed with the sodium bicarbonate rather than the tartaric acid? (*e*) Why store in well-closed containers, in a dry place? (*f*) Why are the contents of a blue paper and those of a white paper dissolved in separate 60-ml. portions and the solutions mixed rather than adding to powders the a single 120-ml. portion?

34. Why are effervescent mixtures of value in pharmaceutical practice?

35. Select a proprietary preparation which is based on an effervescent mixture as administration mode. Identify: carbon dioxide source, acid, diluent, medicinal agent.

PRACTICE IN EQUATION WRITING

36. Write equations to show each of the following.

(*a*) Why solutions of ammonium salts of weak base anions have an acidic reaction.

(*b*) Why solutions of sodium tetraborate have an alkaline reaction.

(*c*) Why an aqueous suspension of aluminum hydroxide is unstable in a medium whose pH is 4.

(*d*) Show the reactions taking place if a bottle of sodium hydroxide is left standing unstoppered.

(*e*) Show reactions taking place when ammonia solution is used as a base.

(*f*) Show the neutralization reaction of ascorbic acid with sodium bicarbonate.

(*g*) Using RNH_2 as an amine present in a crude drug, show why addition of hydrochloric acid to the extracting menstruum improves the efficiency of the extraction.

37. Write equations for the reactions suitable for identification of the following: an ammonium salt, hydrochloric acid, phosphate buffer, carbonate.

---

[15] *National Formulary (N.F.) XII,* American Pharmaceutical Assn., Washington, D.C., 1965, pp. 352-3.

EIGHT · THE SOLUBILITY OF PHARMACEUTICALS AND ITS CONTROL

Solubility and the solubility behavior of compounds is a matter of continuing interest to the pharmaceutical chemist and pharmacist. In addition to its strictly chemical involvement in analytical and synthetic procedures, solubility behavior is extremely important in the preparative aspects of pharmacy. An adequate knowledge of solubility and solubility behavior is essential in order to control the solubility of drugs to achieve suitable doses and to control their assimilation. If possible, insoluble or slightly soluble medicinals must be rendered more soluble when solutions are the desired administration form or if assimilation is poor. On the other hand, if a suspension or gel is to be dispensed, conditions must be selected such that solubilization is avoided. Since so many medicinals are combinations of various products, the possibility of precipitation incompatibilities is always present and must be recognized.

As an introduction to this chapter the fundamental factors affecting solubility and rate of solution will be considered, including certain general properties of solutions important in the control of tonicity. Since soluble substances of pharmacy and medicine so frequently involve alkali metal cations and the closely related ammonium radical and its derivatives, the chemistry of this family of elements will be surveyed. "Rules of thumb," useful in predicting the solubility behavior of inorganics, will follow. The chapter will close with a discussion of the chemical methods available for improvement of the solubility of chemical moieties.

THE PROCESS OF SOLUTION

One of the oldest solubility rules is "like dissolves like." This observation grew from the gradual realization that a polar solvent, such as water will ordinarily dissolve a polar solute such as sodium chloride or ethanol, but not a nonpolar substance such as hexane. This rule provides reasonably useful solubility information, as will be evident from a descriptive consideration of the process of solution. This discussion will be generally biased toward aqueous ionic solutions.

Solution of a solute in a solvent consists of a mutual separation of particles from their initial environments. In order for a solute to dissolve in a solvent it must be able to separate the neighboring molecules of the solvent sufficiently to provide a space, or "hole," for itself in the solvent structure. Simultaneously, the solvent molecules must be able to separate the solute particles from one another. The determining factor in both processes is the field strength of the particles involved. A nonpolar substance will, at best, have a very weak field and will be thus unable to overcome the strong field interactions existing between the particles of a polar substance, whether crystal ions or solvent molecules. This is especially true when water is the solvent; the strong hydrogen bonds between the water molecules must be broken if the solute is to achieve the uniform dispersion of its particles known as a solution.

Properties of the solvent. For good solvent action on ionic or polar compounds a solvent must be a polar substance. Polarity is most conveniently expressed in terms of dielectric constant.[1] Water has a high dielectric constant; this accounts in part for its excellent solvent powers. Ethanol, with a lower dielectric constant, is a less effective solvent. A liquid of very low dielectric constant, such as hexane, has no solvent action for polar solutes.

Interaction between a polar solvent molecule and a solute of suitable polarity implies at least some degree of solvation during the solution process. A solvent will dissolve a given solute when the solvation energy equals or is greater than the lattice energy of the solute. Using water as the solvent, solution is possible when the heat of hydration, which represents the field interaction energy of the water and the solute particles, is equal to or larger than the heat of crystallization, representing the field interaction energy between individual solute particles. This may be expressed mathematically as

$$\Delta H_{\text{solution}} = \Delta H_{\text{hydration}} - \Delta H_{\text{lattice}}$$

If the heat of hydration is greater than the lattice energy, a substance is said to have a positive heat of solution. This is true for most aqueous solutions of electrolytes. The excess heat energy appears as kinetic

[1] Dielectric constant, ϵ, is a measure of the strength of interaction of the field forces of an infinite number of particles with an imposed electrostatic field.

energy; the solution becomes warm or even hot. Occasionally the heat of hydration is less than the lattice energy, but if the difference is small the electrolyte will dissolve. The energy deficit is made up by abstracting equivalent kinetic energy from the system. The temperature of the resulting solution falls; the solution feels cold to the touch. Such a solute, e.g., ammonium nitrate, is said to have a negative heat of solution. The sole source of hydration energy for solution of ammonium nitrate is the hydration of the nitrate ion. Ammonium ions hydrate to a limited extent, if at all. A few electrolytes such as sodium chloride have an almost exact balance between hydration energy and lattice energy. When they dissolve, there is no heating or cooling effect.

High polarity, though essential, does not assure good solvent action for electrolytes. Nitrobenzene has a larger dipole moment than water, but is a poor solvent for electrolytes because its molecules are too large to penetrate the crystal structure of the solute. Only substances having small molecules are effective solvents. For good solvent action the solvent molecules must be able to enter the interstitial spaces of the crystal, literally to attack from all directions. The water molecule is the smallest of the common solvents. It owes its excellent solvent power for electrolytes to this factor as well as to its favorable dielectric constant.

The location of the dipole on the solvent molecule is also important for solvent action. It must be on the surface, not located within the structure where its effect cannot be felt by the solute particles. Alcohols and the corresponding chloro compounds illustrate the importance of dipole location. The hydrogen atom of the hydroxyl group is buried within the electron cloud of its oxygen atom so that the dipole is on the surface of the molecule. In the corresponding chloro compounds the large chlorine atom effectively screens the dipole from external influences. Figure 8.1 illustrates this dipole location on the ethanol and chlorethane. Although they are not solvents for all electrolytes, the low-molecular-weight alcohols do have fair solvent action for these compounds. The corresponding chloro compounds are poor electrolyte solvents.

Properties of the solute. The solubility behavior of a solute depends on the same physical properties involved in complex formation: charge,

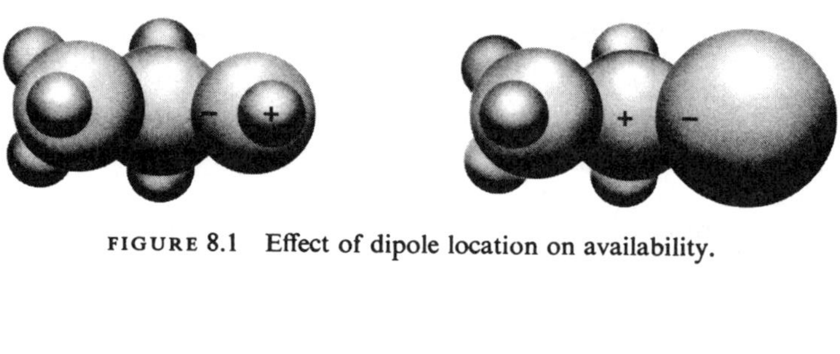

FIGURE 8.1 Effect of dipole location on availability.

size, and valence electron cloud structure. The difference is in degree of application. The extent of interplay of these factors with the solvent molecules determines the solubility behavior of the solute. Soluble electrolytes are those substances one or both of whose ions are able to form complexes with solvent molecules, the so-called solvates. Insoluble electrolytes are those substances whose existing structure is stable with respect to the possible solvate structures.

Simple ions with the inert gas structure tend to be soluble, especially if of low charge. The alkali metal cations almost invariably appear in water-soluble salts. Similarly, the monovalent halogen (and pseudohalogen) ions are soluble except when combined with an easily polarized cation. Ergo, the alkali metal halides are all soluble.

The alkaline earth cations occur with about equal frequency in soluble and insoluble salts. Here the charge effect makes its appearance. The double positive charge retains these cations securely within their crystal lattice; greater hydration energy is needed to release them from the crystal. Salts of alkaline earth cations with divalent anions are almost always insoluble, the interaction between the double positive and double negative field being too strong to be overcome by solvation energy effects. On the other hand, except for the fluorides, their salts with monovalent anions tend to be soluble. Because of its small ionic radius the fluoride ion has a very intense negative field. Consequently, its electrostatic interaction with the divalent positive field of an alkaline earth cation is as strong or stronger than that produced by divalent anions.

The effect of relative ionic charge is well illustrated by reconsidering the insoluble alkaline earth divalent anion salts. If the charge of the divalent anion is reduced by addition of a proton, the resulting compound's solubility is improved. The net electrostatic field interaction of the divalent cation with two monovalent anions is less than its field interaction with the divalent anion. Insoluble calcium carbonate and calcium sulfate yield sparingly soluble calcium hydrogen carbonate and calcium hydrogen sulfate on reaction with hydronium ion.

The effect of valence shell structure on the solubility of salts is illustrated by comparing Group IA salts with the corresponding Group IB salts: e.g., potassium chloride, soluble; silver chloride, insoluble. The greater polarizing ability of the noninert gas shell results in a greater polarization of the anion and a corresponding increase in lattice energy. The hydration energy is too small to overcome this effect.

The solubility behavior of complex anions is generally related to the strength of their conjugate Brønsted acid. The attraction of these anions for a small cation parallels their attraction for the proton. Anions which have strong conjugate acids will usually have soluble salts, whereas those of weak acids usually have insoluble salts. If the series $AgClO_4$, Ag_2SO_4, Ag_3PO_4 is considered, the first and last are extremely soluble and insoluble, respectively, with the sulfate moderately soluble—an order of solubility exactly paralleling acid strength. This behavior is reversed with corresponding salts of large cations. The large cation has its center of positive charge sufficiently removed from the positive center of the complex ion to reduce the mutual repulsive effects sufficiently to allow interaction between the positive charge of the cation and the electron cloud of the anion. This is illustrated by the cesium ion; $CsClO_4$ is insoluble, Cs_2SO_4 is moderately soluble, and Cs_3PO_4 is soluble.

METHODS FOR EXPRESSION OF SOLUBILITY

The qualitative expression of the solubility of compounds is relative and subsequently can be misleading. The classification adopted by the compendia, e.g., *U.S.P. XXI*[2], is a useful attempt to establish order to the field. The scheme is essentially reproduced in Table 8.1.

TABLE 8.1 A Scheme for Expressing Relative Solubilities

Descriptive Term	Ratio Solute:Solvent
Very soluble	<1:1
Freely soluble	1:1 to 1:10
Soluble	1:10 to 1:30
Sparingly soluble	1:30 to $1:10^2$
Slightly soluble	$1:10^2$ to $1:10^3$
Very slightly soluble	$1:10^3$ to $1:10^4$
Insoluble	$>1:10^4$

There is no recognized method for the quantitative reporting of solubility. Generally the concentration at saturation is given either as grams of solute to saturate one hundred grams of solvent (w/w), or grams of solute dissolved in one hundred milliliters of saturated solution. Solubility is also reported as the milliliters of solvent necessary to dissolve one gram of solute (w/v). Molar concentrations are rarely used.

Since the numerical value for the solubility of a given compound is unique for each method of reporting, the mode of reporting must be specified. In addition, the solubility given is valid only for the saturated solution of the salt in the *indicated* solvent at the *specified* temperature.

[2] *U.S.P. XXI*, p. 1441.

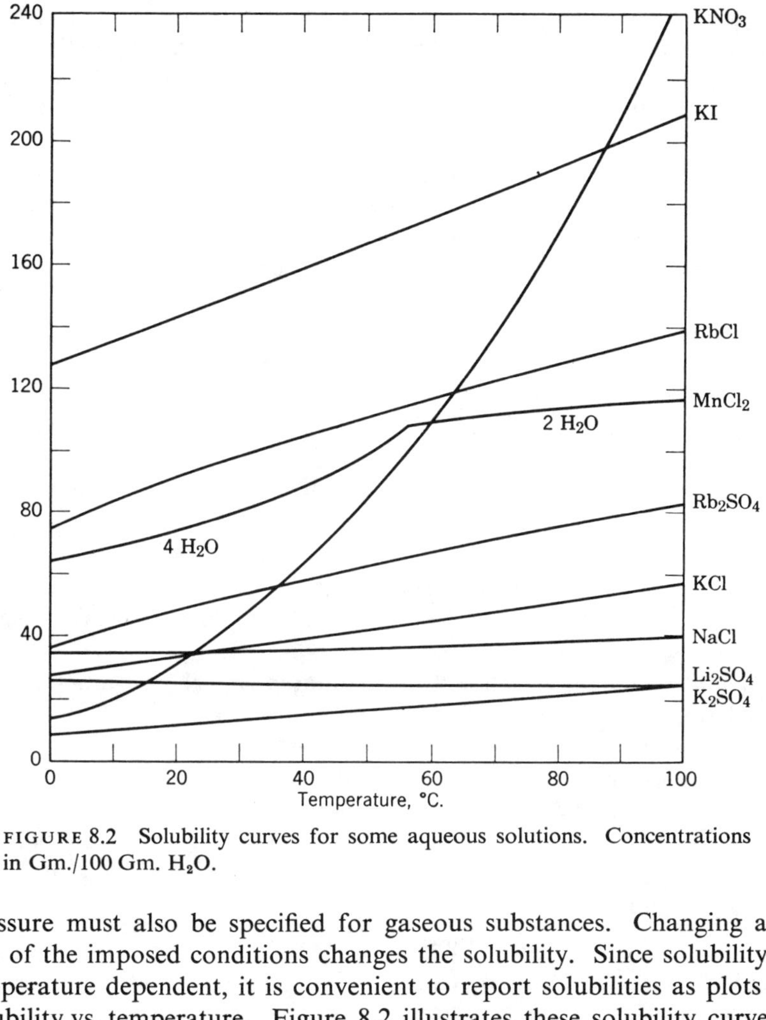

FIGURE 8.2 Solubility curves for some aqueous solutions. Concentrations in Gm./100 Gm. H_2O.

Pressure must also be specified for gaseous substances. Changing any one of the imposed conditions changes the solubility. Since solubility is temperature dependent, it is convenient to report solubilities as plots of solubility vs. temperature. Figure 8.2 illustrates these solubility curves.[3] The salts have been selected to show the effect of the charge and structural features of the ions involved.

Factors affecting solubility. The solubility of a given salt is affected by other dissolved substances. The effects are especially pronounced when high concentrations of the foreign substance are present. If the foreign salt has a common ion, the solubility of a given salt may be

[3] Data for construction of these curves and other solubility data may be found in *Handbook of Chemistry and Physics,* CRC Press, Inc. Boca Raton, Florida.

decreased markedly. Molecular compounds and salts without a common ion also affect solubility. Molecular substances (e.g., ethanol) act by changing the properties of the solvent water, particularly the dielectric constant which is usually reduced. The solubility of a given salt as a result is decreased by addition of a molecular substance.

Foreign salts will change the character of the solvent water by changing the dielectric constant. Solubility can also be changed by field interactions giving rise to ion-pairs, a process which is frequently overlooked. Ions of opposite charge can approach one another closely; their mutual electrostatic attraction causes them to remain with one another. The degree of association of ion-pairs appears to be related to cation field strength as mediated by its ionic radius, extent of solvation, and the electron structure. Ion-pair formation in calcium sulfate solutions has been studied most extensively. Ion-pair possibilities exist between the calcium and sulfate ions, and between calcium ions and/or sulfate ions and other anions and cations, respectively, present in the solution. In a 0.04*M* calcium sulfate solution 70% of the compound is associated as $[Ca^{2+}SO_4^{2-}]$ ion-pairs. If potassium nitrate is present, KSO_4^- and $CaNO_3^+$ are also present; their dissociation constants have been reported as 0.11 and 1.90, respectively. In the presence of chloride ion $CaCl^+$ ion-pairs form.

Solubility product. Although the common ion effect cannot be predicted from simple solubility data, such data may be used for the calculation of the solubility product, K_{sp}, of a salt. This constant is a valuable tool for comparison of solubilities, calculations involving solubility, and in the evaluation of the common ion effect on solubility.

The solubility product is derived by assuming an excess of an insoluble or slightly soluble salt, M_aX_b, suspended in excess in water to give a saturated solution:

$$[M_aX_b](\text{solid}) \rightleftharpoons [M_aX_b](\text{dissolved})$$

Since the salt is only slightly soluble, the concentration of the solution is sufficiently small to assume 100% ionization of the dissolved phase,

$$[M_aX_b](\text{dissolved}) \rightarrow aM^{b+} + bX^{a-}$$

The equilibrium expression may be written

$$K = \frac{[M^{b+}]^a[X^{a-}]^b}{[M_aX_b]_{(\text{dis})}}$$

Since the solid phase is in excess, the concentration of the dissolved phase may be considered a constant:

$$K[M_aX_b]_{\text{dis}} = [M^{b+}]^a[X^{a-}]^b$$

The new constant term is the solubility product, K_{sp},

$$K_{sp} = [M^{b+}]^a[X^{a-}]^b$$

Since $[M^{b+}]$ and $[X^{a-}]$ refer to the total molar concentration of each ion, irrespective of source, it is possible to calculate the common ion effect on the solubility of a salt if its solubility product is known. However, temperature and solvent must still be specified. Other ions present in the solution, but not appearing in the K_{sp}, will affect calculations. Some of the problems in solubility product calculations and the variables neglected in simple derivations, as above, have been discussed by S. Lewin.[4]

COLLIGATIVE PROPERTIES OF SOLUTIONS

Identification and relationships. The colligative properties of solutions are independent of the chemical and physical properties of the solute. Their magnitude is directly related to the number of independent solute particles present in the solution. The greater the number of particles dissolved, the greater the magnitude of a given colligative property. In this aspect of solutions a truly democratic status of solute particles is approached; the small lithium ion is the equal of the large cesium ion, a uranium hexafluoride molecule, a protein molecule, or a colloidal particle. In practice, the fields accompanying particles do modify this equality to some degree, particularly in higher concentrations.

Colligative properties of solutions are usually given as the vapor-pressure lowering, the freezing-point depression, the boiling-point elevation, and the osmotic pressure. In effect colligative properties are all related to the vapor-pressure lowering—the ability of solute particles to depress the vapor pressure of the solution below that of the pure solvent at the same temperature. This relationship is best shown graphically, as in Fig. 8.3. The vapor pressures of pure water and ice and of an aqueous solution are plotted as a function of temperature. Since a curve representing the solution falls beneath that for pure water at all times, it is evident that at a given temperature the vapor pressure of pure solvent water is always greater than that of any of its solutions.

VAPOR-PRESSURE LOWERING: Raoult's law is the mathematical representation of this fact:

$$p_a = N_a p_a^0$$

where p_a is the vapor pressure of the solution, p_a^0 the vapor pressure of

[4] S. Lewin, *The Solubility Product Principle*, Interscience, New York, 1960.

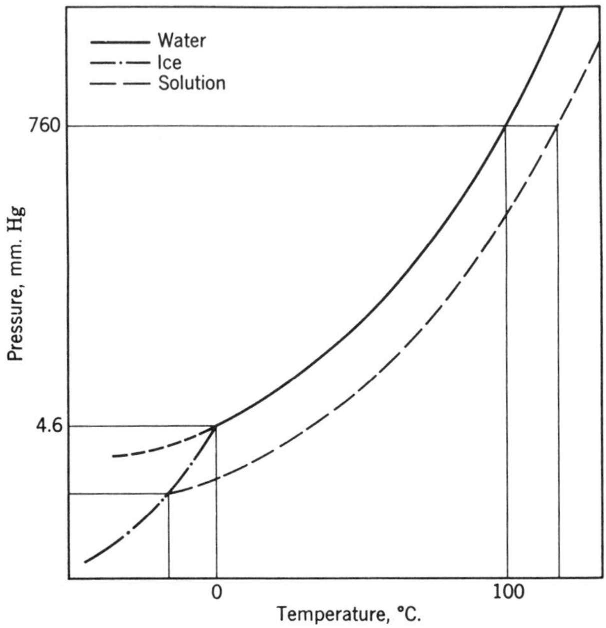

FIGURE 8.3 Comparative vapor-pressures curves for water and an aqueous solution. (Not drawn to scale.)

the pure solvent, and N_a the mole fraction[5] of solvent present; all values are at a specified temperature. For most work with solutions it is more convenient to express Raoult's law in terms of the solute:

$$(p_a^0 - p_a) = N_b p_a^0$$

An examination of Raoult's law reveals that all variables are independent of the physical or chemical properties of the solute.

Although lowering of the vapor pressure should theoretically be directly proportional to the number of particles (mole fraction) of solute present, in practice there is a progressively larger deviation from theory with increasing concentration. This arises from an increasing interference with the independent character of the individual solute particles by field

[5] Mole fraction is defined as the fraction of the total moles present in a system represented by a given substance. If there are n_a moles of substance a, n_b moles of substance b, and n_c moles of substance c, the mole fraction of substance a is

$$N_a = \frac{n_a}{(n_a + n_b + n_c)}$$

At low concentrations the mole fraction of solute and the molality of a solution approach equality.

interactions of neighboring particles. Since increasing concentration has the effect of decreasing interparticle distances, and since the strength of the field forces of a particle is inversely proportional to distance from the particle, it is evident why interactions must become stronger with increased concentrations. Consequently fewer particles are independent in the true sense as concentration increases and their ability to lower the vapor pressures decreases. Figure 8.4 shows this concentration effect for aqueous solutions of several electrolytes, selected to show the effect of charge, ionic diameter, and valences shell structure. In order to show the mean effect of each particle the experimental vapor-pressure lowering has been divided by the number of ions theoretically introduced per molecular weight of salt. A similar concentration effect exists for each of the other colligative properties.

BOILING-POINT ELEVATION AND FREEZING-POINT DEPRESSION: Examination of Fig. 8.3 reveals why the boiling point of a solution must be higher than that of the pure solvent. The normal boiling point of a liquid system is that temperature at which the liquid's vapor pressure is 760 mm. of mercury. Since the vapor pressure of a solution is less than that of the pure solvent at any temperature, the solution can reach a vapor pressure of 760 mm. of mercury only at a temperature above the boiling point of the solvent.

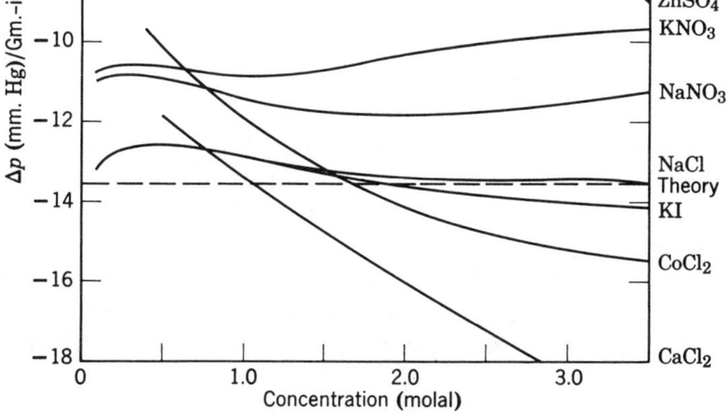

FIGURE 8.4 Effect of ionic properties on vapor pressure lowering.

Thermodynamically it can be shown that the elevation of boiling point, ΔT_b may be expressed as

$$\Delta T_b = mK_b$$

where m is the molality of the solution, and K_b is the molal boiling-point constant. K_b may be defined as the boiling-point elevation of a one molal solution in which the solute exists as undissociated and, for practical purposes, unsolvated molecular particles. For water K_b has a value of 0.512°C.

A liquid and its solid phase can exist in equilibrium only if the vapor pressure of the two phases is identical. Reference to Fig. 8.3 reveals that since the vapor pressure of the solution is less than that of the pure solvent, the temperature at which the vapor pressures of the ice and the solution become identical must be less than the freezing point of the pure solvent. Thermodynamically it can be shown that the freezing point depression, ΔT_f, may be expressed as

$$\Delta T_f = mK_f$$

where K_f is the molal freezing-point constant and is defined similarly to K_b. The K_f for water is 1.86. The molal constants for several other solvents are found in Table 8.2.

TABLE 8.2 MOLAL PROPERTIES OF SOLVENTS

Compound	K_b	Boiling Point, °C.	K_f	Freezing Point, °C.
Acetic acid	3.07	118.5	3.9	16.9
Benzene	2.53	81.2	5.12	5.5
Ethanol	1.22	78.3	—	(−112)
Phenol	3.56	181.2	7.27	42.0
Water	0.512	100.0	1.86	0.0

OSMOTIC PRESSURE: The colligative property of greatest direct interest in pharmacy is osmotic pressure. If two solutions of different concentration (or a solution and its pure solvent) are separated by a semipermeable membrane,[6] the solvent tends to flow through the membrane from the dilute solution to the concentrated solution until equal concentrations are established. Osmotic pressure is the pressure gradient which exists across the membrane. It is defined more exactly as the pressure which must be applied to the more concentrated solution to prevent osmosis. Numerical values for osmotic pressure may be unexpectedly large, e.g., 100 atm.

[6] A membrane permeable to solvent molecules but impermeable to the solute particles.

J. H. van't Hoff first pointed out in 1885 that osmotic pressure, Π, can be related to concentration through an equation resembling the ideal gas law:

$$\Pi = mRT$$

where m is the difference in the molal concentration of the two solutions,

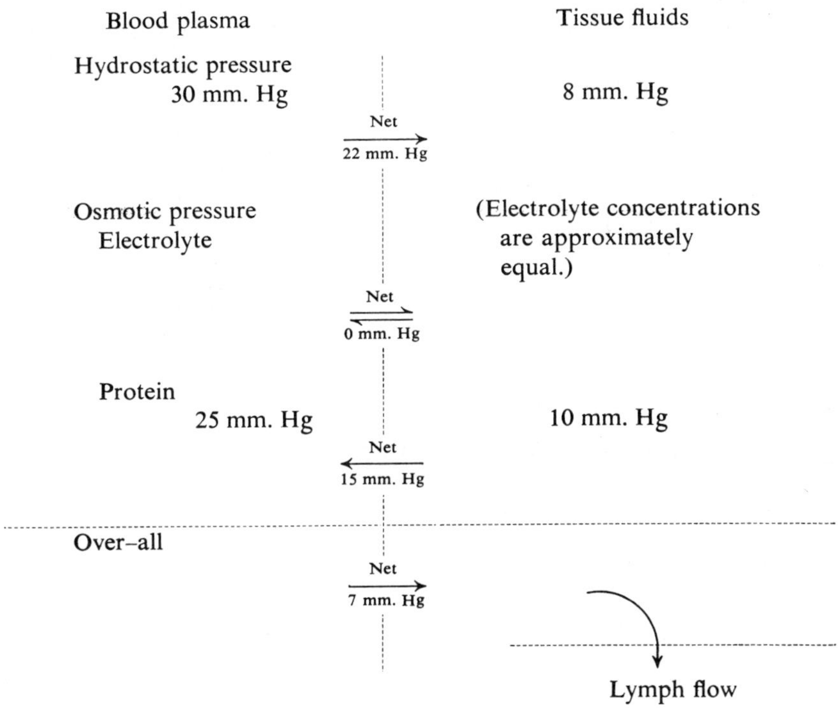

FIGURE 8.5 Fluid balance in capillary net.

R is the gas constant, and T is the absolute temperature. More sophisticated derivations have since been made. The previous equation is really a special case of a more exact equation,

$$\Pi = \frac{RT}{\bar{V}_a} \ln \frac{p_a^0}{p_a}$$

where $\bar{V}_a$ is the partial molal volume of the solvent. The molar volume of the solvent may usually be substituted for $\bar{V}_a$.

Osmotic pressure and the body fluids. Osmotic pressure is important for regulation of fluid flow between physiological fluids for maintenance of isotonic conditions. The role is best illustrated by considering the fluid balance existing between the blood plasma and tissue fluids in the capillary beds. A delicate balance between osmotic pressure and hydrostatic pressure allows for a very small net pressure favoring the tissue fluids (compensated by lymph flow). This is shown in Fig. 8.5. Although the net values are not as easy to establish, a similar fluid balance exists between the extracellular and intracellular tissue fluids.

Osmotic pressure figures importantly in various abnormal physiological conditions. It is of especial importance to the pharmacist because of the

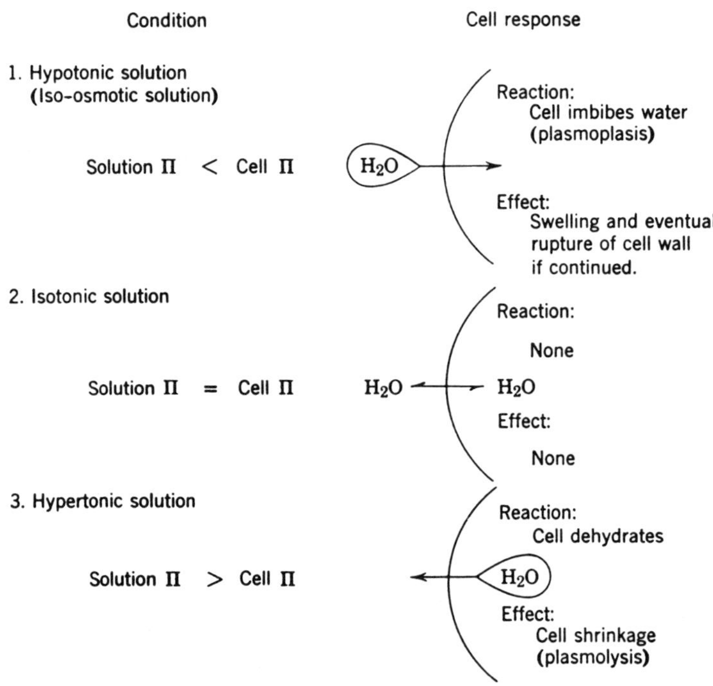

FIGURE 8.6 Cell response to osmotic pressure of solutions.

desirability for maintaining isotonic conditions at the site of administration of medicinals. This is true not only for injectables but also for collyria and nasal sprays. The cell walls and the various membranes (e.g., nasal, capillary) are semipermeable membranes. Drug solutions with osmotic pressures differing from those normally in contact with these membranes disturb the normal osmotic balance; as a result irritation and inflammation may occur. The physical effect of hypotonic and hypertonic solutions is best illustrated diagrammatically, as shown in Fig. 8.6.

Experimental determination of osmotic pressure is difficult, requires special equipment, and is time consuming. Fortunately, osmotic pressure is directly related to all other colligative properties:

$$\frac{\Delta T_f}{K_f} = \frac{\Delta T_b}{K_b} = \frac{p_a{}^0 - p_a}{p_a{}^0} = \frac{\Pi}{RT}$$

Any one of the other three colligative properties may be determined experimentally and used to calculate the corresponding theoretical osmotic pressure. Freezing-point depression measurements are easiest, requiring only an adequate thermometer (or other temperature-measuring device) and a suitable cooling bath.[7] Experimental freezing-point depressions can establish the concentration of a given medicinal agent which is iso-osmotic with the various body fluids. The freezing point of blood plasma and the lachrymal fluids has been established at −0.52°C., equivalent to a 0.9% solution of sodium chloride.

Tonicity control. Experimental freezing-point depressions, although used infrequently for iso-osmotic and isotonic calculations of pharmaceutical solutions, are the experimental basis for all methods of calculation. The freezing points of some 1% w/v solutions are given in Table 8.3.[8] The calculations based on freezing-point depression values are straightforward. The two following examples illustrate:

1. How many grams of $CaCl_2 \cdot 2H_2O$ must be dissolved in 100 ml. of water to render the resulting solution isotonic?

SOLUTION: From Table 8.3 it is seen that a 1% solution of this salt has a ΔT_f of 0.42°C. A direct proportion may be used:

$$\frac{\text{Gm. } CaCl_2 \cdot 2H_2O}{1\% \text{ solution}} = \frac{\Delta T_f \text{ (isotonic)}}{\Delta T_f \text{ (1\% } CaCl_2)}$$

Substituting,

$$\frac{x}{1} = \frac{0.52^\circ}{0.42^\circ}$$

$$x = 1.24 \text{ Gm. required}$$

2. Prepare 1 liter of 0.45% NaCl, rendered isotonic with glucose.

SOLUTION: From Table 8.3 ΔT_f for 1% NaCl and 1% glucose are 0.576 and 0.095, respectively. Thus

$$\Delta T_f(\text{NaCl}) = 0.45 \times 0.576^\circ = 0.259^\circ$$

and

$$\Delta T_f(\text{glucose}) = 0.52^\circ - 0.259^\circ = 0.26^\circ$$

[7] Shoemaker, D.P. *et al, Experiments in Physical Chemistry,* 4th ed., McGraw-Hill, New York, 1981, p. 183.

[8] Taken in part from a compilation by W.J. Husa and O.A. Rossi, "A Study of Isotonic Solutions," *J. Am. Pharm. Assoc., Sc. Ed.,* **31**, 270 (1942).

TABLE 8.3 Freezing-Point Depression of Solutions

Solutions[a] (1 % w/v)	°C.	Solutions (1 % w/v)	°C.
Alum (potassium)[b]	0.11	Potassium iodide	0.22
Ammonium chloride[b]	0.655	Potassium nitrate	0.35
Ascorbic acid[b]	0.105	Procaine hydrochloride[b]	0.14
Atropine sulfate	0.08	Silver nitrate	0.225
Boric acid[b]	0.315	Sodium benzoate	0.26
Caffeine	0.075	Sodium bicarbonate	0.40
Calcium chloride	0.42	Sodium biphosphate	0.265
Calcium gluconate	0.082	Sodium bromide	0.36
Calcium lactate	0.12	Sodium cacodylate	0.175
Chlorbutanol[c]	0.14	Sodium carbonate, monohydrate[b]	0.33
Cocaine hydrochloride[b]	0.11	Sodium chloride	0.576
Cupric sulfate	0.09	Sodium citrate	0.185
Dextrose	0.095	Sodium iodide	0.22
Ephedrine hydrochloride[b]	0.16	Sodium nitrate	0.36
Ephedrine sulfate	0.11	Sodium phosphate (monobasic)	0.29
Glycerin[b]	0.205	Sodium phosphate (dibasic)	0.13
Homatropine hydrobromide[b]	0.11	Sodium salicylate	0.23
Lactose	0.06	Sodium sulfate	0.165
Magnesium chloride	0.245	Sodium tetraborate	0.25
Magnesium sulfate	0.115	Sodium thiosulfate	0.18
Mercuric chloride[c]	0.07	Sodium phenobarbital[c]	0.155
Methenamine[b]	0.14	Sucrose	0.055
Mild protein silver (Argyrol)	0.11	Urea[b]	0.31
Morphine sulfate[c]	0.07	Zinc sulfate[c]	0.09
Phenol	0.20		
Pilocarpine hydrochloride[b]	0.125		
Potassium chloride	0.485		

[a] Official formulas are implied.
[b] Iso-osmotic rather than isotonic.
[c] Information unavailable concerning iso-osmotic or isotonic classification.

Setting up a proportion, as in example 1,

$$\frac{x}{1} = \frac{0.26}{0.095}$$

$$x = 2.74\% \text{ glucose}$$

Converting to a liter basis,

NaCl: $10 \times 0.45 = 4.50$ Gm. required

Glucose: $10 \times 2.74 = 27.40$ Gm. required

These problems demonstrate the basic dependence of isotonicity on freezing-point depression information. Methods of calculation more commonly used by pharmacists, e.g., "sodium chloride equivalents," are based on constants derived from experimental ΔT_f information.[9] It is claimed that these calculations are simpler.

Isotonic solutions prepared using ΔT_f or derived data are subject to two types of error: those of a chemical nature and those of a biological nature. The chemical errors are related to concentration effects. The calculations fail to take into account a change in activity with change in concentration.[10] However, since the solutions are usually quite dilute (e.g., about $0.075M$ for 0.45% sodium chloride), the concentration effect is relatively small and generally insignificant.

An additional concentration effect may be present when preparing buffered isotonic solutions. The buffer salts as well as the medicinal compound contribute to the osmotic properties of the solution. The usual procedure involves the use of buffer solutions which are themselves isotonic or slightly hypotonic, e.g., the phosphate and borate buffers previously described (page 179, 180). Small quantities of drugs having low osmotic effects may be added to these buffers, as the solid, with only a slight resulting hypertonicity. Recent work indicates that slightly hypertonic solutions may be better tolerated than their isotonic counterparts. Larger quantities of drugs will, however, render the buffers excessively hypertonic. *U.S.P. XXI*[11] has an excellent discussion of isotonicity and pH of ophthalmic solutions and of osmolarity of parenterals.

Hemolysis studies, using red blood cells, have recently demonstrated that osmotic concentrations of many medicinals as determined by use of freezing-point data are not related to the true osmotic pressure under physiological conditions. Many medicinals pass freely through the cell membrane, e.g., ammonium chloride, ethanol, methenamine. Others, e.g., sodium carbonate may damage the cell wall due to excessive pH conditions. When a salt can pass through a cell membrane, it obviously cannot exert an osmotic pressure. It is recommended that solutions of this type be termed iso-osmotic, rather than isotonic. Iso-osmotic concentrations may differ markedly from the true isotonic concentration.[12]

---

[9] E. R. Hammarlund and K. Pedersen-Bjergaard, "New Isotonic Solution Values," *J. Am. Pharm. Assoc., Pract. Ed.*, **19**, 38–39 (1958).

[10] ΔT_f is the basis for one method of determining activity coefficients.

[11] *U.S.P. XXI,* pp. 1338-1339, pp. 1256-1257.

[12] A discussion, references, and a table summarizing the percent hemolysis of a large number of medicinals in current use is presented in E.R. Hammarlund and K. Pedersen-Bjergaard, "Hemolysis of Erythrocytes in Various Iso-osmotic Solutions," *J. Pharm. Sci.,* **50,** 24-30 (1961).

Those compounds appearing in Table 8.3 which give iso-osmotic rather than isotonic solutions have been suitably identified.

THE ALKALI FAMILY OF ELEMENTS

Because of the inherent solubility of their salts in aqueous media and fair solubility in alcoholic media, the alkali metal cations find extensive use in the practice of pharmacy. Many otherwise insoluble medicinal agents are successfully administered in anionic form, as a soluble alkali metal salt. Being normal cationic constituents of the body fluids, sodium and potassium ions thus administered have minimal side effects and usually have no elimination problems. They are employed medicinally in fluid and electrolyte replenishers and in physiological salt solutions.

TABLE 8.4 ATOMIC CHARACTERISTICS OF THE ALKALI METALS

Element	Lithium	Sodium	Potassium	Rubidium	Cesium
Symbol	Li	Na	K	Rb	Cs
Atomic weight	6.94	22.99	39.10	85.47	132.91
Atomic number	3	11	19	37	55
Valence electron configuration	$1s^2 2s^1$	$2s^2 2p^6 3s^1$	$3s^2 3p^6 4s^1$	$4s^2 4p^6 5s^1$	$5s^2 5p^6 6s^1$
Atomic radius (Å)	1.52	1.86	2.31	2.44	2.62
Ionic radius (Å)	0.60	0.95	1.33	1.48	1.69
Ionic radius, hydrated (Å)	3.40	2.76	3.32	2.28	2.28
Ionization potential (ev.)	5.39	5.14	4.34	4.18	3.89
Electrode potential (v.)	3.02	2.71	2.92	2.99	3.02
Electronegativity	1.0	0.9	0.8	0.8	0.7

Since sodium is the cation primarily concerned with fluid balance in the intercellular fluids, it is widely used as its chloride for the regulation of the tonicity of pharmaceutical preparations, e.g., injections and collyria.

Group properties. Data pertinent to a discussion of the properties of this group are given in Table 8.4. The ground state electronic structure is a single *ns* electron overlying an inert gas structure. Because of the effective screening of the nuclear charge by this underlying electron shell, the single valence electron is lost with a minimum expenditure of energy; ionization potentials are comparatively low. The ease of ion formation may be demonstrated by a comparison of the ionization potentials of sodium and its three horizontal neighbors; the ionization potentials for the first electron are 21.60, 5.14, 7.64, and 5.98 ev for Ne, Na, Mg, and Al, respectively.

The effect of size on the ease of ion formation is illustrated by comparing atomic radii and ionization potentials within the alkali family (Table 8.4). The larger members lose their single valence electron very easily because it is so far removed from the nucleus. Thus ions of the alkali metals are extremely stable. The reducing environments required for restoration of the ion to its ground state atomic form are not encountered in pharmaceutical preparative practice; redox incompatibilities are nonexistent.

The single positive nuclear charge on alkali metal cations is effectively screened by their spherical, tightly held electron shell. Such cations therefore have little or no polarizing effect on anions or molecules. At the same time, their electron cloud also screens out the polarizing influences of other particles. As a result, the cations of this family are unable to act as the central ion of complexes. Only lithium and sodium, because of their relatively small radii, are capable of existing as hydrates; potassium ion hydrates rarely; and rubidium and cesium ion do not hydrate at all. But a hydrated alkali cation cannot act as an acid. The weak, well-shielded cationic charge is unable to distort the electron cloud of a complexed water molecule sufficiently to allow release of a proton to a stronger base.

The weak fields of these cations impose certain characteristics on their salts. The crystalline forms are ionic. Because of the weak attraction for anions, their normal salts of the oxygen acids (e.g., sodium carbonate) are relatively stable to heat.

Since the alkali cations are chemically inert in pharmaceutical systems, the chemical properties of alkali compounds will usually be determined by the anion present. Sodium hypochlorite is an oxidizing agent because of the hypochlorite ion. Tertiary sodium phosphate is a base due to the tertiary phosphate ion. Sodium hydrogen sulfate is acid because of the hydrogen sulfate ion.

Because of the very weak acidic character of any hydrated cation, the pH of solutions of alkali metal salts depends solely on the base strength of the accompanying anion. If the anion is a very weak base, e.g., nitrate, a neutral solution will result. If it is a strong base, e.g., carbonate, an alkaline solution will result. If it is a very strong base, e.g., hydroxide, a very alkaline solution results. Only the acid salts of strong acid anions, e.g., hydrogen sulfate, are capable of giving an acid reaction to solutions of alkali metal salts.

In conclusion, the characteristic chemical properties of an alkali metal compound are largely dependent upon the anion present. However, many physical properties will be a function of both cation and anion.

The particular physical properties of interest to the pharmaceutical chemist are solubility, largely dependent on the alkali cation, and the molecular properties, dependent on both the cation and anion.

Group differences. LITHIUM DEVIATIONS: Differences in properties, when existing in a given series of alkali metal salts, are usually a matter of degree, albeit usually small. The largest deviations from group behavior are exhibited by the lithium ion. It has a helium-type electron structure (see Table 8.4) rather than the $8e^-$ inert gas shell common to the other members of the family. The lithium ion also is small. These structural elements result in a comparatively strong positive field, giving lithium a much stronger polarizing action on nearby anions or molecules. The effect is shown when the ionic radius of the hydrated lithium cation (the ion of aqueous solutions) is compared with the ionic radius of its unhydrated cation. The water hull associated with the hydrated lithium ion increases the radius by a factor of almost six. In aqueous media the hydrated lithium ion is larger than all other members of the family, including cesium. As a result of hydration such properties as ionic mobility and penetration through semipermeable membranes will be radically changed from normally expected values. Solutions of lithium salts also show strong negative deviations from most other normal physical properties. Freezing points, aqueous vapor pressures, etc. are lower than expected. The application of lithium bromide to absorption refrigeration and lithium chloride to humidity control depends on these unusually large deviations.

Lithium salts generally have a much higher solubility in organic media than other members of the family. Lithium chloride is a valuable conducting salt for electrochemical experimentation in nonaqueous media. The solubility of LiH and $LiAlH_4$ in diethyl ether makes these compounds especially useful as reducing agents. Increased solubility in organic media probably results from the comparatively greater polarizing power of the lithium ion. When acting on large anions and easily polarized linkages, this polarizing power probably accounts for the heat liability of lithium compounds such as the carbonate and nitrate. These salts of the other members of the family are heat stable at comparatively high temperatures.

For comparison purposes lithium resembles calcium much more closely than the other members of its own family. Lithium is a bridge element. Technically its bridge should be magnesium, although the resemblance is somewhat closer to calcium. Apparently the screening effect of the helium electron structure on a single positive nuclear charge is about equal to that of the $8e^-$ shell on the dipositive charge of the calcium ion. This similarity to calcium is especially evident on examination of solubility behavior. Lithium carbonate is almost insoluble in water; lithium hydroxide is only slightly soluble. The acid strength of the hydrated lithium ion is similar to that of the hydrated calcium ion.

Because lithium is considered a member of the alkali metal family, its incorporation into a pharmaceutical preparation may lead to difficulties not predictable on that basis. With properties very strongly reminiscent of calcium, chemical and physiological incompatibilities similar to those of calcium may appear. The fact that lithium may not be substituted *ad liberum* for sodium or potassium in preparations intended for human consumption was tragically demonstrated quite recently when lithium compounds were used in a salt substitute formulation for individuals on salt-free diets. Comparison of lithium and calcium metabolism and of the syndromes of lithium intoxication and hypercalcemic conditions reveals some interesting similarities. This strongly suggests that the cells are unable to distinguish lithium from calcium and unequivocally identify and treat lithium as an alkali metal cation.

Cesium and rubidium have no properties of particular interest to pharmacy.

SODIUM VS. POTASSIUM COMPOUNDS: Of greatest interest to the pharmaceutical chemist is the rationale underlying the selection of a sodium compound versus a corresponding potassium compound, and vice versa. Except for the degree of hydration in aqueous solution the chemical properties of these two ions are quite similar, as are most physical properties (except those due strictly to mass).

Unless certain advantages accrue from use of a potassium salt, economics usually dictate the use of the sodium salt of a given anion. Most common sodium compounds are available at much lower cost than the corresponding potassium compound simply because of the relative abundance of natural sources of sodium ion as compared to the relative scarcity of deposits of potassium salts. The greater number of equivalents per unit weight skews the cost differential even more in favor of the sodium compound. With 500-gram portions of each hydroxide as a basis for calculation, the sodium hydroxide sample has 500/40.0 or 12.5 equivalents, whereas the potassium hydroxide sample has only 500/56.0 or 8.9 equivalents. In other words a given weight of sodium hydroxide has 40.5% more reacting power as a base than the same weight of potassium hydroxide.

Despite the chemical and physical similarity of their compounds and the cost differential, subtle differences favor potassium. Potassium salts generally tend to have greater solubility in polar organic liquids: potassium hydroxide dissolves in 3.0 parts ethanol, sodium hydroxide in 7.2; potassium chloride in 250 ml. ethanol, sodium chloride is insoluble. This accounts for the popularity of alcoholic KOH and the rare use of alcoholic NaOH.

Potassium salts generally are less deliquescent (hygroscopic) than corresponding sodium compounds. Hence potassium permanganate is universally used over the highly deliquescent sodium permanganate. Extremely hygroscopic sodium oxide and sodium hydroxide are unsuitable as carbon dioxide absorbers (page 593). Potassium nitrate formerly was a component of black powder, but at present sodium nitrate is used crystallized in a form which is not hygroscopic!

Potassium compounds are selected to impart properties not given by a sodium salt. Potassium hydroxide must be used in the preparation of "soft soaps." Potassium bicarbonate is used as an ingredient in permanent wave formulations for a pronounced softening effect on the hair.

The living cell is able to distinguish between sodium and potassium ions. Potassium is the dominant cation of the intracellular fluids, whereas sodium plays a similar role in all extracellular fluids. This biochemical differentiation may be responsible for the preferential use of one or the other salt in a given pharmaceutical preparation, e.g., with hypertensive patients the potassium salt is usually preferred, especially if the drug is to be taken frequently or in large quantities.

Analytical problems. The paucity of suitable insoluble compounds makes difficult direct quantitative and qualitative analysis of all members of the family, except possibly lithium. Interferences from other members of the family as well as from the ammonium ion frequently impose additional difficulties. Sodium may be detected in the presence of both lithium and potassium ions by precipitation as the pyroantimonate and sodium zinc uranyl acetate or sodium magnesium uranyl acetate:

$$Na^+ + 3UO_2(OAc)_2 + Zn(OAc)_2 + HOAc + 10H_2O$$
$$= NaZn(UO_2)_3(OAc)_9{\cdot}9H_2O\downarrow + H_3O^+$$

Potassium ion may be precipitated as the hexachlorplatinate(III), acid tartrate, fluosilicate, perchlorate, and the cobaltinitrite. The last named is least soluble. Ammonium ion interferes with each of the potassium tests.

The so-called insoluble salts of sodium and potassium are only relatively insoluble; most other cationic salts of similar solubility would be considered slightly soluble. This is evident by a consideration of the solubilities, expressed in grams dissolving in 100 Gm. water, as given in Table 8.5.

Spectroscopic analysis is a valuable analytical procedure for these elements. Flame tests provide a useful qualitative means for detection and identification of the elements of the alkali and alkaline earth family (page 315). These tests are best performed by heating a small portion of a crystal on a platinum wire loop in the outer cone of the Bunsen flame.

TABLE 8.5 Insoluble Sodium and Potassium Salts

Anion	Cation	
	Sodium	Potassium
Cobaltinitrite	Very soluble	0.09 (0°C.)
Hexachlorplatinate(III)	Soluble	0.74 (0°C.)
Hexafluosilicate	0.65 (17°C.)	0.12 (17°C.)
Hydrogen tartrate	11.1 (0°C.)	0.37 (0°C.)
Perchlorate	170. (0°C.)	0.75 (0°C.)
Pyroantimonate	0.03 (12.3°C.)	2.82 (20°C.)
Tetrafluoborate	108. (27°C.)	0.44 (20°C.)
Zinc uranyl acetate	Insoluble	Soluble

The chlorides, generally more volatile, are the most satisfactory test crystals. The colors imparted to the flame are given in Table 8.6. The sodium flame test is very sensitive. Most substances will give a transient sodium flame due to trace impurities of this cation. A persistent yellow flame is a positive test for sodium. A few other elements principally copper and boron, give flame tests similar to barium under certain conditions.

AMMONIA AND ITS DERIVATIVES

Because many of its chemical and physical properties are similar to those of the alkali metal cations, it is advisable to consider ammonium ion at this time. The wide-spread use of substituted ammonium compounds in pharmaceutical preparations, because of their excellent solubility characteristics, necessitates a study of these compounds and their parent, ammonia.

General chemistry of ammonia and its compounds. Structurally the

TABLE 8.6 Flame Tests

Element	Flame Color	Masked By
Barium	Yellow-green	—
Calcium	Brick-red	Ba
Lithium	Carmine	Ba, Na
Potassium	Violet	Li, Na[a]
Sodium	Yellow	—
Strontium	Crimson	Ba
Thallium	Green	—

[a] Blue glass will mask the sodium color.

ammonia molecule is a complex with a nitrogen atom as its central atom. The nitrogen, in a minus three oxidation state, has four sp^3 hybrid orbitals in the usual tetrahedral array. Since only three hydrogen atoms are available for the formation of σ bonds with these hybrids, the fourth tetrahedral position is occupied by an unshared electron pair. The repulsive electrostatic effects of this unshared pair cause a slight distortion of the tetrahedral angle.

The visualization of the structure and reactions of the ammonia molecule and its derivatives can be aided by a comparison with their carbon analogs. The unsubstituted compounds of the methane and the ammonia series are shown in Fig. 8.7. The charge differences, which are the most obvious differences, are inherent from the ground state electron structure of the carbon and nitrogen atoms: $1s^22s^22p^2$ and $1s^22s^22p^3$, respectively.

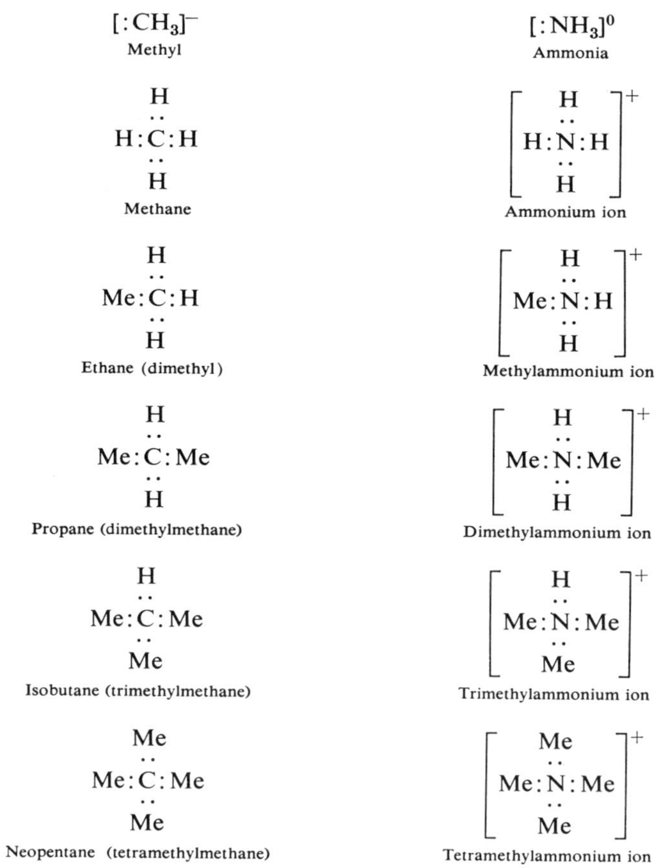

FIGURE 8.7 Substituted ammonia compounds and their carbon analogs.

The accompanying increase in atomic number results in a rather significant increase in the electronegativity of the nitrogen atom—3.0 compared to 2.5 for carbon. Since atomic radii are not significantly different, 0.77 Å vs. 0.70 Å, and the hybridization achieved is identical for both atoms, it is the difference in electronegativity which is largely responsible for the difference in reactivity and properties.

It is possible to replace one or more of the hydrogen atoms of ammonia with other substituent groups, e.g., alkyl, hydroxy, and amine groups, giving rise to primary, secondary, and tertiary amines and compounds such as hydroxylamine and hydrazine. Because of its unshared pair the ammonia molecule is able to form a coordinate covalent bond with a proton or any other suitable electron pair acceptor. Ammonium salts, amine salts, and quaternary ammonium salts result from such coordination:

$$NH_3 + HCl = NH_4^+ + Cl^-$$
$$NH_3 + BF_3 = H_3N{:}BF_3$$
$$CH_3NH_2 + H_3O^+ = CH_3NH_3^+ + H_2O$$
$$(CH_3)_3N + CH_3I = (CH_3)_4N^+ + I^-$$

These reactions illustrate the two ammonia-based conjugate pairs:

$$NH_3 + (H^+) = NH_4^+$$
$$NH_2^- + (H^+) = NH_3$$

Ammonia is a weak base. Its pK_b is 4.86. The assignment of the name "ammonium hydroxide" to its solutions is in error. Ammonia solutions consist almost entirely of ammonia molecules in the solvent water. Ammonia water, or ammonia solution, is a more appropriate name for these solutions.

The ammonium salts are acids. The inorganic amides, imides, and nitrides are bases. The ability to form stable amides with strongly electropositive elements, as compared to the instability of corresponding methides, is due to the stronger electronegativity of the nitrogen atom.

Ammonia is a polar molecule. Thus electrostatic field interactions are possible. The dipole interactions of the ammonia molecule make it an excellent ligand, giving rise to the ammine series of complexes; ammines form with neutral molecules as well as with cations, e.g., $PCl_3{\cdot}5NH_3$, $SbCl_5{\cdot}6NH_3$. The presence of an amine group in its structure usually renders a molecule an excellent ligand. The polarity of the ammonia molecule is also responsible for the excellent electrolyte solvent action of liquid ammonia.

The over-all chemistry of the ammonia molecule is summarized in Fig. 8.8. The ammonium salts and the amines and their salts have the widest application in pharmaceutical practice and medicine.

All cationic derivatives of the ammonia molecule have relatively large ionic radii. Their single positive charge, even in the ammonia ion, is effectively screened by the electron clouds of the molecular bonds within the ionic structure. The net result is that the electrostatic field associated with any one of these cations is extremely weak. Consequently all are highly soluble. It is impossible to achieve effective interaction with anions, an interaction necessary for easy crystallization from solution into an ionic lattice. Even solvation of these cations is minimal. The high solubility of these cations is potentiated by the high dielectric constant

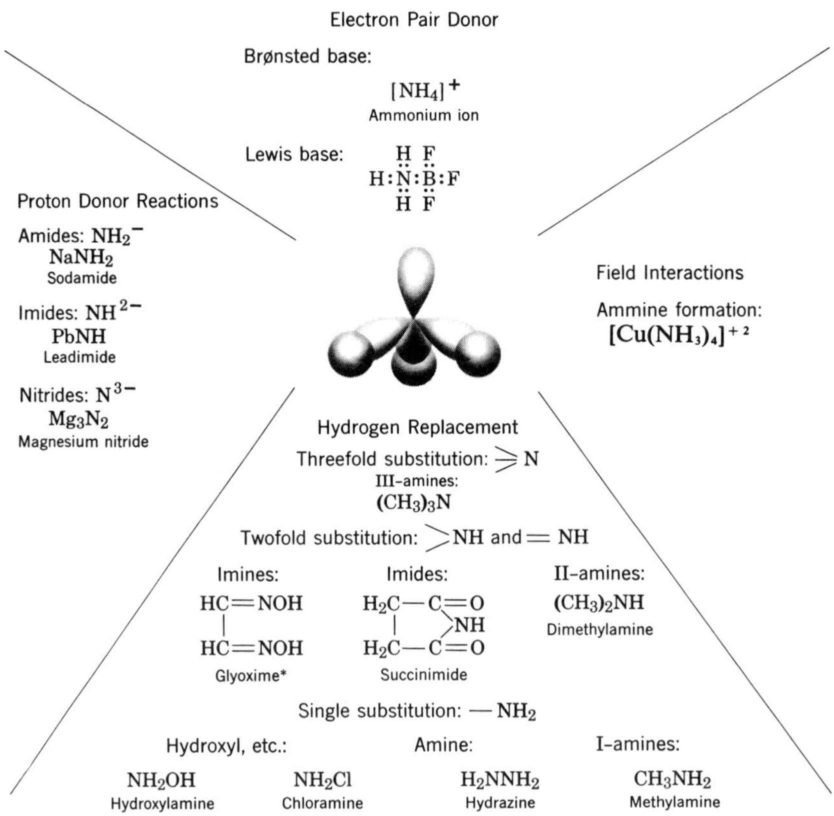

FIGURE 8.8 The ammonia-based family of compounds. A single ammonia moiety may undergo substitution, act as an electron pair donor, and act as a ligand. The starred substance, glyoxime, is a hydroxyl-substituted imine.

of water; it effectively aids in reducing interaction between the weak cation field and anions present in solution.

Ammonium ion and its properties. The unsubstituted ammonium ion resembles potassium ion in many of its physical properties, e.g., conductivity, freezing-point depression. The solubility behavior of ammonium salts is intermediate to that of the corresponding potassium and rubidium salts.

Ammonium salts differ from the corresponding potassium salts in several ways important to pharmacy. Because ammonium ion is a weak acid, the stability of its compounds is a function of the acid strength of its anion's conjugate acid. Ammonium salts of very weak acids, e.g., boric, are unstable even at room temperature. Those of moderately weak acids, e.g., carbonic, decompose slowly at room temperature. Salts of strong acids, e.g., hydrochloric, are stable at room temperature but decompose at high temperatures (page 248). The acid salts of a polyprotic acid are generally more stable than the normal salt.

Because the ammonium ion is an acid ($pK_a = 4.86$), its salt solutions will be acid to neutral rather than neutral to alkaline as the potassium salt solutions are. The anion present will determine the ultimate degree of acidity; depending on its base strength, it will to some degree counteract the hydronium ion generated by reaction of the ammonium ion with water:

$$NH_4^+ + H_2O = NH_3 + H_3O^+$$

If the anion is a strong base, e.g., acetate, it will generate hydroxyl ion by reaction with water:

$$OAc^- + H_2O = HOAc + OH^-$$

Since the ion product of water cannot exceed 10^{-14}, H_3O^+ and OH^- in excess of the ion product must react. Consequently, the pH of such a solution will approach 7.[13] If, on the other hand, the anion of the ammonium salt is a weak base, e.g., chloride, virtually no hydroxyl ion is generated. The hydronium ion now predominates, and the solution has an acid pH of about 5.0.

Amine salts. The mono-, di-, and trisubstituted ammonium ions are the so-called amine salts. Good water solubility is the property of interest to the pharmacist and pharmaceutical chemist. The free amine bases, except those of low molecular weight, are insoluble in water. Conversion

[13] Ammonium acetate solutions will have a pH of 7 only with fresh solutions prepared from fresh, pure salt. If an old, partly decomposed salt, is used, the solution will have an acid pH.

to the amine salt is usually straightforward and makes the amine available in a soluble form (page 224).

The partially substituted ammonium cations have one or more nitrogen-coordinated hydrogen atoms in their structure. One is available for acid-base reactions, making these cations, but *not* the tetrasubstituted cation—a weak acid by the Brønsted definitions. The respective conjugate base is the corresponding free amine. This acidic character of the amine salts is important with aqueous solutions, since partial neutralization of the amine cation with solvent water molecules always results in an acid solution. All amine salts decompose in basic media:

$$C_6H_5NH_3^+ + OH^- = C_6H_5NH_2\downarrow + H_2O$$

Quaternary ammonium salts. The tetrasubstituted ammonium ion is of particular interest in pharmacy for several reasons, With all hydrogens replaced from its structure, it has a number of properties which differ from those of ammonium ion and its three partially substituted derivatives. It has no available proton; the tetrasubstituted ion is incapable of acting as an acid according to the Brønsted, Franklin, or Lewis definitions. The ability to act as an acid is peculiar to the ammonium ion and its partially substituted derivatives only.

The properties of the tetrasubstituted ammonium ion approach those of the larger alkali metal cations very closely. The common salts are strong electrolytes and are very soluble in aqueous media. Having nearly a 100% ionic character, the hydroxides, e.g., $[(CH_3)_4N]^+(OH)^-$, yield some of the strongest aqueous bases known. This strong basic property is not due to the tetrasubstituted ammonium ion, but to the accompanying ionic hydroxyl ions.

The tetrasubstituted ammonium ions are cationic surface active agents. One aspect of this property is a strong germicidal action. Quaternary ammonium compounds such as benzalkonium chloride are used in pharmacy for both properties.

PRACTICAL SOLUBILITY RULES

Although theoretical interpretation of solubility behavior provides a valuable understanding of the fundamental mechanisms involved, a descriptive knowledge is of equal importance for practical work. Ideally each solubility problem should be treated in terms of the ionic radius, valence shell type, etc. However, the time element in practical work does not allow for an introspective perusal of all factors. In addition, the values of many important parameters are unknown or not accurately known. In the dispensing laboratory, in the control laboratory, or in the research laboratory solubility problems must be anticipated and, if present, corrected immediately. A good back ground in qualitative analysis is valuable here.

In the following discussion exact solubility data will not be given, since they are available in the compendia and chemical handbooks. Rather, predictive solubility rules will be reviewed. For convenience the discussion will be divided into two categories based on the ions present in a given salt: (1) Cations and anions either of which if present in a salt usually render the salt soluble. (2) Cation and anion combinations whose presence in a salt will usually render it insoluble. Conditions which affect normal solubility behavior will also be discussed.

Soluble compounds. All common salts of the alkali metal cations are water soluble. Exceptions, given in Table 8.5, are not of direct importance to pharmaceutical practice. The solubility behavior of lithium salts more closely approximates that of the calcium ion than the alkali cation group. Ammonium salts, substituted ammonium salts (amine salts), and quaternary ammonium salts are also water soluble.

Nitrates, nitrites, acetates, and chlorates are usually soluble, as are most gluconates, lactates, and sulfonates. Exceptions are found among the acetates. Silver(I) and mercury(I) acetates are considered soluble but actually have solubilities approximating lead(II) chloride, usually considered insoluble. This solubility is in the order of a 1 % solution at 25°C., which is adequate for most pharmaceutical purposes. Acetates of high-valenced cations, e.g., aluminum, hydrolyze in solution and may precipitate as basic acetates. Some complex ions in which nitrite ion acts as a ligand are insoluble. Some basic nitrates, e.g., bismuthyl, are insoluble.

Sulfates, sulfites, and thiosulfates are usually soluble except for their alkaline earth and lead(II) salts. The order of solubility for a given alkaline earth cation is the order just cited, the sulfate always being least soluble. Within the alkaline earth family the solubility of the salts of any one of these anions decreases with increasing atomic weight of the cation: e.g., magnesium sulfate is soluble; barium sulfate is insoluble. The hypophosphites of the alkaline metal earths are only moderately soluble.

Halide (except fluoride) and pseudohalide salts are usually soluble. The insoluble exceptions are those of the monovalent coinage cations and mercury(I). Fluoride ion, because of its unusually small ionic radius does not follow the group solubility behavior. The solubility of its salts approximates that of the corresponding sulfate. Some apparently anomalous solubility behavior of the halides and pseudohalides results from their excellent ligand properties. Their insoluble simple metallic salts will usually dissolve in an excess of the anion, forming a soluble complex. The following equations illustrate this behavior:

$$Hg^{2+} + 2I^- = HgI_2 \text{ (insoluble)}$$
$$HgI_2 + 2I^- = [HgI_4]^{2-} \text{ (soluble)}$$

Oxy halide salts are insoluble, e.g., antimonyl chloride.

The acid salt(s) corresponding to an insoluble normal salt will be at least slightly soluble. Calcium carbonate is insoluble, but calcium bicarbonate is somewhat soluble (e.g., as temporary hardness of water). Determination of lead(II) as a sulfate can be complicated by the comparative solubility of the bisulfate.

Most common hydrates are soluble. Ammines and most large complex ions tend to be water soluble.

Insoluble compounds. Except for those of the alkali metals, salts of anions not discussed in the foregoing section are usually insoluble. The more common of these anions follow.

The oxides and hydroxides are insoluble in water. However, in addition to the soluble alkali compounds the barium compounds of these anions are soluble; the calcium and strontium compounds are slightly soluble (page 319). For the soluble oxides, the oxide ion, being a stronger base than the hydroxyl ion, dissolves as the corresponding hydroxide:

$$BaO + H_2O = Ba(OH)_2$$

The solubility of all other oxides and hydroxides is a function of the pH of the medium in which they are placed. Metallic oxides and hydroxides, e.g., iron(II) oxide, dissolve in acid media; nonmetallic, e.g., chromium(VI) oxide, dissolve in basic media; amphoteric, e.g., zinc oxide, dissolve in both acidic and basic media. Basic salts tend to be insoluble. The hydroxy and oxy bismuth compounds are good illustrations.

The normal phosphates, arsenates, carbonates, silicates, borates, and hypochlorites are insoluble (except for their alkali salts). However, each of these anions is a strong Brønsted base. They will, therefore, dissolve if in the presence of acids having a weaker conjugate base, e.g., H_3O^+.

$$FePO_4\downarrow + 2H_3O^+ + 4H_2O = [Fe(H_2O)_6]^{3+} + H_2PO_4^-$$

The resultant acid anions are more soluble than the corresponding normal salt.

The sulfides are insoluble except for those of the alkali and alkaline earth cations. The latter, however, are hydrolyzed in aqueous media.

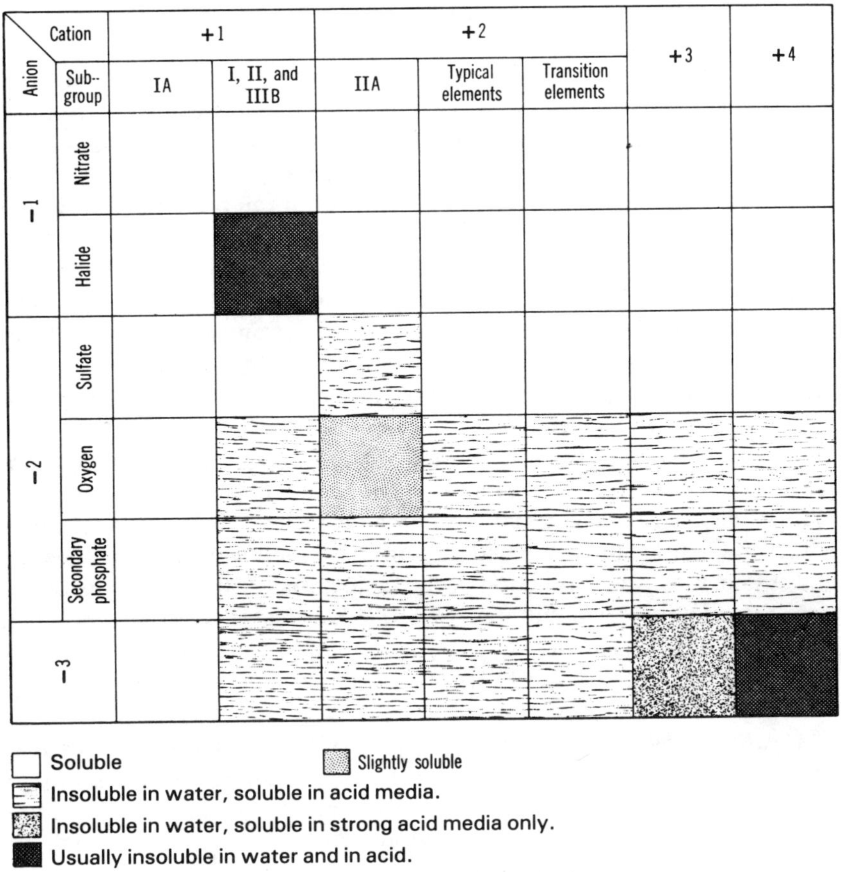

FIGURE 8.9 Solubility trends in electrolytes.

Group IA: alkali cations and ammonium ion, quaternary ammonium ions, and most protonated amines.

Nitrate subgroup: nitrate, nitrite, chlorate, acetate, gluconate, lactate.

Halide subgroup: chloride, bromide, iodide, cyanide, cyanate, thiocyanate, sulfide, sulfhydryl, salicylate.

Sulfate subgroup: sulfate, sulfite, thiosulfate, chromate, hypophosphite, oxalate, tartrate, fluoride.

Simple oxygen anions: oxide, peroxide, hydroxide, basic salts.

Secondary phosphate subgroup: secondary phosphate and arsenate, arsenite, carbonate, borate, hypochlorite, citrate.

Phosphate subgroup: tertiary phosphate, tertiary arsenate.

The sulfides of iron(III), chromium(III), and aluminum(III) are soluble, but undergo reaction with water to precipitate the corresponding insoluble hydroxides. (Iron(III) simultaneously reduces to iron(II).)

Except for those of the alkali metals (and, infrequently, calcium) salts of most organic acids are insoluble. However, salts of carboxy acids of low molecular weight are generally soluble. Most salts of sulfonic acid are soluble. Most heavy metal citrates and tartrates are insoluble, but the picture is confused by the tendency of these two acids and their anions to form soluble complexes with the same metals.

All heavy metals give insoluble substances in the presence of proteins and proteinates (page 464).

Although it is impossible to display the solubility of electrolytes in a comprehensive, definitive way, solubility behavior is sufficiently systematic to allow construction of a solubility chart based on the solubility trends of specific combinations of cations and anions. One such chart has been attempted in Fig. 8.9. This chart is not intended as a substitute for accurate specific knowledge of the solubility behavior of salts. It serves as a general guide only because of the inaccuracies inherent in the simplification procedure. Examples of such inaccuracies are: a single member of a family of salts may deviate from the norm, e.g., silver fluoride, which is very soluble; solubility behavior within a family can vary significantly and systematically as a function of the ionic radius of the cations involved, e.g., the alkaline earth sulfates (page 319); some salts, though soluble, decompose in certain media, e.g., aluminum acetate in water and all carbonates in acid media. Further examples are given in the following section.

Apparent solubility and insolubility. In practice, solubility rules may prove misleading or even valueless unless the variables affecting solubility are considered. The solubility rules given are for "room" temperature. The solubility behavior of a given compound may be considerably different at freezing and boiling temperatures, depending upon its temperature coefficient of solubility (see Fig. 8.2).

The effect of pH is probably the most frequently overlooked or neglected variable in solubility work. It is easy to assume the neutrality of a solution, especially by the inexperienced. Most aqueous solutions are acidic or alkaline in reaction because of interaction of the cations and/or anions with water or one of its conjugates. Consider the hypothetical salt, $[M(H_2O)_6]^{3+}X_3$; the reactions possible are

$$[M(H_2O)_6]^{3+} + H_2O = [M(H_2O)_5(OH)]^{2+} + H_3O^+$$

etc., and

$$X^- + H_2O = HX + OH^-$$

Since oxonium ion and hydroxyl ion concentrations in an aqueous system must satisfy the condition

$$[H_3O^+][OH^-] = 1 \times 10^{-14}$$

the predominant ion of the two is determined by the comparative acid-base strength of the conjugate pairs involved. If the hydrated cation is stronger as an acid than the anion is as a base, the oxonium ion predominates, $[H_3O^+] > 10^{-7}$, and an acidic pH is established. If the reverse is true the hydroxyl ion predominates, $[OH^-] > 10^{-7}$, and a basic pH results. If about equal, $[H_3O^+] \sim [OH^-]$, an approximately neutral solution is maintained. The resultant pH may radically change the course of a reaction, as illustrated by a solution of aluminum sulfate. Summing the three steps in the reaction of the hydrated aluminum cation as a single equilibrium,

$$[Al(H_2O)_6]^{3+} + 3H_2O \leftrightarrows [Al(H_2O)_3(OH)_3]\downarrow + 3H_3O^+$$

the effect of the resultant pH on the solubility behavior of the aluminum cation may be developed. If a concentrated solution is prepared, the concentration of oxonium ion builds up to give a pH of about 3. At this pH the aluminum is soluble as the hexa- and pentaquo cations; no precipitate forms. If, however, a very dilute solution is prepared, the concentration of oxonium ion cannot approach the above value because of the dilution effect of the large quantity of water present. The pH cannot fall below 6–7; the hydroxide precipitates.

Double salts. In solutions containing certain combinations of two simple salts neither salt will crystallize as long as both are present. A double salt—a new crystalline phase of constant composition—separates from the solution. Only after one of the simple salts has been completely crystallized as the double salt will the other simple salt crystallize.

The alums and schönites are examples of double salts. The alums have the general formula $M_2{}^{I}M_2{}^{III}(SO_4)_4 \cdot 24H_2O$, where M^{I} is a monovalent cation (e.g., Na^+, K^+, $NH_4{}^+$, Tl^+) and M^{III} is a trivalent cation (e.g., Al^{3+}, Cr^{3+}, Fe^{3+}, Tl^{3+}). Potassium alum, $K_2Al_2(SO_4)_4 \cdot 24H_2O$, is the prototype. The schönites (Tutton Salts) have the general formula $M_2{}^{I}M^{II}(SO_4)_2 \cdot 6H_2O$, where M^{I} is a monovalent cation, as above, and M^{II} is a divalent cation (Cu^{2+}, Mg^{2+}, Fe^{2+}, Co^{2+}, Ni^{2+}, Zn^{2+}). The ions of each respective charge must have approximately the same ionic radius; e.g., $M^{II} \sim 0.7$ Å, $M^{III} \sim 0.5$ Å.

The reason for the solubility behavior of double salt systems is evident from an examination of the solubility curves of the three substances involved. The curves for potassium alum and its parent salts, potassium

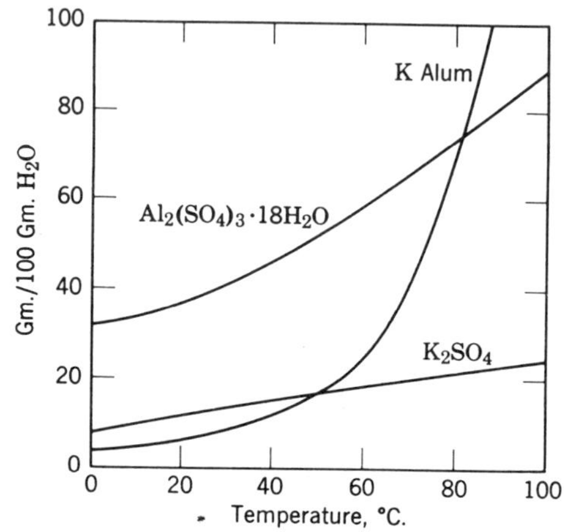

FIGURE 8.10 Crystallization of a double salt.

sulfate and aluminum sulfate, are shown in Fig. 8.10. At lower temperatures the alum is the least soluble substance present in a solution. It must therefore preferentially crystallize.

Double salts may be described as intermediate between solid solutions (page 382) and complex ions (page 63). They resemble solid solutions in that on solution their crystals give rise to the constituent, free, independent ions. They are unlike solid solutions, and like complexes, in that they have constant composition and a crystal structure of their own. Double salts are unlike complexes in that the structural units of the crystal do not persist in solution. This comparison is summarized in Table 8.7.

SOLUBILITY CONTROL

One of the problems often facing a pharmaceutical chemist is the limited solubility of a therapeutic agent. A useful compound may be

TABLE 8.7 COMPARISON OF THE THREE TYPES OF CRYSTALLINE SOLIDS

	Solid Solution	Double Salt	Complex Ion
Composition ratios	Ratio of precursor ions can vary	Constant composition	Constant composition
Crystal structure	That of the component compounds	Unique	Unique
Restrictions	Component salts must be isomorphous	—	—
Behavior on solution	Yields precursor ions	Yields precursor ions	Complex ions persist

TABLE 8.8 SOLUBILITY IMPROVEMENT OF MEDICINALS[a]

Insoluble Compound				Soluble Product		
Name	Functional Group	Solubility	Suitable Reagent	Solubilizing Groups	Solubility	Name
Thiamine	$—NH_2$	Insoluble	HNO_3	$[—NH_3]^+NO_3^-$	35	Thiamine nitrate
Epinephrine	$—NH(CH_3)$	3300	$H[HC_4H_4O_6]$	$[—NH_2(CH_3)]^+ [HC_4H_4O_6]^-$	3	Epinephrine bitartrate
Tetracycline	$—N(CH_3)_2$	2500	HCl	$[—NH(CH_3)_2]^+Cl^-$	10	Tetracycline hydrochloride
Tribasic calcium phosphate	PO_4^{3-}	40,000	HCl	$[HPO_4]^{2-}$	5000	Dibasic calcium phosphate
Salicylic acid	—COOH	460	NaOH	$Na^+[—COO^-]$	10	Sodium salicylate
Cyclohexane-sulfamic acid	$—NHSO_3H$	Insoluble	$Ca(OH)_2$	$Ca^{2+}[—NHSO_3^-]_2$	4	Calcium cyclamate
Sulfadiazine	$—SO_2NH$	13,000	NaOH	$Na^+[—NSO_2—]^-$	2	Sodium sulfadiazine
Iodophthalein	—OH and —COOH	Insoluble	NaOH	$Na^+[—O]^-$ / $Na^+[OOC—]^-$	7	Sodium iodophthalein

[a] Solubilities are expressed as milliliters of water required to dissolve one gram of substance.

insoluble or have so limited a solubility that the volume of aqueous solution necessary for an effective dose becomes unreasonably large. A knowledge of standard methods for improvement of solubility is valuable and may avoid the necessity of resorting to a nonaqueous media.

Role of functional groups. A survey of the solubility rules developed in the preceding section reveals that a salt which has as one of its ions an alkali metal cation, the ammonium ion or one of its substituted derivatives, or a protonated anion (an acid salt) is invariably soluble. The chemistry of solubility improvement therefore centers about the conversion of the insoluble or slightly soluble compound into a salt which includes one of these ions. A knowledge of the functional groups is essential, and a good working knowledge of the Brønsted theory of neutralization is required since the suitable functional groups are invariably acidic (e.g., —COOH) or basic (e.g., PO_4^{3-}, —NH_2).

Neutralization techniques. Basic groups present in insoluble compounds include primary, secondary, and tertiary amines in addition to all normal anions of insoluble salts. Any acid sufficiently strong will render these bases soluble as a substituted ammonium salt (amine salt) for the former, and as the acid salt, for the latter. To obviate problems posed by the relative acid strength of the substance involved, one of the strong mineral acids (hydrochloric, nitric, sulfuric, phosphoric, or hydrobromic) is used for the neutralization. However, a few organic acids (citric, maleic, or tartaric) may be used with some bases. Since the insoluble medicinal bases involved are invariably relatively strong (i.e., their conjugate acids are weak acids), a strong acid will neutralize all. Typical general reactions of this type are

$$RNH_2 + H_3O^+Cl^- = RNH_3^+Cl^- + H_2O$$

$$R_1R_2NH + H_3O^+Br^- = R_1R_2NH_2^+Br^- + H_2O$$

$$BiO^+ + xsH_3O^+Cl^- = Bi^{3+} + xsH_3O^+ + H_2O + Cl^-$$

$$Ca_3(PO_4)_2 + 2H_3O^+NO_3^- = 2Ca^{2+}HPO_4^{2-} + Ca^{2+}(NO_3^-)_2 + 2H_2O$$

Some medicinals rendered soluble by neutralization procedures are given in Table 8.8.

Insoluble substances with acidic functional groups may be converted into soluble sodium, potassium, or calcium salts by a suitable strong base. Because of the acidic character of the ammonium ion, ammonium salts though soluble are usually unsatisfactory for solubility improvement. The acidic functional groups include the carboxy, phenol, amide, and sulfonic acid groups as well as the insoluble inorganic acids. Since sodium salts are most popular, sodium hydroxide, sodium carbonate, or sodium bicarbonate usually serve as the source of base ion. The relative advantages and disadvantages in the use of these three bases have been discussed (page 164). If the potassium salt is desired, the corresponding potassium salts may be used. Calcium hydroxide is generally the only suitable calcium base. Typical neutralizations illustrating these reactions are

$$RSO_3NH_2 + Na^+OH^- = RSO_3NH^-Na^+ + H_2O$$
$$2PhOH + Na^+{}_2CO_3{}^{2-} = 2PhO^-Na^+ + H_2O + CO_2\uparrow$$
$$RCOOH + Na^+HCO_3{}^- = RCOO^-Na^+ + H_2O + CO_2\uparrow$$

Some medicinals rendered soluble by conversion to soluble alkali metal salts are shown in Table 8.8.

Preparation of soluble salts of insoluble compounds is not without problems, however. These may be chemical, pharmaceutical, or medicinal and will vary from substance to substance.

Since the substituted ammonium cations are weak acids (the accompanying anion is invariably that of a strong acid) and the alkali cation salts have anions derived from weak acids (except sulfonate), solutions of these compounds are not neutral. On hydrolysis, the former will be acidic and the latter basic. The actual equilibrium pH will be dependent upon the pK_a of the acids involved. If the pH is extremely acidic or basic or even moderately so, discomfort, irritation, and corrosion in extreme cases will accompany administration. For this reason buffering may be essential. But buffering at an improper pH can initiate reprecipitation of the insoluble form of the compound involved. If the pH of a solution including a substituted ammonium derivative is shifted too far toward the basic side, the insoluble amine will begin to precipitate. Similarly, the usual sodium salts of the weak acids cannot tolerate a shift toward acidic pH values without formation and precipitation of the original insoluble acid. Typical general equations for these incompatibilities may be written:

$$RNH_3{}^+Cl^- + \boldsymbol{B}^- = RNH_2\downarrow + HB + Cl^-$$
$$Na^+OOCR^- + \boldsymbol{HB} = HOOCR\downarrow + B^- + Na^+\cdot$$

HB and B^- have the usual significance.

An excellent example to illustrate the chemistry and reasoning in these procedures is found in the phenolsulfonphthalein molecule. The free acid is soluble, 1 Gm. in 1300 parts water—a solubility requiring unworkable volumes for introduction of the usual dose via injection. The sodium salt is freely soluble. Phenolsulfonphthalien Injection *U.S.P.*

XXI[14] names sodium hydroxide and sodium bicarbonate as suitable neutralizing reagents. Sodium carbonate, another suitable base, is not named in the monograph, but will be included in this discussion.

TABLE 8.9 pK_a OF ACIDS INVOLVED IN PHENOLSULFONPHTHALEIN NEUTRALIZATION

Acid	pK_a
Oxonium ion	−1.74
Phenolsulfonphthalein	7.9
Carbonic acid	6.4
Bicarbonate ion	10.2
Water	16.

The pK_a values for each possible acid-base conjugate pair in this neutralization are found in Table 8.9. Equations for each possible reaction may be written

$$H[Ps] + OH^- \rightarrow HOH + Ps^- \quad (1)$$

$$H[Ps] + CO_3^{2-} \rightarrow HCO_3^- + Ps^- \quad (2)$$

$$H[Ps] + HCO_3^- \rightleftharpoons H_2CO_3 + Ps^- \quad (3)$$

$$H[Ps] + H_2O \leftrightharpoons H_3O^+ + Ps^- \quad (4)$$

$$H_2CO_3 \rightarrow H_2O + CO_2 \quad (5)$$

H[Ps] represents the unneutralized phenolsulfonphthalein molecule, Ps^- its anion. The hydrated sodium ion is too weak as an acid to have any influence on the course of the reaction. Having favorable base strengths, sodium hydroxide and sodium carbonate can react rapidly and almost completely with the H[Ps] molecule, as in Equations 1 and 2. The solubility of the H[Ps] in sodium bicarbonate is not as obvious, since the pK_a values for H[Ps] and carbonic acid seem to preclude reaction between the two. When acid strengths are nearly equal, an equilibrium is set up between the two conjugate pairs (Equation 3). The forward reaction proceeds until the equilibrium concentration of each component is achieved. The net result is a reduction of the quantity of solid H[Ps] to levels allowing its solution. Both the acid and its anion are present in the finished solution when bicarbonate is used as the base.[15] Hydronium ion is too strong an acid to form by direct action between water and H[Ps], as in Equation 4.

[14] *U.S.P. XXI*, p. 821.

[15] The *U.S.P. XXI* specifies a pH of 6.0-7.5. This means that even when sodium hydroxide (or sodium carbonate) is used as the neutralizing agent, some of the acid form is present in the solution.

The pharmaceutical chemist must use restraint in a zeal to render all medicinal substances soluble. There are exceptions to the general rules of solubility; proclorperazine ethanedisulfonate[16] is soluble, 1 Gm. in 1 ml. of water, whereas the corresponding maleate is virtually insoluble in water. It is not always advantageous to increase solubility of a compound. An important attribute of an insoluble compound in contact with an aqueous medium is an automatic control of the quantity of medicant in actual solution. With corrosive or poisonous substances, it is preferable to use an insoluble form if available and sufficiently soluble to maintain an effective concentration. The manipulations necessary to produce a solution of proper concentration (with its inherent chance of error) are eliminated. Hexylresorcinol, because of its phenolic functional group, can be rendered more soluble by conversion to its sodium salt. But the solubility of the compound is such—1 Gm. dissolves in 2000 ml. water—that when used as an anthelmintic an effective concentration is maintained by use of the phenol itself; increased solubility is disadvantageous. A similar situation exists with tribasic calcium phosphate. Neutralization to give the dibasic salt increases solubility by a factor of 10, but the effective antacid activity is reduced by a factor of 10^5, the difference in K_a of the two phosphate ions.

Physiological and pharmacological factors must also be considered. It is essential that the desired medicinal action remain unchanged, and the activity remain constant or nearly so. The change in structure resulting in conversion to the ionic form may bring about a loss of or change in the action of the drug. The calcium phosphates cited in the previous paragraph illustrate a decrease in activity by conversion to a more soluble form.

The physiological activity of the introduced cation or anion must be minimal; hence the popularity of the sodium ion or chloride, dihydrogen-monohydrogen phosphate, and certain organic anions.

Complexation techniques. Complexation can at times be useful for increasing the concentration of an inorganic ion in aqueous solution. Soluble Ferric Phosphate is actually a misnomer for the preparation it names. Insoluble ferric phosphate dissolves because of the chelating action of the added citrate ion for iron(III) ion. The actual concentration of free iron(III) in the presence of citrate is less than the concentration of that ion allowed by the solubility product of the insoluble phosphate. Citrate, tartrate, and versenate[17] are the most popular ligands

[16] Patented.

[17] The disodium monocalcium ethylenediaminetetraacetate is used in pharmaceutical practice. The tetrasodium salt is not used because of the strong complexing action for calcium ion. Its use can lead to a depletion of free calcium ion.

for this type of solubility control. Glycerine, sucrose, salicylate, and 8-hydroxyquinoline are also used.

IMPORTANT TERMS. Define or identify in a single concise sentence if possible.

Alkali elements, alkaline earth elements, bridge elements, colligative property, common ion effect, dielectric constant, freezing point, halide, heat of hydration, hemolysis, hydration, hypertonic, hypotonic, ion-pair, iso-osmotic solution, isotonic solution, lattice energy, molal freezing depression constant, molal solution, mole fraction, osmotic pressure, tonicity, plasmolysis, pseudohalide, quaternary nitrogen, Raoult's law, salting out effect, saturated solution, solubility curve, solvation, temperature coefficient of solubility, vapor pressure, w/w solution.

NOMENCLATURE OF INTEREST. Identify each of the following.

Ammonium hexachlorplatinate(III), dimethyl diethyl ammonium iodide, hexaamminecobalt(II) bromide, hydrogen dicyanoargentate(I), potassium fluosilicate, potassium hexacyanoferrate(II), potassium monosodium cobaltinitrite, sodium fluoborate, sodium pyroantimonate, sodium tetrachloromercurate(II), tetramethyl ammonium hydroxide.

SELF-APPRAISAL QUESTIONS

1. Identify the chemical and physical properties which make sodium ion valuable in pharmaceutical practice. Do the same for chloride ion.

2. Lithium ion has some properties similar to those of sodium ion and some similar to calcium ion. Identify the similarities in each case. Explain.

3. Identify the similarities and differences of ammonium ion as compared to potassium ion.

4. Identify the characteristic group properties of the alkali metals and their cations. How is the electronic structure involved? How are atomic and ionic radii involved?

5. Examination of radius data (Table 8.4) reveals that the hydrated ionic radius of lithium ion is completely out of line from the other members of the family. Explain this extreme deviation. The standard electrode potential of lithium is also out of step with the group values. Is this related to the hydrated radius deviation? Explain.

6. Explain the popularity of the sodium salts of desired anions over the corresponding potassium salts on the basis of (*a*) chemistry, (*b*) physiology or biochemistry, (*c*) economics.

7. Indicate those of the following operations which would find the potassium reagent more efficacious than the corresponding sodium reagent.

Explain fully. (*a*) The oxide used as a carbon dioxide absorber for respired air, (*b*) standard hydroxide solution for neutralization of a solution to be administered as an injection, (*c*) permanganate tablets to be dissolved in water as a germical solution, (*d*) titration of an amine hydrochloride in ethanol solution.

8. Explain on a theoretical basis why the hydrated sodium ion is such a weak acid.

9. Explain on a structural basis why degree of hydration of the alkali cations is inversely related to their atomic weight.

10. Describe the possible chemical incompatibilities which can arise when a solution of a given sodium salt is used as a medium for a medicinal.

11. Write the electronic structure of the ammonia molecule. Indicate the various sites at which reactions can take place. Indicate the resultant products for each reaction type.

12. Differentiate between amide, amine, ammine, nitride. Give an example of each.

13. How are the chemical and physical properties of ammonium salts, primary, secondary, and tertiary amine salts, and quaternary ammonium salts similar? How do they differ?

14. Why is chloride usually the pharmaceutically preferred form of an amine salt?

15. Identify one each of mono-, di-, tri-, and tetrasubstituted ammonium salts in the current compendia.

16. Name the variables which must be specified when giving the concentration of a saturated solution of a given salt.

17. You wish to prepare a saturated solution of sodium chloride as quickly as possible. Outline a suitable procedure and give the basic reasons for each step. (Finished solution temperature, 25°C.) How would you change the procedure if the salt were calcium hydroxide?

18. How are polarizability, dipole moment, and dielectric constant related?

19. Name three structural requirements for a molecule if it is to be a good solvent for electrolytes. Does water meet these requirements? Explain.

20. A hypothetical solute exists as the deca-, hexa-, and dihydrates and as the anhydrous salt. How should the water solubilities compare, expressed as Gm. anhydrous salt per 100 Gm. water? Assume a molecular weight of 100 for the anhydrous salt.

21. Explain on the basis of the properties and structure of the ions involved why you would expect $RaSO_4$ to be highly insoluble, but FrCl to be very soluble.

22. Given the following uncommon compounds, predict those which should be at least moderately soluble in water. Why? Cs_3PO_4, SrF_2, $CaCl_2$, TiS_2, $Mn(ClO_3)_2$, $Hg_2(CNO)_2$, $TcBr_2$, V_2O_5, $Mn(OH)_2$, RbClO, $Mg(HSO_3)_2$, $Mo[B(OH)_4]_2$, $[Co(NH_3)_6](OH)_2$, $K_2Na[Co(NO_2)_6]$, $NiSO_4$, WBr_2, $Ni(H_2PO_4)_2$, $Zn(CN)_2$, $Ag_3[Fe(CN)_6]$, $SrPO_4$.

23. Referring to the compounds listed in question 22, indicate those which should be insoluble in water but soluble in strong acid media.

24. You have a suspension of excess calcium sulfate in water. How will the concentration of dissolved calcium sulfate (Gm./100 ml. solution) change if a fresh sample of the suspension is treated with each of the following? (*a*) Ethanol is added to give a 50% ethanol content; (*b*) enough calcium chloride is added to make the solution 0.05*M* in calcium chloride; (*c*) sodium chloride is added, making the solution 0.1*M* in sodium chloride; (*d*) the quantity of suspended calcium sulfate is doubled; (*e*) the quantity of water is doubled; (*f*) silver nitrate is added to make the solution 0.05*M* in nitrate ion.

25. Distinguish between solid solution, double salt, and complex salt.

26. Why do the common complex ions tend to be soluble in water?

27. You are to prepare if possible 0.5*M* aqueous solutions of each of the following: $Na_2CO_3{\cdot}H_2O$, K_3PO_4, CaO, $FePO_4$, $NaHCO_3$, $Al(OAc)_3$, $[(C_2H_5)_4N^+](OH^-)$, NH_3, $C_6H_{13}NH_2$, $Ca(HSO_4)_2$, $BaSO_4$, $FeCl_3{\cdot}6H_2O$, AgI, CaF_2, Li_2CO_3, $KHSO_3$, $(NH_4^+)(OH^-)$, $(C_2H_5)_2NH_2^+Cl^-$. Classify the solubility of each compound on the basis of Table 8.1. Indicate those solutions which cannot be prepared and explain why. Classify each 0.5*M* solution or saturated solution as to its pH, using the following classification: pH 1–4, acid; 4–6.5, weak acid; 6.5–7.4, neutral; 7.5–9, weakly basic; 9–12, basic; >12 strongly basic.

28. Distinguish between permeable and semipermeable membranes. Are the cell walls truly semipermeable?

29. Distinguish between iso-osmotic and isotonic.

30. Name the colligative properties of solutions.

31. Each of the following when added to 1000 Gm. water in the quantity given should theoretically give a solution having a freezing point of $-1.86°$C.: (*a*) ethanol, 46.1 Gm.; (*b*) sodium chloride, 29.2 Gm.; (*c*) K_2SO_4, 58.1 Gm. The actual freezing points, in this order, are -1.82, -1.68, $-1.34°$C. Explain.

32. Given the prescriptions:

a.Rx			b.Rx		
	Arg. Nitrate	1%		Procaine hydrochloride	1.0 Gm.
	Sod. Nitrate qs			Chlorbutanol	0.5 Gm.
	Aq. Dist. qs ad	30 ml.		Sodium chloride qs	
	M. Ft. isotonic solution for eyes			Purified water qs ad	100 ml.
				M. Prepare isotonic solution	

Perform the necessary·calculations for each, using freezing point depressions. Classify each finished solution as isotonic or iso-osmotic (a/0.246g.)

PRACTICE IN EQUATION WRITING

33. Show why the pH of freshly prepared ammonium acetate solution should be approximately neutral. Show why a solution of this salt becomes acid on long standing.

34. Show the reaction taking place on dissolving aluminum sulfide.

35. Write the reactions taking place during the following qualitative tests: (*a*) uranyl acetate test for sodium ion, (*b*) cobaltinitrite test for potassium ion, (*c*) a series of tests to prove an amine salt is a chloride rather than a bromide.

36. Show equations for the reactions necessary to prove qualitatively that you have $NaHCO_3$ rather than $Na_2CO_3{\cdot}H_2O$.

37. Show why a suspension of calcium carbonate will dissolve if carbon dioxide is bubbled through the suspension for some time.

38. Refer to question 22. Some of the compounds listed probably dissolve but undergo a reaction with water or its ions. Write equations for each case. Write equations for the reactions described in question 23.

39. Write the equations for the reactions by which you judged the pH of the aqueous solutions of question 27.

40. A yellow precipitate formed on addition of a drop of sodium hexachlorplatinate(III) solution to a drop of a solution of a medicinal salt whose cation did not precipitate in any of the qualitative cation groups. Write equation(s) by which it is possible to prove the identity of this cation.

NINE · PRESERVATION, STABILIZATION, AND STERILIZATION OF PHARMACEUTICAL PREPARATIONS

Chemical reactions are generally planned and considered as taking place during the preparative procedures, under the control and observation of the operative. But one aspect of reactions may sometimes be overlooked. Unless a substance or preparation exists in its lowest possible energy form for the environment in which it is placed, it tends to change into the more stable low-energy form, as evidenced by chemical or physical change or by bacterial or mold growth. Hence the worker must be ever aware of the nature of these unwanted reactions and must also be familiar with preventive procedures and counteractive measures.

Essentially three processes assure the ultimate suitability of a pharmaceutical product. Preservation includes operations or techniques intended to maintain a given product in the initial form—e.g., stable storage conditions for ammonium carbonate, inclusion of sulfur dioxide to maintain the sterility of a preparation. Stabilization includes operations intended to prevent or counteract undesirable physical or chemical changes—e.g., inclusion of hypophosphorous acid in hydriodic acid. Sterilization includes those physical and chemical procedures intended to produce products free of contamination by microorganisms. The concern of the pharmaceutical chemist with sterilization arises from the possibility of chemical and physical instabilities during the sterilization period. Although the chapter title lists the three procedures separately, the interrelations among them make it inexpedient to attempt a separate discussion of each.

STABILITY PROBLEMS ASSOCIATED WITH ATMOSPHERIC EXPOSURE

Moisture problems and their control. Living in a sea of air, man ordinarily considers the atmosphere chemically and physically inert. Except for nitrogen and the inert gases, however, this is not true. Given favorable environments the water vapor, oxygen, and carbon dioxide present give rise to a wide range of undesirable reactions. In addition, air-borne microbes and spores are a constant threat to sterility. The prevention or control of these problems is an important aspect of pharmaceutical practice.

Moisture in the air affects pharmaceuticals in several ways: stoichiometry, deliquescence, problems of hydrolysis, and other chemical reactions. The actual extent and rate of reaction for a given pharmaceutical will be mediated by the relative humidity at the time of exposure, an unpredictable variable which necessitates assumption of maximum humidity when planning counteractive measures.

A less evident and therefore more insidious effect of moisture exposure is a change in stoichiometry. Most crystalline inorganic pharmaceuticals are hydrates, the most stable of which is the common commercial form (e.g., the monohydrate of sodium carbonate, the pentahydrate of copper(II) sulfate). If efflorescent, hydrate water may be lost; if hygroscopic, water may be absorbed to give a higher hydrate. The following equations illustrate the possible reactions:

$$[Ca(H_2O)_2]^{2+}(Cl^-)_2 + 4H_2O \rightarrow [Ca(H_2O)_6]^{2+}(Cl^-)_2 \quad \text{(hygroscopic)}$$

$$[Mg(H_2O)_6]^{2+}[SO_4{\cdot}H_2O]^{2-} \rightarrow [Mg(H_2O)_2]^{2+} + [SO_4{\cdot}H_2O]^{2-} + 4H_2O \quad \text{(efflorescent)}$$

Chemical properties are ordinarily unaffected by change in hydration state.

The decrease or increase in water of hydration is a function of the relationship existing between the aqueous vapor tension of the crystal and the partial water vapor tension (humidity) of the atmosphere. The relationship may be expressed as follows:

$$p_{\text{water}}^{\text{hydrate}} > p_{\text{water}}^{\text{air}} \quad \text{efflorescent crystal}$$

$$p_{\text{water}}^{\text{hydrate}} = p_{\text{water}}^{\text{air}} \quad \text{no change}$$

$$p_{\text{water}}^{\text{hydrate}} < p_{\text{water}}^{\text{air}} \quad \text{hygroscopic crystal}$$

The aqueous vapor tension of a hydrate is directly dependent on the temperature of the crystal.

Although many hydrates are characterized by strong efflorescent properties, all hydrates will lose hydrate water depending on the temperature and relative humidity. Depending on the extent of water loss, a sample weighed out on the basis of the formula given on the label must have a greater than calculated number of equivalents. Consider the usual 15-Gm. dose of Epsom Salt; if the sample taken had lost 50% of its water of hydration, the patient actually received a dose equivalent to 20 Gm. of the fully hydrated salt. This is a 33% overdose.

Dehydration can usually be detected visually. Loss of hydrate water results in a change in crystal structure, the surface taking on a chalky or powdery appearance. With extreme efflorescence the powder may separate and accumulate in the bottom of the container. Epsom Salt is an excellent example; the fully hydrated crystals are transparent, colorless spicules. On exposure to dry air the clear crystalline appearance is rapidly lost, and the crystals gradually become translucent and finally powdery in appearance. A change in color may also accompany dehydration. The beautiful blue crystals associated with copper(II) sulfate exist only in the pentahydrate form. The color is lost on exposure to dry air; exposed crystals have a white, powdery surface. Large quantities of the white, powdery material separate and accumulate in old, much exposed samples.

Unless a salt is already in its highest hydration state, moisture can be slowly absorbed to give increasing ratios of higher hydrates. Stoichiometric errors are opposite to those of efflorescent compounds; weighed samples must always have fewer equivalents than calculated on the basis of the label formula. The exsiccated form of common pharmaceuticals belong to this category, e.g.,

$$FeSO_4{\cdot}H_2O + 3H_2O = [Fe(H_2O)_3][SO_4{\cdot}H_2O]$$

The extreme case of moisture absorption is exhibited by deliquescent substances; the affinity for moisture is so high that the substance ultimately dissolves in its own water of hydration and becomes liquid. Many sodium and lithium salts exhibit this property, e.g., sodium hydroxide. The handling of deliquescent substances is difficult even in low humidities.

Some compounds undergo chemical decomposition if exposed to moisture. Surface hydrolysis can give rise to a coating of insoluble basic salts. Zinc chloride, for example, undergoes the following surface reactions:

$$[Zn(H_2O)_6]^{2+} + H_2O = [Zn(OH)(H_2O)_5]^+ + H_3O^+$$

$$2[Zn(OH)(H_2O)_5]^+ + CO_2 = [Zn(OH)]_2CO_3\downarrow + 2H_3O^+ + 7H_2O$$

$$[Zn(OH)(H_2O)_5]^+ + Cl^- = Zn(OH)Cl\downarrow + 5H_2O$$

A freshly prepared solution of this compound will appear cloudy as a result. Resolution of the insoluble basic salts may be attained by adding a small amount of the acid of the anion of the salt.

$$Zn(OH)Cl + [H_3O^+ + Cl^-] + 4H_2O = [Zn(H_2O)_6]^{2+} + 2Cl^-$$

$$[Zn(OH)]_2CO_3 + 4H_3O^+ + 2Cl^- + 5H_2O = 2[Zn(H_2O)_6]^{2+} + 2Cl^- + CO_2\uparrow$$

The resulting solution will be more acid than usual; an excess of the acid is required to effect complete solution.

Anhydrides will react with moisture of the air to give the corresponding acids or bases.

$$P_2O_5 + H_2O = 2HPO_3$$

$$CaO + H_2O = Ca(OH)_2$$

Although acid or base equivalents per molecule are unaffected, weight stoichiometry is changed.

Caking is a problem associated with hygroscopic substances. On absorption of moisture, particles tend to cement to one another. Many common substances are subject to caking, e.g., potassium chloride, sodium nitrate, ammonium chloride, potassium acetate. Calcium chloride hexahydrate will cement into a solid crystalline mass if exposed to moisture even briefly. Sodium carbonate also cakes, but a chemical reaction is responsible. There is a simultaneous absorption of carbon dioxide; caking of the crystals involves hydration with simultaneous formation of sodium bicarbonate at the contacting interfaces. Caking may also take place when substances with low melting points are subjected to temperatures at or slightly above their melting point. On partial melting and subsequent cooling, the crystals cement to one another. If completely melted, solidification will lead to a solid crystalline mass within the container.

Moisture control must ordinarily be restricted to closure control; only well-fitting, tight closures may be used. Small packets of drying agent, e.g., silica gel, may be inserted into the neck of bottles filled with hygroscopic substances in order to maintain a low humidity. Room humidity control, though effective, is economically unfeasible for everyday stockroom purposes. Only in a limited number of specialized manufacturing procedures is the cost warranted.

Control over the caking of crystals has made advances during the past decade or so. One of two methods is successfully employed. The older and still more common method is the addition of small percentages, 1–5%, of a substance which preferentially absorbs moisture. For pharmaceutical purposes this moisture absorber must be innocuous, must not impart an undesirable taste, and must be inert to all substances present except water. The most common anticaking compounds are magnesium oxide, magnesium carbonate, starch, various clays, diatomaceous earth, amorphous silica, and vermiculite. These compounds are effective by means of two mechanisms. First, chemical reaction may take place between the anticaking compound and the absorbed water, e.g.,

$$MgO + H_2O \rightarrow Mg(OH)_2$$

The requisite for this action is the formation of a substance with an extremely low aqueous tension. Second, the water may be physically adsorbed on the surfaces or absorbed within interstitial channels which exist in the absorber, e.g., amorphous silica. Again the adsorbed moisture must be held very tightly if the anticaking compounds are to be effective.

Recently coating techniques have been used for control of salt caking. Each crystal is coated with an otherwise inert material to a thickness sufficient to prevent access to moisture. High-molecular-weight amines and long-chain quaternary ammonium compounds have proven useful in this application.

Water is a catalytic agent for oxidations, oxygen being available in the atmosphere. Dry crystals of iron(II) sulfate are stable to oxygen of the air. However, in a moist atmosphere the salt is sufficiently hygroscopic to accumulate a water film on the surface of each exposed crystal. Oxygen dissolves in the moisture film and attacks the iron(II) ions, oxidizing them to iron(III). A surface film of a basic iron(III) sulfate slowly accumulates on iron(II) sulfate crystals after repeated exposure to moisture.

$$[Fe(H_2O)_6]^{2+}[SO_4{\cdot}H_2O]^{2-} + (aq) = [Fe(H_2O)_6]^{2+} + [SO_4{\cdot}H_2O]^{2-} \quad \text{(surface solution)}$$

$$[Fe(H_2O)_6]^{2+} + \tfrac{1}{2}O_2 + H_2O = [Fe(H_2O)_6]^{3+} + 2(OH)^- \cdots$$

On the other hand, arsenic(III) oxide is stable to oxygen even in moist atmospheres.

Oxidative problems and their control. Oxygen, being a fair oxidizing agent, presents storage problems for easily oxidized substances, especially in aqueous media. Arsenic(III) oxide, though resistant to oxidation by oxygen in the crystalline state, is oxidized quite readily to arsenic acid by dissolved oxygen in aqueous media. In addition to iron(II) compounds, iodides and, to a lesser degree, bromides are susceptible to air oxidation. Inert atmospheres and antoxidants are used to control air oxidation.

Nitrogen is usually employed as an inert atmosphere. The strength of the triple bond, 1-σ and 2-π, binding of the atoms makes it a very stable molecule. It is used as an inert gas over various oils (e.g., cod liver oil, castor oil, olive oil) during storage to retard oxidation and thus prevent rancidity and vitamin loss. Nitrogen replaces the air in containers for parenterals and other oxidation-prone solutions. It is also used as an inert atmosphere during synthesis and analysis.

Helium is an even less reactive inert atmosphere, but because of its high cost it is used only in special cases. Argon is a suitable substitute for helium, although it has not been employed to any great extent in the chemical or pharmaceutical industry.[1] Carbon dioxide may also be used as an inert atmosphere provided that no strong base is present.

In general pharmaceutical practice inert atmospheres are of value only in the unopened container. In a single-dose container the purpose is accomplished. But with multiple-dose containers or with bulk chemicals or solutions, oxygen enters the free space on opening. In the laboratory, elaborate apparatus may be designed to allow performance of entire synthetic or analytical procedures under a blanket of inert gas.

The use of antoxidants in pharmaceutical practice is a more practical procedure. An antoxidant compound must have two attributes. First, the antoxidant and its reaction products must be physiologically innocuous or present in such low concentrations that it is practicably innocuous. Second, it must be effective in reasonably low concentrations.

Antoxidants may act through competitive oxidation, i.e., the oxygen of the air preferentially reacts with the antoxidant. Sulfur dioxide and the sulfites act in this capacity. On the other hand, an antoxidant may not be directly oxidized by the oxygen, but may act effectively to reduce to its original form any of the oxygen-sensitive pharmaceutical which may have been oxidized. Thiosulfates and hypophosphorous acid and its salts act in this manner.

Sulfur dioxide and its family of salts are probably the most widely used antoxidants. The general effectiveness of sulfur(IV) compounds as antoxidants as well as preservatives accounts for this popularity. The compounds are reasonably stable in storage before use, and excellent analytical procedures are available for their determination. Although their chemistry is well known, unfortunately they are too often used on a strictly empirical rather than a sound scientific basis.

Sulfur dioxide and its derived compounds constitute a family with very diverse reactivity, being able to participate in almost all types of chemical reactions. In addition to acting as reducing agents, these compounds act as oxidizing agents, constitute a well-defined acid-base system, and complex readily as ligands. Except for the alkali salts, sulfites have limited

[1] Large volumes of argon are being used in the metallurgical industry for procedures in which helium is too expensive and nitrogen is unsuitable because of nitride formation. The gas is available as a by-product of various air liquefication processes, e.g., tonnage oxygen for steelmaking. At the close of 1962 the average cost per 100 cubic feet of helium and argon were \$5.00 and \$2.35, respectively.

solubility in neutral and alkaline media. Liquid sulfur dioxide is an excellent solvent for most electrolytes.

Together with the parent acid anhydride and sulfurous acid, the sulfur(IV) family includes the anions hydrogen sulfite (bisulfite), sulfite, and metabisulfite (pyrosulfite). Metabisulfite (page 522) is the actual solid used when a bisulfite salt is specified. The sodium salts are commonly used; the ammonium and potassium salts find occasional use. The calcium salt is rarely used because of poor solubility.

The antoxidant and preservative action of sulfur(IV) compounds is independent of the particular compound used. The choice of compound is made on the basis of the pH of the solution to be protected and the final concentration of antoxidant desired. The role of pH is best understood by considering the acid-base interrelations among the various members of the family. When the salts are used, the cation is without effect on these equilibria. The equilibria which exist in aqueous solutions of the sulfur(IV) system are diagrammed in Fig. 9.1. It is apparent from the equilibria and pK_a values that the sulfite ion is the anion of alkaline media. The bisulfite ion predominates under moderately acid conditions. At intermediate pH values the two anions coexist, the relative ratios being a function of the pH. The pH can be estimated by the equation

$$\mathrm{pH} = 7.2 + \log \frac{[SO_3{}^{2-}]}{[HSO_3{}^{-}]}$$

Sulfur dioxide molecules exist only in acid media of at least moderate strength; at higher pH it is completely converted to bisulfite and sulfite ions.

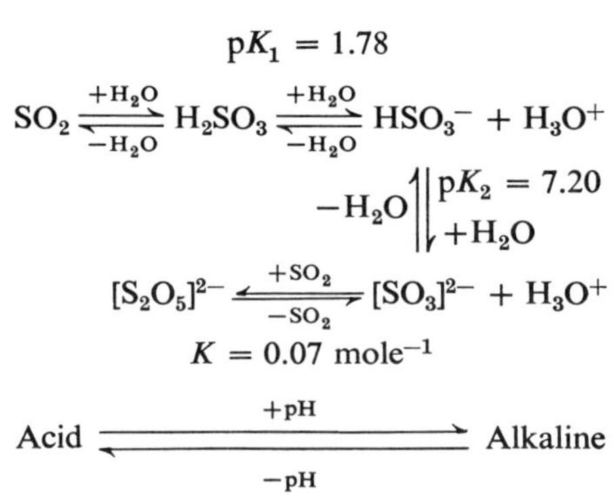

FIGURE 9.1 Equilibria in solutions of sulfur dioxide and its derivatives. The solubility of sulfur dioxide gas in water: 22.8, 4.5, and 0.58 Gm./100 ml. at 0, 50, and 90°C., respectively.

If the initial pH of a medicinal solution is to be maintained, the sulfur(IV) compound stable at that pH should be used, especially in unbuffered solutions. For example, a normal sulfite salt should not be used in an unbuffered solution to be maintained at a pH of, say, 4, since it will cause a drift toward a higher pH.

$$SO_3^{2-} + H_3O^+ \rightarrow HSO_3^- + H_2O$$

Similarly, sulfur dioxide will cause an acid shift if added to an unbuffered solution in the alkaline pH range.

$$SO_2 + OH^- \rightarrow HSO_3^-$$

TABLE 9.1 SULFUR DIOXIDE EQUIVALENTS OF COMMON SULFUR(IV) ANTOXIDANTS

Compound	Per cent
SO_2	100
$Na_2S_2O_5$	67.4
NH_4HSO_3	64.6
$NaHSO_3$	61.6
$K_2S_2O_5$	57.6
$KHSO_3$	53.3
$(NH_4)_2SO_3$	47.8
Na_2SO_3	50.8
$K_2SO_3{\cdot}2H_2O$	33.0

If the solution is sufficiently alkaline, this could be followed by the reaction

$$HSO_3^- + OH^- \rightarrow SO_3^{2-} + H_2O$$

In buffered solutions the effects are not as pronounced.

There is no agreement as to the effective quantities of this series of compounds. Trial and error methods are frequently used to determine the concentration necessary in a given preparation. The concentrations will fall within the 0.01–1.0% range, 0.1% being most common. When parenterals are to be dispensed in multiple-dose vials, higher quantities of antoxidant should be added than for single-dose volumes since each withdrawal introduces an equal volume of air into the vial.

The percentage of antoxidant will vary somewhat depending upon the sulfur(IV) compound. The relative effectiveness of the compounds is often expressed in terms of the equivalent quantity of sulfur dioxide. Table 9.1 gives the percentage of sulfur dioxide in the commonly used compounds of this series.

Qualitative judgement still enters into an evaluation of the continuing effectiveness of sulfur(IV) compounds in this application. If the solution so protected is colorless, the antoxidant is presumed effective since oxidative degradation of the medicinal usually results in the appearance of color.

The value of sulfur dioxide and the various sulfites as antoxidants depends upon their avidity for an additional oxygen atom. The net reaction taking place is represented electronically in Fig. 9.2. But mechanism of the reaction is much more complicated than indicated. The diatomic oxygen atom must be broken down in a series of processes which probably involve one or more free radicals. Several mechanisms have been postulated on the basis of available evidence. Thiosulfate, an impurity in all sulfites, acts as a catalyst for the reaction. Traces of transition element cations probably also have a catalytic affect on the reaction.

The electrode potential for the oxidation of sulfur(IV) depends on the pH of the medium (see Table 4.2). In acid media at 25°C.

$$SO_3^{2-} + 3H_2O = SO_4^{2-} + 2H_3O^+ + 2e^- \qquad \mathscr{E}^0 = -0.17$$

The electrode reaction for oxygen in acid media may be written

$$\tfrac{1}{2}O_2 + 2H_3O^+ + 2e^- = 3H_2O \qquad \mathscr{E}^0 = +1.23$$

These two equations may be summed and the $\mathscr{E}^0$ for this over-all reaction determined algebraically by adding the $\mathscr{E}^0$ values for the two half-reactions:

$$SO_3^{2-} + \tfrac{1}{2}O_2 = SO_4^{2-}$$
$$\mathscr{E}^0 = +1.06$$

Thus under standard conditions this reaction is spontaneous, $\mathscr{E}^0$ being positive. The Nernst equation for the cell reaction may be written

$$\mathscr{E} = 1.06 - \frac{0.0591}{2}\log\frac{[SO_4^{2-}]}{[SO_3^{2-}][O_2]^{1/2}}$$

Because of the large positive value of $\mathscr{E}^0$, sulfite is an effective antoxidant even when the reaction product, sulfate, predominates. Its effectiveness is best shown by calculations assuming different fractions of sulfite consumption, as has been done in Table 9.2. For example, even after 99.95% of an original 0.1% sulfite has been consumed, good protective action remains as is evidenced by the $\mathscr{E}$ value, +1.06. The other sample calculations are equally enlightening. (The oxygen contribution in these

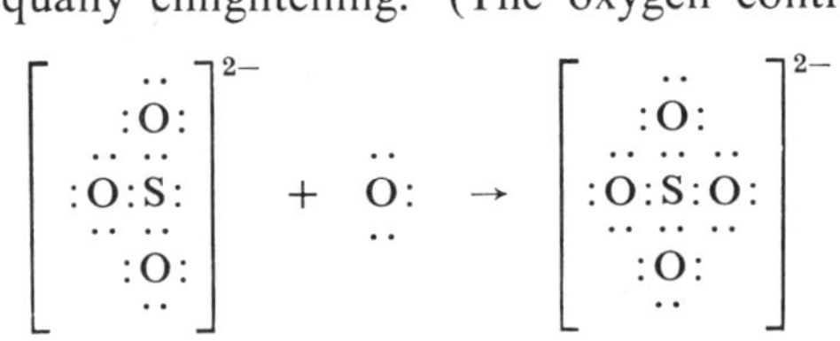

Sulfite Oxygen Sulfate

FIGURE 9.2 Sulfite as a reducing agent.

calculations has been assumed to be a constant. If included, the calculated $\mathscr{E}$ values would be somewhat less positive.)

Sulfur dioxide and the sulfites have a relatively wide range of possible incompatibilities. This follows from their diverse chemical properties. In the concentrations ordinarily employed calcium ion will be precipitated from neutral or alkaline solution (solubility = 0.004 Gm./100 ml.) If calcium ion is required for medicinal reasons, the pH must be adjusted to the acid side if the solution is to be protected by sulfite. Dyes are decolorized when sulfur(IV) compounds are added. Many aromatic anions, e.g., phenolates, inhibit their antoxidant action. Alkyl halides and compounds with alkene links will react to give sulfuric acid derivatives.

TABLE 9.2 EFFECTIVENESS OF SULFUR(IV) AS AN ANTOXIDANT

Per cent NH_4HSO_3	Initial $[SO_3^{2-}]$	Concentrations after Antoxidant Action $[SO_3^{2-}]$	$[SO_4^{2-}]$	$\frac{[SO_4^{2-}]}{[SO_3^{2-}]}$	$-0.03 \log Q_a$	$\mathscr{E}$
0.01	0.001	0.0009	0.0001	0.11	+0.029	+1.09
		0.0005	0.0005	1.0	0	+1.06
		0.0001	0.0009	9.0	−0.029	+1.03
		0.0_55	0.0_3995	199.0	−0.069	+0.99
0.1	0.01	0.0099	0.0001	0.0101	+0.060	+1.12
		0.005	0.005	1.0	0	+1.06
		0.0001	0.0099	99.0	−0.06	+1.00
		0.0_55	0.009995	1999.	−0.096	+0.96
1.0	0.10	0.09	0.01	0.11	+0.029	+1.09
		0.05	0.05	1.0	0	+1.06
		0.001	0.099	99.0	−0.060	+1.00
		0.0001	0.0999	999.0	−0.090	+0.97

The packaging of solutions protected by the sulfite group of compounds also presents problems. Low-density plastic vials and bottles are not impervious to penetration by gaseous molecules, including oxygen. The slow diffusion of oxygen through the plastic walls may exhaust the sulfite present. High-density plastics where gaseous diffusion is negligible are preferable. Sulfite penetrates rubber stoppers and changes the sorption properties of rubber, resulting in increased solvent loss and a concentration effect on the contents.

Sulfur(IV) compounds are used as antoxidants and preservatives in a wide range of pharmaceutical preparations, e.g., penicillin suspensions, ascorbic acid, *p*-aminobenzoic acid, sodium aminosalicylate, steroids, and alkaloid salt solutions. Sulfur(IV) compounds also find use in the food industry for preservation of dried fruits and fruit juices and for blanching of vegetables. Sulfur dioxide is employed as a gaseous fumigant, e.g., for grains.

Hypophosphorous acid, the hypophosphites, and the thiosulfates are also inorganic antoxidants. They are used principally for the prevention of oxidation of iodide compounds and iron(II) compounds. They act by reducing the oxidized product back to its original reduced state. The chemistry of the hypophosphite series of compounds and its role in the phosphorous series of acids will be considered later (page 323). On oxidation of hypophosphites, phosphorus(I) is oxidized to phosphorus(V). Its antoxidant action on iodine, formed by oxidation of iodide by atmospheric oxygen, may be expressed by the reaction

$$(H_2PO_2)^- + 2I_2 + 6H_2O = 4H_3O^+ + 4I^- + H_2PO_4^-$$

Hypophosphorous acid is used as an antoxidant in hydriodic acid, hydriodic acid syrup, and ferrous iodide syrup. Ammonium hypophosphite is used with ammonium iodide for the same purpose. The concentration range is usually 0.5–1.0%. Sodium hypophosphite is used as an antoxidant preservative for food products, e.g., meats, flour, as it inhibits rancidity.

Because of the strong reducing properties of the hypophosphite series of compounds care must be exercised in the presence of oxidizing agents. Explosive reactions will result on mixing the dry salts or combining concentrated solutions of oxidizing agents with concentrated solutions of hypophosphites. Hypophosphites are heat liable (page 325).

Sodium thiosulfate (page 525) may be used as an antoxidant in solutions of iodides; e.g., Potassium Iodide Solution, *N.F. XI*, uses 0.05% "hypo" as an antoxidant.

Acid-base and solubility problems due to carbon dioxide. Dissolved carbon dioxide in aqueous media leads to an acid shift in pH. Distilled water, prepared under properly controlled conditions, should be neutral; however, exposure to the air will cause a gradual shift owing to the equilibria

$$CO_2(\text{dissolved}) \xrightleftharpoons{+H_2O} H_2CO_3 \xrightleftharpoons{+H_2O} H_3O^+ + HCO_3^-$$

Water in equilibrium with atmospheric carbon dioxide will have a pH of about 5.5. Some dissolved medicinals especially sensitive to slightly acid media may undergo color changes, oxidation, or separation of insoluble substances. If a medium is initially alkaline, the absorption of carbon dioxide will be rapid:

$$CO_2 + OH^- \rightarrow HCO_3^- \xrightarrow{+(OH)^-} CO_3^{2-} + H_2O$$

All basic solutions for this reason must be protected from atmospheric carbon dioxide if their pH is to be maintained.

Since dissolved carbon dioxide gives rise to carbonate ion, neutral or alkaline solutions of all except alkali cations will precipitate insoluble carbonates. Such solutions as lime water must be protected from the atmosphere at all times.

$$Ca^{2+} + 2(OH)^- + CO_2 \rightarrow CaCO_3\downarrow + H_2O$$

The usual method of maintaining a supply of lime water by periodic addition of purified water to a bottle having a large excess of solid calcium hydroxide is unrealistic. Carbon dioxide has access to the solution after each withdrawal of solution and on each addition of fresh water. The excess white residue slowly changes from slaked lime to precipitated calcium carbonate. Periodically a sample should be withdrawn for checking its pH, which should be about 13.5, and its ability to form a voluminous precipitate of fluffy calcium carbonate on addition of a soluble carbonate. Barium hydroxide, a convenient analytical reagent, is similarly affected.

The simplest solution to this problem is the maintenance of tight closures and minimum exposure during the withdrawal of samples. Carbon dioxide-free air may be used over solutions. When frequent withdrawals are made from stock bottles of solutions, a pressure system, as shown in Fig. 9.3, may be devised.

Microorganism contamination. Although this aspect of pharmaceutical practice is the province of the microbiologist, the contamination of preparations by air-borne microorganisms and spores is of indirect interest to the pharmaceutical chemist. Preparative procedures must be carried out under conditions which minimize such contamination. Of equal importance is an understanding of the thermal and chemical stability of a preparation during heat and chemical sterilization procedures. These topics will be discussed later in the chapter.

TRACE ELEMENTS AS CONTAMINANTS

During the commercial manufacture of bulk as well as most fine chemicals, it is impossible to prevent contamination of the product by traces of various elements. Contaminants may be introduced as trace impurities in the raw materials in the manufacturing process or they may arise from the walls of the reaction vessel. Glass and metallic alloys are the material of industrial reaction vessels, and thus sodium, aluminum, iron, chromium, copper, manganese, etc. will be present in varying amounts in all chemicals.

The compendia require *limit tests* to prevent use of products with excessive contamination. Although the same test is usually used for a given trace impurity, e.g., arsenic, the actual limit tests prescribed vary from entry to entry. The tests specified for a given entry depend on possible sources and types of contamination, usually centering about

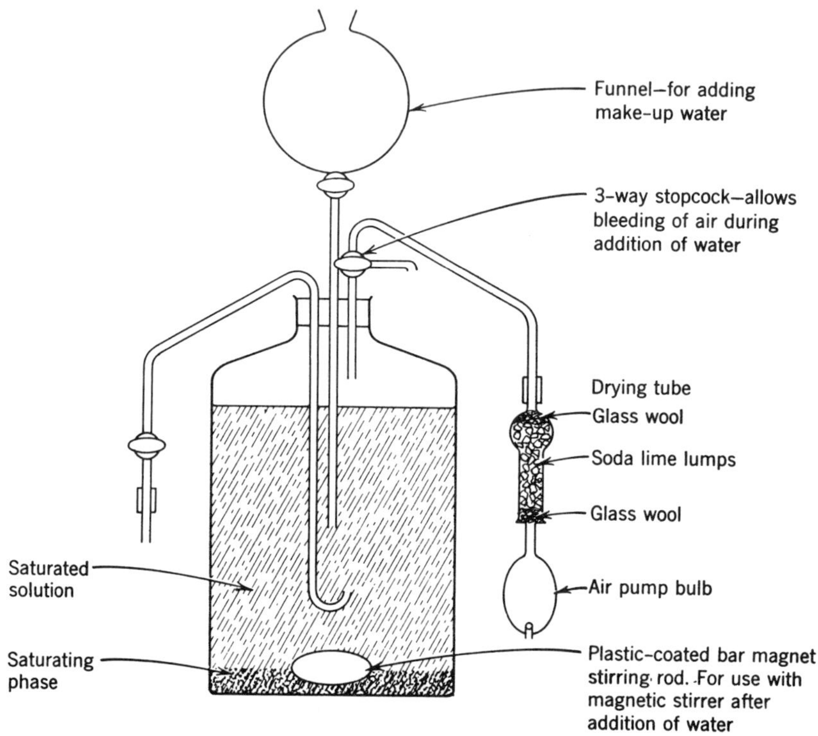

FIGURE 9.3 Apparatus for the preparation, storage, and dispensing of carbon dioxide-sensitive solutions.

raw materials, method of manufacture, and handling. Typical limit tests are for arsenic, lead, iron, heavy metals, chloride, sulfate, and ammonium salts.

Because of their catalytic property, the transition elements may lead to storage problems and other incompatibilities. They may be removed or rendered innocuous by any one of several procedures. Sequestering agents will complex these ions; in the complexed form they lose their catalytic properties. Cation exchangers (page 143) may be used to effect their removal. Hydrogen peroxide may serve as an example of a substance extremely sensitive to catalytic decomposition (page 442).

PHYSICAL FACTORS IN INSTABILITY

Physical as well as chemical conditions will produce instability in pharmaceutical systems. Light and heat are the only energy conditions of importance. Heat has two aspects; some systems are heat labile; others are sensitive to freezing temperatures.

Control of photodegradation. Numerous substances of importance in pharmacy and medicine are light sensitive, including such inorganics as silver, gold, and mercury salts, the iodides, and hypochlorites.

Although the underlying laws of photochemistry were known half a century ago, progress in the field was slow. However, within the past two decades progress has been accelerated by the application of new spectrochemical theories to atomic and molecular phenomena and by the introduction of new and improved experimental techniques. It is entirely probable that in the near future a textbook will discuss photosensitive pharmaceutical systems on a fundamental rather than an empirical basis.

To induce a photodecomposition an absorbed photon of light must have sufficient energy to cause excitation of one or more of the bonding electrons. Such excitation may cause the electron to move into a nonbonding orbital or to leave the compound and be subsequently ionized. Only light of the shorter wavelengths—ultraviolet, violets and blues—have sufficient energy to initiate electron excitation. Photons of longer wavelengths—the yellows, reds, and infrared—do not have sufficient energy to initiate chemical reactions.

Pharmaceutical control of photodegradation must therefore be designed to prevent absorption of those energetic wavelengths which can initiate a photoreaction. In one method the container itself filters out the light responsible for the photodecomposition. Dark glass bottles have been the traditionally effective means of preventing high-energy photons from penetrating the photosensitive system. A clear or translucent glass container may be used if made light-resistant by applying an opaque coating. Protection from light may also be provided by means of an opaque covering for clear containers; a statement that the covering must remain in place should appear on the label. Metallic ointment tubes as well as storage in the dark accomplish the same objective. Although not sophisticated, these methods accomplish their purpose.[2]

A newer mode of attack has been the use of screening agents. Here a substance is included in the preparation which will preferentially absorb the critical light. Application of screening agents requires a knowledge of the light absorption characteristics of both the screening agent and the compound being protected. To be effective, the *absorption envelop*[3] of the screening agent must overlap that of the light-sensitive compound. Also its *absorbance index*[4] must be greater than that of the protected medicinal

---

[2] *U.S.P.XXI*, p. 8

[3] The band(s) of wavelengths absorbed. For an elementary background in the fundamental laws and terminology of photochemistry, see Daniels and Alberty, *Physical Chemistry*, Wiley, New York, 1955, pp. 66–71, 574–596.

[4] A photochemical term whose numerical value expresses the relative efficiency of light absorption at a given wavelength. A small absorbance index, e.g., 250, indicates poor absorption efficiency; a high value, e.g., 20,000, indicates an efficient absorber.

in the overlapping absorption envelops. In summary, an effective screening agent preferentially absorbs those wavelengths of light causing photodegradation of the medicinal. In effect the action of a screening agent is analogous to that of an antoxidant. Just as an antoxidant preferentially reacts with any oxygen present or entering a solution, the screening agent preferentially absorbs the destructive wavelengths of light which enter the solution.

Since the screening agent must be present in effective concentration, it must be innocuous to the patient and chemically and physically compatible with all other compounds present. The screening agent may also be included in the coating on a capsule or tablet or on the surface of the container, or may be directly included in the glass or plastic used for manufacture of the container itself.

Some of the ultraviolet absorbers developed for application on space craft conceivably can be of value in pharmaceutical applications, especially as coatings, providing toxicity problems do not exist. Ferrocene and its derivatives—1,1′-ferrocenedicarboxylic acid, benzoylferrocene, dibenzoylferrocene, 2-hydroxybenzoylferrocene, 2-hydroxy-4-methoxybenzoylferrocene, and 2-methoxyferrocene (page 76)—are highly colored red to orange substances which are strong absorbers in the 3500–5000 Å range. These compounds have been used in space work in 1% melamine alkyd resins. The usual commercial screening agents—benzophenone, benzotriazol, phenylsalicylate—were found to be less effective in this application.

Fluorescent lamps have higher ultraviolet outputs than filament lamps. Photosensitive substances should be protected from direct exposure.

Thermal degradation of pharmaceuticals. Thermal liability of compounds is important under three conditions: during preparative stages requiring heating, during sterilization procedures depending on heat, and during storage. Heat liability problems have two aspects. If a new chemical compound(s) results, the medicinal efficacy of the original substance is lost. In addition the decomposition products may be toxic or have dangerous side effects. Where actual decomposition may not take place, stoichiometry may be affected, e.g., partial dehydration of a hydrate or the escape of a gas from solution.

Heat does not affect the bonding electrons directly. Rather it causes the atoms of a given structure to vibrate with increased frequency and amplitude within their sphere of influence. As the excursions of a pair of thermally agitated atoms moves them successively further from one another, a separation is finally attained at which bond rupture takes place.

Thus, when vibrating sufficiently, a hydrate water molecule can overcome the electrostatic attraction of the charged central atom.

General rules exist which predict inherent heat liability of inorganic compounds. If the elements of water are present, especially as hydrate water or hydroxyl groups, water will split out on heating. If the elements of carbon dioxide are present, especially if a bond already exists between carbon and oxygen, carbon dioxide will be evolved on heating. Finally, if the elements of ammonia are present, ammonia gas may be given off on heating.

All hydrates are heat liable, many at temperatures below the boiling point of water. The required portions of hydrates should be weighed if possible before heat sterilization. Hydrates should never be stored where they become excessively hot—near steam lines, over radiators, under low roofs. On heating, hydrated halide salts tend to convert to an oxy salt by splitting out hydrogen halide (page 335, 358).

Hydroxy compounds tend to split out water on heating, forming the corresponding oxides. This includes the hydroxy acids as well as the bases. The temperature of decomposition is directly related to the strength of the particular compound as an acid or base. The alkali hydroxides are stable at all ordinary temperatures; sodium hydroxide melts without decomposition at 318°C. On the other hand, hydroxides which are very weak bases, such as mercury(I) hydroxide, will disproportionate at room temperature on precipitation in aqueous solution. Strong acids such as sulfuric are stable at relatively high temperatures (300°C.), whereas weaker acids such as carbonic are dissociated almost completely into anhydride at room temperature. Acid and basic salts, e.g., $NaHSO_3$, Bi(OH)Cl, dehydrate on heating. But for the most part the temperatures involved are higher than those met in sterilization or in storage.

Carbonates, except for barium carbonate and the alkali metal carbonates, are heat liable even at the temperature of boiling water. Because sodium bicarbonate and, to a lesser extent, potassium bicarbonate are so widely used for pH adjustment, as a neutralizer, and as a buffer in injectables, their heat liability presents serious problems. Both the dry salt and its solutions decompose completely at the temperatures maintained during heat sterilization.

Preparations which include bicarbonate may be heat sterilized if the solution is first saturated with carbon dioxide for a minimum of one minute and the ampule then sealed, gastight, before autoclaving. The excess pressure of the carbon dioxide effectively prevents decomposition. Although this procedure does not prevent a shift of the equilibrium in favor of carbon dioxide during the sterilization process, on cooling the equilibrium shifts back to favor the bicarbonate. This means, however, that the ampules must be cooled to room temperature and allowed to equilibrate for at least two hours before opening. It is also essential that there is no leakage of carbon dioxide. The glass must be heavy enough to withstand the high gas pressures attained during heating. The equilibria may be represented by

$$2HCO_3^- \underset{\text{room temperature}}{\overset{\text{sterilization temperature}}{\rightleftharpoons}} CO_3^{2-} + H_2O + CO_2$$

Bacteriological filtration (ultrafiltration) of bicarbonate solutions has also been used. (A final method consists of weighing reagent-grade sodium bicarbonate into warm sterile water for injection. However, this method may result in nonsterile preparations.)

Ammonium salts (and most substituted ammonium salts) are heat liable. The temperature at which decomposition takes place is a function of the anion base strength or, conversely, the strength of the anion's conjugate acid. Ammonium salts of strong acids are comparatively stable, requiring temperatures of 400–500°C. for decomposition. Salts of acids with $pK_a > 5$ decompose slowly at room temperature and are unstable at the temperature of boiling water. Ammonium salts of very weak acids, $pK_a > 10$, are unstable at room temperature. This is summarized in the following equations.

$$NH_4^+Cl^- \xrightarrow{500°C.} NH_3\uparrow + HCl\uparrow$$

$$NH_4^+OAc^- \xrightarrow{100°C.} NH_3\uparrow + HOAc\uparrow$$

$$(NH_4^+)_2CO_3^{2-} \xrightarrow{18°C.} NH_3\uparrow + H_2O + CO_2\uparrow$$

Ammonium carbonate is the principal heat-liable ammonium salt of pharmaceutical interest. If a container with the dry salt is not tightly sealed, slow decomposition will result in a continuing loss of the compound. If stored under conditions of slightly elevated temperatures, loss will be rapid. This is also true of ammonium carbonate solutions, e.g., aromatic ammonia spirit.

Peroxides and the simple oxides of the inactive metals are also heat liable at relatively low temperatures.

The solubility of a dissolved gas is inversely proportional to the temperature of its solution. Everyone has observed the bubbles of air forming on the inside surface of a glass filled with cold water as it gradually warms to room temperature. Solutions of gases must be protected from elevated temperatures and provided with well-fitted closures. Solutions having dissolved gases formed as a result of hydrolysis (e.g., aromatic ammonia, sulfurated lime) must be similarly protected.

Freezing temperatures also cause difficulties. Suspensions tend to precipitate on freezing. The formation of ice removes water essential for maintenance of the suspension. Since the quantity of a given suspended substance which may remain in suspension is directly related to the quantity of suspending agent (water), precipitation results. Magnesia Magma, Milk of Bismuth, and Mild Silver Protein are typical preparations of this type.

In dealing with preparations to be sterilized or which may be subjected to higher temperatures, it must be remembered that solubility, pH, ionization, the position of an equilibria, and reaction rates are a function of temperature. The following rule of thumb is often applied: the rate of reaction doubles for every 10° increase in temperature (Celsius scale). Although a system may be stable at room temperatures, it may become unstable at sterilization temperatures. Degree of hydrolysis and pH change quite radically with increased temperature.

STERILIZATION

Irrespective of the method used, the pharmaceutical chemist is interested in sterilization procedures because of possible decomposition and untoward reactions within the system being sterilized. The method of sterilization selected is determined largely by the chemical and physical properties of the system; the conditions must be such that all components are stable, but any microbes and their spores are liable.

Sterilization methods may be classified as physical or chemical. Heat, radiant energy, and filtration constitute the physical methods. Chemical sterilization depends upon the action of compounds lethal to microbes.[5]

Physical methods of sterilization. Decomposition problems arising during heat sterilization procedures have been discussed in an earlier section of this chapter. Under moist heat conditions hydrolysis problems are aggravated. Dry heat, on the other hand, will increase oxidation and dehydration rates.

Short wavelengths of radiant energy are lethal to all living matter. Ultraviolet (2400–2700 Å is most effective) and gamma rays are used for sterilization procedures. Ultraviolet cannot be relied upon to give 100% kills; its low penetrating power prevents it from reaching the interior of most solids. Ultraviolet is used primarily to eliminate air-borne organisms in sterile areas. (Many compounds are also photosensitive to these high-energy radiations.)

Gamma rays are obtained from one of two sources: the van de Graaf

[5] Sterilization sometimes designates physical procedures, while the term disinfecting refers to chemical procedures.

electron accelerator or the synthetic radioactive isotope of cobalt, CO^{60} (page 557). Unlike ultraviolet, these rays have high penetrating power; solids are optically void to γ rays. Although γ radiation holds great promise for the future, many problems remain unsolved at present. Because of their high energy, these rays not only are lethal to microbes but also readily induce chemical degradation. Loss of potency, introduction of untoward tastes and odors, and appearance of color are typical results. Clear glass becomes yellow on γ irradiation. Of equal importance, radiation hazards exist for personnel. Special training is required to deal with these hazards. At the present surgical sutures and a few similar products are the only pharmaceutical items being commercially sterilized by gamma-radiation techniques. It has been found that gamma radiation from cobalt-60 acts synergistically with chlorobutanol.

Filtration techniques are frequently applied to substances which cannot withstand sterilization. They should ordinarily be restricted to those parenterals which are thermoliable, e.g., some glandular extracts. A bacteriostat or bacteriocide should usually be included.

A suitable filter has pores small enough to trap microorganisms physically. (Viruses pass through these filters.) A wide variety of materials have been used to make the filters: asbestos and asbestos-nylon pads, sintered glass, sintered stainless steel, unglazed porcelain or infusorial earth with clay and asbestos binders, porous clay, unglazed porcelain plates, nitrocellulose membranes. Although these materials are chemically inert, they can undergo physical reaction with many compounds found in pharmaceutical systems. Adsorption of drugs by these materials can be annoying, especially when present in low concentrations, and filter material must always be selected with this possibility in mind. Chemical interactions may be possible since some of these materials possess catalytic properties.

These extraneous properties of the filter media vary with pH and temperature. For a given pharmaceutical a given filter media may not adsorb or have catalytic properties at one pH or temperature, but be completely unsuitable with a change in pH or temperature. Adsorption properties of the silicate materials used as filter media (diatomaceous earth, infusorial earth, glass, china, clays, asbestos) will be discussed in Chapter 10.

Chemical methods of sterilization. In chemical sterilization techniques the significant variables are concentration of sterilizing agent, exposure time, temperature, humidity, wetness of solid materials, residue of agent remaining and its decomposition products and their effects, and chemical interaction between the pharmaceutical to be sterilized and the sterilizing agent. Gaseous substances, sometimes called fumigants, include sulfur dioxide, ozone, formaldehyde, ethylene oxide, and propylene oxide. The three organic gases give explosive mixtures in air. For this reason they

TABLE 9.3 SUBSTANCES WITH STORAGE PROBLEMS

Substance	Symbol and Temperature, °C.[1]
Aluminum chloride (anhydrous)	M, 30
Aluminum subacetate solution	R, 0–30
Ammonia water	V, 30
Ammonium acetate	V, M, 30
Ammonium carbonate	V, 30
Ammonium carbonate, aromatic	V, 30
Ammonium nitrate	M, 30
Ammonium thiocyanate	M, 30
Antimony trichloride	R, M, 30
Bismuth magma	R, 0–30
Bismuth trichloride	R, M, 30
Calcium chloride	M.P., M, 30
Calcium nitrate	M.P., M, 30
Calcium oxide; hydroxide	R, M
Cobalt(II) acetate; chloride	M, 30
Cobalt(II) sulfate	T.T., M, 30
Gold(III) halides	M, 30
Hydriodic acid	R, 18
Hydrochloric acid	V, 18
Hydrogen peroxide	R, 18
Iodine (elemental and all solutions)	V, 18
Iron(III) ammonium sulfate	M.P., R, M, 18
Iron(III) chloride	M.P., M, 18
Iron(III) nitrate	M.P., M, 30
Iron(II) sulfate	R, 30
Iron, peptonized	M, 30
Magnesium chloride	M, 30
Magnesium magma	0–30
Manganese chloride	M, 30
Mercurial ointment, strong and mild	M.P., 0–30
Mercury(II) nitrate	M, 30
Mercury(I) nitrate	R, 30
Potassium fluoride; nitrite	M, 30
Potassium hydroxide	M, R
Potassium thiocyanate	M, 18
Sodium hydroxide	R, M
Sodium perborate	R, M, 30
Sodium phosphate, dibasic, heptahydrate	M, 30
Sodium sulfate	M.P., 30
Sodium sulfide	M.P., T.T., R, M, 18
Sulfurated lime solution	R, V, 30
Sulfurated potash	R, M, 30
Tin(II) fluoride	R, M
Zinc chloride	M.P., M, 30
Zinc nitrate	M.P., M, 30
Zinc perborate	R, 18

[1]The letters used here are defined on p. 252.

are frequently mixed with an inert gas before use, e.g., ethylene oxide 10% and carbon dioxide 90%. If carbon dioxide gives incompatibilities, nitrogen gas may be substituted as a gaseous diluent.

Chemical sterilization using liquids or solids is accomplished with the sulfur(IV) compounds, metallic compounds (especially mercurials), quaternary ammonium compounds, alcohols, acids, phenols, and halogenated compounds. The usual limiting concentrations are: sulfur(IV) compounds, 0.1% SO_2 or its equivalent; mercurials, 0.01%; quaternary ammonium, 0.01%; benzyl alcohol, chlorbutanol, 0.5%; phenol or cresol, 0.5%. The quantity used is dictated by the toxicity of the sterilizing agent and the volume of solution to be administered to the patient.

Each protected system represents its own problems with respect to chemical and physical interactions between the germicide used and the other components present. Each must be considered in its own light.

CHEMICALS WITH STORAGE PROBLEMS

A list of heat-sensitive and otherwise unstable inorganic chemicals and pharmaceutical preparations is given in Table 9.3. The following symbols are used in the table to indicate why these substances must be protected: M.P., the substance has a melting point within ordinary temperatures or it softens unduly on being subjected to higher temperatures, giving rise to caking problems; T.T., a transition temperature exists within the range of temperatures ordinarily encountered and may result in a change in appearance of the material; V., the substance is highly volatile; R., the substance is sensitive to chemical decomposition through oxidation, reduction, hydrolysis, or neutralization by carbon dioxide; M., the substance must be protected from moisture. Maximum and minimum storage temperatures (in °C.) are also indicated: a single number represents the maximum safe temperature; two numbers represent the safe range for substances which decompose at too low or too high temperatures. Unless otherwise specified the data refer to all forms of the given compound. Inorganic compounds or preparations omitted from this list are not necessarily stable and without storage problems. For example, data on light sensitivity, efflorescence, and microbe contamination have not been included.

STUDY AND SELF-APPRAISAL AIDS

IMPORTANT TERMS. Define or identify each of the following in a single concise sentence if possible.

Absorbance index, absorption band, acid anhydride, activity quotient, adsorption, antoxidant, angstrom unit, dehydration, deliquescent, $\mathscr{E}$, $\mathscr{E}^0$, effervescent, efflorescent, exsiccated, fumigant, γ ray, humidity, hydrate, hygroscopic, inert gas, infrared, i.r., lethal, microbe, millimicron, mμ, Nernst equation, optically void, photodegradation, photon, Q_a, relative humidity, screening agent, sphere of influence, standard electrode potential, preservative, sterilization, sulfur(IV), ultrafiltration, ultraviolet, u.v., wavelength.

NOMENCLATURE OF INTEREST. Identify each of the following.

Ammonium carbonate, argon, Aromatic Ammonia Solution, chlorbutanol, ferrocene, helium, hypo, nitrogen, potassium hydrogen trioxosulfate(IV), potassium pyrosulfite, sodium bisulfate, sodium hypophosphite, sodium metabisulfite, sodium thiosulfate, sulfur(IV) oxide, sulfurated lime solution, sulfurous acid.

SELF-APPRAISAL QUESTIONS

1. Write the electronic structures of hypophosphorous acid, thiosulfate ion, sulfur dioxide, nitrogen, sodium sulfite, ammonium sulfate, trimethyl duodecyl ammonium chloride.

2. A sample of Glauber's Salt lost 50% of its water of hydration. A 15.0 Gm. cathartic dose is weighed out. How many grams of the fully hydrated salt is this dose equivalent to? (20.8g)

3. Single-dose portions of a hydrated salt are weighed out and sterilized in unsealed ampules. Will the intended concentration of the salt be obtained on adding the required volume of sterile water for injection?

4. Lime water should have a pH $>$ 12.5 and give a copious precipitate when a solution of a soluble carbonate is added. Explain.

5. Why is it possible to disregard the cation in the acid-base equilibria of alkali sulfite solutions? Are any sulfites in use which require consideration of the cation. Explain.

6. Sodium hydrogen sulfite is added to solutions adequately buffered to give pH's of 3, 5, 7, 9, respectively. Which of the sulfur(IV) compounds or ions predominates in each finished solution? What is the ratio of acid to normal salt at pH5? (At pH = 3 the HSO_3^- to H_2SO_3 ratio is 30/1.)

7. Recalculate the values of $\mathscr{E}$ for the 0.01% ammonium sulfite solutions given in Table 9.2 by including the oxygen pressure in the evaluation of Q_a. The partial pressure of oxygen may be taken as 0.2 atm.

8. A liter of an injection is to be stabilized with the equivalent of 0.15% sulfur dioxide. How many grams sodium metabisulfite must be added per liter to achieve this level? (2.23 g)

9. If Calcium Chloride Injection, 1 Gm./10 ml., is stabilized with

0.1% sodium bisulfite, will calcium sulfite precipitate? (yes)

10. Thioglycollate is used as the iron limit test in the *B.P.* (*British Pharmocoepeia*). It gives a purple color for iron(II) and iron(III) in alkaline media, and for iron(III) in acid media. Iron(II) gives no color with this reagent in acid media. What is the structure of this reagent, and thiocyanate and hexacyanoferrate(III) reagents? Why is the thioglycollate a more effective reagent for iron than either of the other two reagents, even in the presence of substances such as citrate?

11. Vitamin C is unstable in the presence of traces of transition metal cations. How would you proceed to stabilize a solution of this compound? (It is also unstable in alkaline media.)

12. Why is nitrogen, rather than helium, ordinarily used as an inert atmosphere?

13. A light-sensitive compound gives an orange-colored solution. Approximately what wavelengths of white light (expressed as color) are being absorbed?

14. Ferrocene is described as a highly colored red to deep orange compound, yet it is an effective blue and u.v. absorber. Explain.

15. Which of the following solids are unstable at room temperature? NH_4Cl, $Hg(OH)_2$, $NaHCO_3$, $NH_4[B(OH)_4]$.

16. Predict the effect of heat on a solution of a stable acid salt, e.g., $NaHSO_4$.

17. Give as many different methods as possible for preparing a sterile sodium bicarbonate injection. Explain the problems encountered with each method and the precautions taken to prevent them.

18. You have freshly prepared solutions of each of the following: 5% sodium sulfite, saturated potassium bicarbonate, 0.001*M* aluminum chloride, 0.05*M* iron(II) sulfate, 0.4*N* sodium hydroxide, 3% hydrogen peroxide, 2% potassium tertiary phosphate. Which will be stable on standing? Indicate the reason for instability of the others.

19. How would you overcome the instability problems of question 18?

20. Which of the following solids have storage problems? Indicate the nature of the problem and the methods by which it may be overcome or minimized. Tin(II) fluoride, ammonium carbonate, calcium chloride dihydrate, iron(II) chloride, iron(III) chloride, sodium hydroxide, magnesium sulfate heptahydrate, zinc chloride.

21. Unstable medicinal-preservative pairs follow. Indicate the reason for the instability and give the reason why the preservative (or stabilizer)

is effective. Hydrogen peroxide, 8-hydroxyquinoline; epinephrine, nitrogen; ammonium iodide, ammonium hypophosphite; diluted hydriodic acid, hypophosphorous acid; iron(II) sulfate, sodium bisulfite.

22. You have prepared a solution of diluted hydriodic acid and find that the usual antoxidants are out of stock. How could you maintain this solution until a new stock of antoxidant is obtained?

PRACTICE IN EQUATION WRITING

23. Write equations demonstrating the effect of moderate heat (100–200°C.) on the following salts: $NaHSO_3$, $KHCO_3$, $NH_4^+COONH_2^-$, NH_4Cl, KH_2PO_4, $BaCO_3$, $CaSO_4 \cdot 2H_2O$.

24. Write equations for the decomposition of boiling aqueous solutions of the following: sodium bicarbonate, sulfurous acid, hydrogen peroxide, aluminum sulfate.

25. Write equations demonstrating the reasons why the following solutions must be adequately stoppered: lime water, ferrous chloride, sodium sulfite.

26. Write equations for the reactions taking place on the surface of lead acetate when this substance is exposed to moist air.

27. Write equations for the reactions taking place on decomposition of all compounds designated by R in Table 9.3.

28. Show the reactions for the anticaking action of magnesium oxide; amorphous silica.

29. Write equations for all reactions taking place during the limit tests for: lead, arsenic, chloride, sulfate, heavy metals.

30. The reagent for the iron limit test for aluminum chloride is thiocyanate, and for aluminum sulfate it is ferrocyanide (*U.S.P. XXI*). Is there any advantage for one or the other of these reagents? Write equations for the reactions.

TEN · PHYSICAL PHENOMENA IN PHARMACEUTICAL PRACTICE I: *amorphous silica, insoluble hydrous oxides, soluble silicates*

Physical phenomena rather than chemical are the basis for an extensive group of operations and techniques important in the practice of pharmacy. A well-defined relationship exists between each and the type of structure a compound must possess to be suitable for its execution. Table 10.1

TABLE 10.1 CLASSES OF SUBSTANCES SERVING AS PHARMACY AIDS

Diluent, Extender, Filler.
Distributing Agent, Excipient, Inert Carrier.
Adsorbent, Clarifier, Filter Aid, Whitening Agent.
Emulsifier, Suspending Agent, Thickener.
Absorbent, Anticaking Agent, Moisture Absorber.
Lubricant, Abrasive.
Protective, Covering Agent, Dusting Powder, (Cataplasm).

displays the classes of substances involved. One or more silicon compounds have the prerequisite structural features for successful employment in each class. Presentation is therefore expedited by using silicon and its compounds as a basis for the elucidation of the relationship existing between structure and the various pharmaceutical operations and techniques in which these substances serve.

For convenience discussion is divided into two chapters. Chapter 10 is devoted to the general background chemistry of silicon, silicon dioxide, silicic acid and its polymers, and some soluble synthetic silicates. Chapter 11 is devoted to the insoluble silicates. Substances other than the silicon compounds with analogous or equivalent structures, e.g., the aluminas, and applications are introduced whenever appropriate.

SILICON—ITS GENERAL CHEMISTRY

Silicon is the second most abundant element, making up 28 per cent of the earth's crust. Other than the limestones and dolomites, all rocks are siliceous. A few are simple silicates, e.g., beryl, but most are aggregates of two or more silicates, e.g., granite, which is a composite of quartz, feldspar, and mica.

Silicon chemistry is modern chemistry. Because of the infusibility and nonreactivity of silicon compounds worthwhile progress was meager until the present century. Even then real progress was not evident until the late twenties when W. L. Bragg and L. Pauling initiated studies applying x-ray techniques to the elucidation of the structural features of the silicates. With this groundwork progress has since been rapid.

TABLE 10.2 Comparison of Ionic Character of Similar Silicon and Carbon Compounds

Bond	% Ionic Character	Bond	% Ionic Character
$\overset{\delta+}{Si}$—$\overset{\delta-}{C}$	12	C—C	0
$\overset{\delta+}{Si}$—$\overset{\delta-}{Br}$	22	$\overset{\delta+}{C}$—$\overset{\delta-}{Br}$	2.5
$\overset{\delta+}{Si}$—$\overset{\delta-}{Cl}$	30	$\overset{\delta+}{C}$—$\overset{\delta-}{Cl}$	6.0
$\overset{\delta+}{Si}$—$\overset{\delta-}{O}$	51	$\overset{\delta+}{C}$—$\overset{\delta-}{O}$	22.0

The element silicon occupies the same role in inorganic chemistry held by carbon in organic chemistry. Examination of the ground state electron structures, ns^2np^2 (see Table 17.5), indicates that silicon chemistry should closely approximate carbon chemistry. Both elements show a covalency of four, giving rise to sp^3 hybrids, with the familiar tetrahedral arrangement.

The chemistry of the two elements differs because of several factors. Although the valence electrons are similar, those of silicon, being in the next higher major quantum group, are further removed from the influence of the positive electrostatic field of the nucleus and consequently are held less securely. This is indicated by a comparison of the covalent radii, 0.77 Å for carbon and 1.18 Å for silicon. The lesser screening power of the helium core of electrons (carbon) may be contrasted to the greater screening effect of the neon core (silicon). Thus the silicon nucleus has a lessened attraction for electrons, and bonds between silicon and other atoms have greater electrostatic character than the corresponding carbon links. The ability of electronegative elements to attract more effectively the bonding electron pair from the silicon atom and thus impose a polarity on the bond is clearly shown in Table 10.2.

The greater covalent radius and its effect on silicon bonds give rise to two other deviations from carbon chemistry. The silicon atom is incapable of forming multiple bonds; unshared electron pairs at each side of a suitable σ bond are too far apart to give the overlap necessary to form stable π bonds. There is no silicon analog of ethylene or of the carbonyl moiety. Catenation, which reaches its acme with carbon, is unknown in natural silicon compounds. Silicon hydrides (silanes) up to Si_6H_{14} have been synthesized, but all, particularly the higher members, are unstable.

These differences in bonding lead to some striking differences in the respective compounds of silicon and carbon. Carbon dioxide and carbon disulfide occur as independent molecules. In the solid state the crystal binding forces are van der Waals. The silicon analogs do not exist as molecules; SiO_2 and SiS_2 are covalently bonded polymeric substances. There is no discontinuity in the bonds between the atoms of a single silica crystal, as there is no discontinuity in the bonding of the carbon atoms of a diamond crystal. The simple carbon hydrides, halides, and sulfides (e.g., CH_4, CCl_4, CS_2) are insoluble in water and do not hydrolyze at room temperatures. The corresponding silicon compounds are somewhat soluble in water and hydrolyze readily. (Silane requires a trace of base to initiate the hydrolysis.)

Another structural feature further differentiates the silicon atom from the carbon atom. Unlike the second major quantum group the third has *d* orbitals available for occupancy by electrons. Although the *d* level is unoccupied in the ground state silicon atom, it is nonetheless available for bonding. As a result sp^3d^2 hybridization is possible, giving rise to silicon compounds with the familiar octahedral arrangement. The octahedral hexafluosilicate ion, $[SiF_6]^{2-}$, has been known for many years. Recently six coordinate compounds of silicon and bidentate organic ligands have been synthesized.

Since silicon has a position in the periodic table midway between the metals on the left and the nonmetals on the right, it can be expected to exhibit both basic and acidic properties. Being the second lightest member of its family, it has a comparatively small electropositivity. Consequently, though silicon tends to exhibit greater basicity in its compounds than carbon, the acidic tendencies still predominate.

SILICA, SILICIC ACID, AND AMORPHOUS SILICA SYSTEMS

Silicon dioxide (silica). Silica is widely distributed in nature. Rock crystal (quartz),[1] tridymite, and cristobalite are pure forms. Sand, quartzite, sandstone, and flint are comparatively impure forms. The gems—amethyst, agate, onyx, and opal—are silica. Silica occurs in the ashes of many grains, tobacco, and bamboo. Sponges and feathers of some birds are rich in silica. Diatoms and infusoria have siliceous skeletons, which have been laid down in tremendous beds as diatomaceous earth and infusorial earth.

Silicon dioxide is extremely resistant to most chemical reagents. Of the halogens only fluorine and hydrofluoric acid will attack it at room temperature.

$$(SiO_2)_n + 2nH_2F_2 = nSiF_4\uparrow + 2nH_2O$$

Strong alkalis attack it, as will the alkali metal carbonates.

$$(SiO_2)_n + 4nNa^+OH^- = nNa^+{}_4SiO_4{}^{4-} + 2nH_2O\uparrow$$
$$(SiO_2)_n + 2nNa^+{}_2CO_3{}^{2-} = nNa^+{}_4SiO_4 + 2nCO_2\uparrow$$

The latter reactions, carried out in the fused state, are the basis for the preparation of soluble silicates. The reactions proceed to completion because of the nonvolatility of silica as compared to very volatile water and carbon dioxide.

It is apparent that silicon dioxide is a true acid anhydride. However, because of the insolubility of the compound in water, the reaction

$$2nH_2O + (SiO_2)_n \rightleftharpoons nH_4SiO_4$$

is of academic interest only. The actual solubility in water is a function of particle size. Pure amorphous silica will dissolve to give a 0.01% solution, whereas coarse particles will give a 0.0006% solution.

Silicic acid. This substance is an extremely weak acid, $pK_1 = 9$, $pK_2 = 13$. It is unknown in the crystalline form. Dilute solutions are prepared by hydrolysis of its esters or halides, by cation exchange techniques involving soluble silicates, and by displacement reactions.

Silicic acid solutions are unstable. Monomeric silicic acid polymerizes in steps to give dimers, trimers, etc., until polymeric silicic acid or silica gel precipitates. Stability is a function of pH; Table 10.3 summarizes

---

[1] In addition to inertness, crystal quartz has several interesting properties. It is optically active, it is transparent to ultraviolet light, and its coefficient of expansion, 0.0_645, is only one eighteenth that of glass.

TABLE 10.3 The Effect of pH on Gelling of Silicic Acid Solutions

pH	Gelation Tendency	Factors
6–7 and >7	Retarded	Strong negative charge on particles retards gel formation. (Presence of soluble salts increases gelation at a given pH in this range.)
5–6	Rapid	Gel formation is most rapid in this range.
3-5	Low	A function of $[OH^-]$.
2-3	Very low	Maximum stability range of silicic acid solutions.
<2	High	Rate is proportional to H_3O^+ and F^-. (F^- induces gelation if present in only a few p.p.m.)

the gelling tendency of silicic acid solutions as a function of pH. Stability is also affected by the age of the solution and electrolyte impurities present.

Evidence for the existence of silicic acid is indirect: its ability to pass through dialysis membranes and its inability to precipitate albumin indicates low molecular weight. This is affirmed by freezing-point, potentiometric, and conductivity measurements. The preference for the formula H_4SiO_4 is based on the existence of esters, $Si(OR)_4$, and Raman spectra. Both indicate a tetraprotic acid rather than the dihydrate, $H_2Si(OH)_6$, or the meta acid, H_2SiO_3. The tetrahedral arrangement characteristic of silicates also lends support to H_4SiO_4 as the correct formula for the simple acid.

Amorphous silica. NATIVE: Diatomaceous earth (terra silicea purificata, purified siliceous earth, purified kieselguhr, fossil flour, tripoli, kieselerde, diatomite, celite) consists of spicules, rods, and stars of amorphous silica in various stages of crystallization. As a deposit ages, the silica slowly but continuously changes from amorphous silica to quartz. The term infusorial earth is also used. Technically it refers to siliceous skeletal material arising from the infusoria, members of the animal kingdom, whereas diatomaceous earth is the siliceous skeletal debris of diatoms, members of the plant kingdom.

Crude diatomaceous earth is purified by boiling with concentrated hydrochloric acid. The acid causes solution of the calcium, iron, and similar compounds usually associated with the deposits as impurities. The dissolved impurities are then removed by thorough washing with water.

Finally, volatile and carbonaceous materials are removed by calcination. Particle size is generally controlled by sifting, or by air-blowing for extremely fine particles.

In common with all substances with complete covalent bonding diatomaceous earth is a hard, chemically inert substance. Because of the gross open physical structure resulting from the matting of the spicules, the material has excellent absorbent properties. Siliceous earth retains its dry powdery characteristics even after absorbing its own weight of liquid. It will absorb up to four times its weight of water. Thus siliceous earth has found use as a carrier for liquids, both aqueous and nonaqueous. This makes possible the dispensing of fluid extracts in powder form. Glyceryl trinitrate (trinitroglycerine) was formerly absorbed on this substance to form dynamite; wood meal is now used. The moisture-absorbent ability, together with the shape of its particles and its chemical inertness, has made siliceous earth a valuable anticaking agent. It has also been used as a dusting powder, but has an objectionable grittiness if too coarse particles are used. The air-blown form is used in face powders.

The chemical inertness, open structure, matting qualities, and the absence of adsorptive properties make siliceous earth an excellent filter aid and clarifying agent, especially for fats, oils, syrups, and serums. A mixture of the discoid and pennate forms is best for filtration. Because of the covalent bonding within the particles, the structure is strong. This strong structure imparts hardness which makes siliceous earth a useful mild abrasive. As such it finds use in tooth pastes; however, it may scratch the enamel of the teeth. It is also used in polishes and scouring soaps. The chemical inertness of siliceous earth renders it an excellent filler and extender. Oral preparations may include this substance since it passes through the intestinal tract unchanged. The substance is useful as an excipient for volatile oils and mixtures which liquefy, e.g., phenol, camphor.

SYNTHETIC: There are two types of manufactured amorphous silica. The first, "silica fume," is prepared by condensation of silica from its vapor phase. Fume may be formed by the vaporization and recondensation of silica itself, or by the burning of silicon, silicon tetrachloride, silicon disulfide, or silicic acid esters, e.g., ethyl silicate in air. Typical reactions are

$$(SiO_2)_n \xrightarrow{2230°C} nSiO_2 \xrightarrow{\text{cool}} m(SiO_2)_p, \qquad \text{where } mp = n.$$

$$Si_x(\text{solid}) \xrightarrow{\text{heat}} ny\text{Si(gaseous)} \xrightarrow{+O_2(\text{air})} y(SiO_2)_n, \qquad \text{where } ny = x$$

$$nSi(OC_2H_5)_4 \xrightarrow{O_2(\text{air})} nSiO_2\downarrow + 8nCO_2\uparrow + 10nH_2O\uparrow.$$

The exhaust vapors are chilled and the fine silica powder is collected.

The second type of synthetic amorphous silica is "silica gel," a hydrated form. It is prepared by precipitation from aqueous media,

$$x[Na^+_4(SiO_4)^{4-}] + 4x[H_3O^+ + Cl^-] = 4xNa^+Cl^- + (H_4SiO_4)_x\downarrow + 4xH_2O$$

$$xSi(OR)_4 + 4xH_2O \xrightarrow{\text{boil}} (H_4SiO_4)_x\downarrow + 4xROH$$

$$x[Na^+_4(SiO_4)^{4-}] + R(SO_3^-)_x\,[H_3O^+]_x = (H_4SiO_4)_x\downarrow + xH_2O + R(SO_3^-)_xNa^+_{4x}$$

where $R(SO_3^-)_x$ represents the anion skeleton of a cation exchange resin. The conditions selected are favorable for polymerization of the monomeric acid as it forms.

$$(H_4SiO_4)_x\downarrow = [(SiO_2)_x(H_2O)_y] + (2x - y)H_2O$$

The purest gels are obtained by ion exchange methods. The acid precipitation products are least pure; dialysis can remove the anions of the by-product salt, but complete removal of the cations is not possible. The physical appearance and handling qualities of these gels are a function of their water content, shown in Table 10.4.

TABLE 10.4 Effect of Water Content on the Physical Condition of Silica Gels

300 H_2O/SiO_2	Maximum water content (some water can be separated by mechanical pressure)
30–40 H_2O/SiO_2	Easily cut (exhibits syneresis)
20 H_2O/SiO_2	Stiff
10 H_2O/SiO_2	Friable, crumbly
6 H_2O/SiO_2	May be triturated to a fine powder

Most silica gels are xerogels—gels which have been freed of excess liquid phase. During the required evaporation procedure the surface tension of the water or solution in the pores of the gels exerts a strong compression action on the gel mass. This compressive force is resisted by the strength of the gel framework, which is a function of the precipitation conditions, the concentration of soluble silicate, and the silica/alkali ratio of the soluble silicate used as the raw material. The final balance between surface tension and the strength of the gel framework will determine the properties of the finished xerogel. For a pulverent, low-density, finely divided gel, the freshly precipitated gel is first treated with a water-miscible organic liquid, e.g., ethanol or methyl ketone, and then dried.

The final physical form determines the properties of the given gel and its suitability for a given application. If intended as a filler, a light, friable, easily powdered product is desired. Use as an adsorbent requires a porous gel with a high specific surface. Use as a carrier requires a strong, hard gel which will not shrink or crack when heated. Certain of the properties used for the evaluation of silica gels, together with their usual range of values, are given in Table 10.5.

PHYSICAL STRUCTURE OF SILICA GELS: X-ray studies indicate that freshly prepared gels possess the cristobalite[2] structure. Magnetic data indicate the presence of silicon dioxide and water, rather than reaction products of these two compounds. Electron microscope studies indicate that the individual particles are not compact but spongy in character, probably aggregates of smaller silica units.

TABLE 10.5 IMPORTANT PROPERTIES OF SILICA GELS

Property	Units	Range
Area	$m.^2$/Gm.	20–900
Gas adsorption at p_0	cc./Gm.	negligible to 2500
Pore volume	cc./Gm.	negligible to 400
Average pore radius	Å	negligible to 400

Structurally the particles range from a meshwork of chains and rings to highly condensed three-dimensional units. Because of the nature of this structure, each particle may be considered a meshwork of primary and secondary channels. The primary channels are fine, nearly of molecular dimensions. They have diameters in the range of 5–10 mμ, are about 45 mμ in length, and have a tortuous shape. Primary channels are not destroyed by pulverization. Any water present is adsorbed water. The secondary or coarse channels obey capillary theory. They fill with water only at high humidities. Their diameters range from 260–440 mμ. Pulverization destroys these channels.

The size, distribution, and ratio of primary and secondary channels determine the physical strength, absorption capacity, and adsorption capacity of a gel. They also determine the limiting size of molecules which may enter. However, it is the nature of the exposed surfaces which determine the relative reactivity or inertness of a given gel. The bulk of a given silica particle consists of interior SiO_2 units; each silicon atom in the interior is tetrahedrally surrounded by four oxygen atoms. Each interior oxygen atom is in turn shared by two different silicon atoms. A polymeric, continuously bonded internal structure results. At the surface the coordination forces of a given SiO_2 unit cannot be fully satisfied since

[2] Silicon dioxide appears in one of three crystalline modifications: quartz, tridymite, and cristobalite.

it is not surrounded on all sides by silica units. Two types of surface groups are possible; the silanol and the siloxane. Both are shown diagrammatically in Fig. 10.1. A given silica gel surface may be partly silanol and partly siloxane. Siloxane groups hydrate to some extent when exposed to very high humidities.

The final surface structure achieved by a given gel is largely determined by the temperatures to which it was exposed during the manufacturing process. Silica fume, because of the high temperatures required for preparation, has siloxane surfaces. If a precipitated gel is heated carefully during drying, a predominantly silanol surface is obtained; if heated to 800°C. in a vacuum, all silanol groups dehydrate to siloxane. Electrolyte impurities lower the temperature for dehydration of the silanol groups.

SURFACE REACTIONS OF AMORPHOUS SILICAS: Siloxane groups tend to be inactive. Silanol groups are comparatively reactive. These bound hydroxo groups are weakly acidic and undergo chemical reactions typical of acidic hydroxo functional groups. Primary and secondary alcohols (but not tertiary) of two or more carbons esterify easily. The reacting alcohols may be substituted alcohols. The surface of a gel particle covered with these esterified silanol groups takes on the general behavior of the substituting group. Esterified with an unsubstituted alcohol, a gel particle assumes the properties of a hydrocarbon, i.e., it becomes hydrophobic. Esterified with an amine-substituted alcohol, the gel particle assumes anion exchange properties.

Heats of adsorption of a given substance favor adsorption on a silanol surface rather than on a siloxane surface. Methyl red absorbs readily on a silanol surface but with difficulty on a siloxane surface. Because of its acidic nature, a silanol surface has a strong affinity for basic substances, e.g., fuchsin, methyl violet, ammonia. Water will not leach such substances when adsorbed. On the other hand, acidic substances are weakly adsorbed

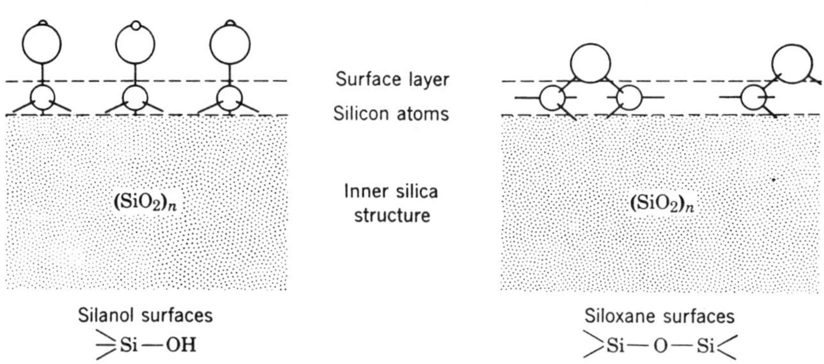

FIGURE 10.1 Surface atomic arrangements of silica gels.

and may be removed by water leaching. Aluminum hydroxide gel (page 268) has surface structures similar to those of the silica gels. However, its hydroxo groups have a basic character; the adsorption behavior on alumina is the reverse of that described for silica.

The adsorptive properties of silica gel particles can be effective for a very wide range of substances. Individual gels are commercially available with tailored adsorption properties. One group of gels is especially effective as moisture absorbers. An indicator, cobalt(II) chloride may or may not be included to show whether or not the gel has reached its saturation point and is no longer effectively absorbing moisture.[3] These gels may be regenerated repeatedly by drying in air at 150–200°C. Moisture-adsorbent silica gels are available in small moisture-pervious paper bags. Bag sizes are available which allow insertion in the empty space of a container filled with a moisture-sensitive chemical or pharmaceutical preparation. If replaced after removal of a portion of the contents, a moisture-free atmosphere is effectively maintained.

Such gases as ammonia, sulfur dioxide, formaldehyde, carbon tetrachloride are effectively adsorbed. They are easily removed by heating the gel in a stream of air. Solids, e.g., fatty acids, may be adsorbed.

Because of their moisture-absorbent properties, amorphous silicas are valuable anticaking agents, especially for hygroscopic substances (page 236). Very small percentages ($<1\%$) are frequently effective. Having an open, porous structure, silica gels are valuable as absorbing agents. Because they are able to absorb large volumes of oil, they are used as carriers. As such it is possible to apply oils as dusts and powders.

Amorphous silicas serve as thickeners for both aqueous and nonaqueous systems. Thickening is a function of the type of silica surface, moisture content of the silica, and total silica per unit volume, together with the polarity of the suspending medium. Being unable to interact effectively with the fields of the surface silanol and siloxane groups, a nonpolar medium tends to reject the silica particles. The few silica particles accepted tend to flocculate, giving rise to a three-dimensional fibrous siliceous network which occludes the continuous phase of the suspension. Considerable thickening of the suspension results. Since thickening is related to viscosity, an easily measured property of fluid systems, viscosity is used as a measure of the degree of thickening or gelling of suspensions.

On the other hand, molecules of a polar medium interact with the surface silanol and siloxane groups. The degree of electrostatic association,

[3] Anhydrous cobalt(II) chloride is azure blue; the hydrate is pink. A gel's heat of adsorption for water is such that moisture adsorption takes place preferentially over hydration of the cobalt(II) ion. Cobalt(II) cannot hydrate until the gel has become saturated with moisture. This point is indicated by the color change from blue to pink.

or solvation, is directly proportional to the polarity of the medium. Direct contact or field association is no longer possible between solvated silica particles; no framework forms. The viscosity of a suspension in a polar medium is low, approaching that of the pure suspending medium. The ratio of silanol to siloxane groups determines the polarity of the individual particles of a silica gel; plus silanol populations increase the polarity. In a nonaqueous system moisture increases the polarity of the silica particle.

The effect of an additive, i.e., any dissolved compound present in the gel, is dependent on its polarity. A polar additive is effectively adsorbed on the silica particles to give in effect a more nonpolar silica surface.

TABLE 10.6 THE EFFECT OF ADDITIVES ON THE PROPERTIES OF SILICA GEL SUSPENSIONS

Additive	Effect of Additive on Polarity of Silica Surface	Polarity of Medium	Effect on Viscosity
Polar	Decreases	Polar	+
		Nonpolar	−
Nonpolar	Increases	Polar	−
		Nonpolar	+

This results in an increase in viscosity if a polar medium is used. The effect of the combinations of additives and suspending media on the viscosity of the suspensions is given in Table 10.6.

Because of general inertness, including lack of odor and taste, the use of amorphous silica imposes no serious incompatibility problems. Both silica fume and xerogel find application in pharmaceutical and related fields as thickening and gelling agents. They are effective in both aqueous and nonaqueous media for the preparation of lotions, creams, ointments, salves, emulsions, suspensions, jellies, and toothpastes.

Having a favorable refractive index range, amorphous silica particles can be prepared with refractive indices similar to the suspending medium. This allows the preparation of transparent or slightly translucent gels.[4] This is not possible with many of the other thickening agents; opaque systems result when large differences of refractive indices exist.

In chemical practice amorphous silica products are used as substrates and as inert carriers for active materials. This is accomplished either by distributing the active substance throughout the silica particle or by using

[4] If a suspended particle has the same refractive index as the suspending medium, it is invisible.

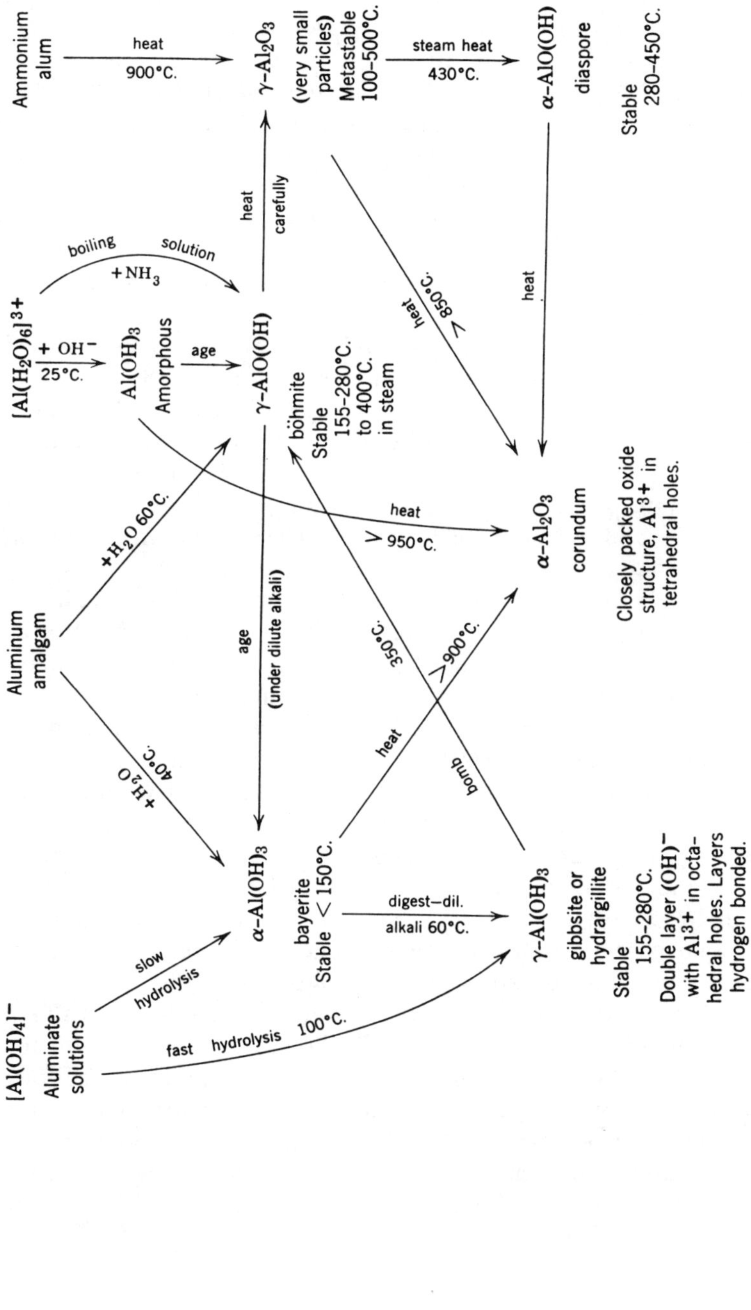

FIGURE 10.2 Alumina and the hydrated aluminas. $Al_2O_3 \cdot H_2O \sim 2AlO(OH)$, and $Al_2O_3 \cdot 3H_2O \sim 2Al(OH)_3$.

the silica particle as a core upon whose surface the active substance is deposited. The coated silica particle acts as though it were a solid having the composition of the coating.

It would seem that applications along these lines should be possible in pharmaceutical practice. Since amorphous silica is inert under the conditions existing in the intestinal tract (except for adsorbing selected trace substances), it is conceivable that this material is suitable as a carrier for small quantities of medicinal agents. The medicinal compound could either be coprecipitated during the formation of the original gel or adsorbed or absorbed on a finished gel. On ingestion the medicinal would be released due to favorable concentration gradients in the intestinal tract or by controlled disintegration of the gel particles. The latter could be accomplished by taking advantage of the relationship between gel stability and pH.

Silicosis. A discussion of silica would be incomplete without a review of silicosis—a lung condition resembling chronic tuberculosis. Only "respirable dust," particles 5 μ and less in diameter, will cause this condition. Silicosis appears only after long exposure, seven years in most cases. Recent studies indicate that only crystalline silica is responsible; apparently amorphous forms are innocuous. Other silicon compounds such as glass and silicates are harmless. The mechanism of action is unknown.

A former treatment consisted of breathing aluminum dust or aluminum-aluminum oxide dust at regular intervals. The mechanism has not been elucidated but has been postulated as related to a decreased solubility of silica. In a recently reported experiment the amorphous silica used gave a 0.017% solution in water (37°C.). Introduction of aluminum oxide reduced the solubility to 0.001%; introduction of aluminum further reduced the solubility to 0.0001%. It seems probable that a thin film of highly insoluble aluminum silicate forms on the silica particles.

Aluminum oxides and hydrated aluminum oxides. Except for the basic nature of the surface hydroxo groups the hydrous aluminas have properties reminiscent of the hydrated silicas. The hydrated aluminum oxides exist in amorphous as well as four well-defined structures: two trihydroxo and two oxomonohydroxo. The relationships existing between these hydrous forms and between the hydrous and the two anhydrous aluminas are shown in Fig. 10.2. The preparative conditions for the pure forms are included. Ordinary absorptive aluminas are prepared by careful heating of the precipitated hydroxide. The temperature and humidity determine the net properties of the resulting hydrous compounds.

Only γ-alumina and the hydrous forms possess the high absorptivity and adsorptivity associated with the activated aluminas. The α-alumina is chemically and physically inert. Its properties are compared with those of γ-alumina in Table 10.7.

In addition to their use as drying agents, absorbing agents, and adsorbing agents in preparative procedures, hydrated aluminas are used as carriers for vaccines and toxoids.[5] For this use the precipitate must be in a reactive form to allow slow solution by the body fluids. The hydroxide may be precipitated in the toxoid or vaccine solution by addition of

TABLE 10.7 Comparison of the Aluminum Oxides

	α-Al_2O_3	γ-Al_2O_3
Structure	hexagonal rhombohedral; well-defined oxide structure	cubic; distorted, badly organized microcrystals
Physical properties		
Specific gravity	4.0	2.5
Hardness	hard	soft
Surface area	–	high in proportion to volume
Chemical properties		
Moisture	inert	hygroscopic
Adsorption	—	+
Absorption	—	+
Acids and bases	insoluble	soluble
General reactivity	nonreactive	very reactive

potassium alum, or the preformed adsorption agent may be added. Aluminum phosphate is also used. The adsorbed forms of these substances provide a slow release of the active drug to the body fluids. Aluminum stearate is used as a gelling agent (Sterile Procaine Penicillin).

Corticotropin may be administered as an adsorbed repository product. The α form of zinc hydroxide is used in this application.

THE SOLUBLE SILICATES

Only the alkali metal silicates are soluble. Even here solubility is limited, those with high silica ratios being insoluble. All are prepared by fusion of silicon dioxide in a suitable alkali base. Since alkali metal oxides and silicon dioxide are mutually soluble over a wide range of ratios, empirical formulae are generally meaningless. Soluble silicates

[5] Adsorbed Tetanus Toxoid, Adsorbed Diptheria Toxoid, Adsorbed Pertussis Vaccine, and combinations thereof.

are usually identified by SiO_2/M_2O ratios. Those soluble silicates with high alkali contents are known as "alkaline silicates"; those of high silica content are identified as "siliceous silicates."

Preparation. Alkaline silicates are prepared by fusion of silica with the hydroxide, MOH, of the desired cation

$$x\text{SiO}_2 + 2yM\text{OH} \xrightarrow{\text{melt}} y(M_2\text{O})\cdot x(\text{SiO}_2) + \tfrac{1}{2}y\text{H}_2\text{O}$$

A relatively pure product results.

Siliceous silicates are also prepared by fusion techniques. American practice uses the carbonate of the desired cation as the base.

$$x\text{SiO}_2 + yM_2\text{CO}_3 \xrightarrow{\text{melt}} y(M_2\text{O})\cdot x(\text{SiO}_2) + y\text{CO}_2\uparrow$$

The carbon dioxide must be completely expelled or the reverse reaction takes place to an appreciable extent on cooling. Moisture will cause hydrolysis to silicic acid gel. European practice substitutes the corresponding sulfate as the base.

$$y\text{Na}_2\text{SO}_4 + y\text{C} + x\text{SiO}_2 \xrightarrow{\text{melt}} y(\text{Na}_2\text{O})\cdot x(\text{SiO}_2) + y\text{SO}_2(\text{dissolved}) + y\text{CO}\uparrow$$

The mechanism is not that given by the indicated stoichiometry. More carbon than indicated must be used. This in turn requires an excess of the sulfate. The resultant product has a lower purity but is more economical than that obtained in American practice. The fused product from either process cools to a clear, greenish blue glass. The color is due to traces of iron compounds.

Soluble silicates may be prepared over a wide range of SiO_2/M_2O ratios. The most common have ratios within the range 2.0–3.2. Those with ratios $\leqslant 2.5$ are soluble in water. Increasing insolubility limits the useful products to ratios < 4.0. The insoluble forms may be "dissolved" by using steam under pressure or by mixing an excess of the finely ground solid with a concentrated solution.

Soluble silicates are usually sold as solutions of highest possible concentration. The limiting factor is viscosity, which is directly proportional to both silicate concentration and silica ratio. Commercial silicate solutions are usually prepared from sodium compounds; some potassium- and lithium-based solutions are available. A few sodium silicates are available commercially as stable crystalline hydrates; a somewhat hygroscopic pentahydrate is available as a free-flowing white crystalline powder melting at 72.2°C. Potassium silicates, though more expensive than the corresponding sodium compounds, are free from efflorescence (bloom) and are generally more soluble in water. Cesium and rubidium silicates have been prepared in the laboratory. Ammonium silicates have not been prepared, although there is evidence for their existence in solutions of concentrated ammonia.

Properties of soluble silicate solutions. Alkaline silicates exist as simple ions in aqueous solutions. The solutions undergo appreciable hydrolysis. Assuming the orthosilicate structure, the first stage may be written

$$SiO_4^{4-} + H_2O \rightleftharpoons OH^- + HSiO_4^{3-}$$

A 1.0M solution of the 1:1 ratio compound is hydrolyzed to about 6%. This is higher than that of corresponding solutions of tertiary sodium phosphate, tetrasodium pyrophosphate, or sodium carbonate.

Siliceous silicates give rise to colloidal systems. Colloidal properties appear at the 2:1 ratio and become pronounced at higher ratios. The simplest test for colloidal silica employs ammonium molybdate solution. Crystalline silicates give an insoluble green-yellow silicomolybdate; colloidal silicates give no color reaction.

Solutions of soluble silicates have complex equilibria reactions. Dilution of a given solution does not bring about an instantaneous change in properties; hours may pass before constant values of pH, conductivity, viscosity, etc. are established. Soluble silicates are precipitated from solution by ethanol, acetone, ammonia (strong solution), and salt brine. These precipitates may be redissolved, but the properties of the new solution differ from those of the original even though the exact concentration is reproduced.

Cations other than alkali cations react with solutions of soluble silicates to give nonstoichiometric precipitates. The products formed are a function of the conditions under which precipitation was effected: the concentration and pH of the silicate solution, the concentration and pH of the solution of the metal cation salt, temperature, and impurities present. With a solid salt the product differs from that obtained with a solution. The precipitates formed are complex mixtures with varying ratios of colloidal silica, basic salts, hydroxides, hydrous oxides, and sometimes colloidal complexes. For example, copper(II) sulfate (0.1M) and a 1:1 ratio silicate solution give a precipitate whose color varies from light blue-green to deep blue. X-ray analysis indicates varying proportions of $4CuO \cdot 3H_2O$, $CuO \cdot SO_3 \cdot 2H_2O$, and amorphous silica.

Magnesium trisilicate belongs to this class of preparations. It is a mixture of hydrous magnesium oxide and hydrous silica. The term trisilicate is a misnomer, as no trisilicate ions are present. It has its origin in the analytical composition of the preparation, expressed as an oxide ratio, $2MgO:3SiO_2$. Magnesium trisilicate is prepared by adding stoichiometric volumes of magnesium chloride or sulfate solution to solutions of the 1:1 sodium or potassium silicate. Because of the alkaline character of the silicate solution, the magnesium ion is precipitated as the hydroxide and hydrated oxide. By arbitrarily selecting the orthosilicate anion, these reactions may be written

$$(SiO_4)^{4-} + 4H_2O \xrightarrow{\text{stepwise}} H_4SiO_4 + 4OH^-$$

$$Mg^{2+} + 2OH^- = Mg(OH)_2\downarrow$$

$$Mg(OH)_2\downarrow \xrightarrow{+H_2O} MgO \cdot nH_2O\downarrow$$

As the hydroxyl ion is consumed, the alkalinity is reduced to pH values favorable for the coagulation of the silicic acid.

$$xH_4SiO_4 = (SiO_2)_x \cdot yH_2O + (x - y)H_2O$$

The precipitates carry down soluble by-product salts, e.g., Na_2SO_4, by occlusion and adsorption. They must be removed by washing. The final moist produce is dried and powdered. Its use as an antacid will be discussed (page 415, 425). It is also used as an emulsifying agent for oils.

The soluble silicates have no direct application in pharmacy. They do serve, however, in manufacturing processes as a source of silicate ion, e.g., magnesium trisilicate, and as an inexpensive source of strong base. They are essential for the preparation of silica gels. The soluble silicates are excellent and inexpensive adhesives and are thus widely used. They also find application in fireproofing and waterproofing techniques.

STUDY AND SELF-APPRAISAL AIDS

IMPORTANT TERMS. Define or identify each of the following in a single concise sentence if possible.

Abrasive, air-blown, amorphous, bidentate, calciferous, catenation, coefficient of expansion, distributing agent, excipient, extender, ferruginous, filler, ground state, octahedron, optical activity, polymer, pore diameter, refractive index, Raman spectrum, silanol, silicosis, specific surface, viscosity, xerogel.

NOMENCLATURE OF INTEREST. Identify each of the following.

α-alumina, aluminum stearate, böhmite, cellite, corundum, diatomaceous earth, disilane, ethylsilicate, hydrargillite, kieselguhr, orthosilicic acid, potassium silicomolybdate, purified siliceous earth, quartz, silane, silica, silica fume, silicon(II) sulfide, tripolite, tridymite.

SELF-APPRAISAL QUESTIONS

1. Using electron structures, account for the similarities and differences in the properties of carbon and silicon atoms.

2. Account for the ease of hydrolysis of silicon tetrachloride as compared to the resistence of carbon tetrachloride to hydrolysis.

3. Explain why silicon and its compounds tend to have acidic rather than basic characteristics.

4. Explain on the basis of structure why diatomaceous earth is useful as filter aid, anticaking agent, excipient, absorptive agent.

5. Outline the two general methods for preparation of amorphous silicas. How do the products differ?

6. Distinguish between primary and secondary channels of silica gels.

7. Explain why a silica gel particle with a silanol-rich surface is more reactive than one which is not.

8. How is it possible to prepare silica gel particles having ion exchange properties? Explain.

9. Give the pharmaceutical uses of hydrated silicas; hydrated aluminas. Explain on the basis of structure and properties why each application is successful.

10. Compare the acid strength of the insoluble hydroxo compounds of aluminum and silicon. Account for the differences.

11. Distinguish between the absorptive properties of silica and alumina gels.

12. Distinguish between silanol group, siloxane group; diatomaceous earth, infusorial earth; α-alumina, γ-alumina; absorbent, adsorbent; alkaline silicate, siliceous silicate.

13. Will an alumina prepared at 1000°C. be active? Explain.

14. Explain why a soluble silicate prepared using salt cake as the source of base ion is less pure than one prepared using soda ash for this purpose. Give some possible impurities.

15. Describe the preparation of medicinal magnesium trisilicate. Explain why it is impossible to obtain a true silicate using this method.

16. Describe a synthetic method which would yield a true magnesium silicate. Why is this product unsuitable as a protective antacid?

PRACTICE IN EQUATION WRITING

17. Write the equations for reactions taking place during purification of siliceous earth.

18. Write equations representing the preparation of each of the following: silica fume from silicon(IV) sulfide; bayerite from sodium aluminate; silicic acid by hydrolysis of ethyl silicate; silica gel from

potassium silicate (1:2 ratio); γ-alumina from böhmite; 1:1 ratio sodium silicate from the 2:1 product (sodium hydroxide is used).

19. Write the equation for the reaction taking place when cobalt(II) chloride acts as a moisture indicator.

20. Write and balance the equation for preparation of 1:2 ratio potassium silicate by the American procedure.

21. Write the equation for hydrolysis of sodium orthosilicate; for sodium disilicate.

22. Write an equation showing why ammonium silicates have not been prepared.

23. Write the equations for all reactions taking place in the official assay for each of the following: magnesium trisilicate, both MgO and SiO_2; Purified Siliceous Earth, both acid solubles and sulfate; Sterile Corticotropin Zinc Hydroxide Suspension, for zinc.

24. Using $(SiO_2)_xSiOH$ as the formula for a silica gel particle, write the equation for the reaction with ethanol, water, aminomethanol.

ELEVEN · PHYSICAL PHENOMENA IN PHARMACEUTICAL PRACTICE II: *relationship between structure and use; the insoluble silicates*

SILICATE STRUCTURES—GENERAL CONSIDERATIONS

Anions, in general, are large and diffuse. Cations, especially those of high charge, are comparatively small and compact. Consequently space requirements of cations are negligible compared to the space requirements of anions. As a result anions tend to dominate structural arrangements. The oxo and silicon(IV) unit particles of the insoluble silicates are no exceptions; silicate structures may be considered close-packed arrangements of oxide ions. Certain physical properties of silicates—refractive index, density, hardness—are almost completely determined by the oxygen packing.

Two close-packed arrangements of oxygen, the hexagonal and the cubic, are possible. In each the arrangement of the oxygen in a given layer is identical to its arrangement in all other layers; placement of the layers with respect to one another differs. Hexagonal closest packing, common to several silicates, is shown diagrammatically in Fig. 11.1. If the diagram is examined closely or, better, if the arrangement is duplicated with small spheres, it is evident that two kinds of open space exist between the closely packed spherical anions. One allows a suitable entering cation to make contact with four surrounding oxygens, giving a tetrahedral arrangement with respect to the cation. The comparatively small silicon(IV), aluminum(III), and beryllium(II) ions may occupy these spaces. Larger cations will not fit; to force accommodation would destroy the oxygen arrangement. The second kind of space within the oxygen packing is somewhat larger. An entering cation makes contact with six neighboring oxygens to form an octahedral arrangement with respect to the cation. Magnesium(II), lithium(I), iron(II), iron(III), and aluminum cations have radii which allow them to occupy such octahedral spaces. Aluminum(III),

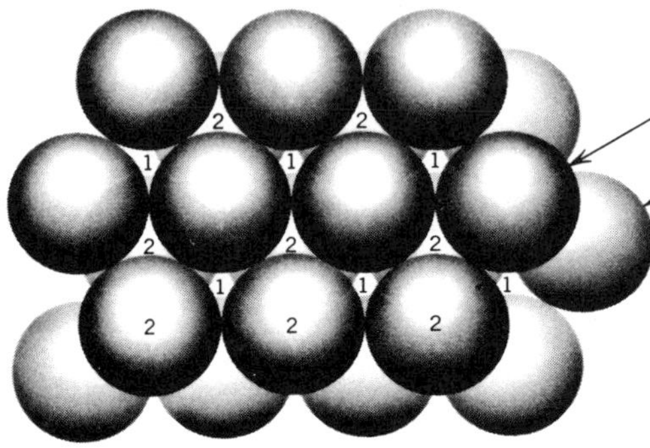

FIGURE 11.1 Hexagonal closest packing of oxide ions. Two layers are shown.

having an intermediate radius, may be accommodated in either the tetrahedral or octahedral holes.

Many common cations are too large to be accommodated in either the tetrahedral or octahedral holes. These include all alkali cations except lithium and all alkaline earths except beryllium and magnesium. When present in a silicate, these cations are accommodated in one of two ways. The silicate may assume a layer structure having slabs of the oxygen-silicon arrangement with alternate layers of the cations. The micas are representative of these structures. On the other hand, the oxide structure may open somewhat and allow formation of large cavities or holes, regularly spaced, within the oxygen-silicon network. Large cations may be accommodated within these holes. The zeolites have this structure. The three methods for accommodating cations are summarized in Table 11.1.

One of the problems which retarded the elucidation of the silicate structures was failure to appreciate that ions of an ideal structure may be substituted to some extent by other ions of approximately the same radius, charge differences notwithstanding. This phenomenon, now known as

TABLE 11.1 ACCOMMODATION OF CATIONS WITHIN THE CLOSE-PACKED OXIDE ARRANGEMENTS OF THE SILICATES

Space Arrangement	Cations Accommodated
Tetrahedral holes	Si, Al, Be
Octahedral holes	Mg, Li, Fe^{II}, Fe^{III}, Al
Layer and cavity structures	Ca, Ba, Na, K, NH_4^+

isomorphous replacement, is widely prevalent among the silicates. Because of isomorphous replacement empirical formulas based on analytical determinations may be meaningless. The substitution of aluminum for silicon is frequent. The net negative charge of the network increases by one unit for every entering aluminum ion. Iron, aluminum, and magnesium freely exchange with one another in the octahedral spaces. In some complex silicate arrangements hydroxyl groups are present. Fluoride ion may isomorphously replace varying numbers of the hydroxy groups.[1]

TABLE 11.2 SILICATE SPACE STRUCTURES

Shared Oxygens[a]/ Tetrahedron	Space Characteristic	Gross Structure	Empirical Formula (Ideal)	Nomenclature	Examples
2	1-dimensional	chains and rings	$[SiO_3]^{2-}$	metasilicate	asbestos
3	2-dimensional	layer or sheet structures	$[Si_2O_5]^{2-}$	disilicate	talcs clays micas
4	3-dimensional	lattice structures	$[SiO_2]^0$	—	feldspars zeolites ultramarines
0	3-dimensional	insular structures	$[SiO_4]^{4-}$	orthosilicate	olivines sodium orthosilicate

[a] Only oxygens shared with other silicon atoms are counted.

Because of the predominance of silicon(IV) in the silicate structures, silicates may be considered arrangements of silicon-oxygen tetrahedra. One, two, three, or all four of the oxygens of a given tetrahedron may be shared with one, two, three, or four other silicon (or aluminum) atoms, respectively. This is known as corner sharing; it is the only sharing found in silicates.[2] The actual number of oxygens per tetrahedron shared with silicon atoms of other tetrahedra depends on the space arrangement of a given silicate mineral. This relationship, together with relevant nomenclature, empirical formulas and examples is shown in Table 11.2.

CHAIN SILICATES

These unidimensional arrangements of silicate tetrahedra are achieved by the sharing of two oxygens per tetrahedron. All chains run parallel

[1] This phenomenon is of particular interest with the insoluble phosphates making up bone and teeth structures. The enamel of the teeth is rich in fluoride as a result of some isomorphous replacement of the normal hydroxy component of the mineral.

[2] Edge and face sharing exist, but are never found in the silicates. In edge sharing two oxygens (or other anions) are shared between the same two cations; in face sharing three anions are shared in common by two cations.

to the same axis. In effect each chain is a long one-dimensional macro-anion, having a negative charge equal to the number of unshared oxygens in the chain; a chain of 1000 silicate tetrahedra would have a net negative charge of 2004. Since the binding forces between the silicon and oxygen atoms making up these chains have about 50% covalent character, it is difficult to break a chain.

The essential electrical neutrality is maintained by lining the prerequisite number of cations between the individual chains; potassium and calcium ions are most common. Binding between adjacent chains is electrostatic, arising from the interaction of the positive field of the cation with the negative fields of neighboring chains. As electrostatic forces are weaker than covalent forces, the silicate chains cleave readily from one another, giving rise to the fibrous structure characteristic of asbestos.

There are two common chain structures. The metasilicates, or pyroxenes, consist of single chains. They are always anhydrous. The metatrisilicates, or amphiboles, are double chains—in effect two pyroxene chains crosslinked by oxygens. The amphiboles include water as hydroxo groups. The two structures are shown in Fig. 11.2. Petrologically the two forms are distinguished by their cleavage angles, 93° and 56°, respectively. The amphiboles are structurally somewhat similar to cellulose.

Asbestos. Chrysotile asbestos (serpentine asbestos), $(HO)_6Mg_6(Si_4O_{11})$ H_2O, is the most distinct form and the most widely used. It consists of very small hollow fibers (in micron range). The other asbestoses, collectively referred to as amphibole minerals, are significantly larger, solid fibers. The latter are commonly found as contaminants in chrysotile, talc, and other minerals.

Chain silicates are chemically inert. Adsorptive powers are minimal. Cation exchange properties exist to a very limited extent. The physical fibrous structure, together with this general inertness, accounts for the popularity of this mineral as a filter medium. The long-fibered amphibole asbestos is preferred. In 1973 the FDA took steps to eliminate the use of asbestos filters in the preparation of parenteral products. The mineral is also heat resistant and a heat insulator.

Asbestosis. This is a disabling lung disorder. It can cause difficulty in breathing, lung cancer, cancer of the membranes of the lining of the chest and abdominal cavity (mesothelioma), and less commonly, cancers involving any part of the gastrointestinal tract. The effects of exposure may be slow in appearing, sometimes not evident for 30 years or longer. The risk of asbestos-related diseases usually increases with the exposure time and the amount of asbestos in the environment. Not everyone exposed to the mineral develops cancer or other complications.

Attapulgite, $Mg_5Si_8O_{20}(OH)_2 \cdot 8H_2O$, has been proposed as a substitute for kaolin (page 288). The material is a double-chain silicate related to the amphiboles. Water molecules are held in the large open spaces which exist between the chains. Isomorphous replacement of magnesium by aluminum in octahedral coordination is extensive.

It is claimed that the adsorptive properties of the material are superior to those of kaolin.[3] There is some evidence for the adsorption of bacteria.

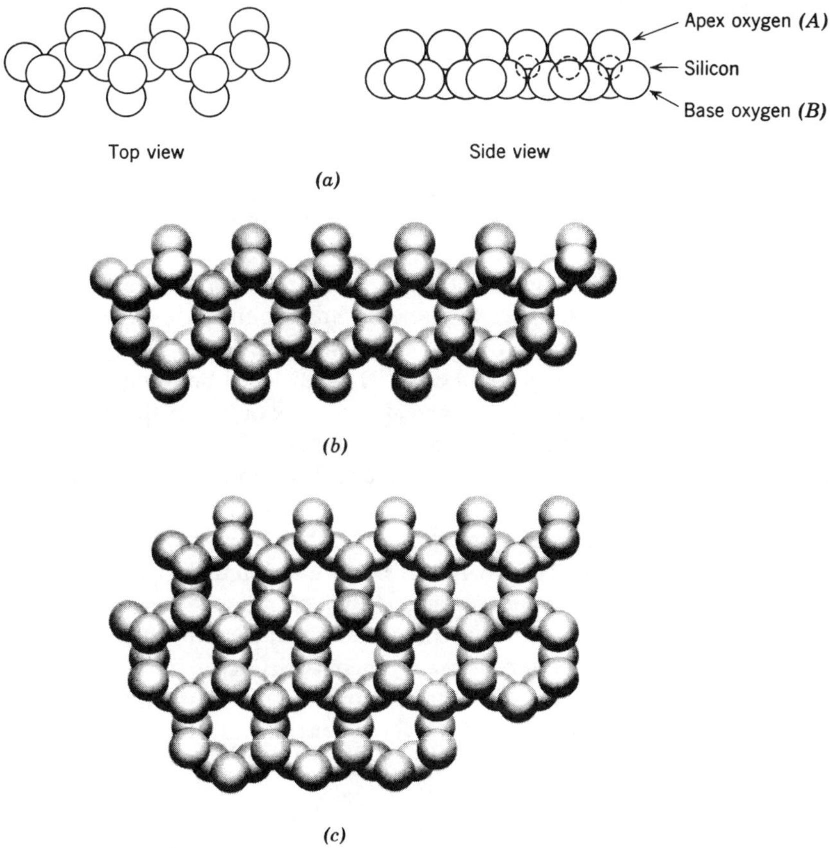

FIGURE 11.2 Basic arrangements of silicate tetrahedra in linear and planar silicate structures. Silicon atoms (small circles) are shown in the single-chain structure. (*a*) Single chains (pyroxenes); (*b*) double chains (amphiboles)—top view; (*c*) sheet arrangement—top view. The side view is identical for all three structures.

---

[3] Martin Barr, "Activated Attapulgite," *J. Am. Pharm. Assn., Pract. Ed.*, **19,** 85–87 (1958).

An aqueous suspension is available for control of diarrhea. Attapulgite clays are used industrially as adsorbants, thickening agents, anticaking agents, and suspending agents. They have also been used as carriers for insecticides and herbicides. This application is interesting in that the carrier-active compound combination can be tailored for sustained release of the active compound in a predetermined pattern.

LAYER SILICATES

General. If the structure of the chain silicates is extended in a second dimension, the layer or sheet silicates result. Each sheet consists of indefinitely repeating six-membered rings of silicate tetrahedra. Only one oxygen per tetrahedron is unshared, as shown in Fig. 11.2. The entire sheet acts as a macroanion. Since the interatomic binding within a silicate layer is by bonds with a strong covalent component, the sheets are mechanically very strong.

The individual silicate sheets of the layer minerals always consist of three planes of atoms: a tightly packed plane, *B*, holding the base oxygen atoms of each tetrahedron, a second oxygen layer, *A*, consisting of the apex atoms of all tetrahedra, and a layer of silicon(IV) sandwiched between occupying the tetrahedral holes (Fig. 11.2). The *A* layer has only one third as many oxygen atoms as the *B* oxygen layer. Consequently there are vacancies in the apex planes of the silicate sheets. These holes may be and frequently are occupied by hydroxide, chloride, and fluoride ions.

With a single exception, the arrangement of the successive silicate sheets in the layer structures is *AB BA AB BA* · · · , found in the micas, talcs, and montmorillonite clays. The exception is the *AB AB AB AB* · · · arrangement of the kaolin clays. For the usual arrangement of the sheets, *AB BA AB BA* · · · , two interfacial alignments exist between the successive silicate sheets, base to base, and apex to apex. The interfacial properties must differ since the density of their oxygen packing is different by a factor of three to one.

The successive sheets of the *AB BA AB BA* · · · arrangements are always firmly bonded between the apex to apex interfaces. The cementing materials are either brucite, $Mg(OH)_2$, or hydrargillite, $Al(OH)_3$. Sufficient hydroxyl ions are omitted to partially or completely compensate the negative charge of the silicate macroanion. The hydroxo groups enter vacant positions in the apex oxo planes of the two silicate sheets involved. The metal cations occupy octahedral holes between the two resultant oxohydroxo planes, as shown in Fig. 11.3. The binding is identical to that existing within the individual silicate sheets; cleavage never takes place between the *A–A* faces of this type of layer silicate structure. The brucite or hydrargillite layers, buried between the silicate sheets of the resultant sandwich-like arrangement, contribute little to the distinguishing properties of the individual silicates.

The cementing material between the *B* faces of the *AB BA AB* · · · arrangements is largely responsible for the distinguishing features of the individual silicates. It specifically determines the ease of cleavage with a given layer silicate. Cleavage always occurs between the *B–B* faces of the sheets, since in all cases the binding forces are weaker than those of the covalent bonds involved in the remainder of the structure.

Talc (*talcum*, *French chalk*, *purified talc*). This is the softest mineral known. In lump form it is known as *steatite* (soapstone). In this family there is no cementing material between the *B–B* faces of successive

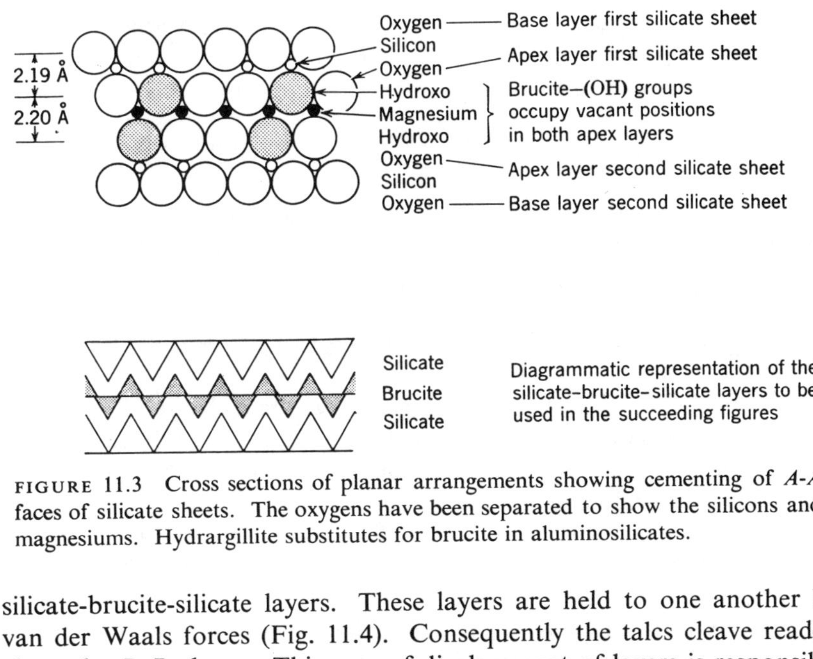

FIGURE 11.3 Cross sections of planar arrangements showing cementing of *A-A* faces of silicate sheets. The oxygens have been separated to show the silicons and magnesiums. Hydrargillite substitutes for brucite in aluminosilicates.

silicate-brucite-silicate layers. These layers are held to one another by van der Waals forces (Fig. 11.4). Consequently the talcs cleave readily along the *B–B* planes. This ease of displacement of layers is responsible for the smooth, unctuous feel characteristic of talcs.

Crude, natural talc is purified by boiling with 2% hydrochloric acid, followed by successive washes with 1% hydrochloric acid. Impurities, largely iron, calcium, and magnesium compounds, dissolve readily as the chlorides and are removed in the washing process. The treated silicate is then dried. The closely bound brucite layer is unaffected.

With both sides of the silicate-brucite-silicate structural groups being closely packed oxide ions, the talcs are characterized by their chemical inertness. They do not react with ordinary acids or bases and are insoluble in all ordinary solvents. They are soluble in hydrofluoric acid. This inertness of the exposed oxo surfaces extends to lack of adsorptive properties. Because of the flat platy character of the mineral, it has good covering power. Since all cations are bound within the covalent network, talc exhibits no cation exchange properties.

Talc has many uses in pharmaceutical practice. Inertness, lack of adsorptive properties, and particle size and shape render talc valuable as filter aid,[4] e.g., Aromatic Elixir, Sweet Orange Peel Tincture, and as a distributing agent,[5] e.g., aromatic waters, Magnesium Citrate Oral Solution. The absence of adsorptive properties is especially valuable since adsorptive losses are nonexistent on filtration of dilute solutions, e.g., alkaloid extracts of crude drugs. Particle size is important when used in either application. If too fine the particles are not removed by ordinary filter materials, but remain suspended to give a cloudy, murky, appearance to the filtrate. The most efficient particles are those passed by a No. 80 sieve but retained by a No. 100 sieve.

Because of its smooth, unctuous feel and good adherence, talc is used as a dusting powder and as a base for medicated dusting powders. For the same reason it finds wide use in cosmetic preparations. At one time it was used as a dusting powder and lubricant for surgeons gloves. Other preparations are now used for this purpose since any unabsorbable talc remaining in the tissues, particularly in the peritoneal cavity, acts as a center of irritation and possible ulceration.

The loose binding between layers makes talc valuable as a lubricant. It serves in foot powders for this purpose. It is used as a lubricating dust for tablet molds,[6] etc. Because of its inertness talc is used as a filler, e.g., soaps, and as a white pigment in cosmetics.

Micas. These layer structures have silicate-hydrargillite-silicate layers, similar to those of talc but cemented by electrostatic forces (Fig. 11.4). This is achieved by including a layer of alkali and/or alkaline earth cations, but not lithium, beryllium, or magnesium, between the *B–B* faces. These cations are too large to be accommodated in octahedral or tetrahedral holes and lie in the depressions existing where superimposed centers of the tetrahedral silicate bases come to lie over one another. Because of the resultant keying effect, micas are elastic.

---

[4] *N.F. XVI,* pp. 1533, 1583.

[5] *U.S.P. XXI,* pp. 1342, 611.

[6] The stearates, usually with varying quantities of palmitates, are also used as lubricants. The sodium (soluble) and magnesium (insoluble) compounds are used most frequently.

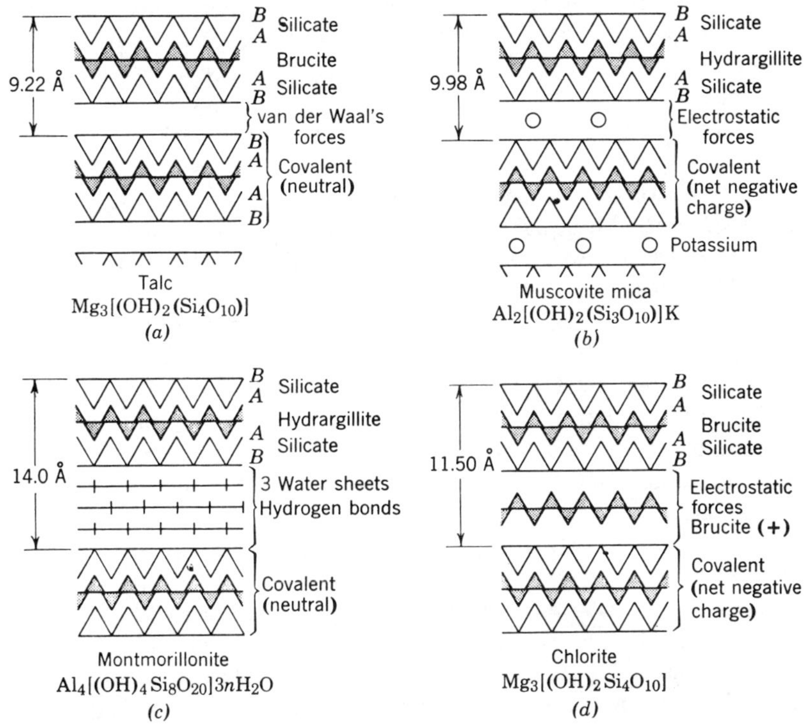

FIGURE 11.4 Cross sections of planar arrangements showing cementing of the *B-B* faces of silicate structures: (*a*) talc, (*b*) muscovite mica, (*c*) montmorillonite, (*d*) chlorite.

The forces involved in cementing the successive layers together are exactly those responsible for an ionic crystal. As electrostatic bonding forces are stronger than van der Waals forces, these layer silicates cleave with greater difficulty than the talcs, but the cleavage along the cation planes is still comparatively easy. Ease of cleavage is directly related to the charge of the cations present.

Like the talcs, the micas are chemically inert and have no pronounced adsorptive powers. They are excellent heat and electrical insulators. They have ion exchange ability, but not to a degree sufficient to be useful. Micas have no important pharmaceutical applications.

Chlorites. This is a group of magnesium silicates having structures and properties very close to the micas. Electrostatic binding of the layers is achieved by replacing the cations of the micas with positively charged brucite sheets, as shown in Fig. 11.4. It may truly be said that this mineral consists of alternate slabs of planar macroanions and planar macrocations.

The silicate layer is rendered more negative by isomorphous replacement of silicon by aluminum, while the brucite layer (which in its pure state is without charge) achieves its positive charge by isomorphous replacement of magnesium by aluminum and by omission of some hydroxo groups. Each aluminum (+3) substituting for a magnesium (+2) gives a single positive charge to the brucite layer. The chlorites are not elastic since the keying action of the individual spherical cations is lost.

Clay Minerals. These substances divide into several families: the montmorillonites (now also known as smectites), the kaolins, the illites, and the vermiculites. Montmorrillonite is the principle constituent (about 90%) of the montmorillonite or bentonite clays. Nonite and saphonite are also clays of this family. Kaolinite, frequently accompanied by halloysite, is the most common kaolin clay. Halloysite is second in abundance in the kaolin family. China clay is relatively pure kaolinite. As found in nature none is even moderately pure. Most deposits are difficultly separable mixtures of the clay with variable amounts of finely divided quartz, feldspar, and mica.

Apart from their role as a fundamental constitutent of the soils, clay minerals find wide use in many industries and have important applications in pharmacy and medicine. The ceramic and paper industries are the largest users of clays. However, the adsorptive and detergent properties lead to many other uses. During World War II European soaps had high percentages of montmorillonite. The gelling and ion-exchange properties also lead to important applications. It should be observed in the following discussions that the uses of these minerals depend almost invariably on surface properties of the otherwise chemically inert particles of clay.

The two main groups of clays—the montmorillonites and kaolins—differ in several important aspects. As explained earlier, the stacking of the layer structures differs: *AB BA AB* · · · and *AB AB AB* · · · , respectively. The montmorillonite group has higher $SiO_2:Al_2O_3$ ratios than the kaolins, and other basic oxides such as Fe_2O_3, MgO, or CaO frequently substituted in varying degrees for the aluminum oxide component. The kaolins never have a magnesium component. The essential features of the two classes of clays are compared in Table 11.3.

THE BENTONITES: The distinguishing feature of these clays is the stacking of one or more water layers between the *B–B* faces of the silicate-hydrargillite-silicate layers. The oxygens of the water molecules are arranged in a nearly coplanar hexagonal pattern, and are hydrogen bonded to one another (Fig. 11.5). The hydrogens not required for bonding the oxygens in their planar network extend above or below the plane. They are available to form hydrogen bonds with oxygens in adjacent

TABLE 11.3 Comparison of the Essential Structural Features of the Two Principal Clays

Class	Bentonite	Kaolin
Principal constituent	Montmorillonite	Kaolinite
Ideal formula	$[Al_4(OH)_4][(Si_8O_{20})]\cdot n(4H_2O)$	$[(OH)_6Al_4][(OH)_2(Si_4O_{10})]$
Charge balance formula	$[Al(OH)]_4^{2+}(Si_8O_{20})^{-8}nH_2O$, where $n = 0, 1, 2,$ or 3	$[Al(OH)_2]_4^{+}(Si_4O_{10})^{-4}$
Stacking of silicate tetrahedra	$\overline{AB}\ \overline{BA}\ \overline{AB}\ \overline{BA}\ \cdots$	$\overline{AB}\ \overline{AB}\ \overline{AB}\ \overline{AB}\ \cdots$
Depth of unit cell along *c* axis	9.13 Å to 14 Å (depending on the number of water layers)	7.1 Å
Miscellaneous	Al may be isomorphously replaced by Mg, Fe(III)	Al is never isomorphously replaced by Mg
Common synonyms	Soap clay, mineral soap bentonite	Bolus alba, white bole, argilla, china clay

layers, either another water layer or the base of a silicate layer. The hexagonal oxygen arrangements of the silicate structures and the water sheets have almost identical dimensions. Since hydrogen bonding is intermediate in effectiveness, these silicates cleave readily into flaky particles. The three common groups having this structure are identified in Table 11.4. The montmorillonite structure is illustrated in Fig. 11.4.

The halloysites are frequently present as minor constituents in the clays used in pharmacy. The vermiculites usually crystallize in soft, pliable, golden yellow or brown lamellae. In structure they are built up of inert talc-like layers separated by double water layers. But unlike the talcs, these minerals have pronounced ion exchange properties. One half of

TABLE 11.4 Layer Silicates with Water Binding

Mineral	Ideal Formula	Charge Equivalency of Units	Water Layers
Halloysite	$[(OH)_8Al_4][Si_4O_{10}]\cdot 4H_2O$	$[Al(OH)_2^{+}]_4[Si_4O_{10}^{4-}]\cdot 4H_2O$	1
Vermiculite	$[(OH)_4Mg_6][Si_8O_{20}]\cdot 8H_2O$	$Mg_4^{2+}[MgOH^{+}]_2[Si_8O_{20}^{8-}]\cdot 8H_2O$	2
Montmorillonite	$[(OH)_4Al_4][Si_8O_{20}]n(4H_2O)$	$[AlOH^{2+}]_4[Si_8O_{20}^{8-}]n(4H_2O)$	n
	$n = 0, 1, 2,$ or 3		

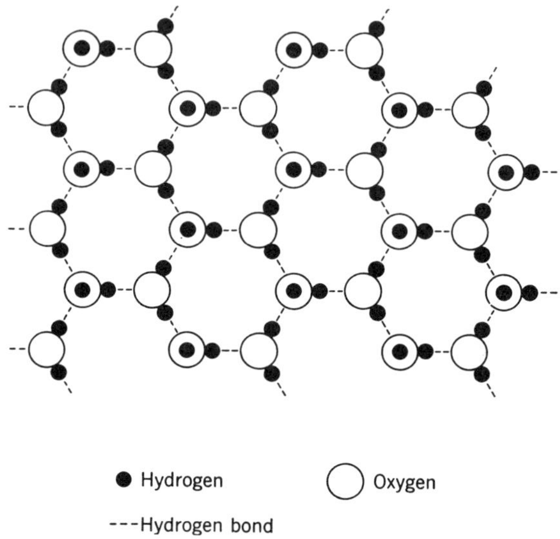

FIGURE 11.5 Water sheets of layer silicates.

the binding water of vermiculite is driven off at 110°C., without exfoliation. If heated above this temperature the remaining binding water is lost, accompanied by exfoliation and loss of ion exchange properties. The flakes expand twenty to thirty times their original thickness along the crystal's *c* axis (perpendicular to the water plane) to give wormlike forms known as *expanded mica.* If vermiculites are heated to 750°C., the layer structure is destroyed.

Expanded mica has excellent shock-absorbing qualities. This, together with its inertness and absorbing properties, makes it an excellent packing material for breakable containers of corrosive chemicals. Because of very low density it is used as an inert, lightweight filler. It is an excellent carrier for driers, e.g., P_2O_5. By mixing the drier with expanded mica, 8–12 or 12–20 mesh, a more porous structure is obtained and the drying action is more efficient. Expanded mica is an excellent sound and heat insulator.

Approximately 1.3 magnesium ions per vermiculite unit cell are replaceable by sodium, potassium, or calcium ions. The molecules of the water layer may be partially replaced by ammonium ions. Glycerine and amines may enter the structure between the layers to form complexes.

The x-ray spectrum of montmorillonite reveals two distinct types of bands. The general bands are not affected by the hydration state of the system. They are characteristic of the nature of the layers making up the particular montmorillonite and are important for identification purposes, since isomorphous replacement is rife within the group. Typical compositions vary widely from the ideal formula given in Table 11.4. The second group of x-ray lines, the basal lines, vary with the state of the system. They are indicative of the spacing between the layers making up the clay. For this reason their spacing will vary with the water content (0, 1, 2, or 3 layers). Variation in the number of water layers is responsible for the "swelling" property of the bentonite clays.

Although the bentonites have the usual chemical inertness of the insoluble silicates, they exhibit many valuable and interesting properties. Ion exchange properties arise from the isomorphous replacement of aluminum by magnesium in the hydrargillite layers during the crystallization of the original clay structure. Since magnesium carries only a double positive charge, the negative charge of the clay layer is increased by one unit for every substitution. An equivalent number of cations must be accommodated in the space between the layers since the tetrahedral and octahedral holes of the structural layers are fully occupied. These cations are exchangeable. The exchange properties vary widely; montmorillonite has exchange capacities ranging from 50–100 me./100 Gm. Bentonites are usually available in the sodium form.

Montmorillonite in the hydrogen form is an excellent cation exchanger. Silver, the alkaline earths, and uranyl ion exchange efficiently from solution up to 300°C. and under intense γ radiation. Most common organic cation exhangers degenerate under these rugged conditions. Montmorillonite has about 1.5 times the theoretical exchange capacity when the substituting cation is very large, e.g., dodecylpropyleneamine; the net charge equivalents of the cations released is less than the total charge of the entering cations. This occurs because sufficient anion, e.g., chloride, is adsorbed to maintain the electrical neutrality of the montmorillonite particles. The exchange balance may be expressed as

$$\begin{aligned}\text{153 me. cation fixed} &= \text{102 me. } Ca^{++} \text{ released} \\ &\quad + \text{1 me. } H^{+} \text{ desorbed} + \text{50 me. } Cl^{-} \text{ fixed}\end{aligned}$$

The montmorillonites are effective adsorption agents. The dipole moment, dipole volume, and degree of asymmetry of the adsorbed phase determine the avidity of the binding. Water, alcohols, amines, ethers, and ketones have the proper parameters of these properties for effective adsorption. The ions present and pH also have an effect.

Alcohols enter between the structural layers, forming flat two-dimensional sheets. In so doing they increase the thickness of the mineral, depending on the number of individual sheets laid down. Amino acids and amines are readily adsorbed, being held by a combination of van der Waals and electrostatic forces. Maximum adsorption occurs as a cation at a pH of 2. The cations, oriented parallel to one another, also arrange in sheets between the structural layers. If the space between these layers is fully utilized, e.g., occupied by an amine, other molecules such as ethanol cannot enter.

Bentonite clays are available in a wide range of purities. The very highly refined grades are white and are available commercially. In the preparation of suspensions the dry clay should be sprinkled on the proper volume of actively agitated water. Once uniformity is attained, the suspension should be set aside overnight to allow complete hydration of the clay particles. Pure suspensions have a pH of 9–10. Stability is best above pH 7, although acid systems may be used. Gelation is enhanced by the presence of a mild alkali, such as MgO. If a formulation is to include electrolytes, addition of the bentonite suspension to the electrolyte solution results in a thinner suspension than if the reverse order is used.

The gelling property of the bentonites make them valuable as suspending and emulsifying agents. A 5% suspension is official in *N.F. XVI* as a suspending agent.[7] Calamine lotion is stabilized. Gels of ammoniated mercury, zinc oxide, sulfur, bismuth subnitrate, etc. have been prepared. Bentonite may be used in the preparation of ointment bases suitable for either water-soluble or oil-soluble drugs. The properties of these systems are modified by the form of bentonite employed. Ammoniated Mercury Ointment prepared using the hydrogen form shows greater antibacterial action than that prepared using the sodium form.

Bentonite clays are detergents. They appear in soap, dentifrices, shaving creams, and cleanser formulations. They also serve as clarifiers and adsorbants. Turbidities are eliminated and dyes and other colored materials may be removed from both aqueous and oil media.

Bentonite has been used in gel form as a bulk laxative. However, continued use in this capacity may produce dietary deficiencies as the gel may adsorb trace elements and certain vitamins from the intestinal fluids. Applied to the skin, topical preparations including 15% bentonite dry to a thin film. This system serves to hold active medicants *in situ*. Bentonite is used as an absorbant powder in dermatology and as a facial pack in cosmetics.

THE KAOLIN CLAYS: The kaolin clays are the one family of layer silicates in which the sheets of joined silicate tetrahedra line up apex to base rather than apex to apex, base to base. The apex spaces of the silicate layers accommodate hydroxo groups of the associated hydrar-

---

[7] *N.F. XVI*, p. 1534.

gillite layer. Because of the negative charge of the silicate sheet, ideally one hydroxo is omitted per aluminum to provide the compensating positive charge. In effect the hydrargillite sheet has the empirical composition, $[Al(OH)_2]^+$, and the resultant silicate-hydrargillite structures are neutral. The successive silicate-hydrargillite layers are cemented by hydrogen bonds and van der Waals forces. The hydrogen bonds are formed by bonding outside hydrogens of the hydrargillite portion of the structure to oxygens of the base of the facing silicate sheet of the next clay layer. The resultant structure is shown in Fig. 11.6.

Kaolins always occur as microparticles of colloidal dimensions, never as large crystals. Symmetrical layers are essential if a layer structure is to grow into large crystals. When, as with kaolin, the two faces of a layer structure are not equivalent there is a tendency toward curvature of the crystal, one face convex, the other concave. Under such conditions it will grow to only a limited extent before the curvature becomes extreme and breakage into particles results. The nonexistence of magnesium analogs

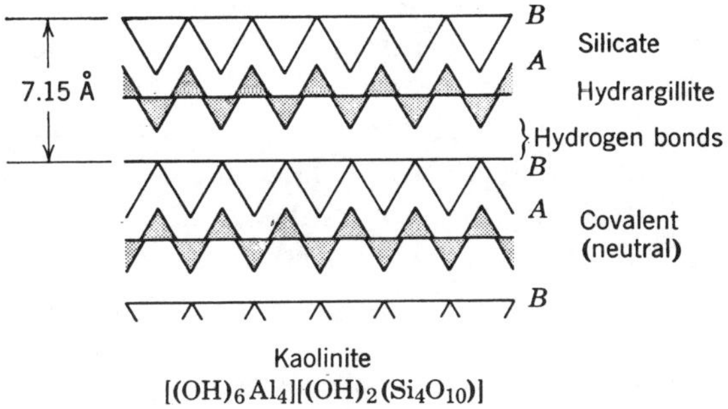

FIGURE 11.6 Structure of kaolinite.

of kaolinite may be attributed to the larger values of the fundamental brucite sheet (5.40 × 9.36 Å) as compared to the silicate sheet (5.05 × 8.71 Å). Integration of the two structures to give the kaolin arrangement would result in extreme curvature and a resultant structural instability.

The properties of kaolins are similar to those of the bentonites. The kaolins are used as clarifying agents. They also are good excipients for salts which are decomposed by organic substances, e.g., silver nitrate. Flocculation of kaolin has been used as a diagnostic aid in rheumatoid arthritis.

Kaolin finds use as an intestinal adsorbant in various enteritis. Its action is twofold: It is thought a protective coating forms on the irritated mucosa, and it acts as an adsorbant of toxic substances, e.g., toxic amines, organic acids of decomposed food, and possibly bacteria. However, the kaolins also adsorb certain vitamins, enzymes, alkaloids, and minerals. Their continued use can lead to failure of medication, to a deficiency of one or more essential trace minerals, and to interference with normal digestive processes. The wide variability of the adsorptive powers of kaolin clays and a tendency to flocculate into sticky masses (due to electrolytes in the intestinal fluids) are disadvantages. Kaolin must not be used when an intestinal obstruction exists.

Externally kaolin finds use as a dusting powder. It is especially useful as a drying agent for moist sores and infections, e.g., weeping eczema and weeping ulcers. A paste in olive oil is used to protect the skin around iliostomy and cecostomy openings from the digestive juices. Kaolin alone and in combination, e.g., with boric acid, finds use as a carrier for both heat and moisture at local inflammations (cataplasm),

Because of the wide variability of the native clays and the range of extraneous materials which may be present, careful preparative techniques are necessary. For therapeutic uses the fine china clays are preferable; they are whiter and have fewer extraneous materials.

The crude clay is treated by elutriation; though quartz, feldspar, mica, etc., are removed the product remains gritty. In order to obtain a finer clay the elutriated material is peptized in water by addition of a suitable electrolyte, e.g., sodium pyrophosphate. The finer particles become sufficiently charged to suspend, the heavier particles drop out. The resultant suspension of fine particles is flocculated either by addition of a large excess of electrolyte or by subjecting it to an electrostatic field. The flocculated clay is washed with water and dried to a soft white, or light yellow powder. The *B.P.* distinguishes between a light kaolin, for internal use, and a heavy kaolin, for external use.

Fuller's earth is a nonplastic clay closely resembling kaolin and having essentially the same properties. It is used as a clarifier and deodorizer for fats and oils.

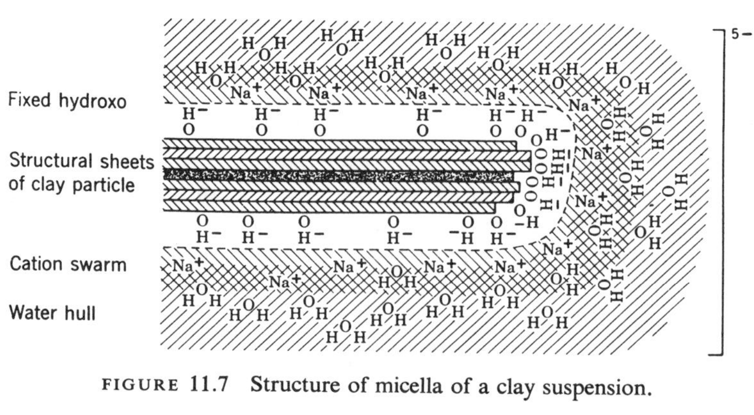

FIGURE 11.7 Structure of micella of a clay suspension.

THE NATURE OF CLAY SUSPENSIONS: The physical and chemical properties of the montmorillonite and kaolin suspensions are sufficiently similar to allow a general discussion. Every suspended clay particle behaves as though it is a colloidal electrolyte, in effect a weak acid macroanion arising by partial ionization of surface hydroxo groups. Because of its negative charge a given clay particle is closely associated with a "swarm" of cations, the *cation hull.* The number of closely associated cations is less than required for neutrality, a net negative particle charge persists. Integral with the cation hull is a shell of associated water, the *water hull.* The clay particle, together with its cation swarm and water hull is known as a *micella.* Figure 11.7 gives a diagrammatic representation of a suspended clay particle. An inverse relationship exists between the size of the cation hull and the water hull.

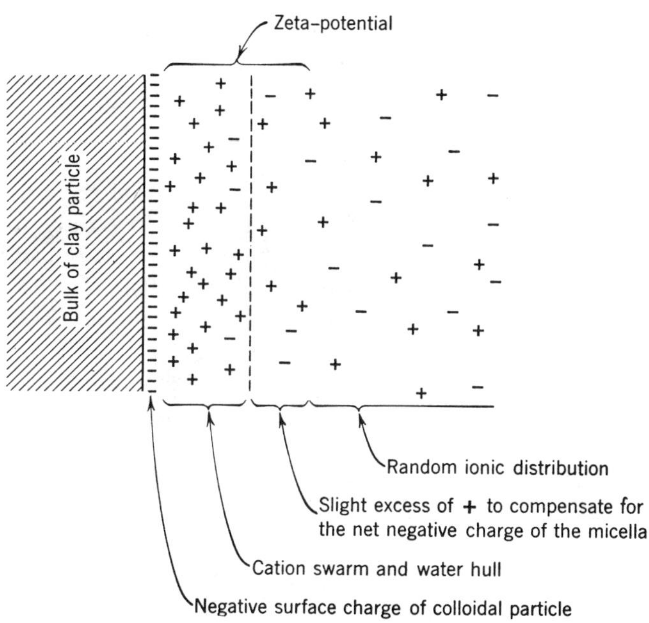

FIGURE 11.8 Charge structure of a micelle and its environment.

A difference in potential, the zeta-potential, exists between the surface of a clay particle and the bulk of the suspending medium. This has been diagrammed in Fig. 11.8. Zeta-potentials of stable suspensions range from 15–70 millivolts. Flocculation takes place at and below the lower limit.

Zeta-potential is a function of the particle size and bulk concentration of the clay together with the ionic radius, polarizability, and hydration of the cations present. The dielectric constant of the suspending medium is also a factor. The effects of these variables are tabulated in Table 11.5.

TABLE 11.5 SUSPENSION VARIABLES AND ZETA-POTENTIAL

Variable	Relationship to Zeta-potential
Clay	
Particle size	Inverse
Bulk concentration	Inverse
Cation	
Charge	Inverse
Radius	Direct
Polarizability	Inverse
Degree of hydration	Direct
Suspending medium	
Dielectric constant	Direct

The cation and water hulls of a clay particle determine the gross physical properties of the suspension. If the water hull is tightly fixed, rigid suspensions result. If the cation is only slightly hydrated or unhydrated, e.g., barium, granular dispersoids result. Highly hydrated cations, e.g., sodium, lithium, give rise to viscous, slimy suspensions. Thixotropy is observed only with highly hydrated cations, e.g., sodium bentonite, 2–3%, forms thick thixotropic systems. Too high cation concentrations diminish the thixotropic effect, if sufficiently high flocculation takes place.

TABLE 11.6 COMPARISON OF LAYER SILICATES

Family	Hardness (Moh Scale)	Layer Thickness (Å)	Particle Size	Interlayer Binding	Cation Exchange Ability	Adsorption Properties
Talc	1–2	9.22	macro	van der Waals	nil	nil
Mica(M^+)	2–3	9.98	macro	electro-static	slight	nil
(M^{2+})	4–6					
Montmoril-lonite	2	9.4–14.0	colloidal	hydrogen	good	good
Kaolinite	2	7.15	colloidal	van der Waals and hydrogen	good	good
Chlorite	2–2.5	14.23	colloidal	electro-static	slight	slight
(Silica)[a]	(7.0)	—	(macro)	(covalent)	none	none

[a] Silica is included for comparison purposes.

ACTIVATED CHARCOAL: Norite, activated carbon, *carbo activatus*, blood charcoal, bone charcoal. A discussion of the uses of clay minerals in pharmaceutical practice and medicine is incomplete without a parallel discussion of charcoal. The two have many interesting points of similarity both in properties and application.

Modern activated charcoals are usually wood charcoals which have been activated by one or more procedures. Lignite, pulp mill char, peat, coal, petroleum coke, and coconut shells are also used. Carbonization is carried out in the absence of air at temperatures less than 600°C. Sometimes the wood is pretreated with a dehydrating agent, e.g., sulfuric acid, zinc chloride, before pyrolysis. Carbonization is followed by a heat and/or chemical treatment to increase porosity and surface area, usually an oxidative process. Steam and carbon dioxide are used at 800–900°C., and air can be used in the temperature range of 300–600°C. Chemicals such as sodium sulfate or dolomite are sometimes mixed with the char. They release oxidizing gases within the mass at high temperatures. Upon completion of this treatment, acid or base is added to adjust the pH to that suitable for the particular application intended. The final preparative step is the removal of solubles with a water wash.

Structurally activated charcoals consist of a series of carbon sheets. Activation brings about a greater separation of these sheets with an accompanying increase of surface area. Charcoals come in two gross forms. Those intended for gas adsorption are coarse, hard, porous particles. This is necessary to allow an easy flow of the gases through the carbon bed. Those intended for adsorption procedures in liquid media are finely ground, 300–350 mesh, to provide maximum surface per unit weight.

Adsorption from gaseous medium is related to the boiling point of the substances in contact with the charcoal. Other variables being equal, the gas closest to its boiling point will be adsorbed first and most tenaciously. This is logical since the heat of adsorption for a given substance is of the same order as its heat of vaporization.

Charcoal is widely used in the practice of pharmacy and medicine. It is used as a filter aid and for adsorption of colored and odorous substances, acids, bases, and neutral salts. It is used as a gastrointestinal adsorbant, frequently in a one to one combination with kaolin. However, the use of activated charcoal for absorption of gases from the gastrointestinal tract is of questionable value; to absorb gases the surface of the charcoal must be dry, an impossible situation in the intestinal tract. Activated carbon is also used as a poultice and dressing for suppurating wounds of various types, odorous, gangrenous, and foul wounds.

Activated carbons find numerous applications in the pharmaceutical industry. They can be used for removal of pyrogens from solutions of inorganic salts. Penicillin is adsorbed from weak solutions. It is recovered from the charcoal in concentrated solution by treatment with a suitable eluent. Streptomycin and some hormones are similarly concentrated. Charcoal is used for odor and taste improvement of potable waters (page 135). Ten to twenty pounds per million gallons of water is usual.

Charcoal is used as an antidote for poisoning by mercury(II) chloride, strychnine, phenol, atropine, oxalic acid, phenolphthalein, and mushroom. It is a component of universal antidotes as an adsorptive.

Molecular sieve carbons have been prepared by pyrolyzing polymers. Structural arrangements arise having properties similar to the synthetic zeolites (page 297). Table 11.6 reviews the distinguishing features of the layer structures studied.

THREE-DIMENSIONAL SILICATES

As shown in Table 11.2 there are two classes of silicates which are covalently bound in all directions within their structure, the "insular" silicates and the "lattice" silicates. The four oxygens of each silicate tetrahedron are shared with four other cations in both. In the insular silicates oxygens are never shared between two silicons of different tetrahedra. In the lattice structures all oxygens are shared between silicate tetrahedra.

The insular structures, mainly ortho- and disilicates, have no direct applications in pharmaceutical chemistry. The oxygen is in hexagonal close packing. Silicon (or aluminum in isomorphous replacement of silicon) occupies the tetrahedral holes, and magnesium, iron(II), iron(III), and aluminum occupy the octahedral holes in the oxygen packing. Because of their completely covalent structure these silicates are characterized by high densities, extreme hardness, and chemical inertness. Even adsorptive ability is absent.

Insular silicates. Some examples of insular ortho silicates are the olivines, Mg_2SiO_4, coffinite, $USiO_4$, the phenacites, $M^{2+}{}_3M^{3+}{}_2(SiO_4)_3$, e.g., $Mg_3Al_2(SiO_4)_3$, and the chondrodites which have alternate slabs of Mg_2SiO_4 and $Mg(OH)_2$. An extremely inert, hard alumino-silicate series includes sillimanite, Al_2SiO_5 and mullite, $Al_6Si_2O_{13}$, suitable for fabrication of extremely inert and abrasion resistant mortars and pestles. The aluminum of sillimanite and mullite is partially in octahedral holes and partially in isomorphous replacement of silicon in tetrahedral holes.

Insular silicates may have disilicate $[Si_2O_7]^{6-}$ or the cyclic anions, $[Si_3O_9]^{6-}$ and $[Si_6O_{18}]^{12-}$. The latter is the basis of beryl, emerald, and aquamarine. The six-membered rings are so arranged that channels exist through their openings, providing space for small molecules within the cages thus formed. A cage plus its entrapped molecule, or "guest," is known as a *clathrate*. If the over-all structure is stoichiometrically exact the resultant silicate is neutral; only neutral molecules, e.g., water, helium, may be guests. If the aluminum:silicon ratio is less than 1:3, small anions such as OH^- and F^- may be guests. If the ratio is greater than 1:3 small cations such as lithium may enter the channels.

Lattice silicates. The lattice silicates have the empirical formula SiO_2 which perforce, is neutral. The network assumes a net negative charge because of extensive isomorphous replacement of silicon(IV) by aluminum(III).

Space utilization is poor in lattice silicates, holes remain within the four- and six-membered rings formed by the constituent tetrahedra. This accounts for their low densities. The cations, necessary to compensate for the negative charge of the anionic network, occupy the holes within the ring structures.

The distinguishing feature of the different lattice silicates is the arrangement of the holes and cavities with respect to one another. If the holes of the four or six-membered rings align, chambers with interconnecting channels result. The zeolites and ultramarines have this alignment of holes. In the feldspars there is no alignment of the openings.

The *feldspars* make up two thirds of all igneous rocks. They are aluminosilicates of sodium, potassium, and calcium, and rarely, barium and lithium. Magnesium and iron are never present. Within the group, orthoclase, ideally $KAlSi_3O_8$, is the most common of all minerals. The structural differences within the feldspar family arise from differences in cation size. The potassium feldspars are monoclinic, the sodium and calcium triclinic.

The densities are within the range 2.5–2.8 Gm./ml., except for a barium feldspar. Because of the covalent bonding of the anionic network, each piece of rock is a single macroanion, with entrapped cations. The hardness is high, 6–6.5 on the Moh scale (silica = 7.0). The feldspars are white, or pale shades of yellow, red, or green.

The entire family is chemically inert. Adsorptive properties are absent. Because the cations are trapped within the closed spaces of the crystal, ion-exchange properties are absent.

Pumice is a porous volcanic rock of low density. It will float on water until waterlogged. Structurally it is a form of obsidian, which is usually found in the vitreous state. Being a three dimensionally linked silicate, it is hard. Pumice is inert both chemically and adsorptively.

These properties make pumice valuable for use as a filtering medium and dispersing agent. It is found in dental preparations as an abrasive.

The *zeolites* and *ultramarines* differ structurally from the feldspars only in that the open spaces enclosed by the four and six-membered rings of silicate tetrahedra are interconnected. The cations are no longer trapped within the cavities but are free to move throughout the crystal. Under suitable conditions the cations may leave the crystals provided an equivalent number of other cations take their place within the macroanion's holes. This freedom of movement of the cations accounts for the excellent cation exchange properties of the zeolites. Sometimes simple anions, e.g., fluoride, find their way into the cavities. This is allowed, provided enough extra cations enter to maintain electrical neutrality.

THE ZEOLITES: The structure of many zeolites, both natural and synthetic, has been elucidated. Chabazite, $(Ca)Al_2Si_4O_{12}\cdot 6H_2O$, consists of parallel hexagonal arrangements of silica-alumina tetrahedra. Eight of these prisms link together to form a partially enclosed cavity. Each cavity is connected to six adjacent cavities through apertures about 3.9 Å in diameter. One cubic inch of chabazite has 5×10^{21} cavities. Each cavity contains twenty four water molecules and four calcium ions, or their equivalent of other cations. Figure 11.9 illustrates a zeolitic structure.

Water is the only neutral molecule found naturally in the cavities, however, other molecules may enter, e.g., ethanol, hexane, mercury. Due to its dipole, water arranges itself on the inner wall of the cavity, the

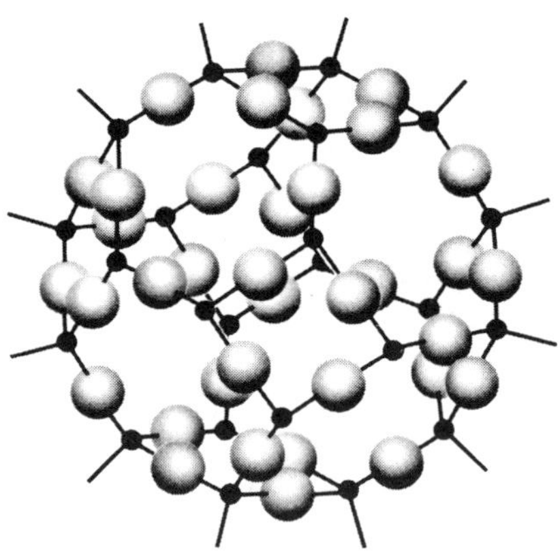

FIGURE 11.9 Open framework silicate structure of the zeolite chabazite.

positive dipole pointing toward the anion net. The electrostatic interaction between the water dipole and the negative charge of the net is so strong that this socalled *zeolite water* is retained above its boiling point. Zeolite water is not hydrate water since there are no breaks in the vapor pressure curves.

Dried zeolites take on water avidly. The reaction is reversible. The zeolite structure is stable at dehydration temperatures. Because of the avidity with which they take on water and the ease of regeneration zeolites are used to scavenge residual quantities of water in gases, even at high temperatures.

Synthetic zeolites, known as *molecular sieves*, are prepared by reaction of suitable silicates, aluminas, and base (Table 11.7). Variation of the raw

TABLE 11.7 RAW MATERIALS FOR PREPARATION OF MOLECULAR SIEVES

Element	Sources
Si	Silica gel, silicic acid, sodium silicate
Al	Alumina, alumina trihydrate, sodium aluminate
Na	Sodium hydroxide (also serves to control pH)

materials selected, and of the reacting ratios, makes possible a wide range of synthetic zeolites. A synthetic zeolite is conveniently prepared by the addition of an aqueous solution of sodium aluminate in sodium hydroxide to a solution of sodium silicate. (Potassium compounds may be substituted.) The resultant mixture is stirred until homogeneous, i.e., until all gels are broken down. Zeolite will crystallize at room temperature, however, crystal growth is faster at higher temperatures ($\leqslant 130°C.$). The crop of crystals is washed with water until the wash water has a pH between 9 and 12. The crystals are dried in a vented oven at 25–150°C.

The following data have been published for Zeolite A:[8] Reaction ratios, $Na_2O:Al_2O_3:SiO_2:H_2O::0.99:1.0:1.85:5.1$. Cubic crystals ($10^{-3}$ mm.) grow over a six-day period at 21°C.; in 45 minutes at 100°C. They are stable up to 700°C. The cavities are approximately spherical, with a volume of 925 $Å^3$, and account for over one half of the crystals' total volume.

To serve as cation exchangers zeolites must satisfy certain conditions. The cations must be small enough to move through the smallest openings in the channels within the crystals. Charge balance must be maintained, e.g., $2Na^+ \approx Ca^{2+}$. Finally, open positions must be available in the cages. The exchange process may be represented as a true ionic equilibrium, controlled by the law of mass action:

$$Na_2[Al_2Si_4O_{12}]2H_2O + 2KCl(aq) \rightleftharpoons K_2[Al_2Si_4O_{12}]2H_2O + 2NaCl(aq)$$

[8] U.S. Patent (Linde) 2882243 and 2882244.

Zeolites are excellent gas adsorbants. Their ability in this field has only recently been studied and applied industrially. A zeolite crystal may be compared to a "jungle gym" through which small children may move easily, but through which larger children, depending on their size, can move only with difficulty, if at all. Molecules, depending on size, may or may not move in and out of molecular sieves.

The adsorption properties of a molecular sieve are much more selective and more readily controlled than those of the more conventional agents, e.g., silica gel, charcoal. The critical pore diameters of the zeolitic material can be tailored exactly for a given application whereas in the ordinary adsorbers the pore diameters vary widely.

Adsorption by a zeolite is dependent upon the cations present in the cavities, the effective aperture or "mesh" of the sieve, and on the entering molecule. The cations are essential; adsorption depends upon the electrostatic charge of the cation for interaction with the dipole, permanent or induced, of the molecule being adsorbed. Consequently divalent cations, because of their stronger fields, are more effective than the univalent. The cations are also important because they mediate the effective diameters of the apertures connecting the cavities within the zeolite.

Molecules with diameters up to 0.5 Å larger than the effective aperture diameter of the zeolite cavities may pass through freely. Ethane with a diameter of 4.0 Å passes through the pores of the sodium form of Linde Type A molecular sieve of pore diameter 3.5 Å, whereas propane, 4.9 Å, cannot. This apparent anomaly is explained by the fact that both the aperture atoms and the atoms of the entering molecule are vibrating with respect to one another. As a result the active effective diameter of an aperture is greater than its "free" diameter. At the same time the kinetic energy of the incoming molecules allows them to "shoulder" their way through the apertures. Since increased temperature increases both the modes of vibration of the aperture atoms and the kinetic energy of the entering molecule, heating has the effect of increasing the mesh of a given zeolite, cooling decreases the mesh.

The cations of a given cavity are stationed either within the cavity or near the aperture. Those near an aperture partially block it. Thus the mesh of a molecular sieve is partially dependent on the size of these blocking cations. To illustrate: The sodium form of the type A sieve has twelve cations per cavity. Four take positions at or near the aperture and partially block it. There are six apertures per cavity and only four sodiums to block them; apparently two apertures are free. However, since a given aperture is common to two cavities enough sodiums are available to block all apertures. If this zeolite is first soaked in a calcium chloride solution a single calcium ion replaces two sodium ions. Only two calcium ions are available per cavity for blocking purposes; two apertures are consequently unobstructed. The free diameter of the aperture is thereby increased from 3.5 Å to 4.2 Å. In practice not all sodium ions need be replaced. Heptane cannot be adsorbed on the sodium form but by replacing one third of the sodiums with calcium it is readily adsorbed. Molecular sieves are ordinarily in the sodium form. Sodium may be replaced in whole or in part by lithium, potassium, ammonium, magnesium, calcium, strontium, barium, and cerium(III) by means of conventional ion exchange techniques.

In addition to its kinetic energy and size the polarizability of a possible entering molecule is important. It must be so constituted that it has a permanent dipole, or a dipole may readily be induced in the environment of the zeolite cavities. Water, with its strong permanent dipole, is so avidly held that the moisture content of a gas may be reduced to 4 ppm. by use of a zeolite as a drying agent. Unsaturated ethylene is readily separated from saturated ethane even though the two compounds differ little in physical properties and the molecules are the same size. Isooctane is separated from normal octane in commercial installations.

The location of the adsorbed molecules varies not only with the adsorbed molecule but also with the zeolite used. With one type of molecular sieve only hydrogen sulfide will be found in the center of a cage or cavity, carbon dioxide, ammonia, and water lie in the center of the hexagonal rings. A slight variation in structure causes the carbon dioxide and hydrogen sulfide to locate in the center of the cage.

Molecular sieves with a range of properties are commercially available. In 1980 80% of the synthetic zeolites produced were used as detergent builders. (They are very efficient in the removal of calcium ions from water and thus replace the phosphates of the usual detergent formulations.) In addition they are valuable for the drying of gases, purification and separation of hydrocarbon mixtures, cleaning of argon, and as carriers for catalysts previously unusable because of high volatility. Although applications in pharmaceutical chemistry are as yet lacking, it would seem that molecular sieves should find application in preparative work and manufacturing, e.g., drying, separations of gaseous intermediates, purification of gases. It is conceivable that they can serve as inert carriers for oral administration of low molecular weight drugs: gaseous, or easily vaporized drugs could be adsorbed on sieves, to be slowly released on passage through the intestinal tract. Possibly the irritating effect of iron(II) could be overcome by administration on a zeolite, the preponderance of other cations in the intestine should favor slow release of the ion from its inert carriers.

Permutites are synthetic siliceous glasses prepared by fusion of soda ash, kaolin, feldspar, and quartz sand. Like the zeolites they are cation exchangers, especially useful for the $Ca^{2+} \rightleftharpoons 2Na^{+}$ exchange. They have found wide application in water softening techniques (as have the zeolites). Since they are unstable in acid media they cannot be used in a hydrogen form.

THE ULTRAMARINES: *Ultramarines* occur naturally, e.g., lapis lazuli, and are prepared synthetically. Their structure is similar to that of the zeolites, with an anion framework having the ideal empirical formula, $[Al_6Si_6O_{24}]^{6-}$. They are cation exchangers. They differ from the zeolites in that water is absent and exchangeable anions as well as cations, are present in the cavities. The anions found are S_2^{2-}, Se^{2-}, Te^{2-}, SO_4^{2-}, Cl^- as well as the negative ion pairs, $NaSO_4^-$, NaS^-. The ultramarines are unstable in acid media.

The ultramarines can be prepared in a wide range of colors. Sulfides have the best colors; the selenides and tellurides would be more useful except for their poisonous nature. It was once thought that sulfur was present in colloidal form, however, such particles would be much larger than the cavities available. It is probable that the sulfur is present as a polysulfide. The colors obtained become more intense and deeper with increasing sulfur content. On the other hand, increasing the silicon ratio from the ideal gives a warmer glow. The cations present modify the color so that practically all colors are obtainable. Table 11.8 lists some of the colors available and the cations and anions present.

Synthetic ultramarines are prepared by calcining a mixture of china clay, soda ash, charcoal, and sulfur. The reaction product is washed, dried, mixed with more sulfur, and roasted in air until the desired tint is obtained.

TABLE 11.8 THE COLOR OF SOME ULTRAMARINES

Cation	Anion	Color
Li^+	S_2^{2-}	Blue or violet
Na^+	S_2^{2-}	Blue, green
Na^+	Se^{2-}	Blood red
Na^+	Te^{2-}	Yellow
K^+	S_2^{2-}	Blue
Tl^+	S_2^{2-}	Violet
Ag^+	S_2^{2-}	Yellow or grey
Zn^{++}	S_2^{2-}	Violet (or colorless)
Mn^{++}	S_2^{2-}	Grey
Ca^{++}	S_2^{2-}	Colorless
Ba^{++}	S_2^{2-}	Yellow-brown

The ultramarines are used as pigments, in both oil and aqueous systems. They are also used as masking agents for yellow tints in sugar, starch, paper, laundry (bluing). Because of their inertness they should be suitable for pharmaceutical preparations as coloring agents or as masking agents.

GLASSES

This type of substance is so common that, like water, it is taken for granted. As a result sight is lost of its valuable properties. It is durable, transparent, it can be colored and shaped, cleaned, and sterilized.

While the term glass is ordinarily applied to the vitreous silicates it is a generic term. Other substances, fluorberylates, certain phosphates, glycerol, methyl methacrylate, polystyrene, cellulose acetate, are vitreous substances. All are amorphous, i.e., they lack a characteristic crystal form, and are isotropic in character.

Vitreous substances possess no sharp melting point but rather soften gradually over a temperature range. Coefficient of expansion, specific heat, refractive index, and dielectric constant exhibit similar gradual changes, known as *anomalies*. The anomalous behavior of the viscosity of a glass is compared with the viscosity changes at the melting point of a crystalline solid in Fig. 11.10. If a substance is capable of existing in either the vitreous or crystalline state, the difference in viscosity of the solid and liquid states is about the same in both forms.

Theories of the vitreous state. Three major theories have been advanced to explain the vitreous state. The oldest considered glass a supercooled liquid. Its inability to predict a given property of a vitreous solid

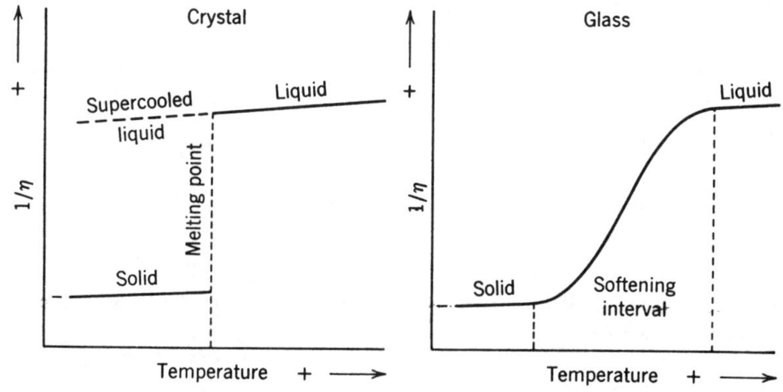

FIGURE 11.10 Fluidity behavior of crystalline and vitreous solids. (Fluidity is the reciprocal of viscosity.)

by extrapolation from the liquid state resulted in its present disfavor. Examination of Fig. 11.10 shows why such extrapolation is impossible.

The crystallite theory of the vitreous state is based on the observation that vitreous solids show diffuse x-ray diffraction patterns resembling those of the same substance in the crystalline state (when available). It postulates that a vitreous solid is built up of crystallites of submicroscopic size, embedded in a disorganized material. It does not explain the ease with which the vitreous state can be realized with some substances whereas others, e.g., the metals and the majority of typical salts, never exist as glasses.

The third theory, the network theory, first outlined by Zachariasen in 1932,[9] appears capable of being developed into a comprehensive theory of the vitreous state. It assumes that glasses have a linking of particles qualitatively similar to that present in the crystalline state. Within very small volumes the structure may be identical to that of the crystal. However, these regions are not discontinuous but are bonded to regions of different symmetry. An overall, disordered, three-dimensional network results.

The silicate tetrahedron is the structural unit of the silicate glasses. However, the geometrical arrangement of the tetrahedra is variable. Instead of the six-membered rings of tetrahedra, common to the crystalline silicates, five-, seven-, or eight-membered rings may exist at random throughout the structure. Figure 11.11 represents such a random structure in two dimensions. Due to the small bond angle deformations imposed by the variable ring structures, bond energies differ slightly. Since a low energy bond breaks at a lower temperature than a high energy bond the glass no longer has a sharp melting point. Rather a progressive softening takes place at higher temperatures, with concomitant higher percentages of bond ruptures.

Chemical composition of glasses. The common glasses (soda glass, soft glass) are insoluble sodium, calcium silicates. Their properties are modified by the inclusion of other cations. Boron, substituted for some silicon, decreases the coefficient of expansion (Pyrex, Vicor). Lead, substituted for some of the calcium, increases the refractive index (flint glass, paste jewelry, optics). Potassium, substituted for sodium, gives dark brown light resistant glasses. Inclusion of praseodynium and neodynium results in a glass (Crooke's) which absorbs strongly in ultraviolet. Traces of manganese dioxide or selenium are sometimes added to neutralize the green or yellow tinge due to iron(II) or iron(III) oxide, present as impurities in the raw materials.

[9] This was predated by Goldschmid's work in the twenties. However, his work was too specific to be generalized into a comprehensive theory.

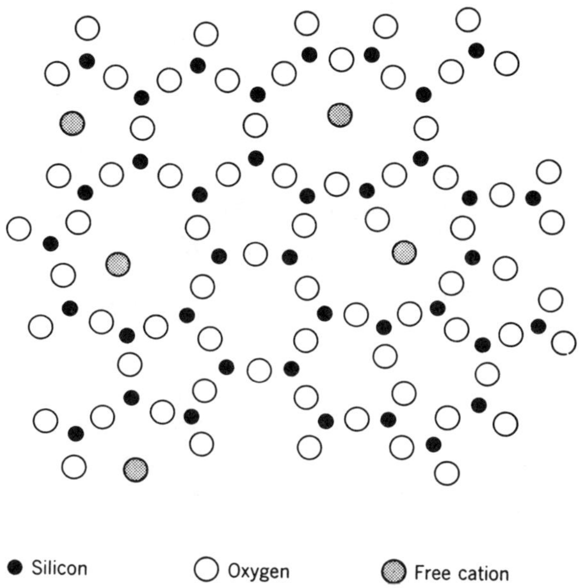

FIGURE 11.11 A disordered silicate network. The fourth oxygen of each silicate tetrahedron lies above or below the plane of the paper. It may be shared with another silicon.

Properties of glasses. While glass is ordinarily considered inert sodium silicate slowly leaches from it when in contact with water. Since the silicate ion is a strong base hydrolysis takes place almost immediately

$$Na^{+}{}_{4}SiO_4{}^{4-} + 3H_2O \rightarrow H_3SiO_4{}^{-} + 4Na^{+} + 3OH^{-}.$$

This explains why solutions stored for some time in glass containers slowly shift to a more alkaline pH. Base sensitive medicinals must be adequately buffered when maintained in contact with glass (page 177). Elevated temperatures, e.g., during sterilization, accelerate this leaching action. The compendia provide testing procedures which measure the rate of attack of glass by water,[10] and classify glasses as Types I, II, and III, intended for parenterals, and Type NP, intended for oral and topical preparations.

As a macroanionic silicate network, albeit somewhat disordered, glass surfaces possess the properties common to silicate surfaces. The adsorptive property is of special interest. Water is strongly adsorbed; very stable films form, not only physical but also by chemisorption to give

[10] *U.S.P. XXI,* pp. 1235.

silanol groups. Grease is readily adsorbed, complicating the cleaning of glass surfaces.

The activity of a given glass surface depends on the pretreatment to which it is subjected. If treated with an acid a surface film of silica gel results

$$SiO_4^{4-} + 3H_2O \rightarrow H_3SiO_4^- + 3OH^-$$

$$H_3SiO_4^- + H_3O^+ \rightarrow H_4SiO_4 + H_2O$$

$$nH_4SiO_4 \rightarrow (H_4SiO_4)_n.$$

Since these reactions take place within the submicroscopic film of solution associated with the glass surface, the silica gel film adheres firmly. Adsorptive properties, e.g., wettability, are greatly increased.

If a glass surface is first treated with sodium fluoride solution, acid treatment no longer produces an active surface. Surface cations also affect the surface properties. Pretreatment with large cations, e.g., barium, lead, results in a less wettable surface than pretreatment with small cations, e.g., tin(II).

In ordinary practice the reactivity of glass surfaces is neglected and rightly so. Since the volume of contained solution is very large compared to the surface of the glass container to which it is exposed, adsorption is minimal. However, there are conditions which may require precautions. If an extremely small container is used for an even smaller volume the exposed surface may no longer be negligible with respect to the volume of solution. This is especially true when working with micro or gamma quantities, e.g., radioactive isotopes (page 570). Appreciable fractions of active material may be lost from solution by adsorption on the glass surface. The same rules for stability of binding as were developed for silica gels apply.

Cleaning of glass surfaces. The selection of the proper reagent for the cleansing of a glass surface should be based solely on its ability to do the given cleaning job most effectively and in a minimum of time. Adequate removal of the cleansing agent from the surface of the glass after cleansing is also important, but is frequently overlooked.

For ordinary cleansing, soaps, detergents, and alkaline cleansing agents (sodium triphosphate) are best. Adherent dirt films which resist are readily removed with scouring powders. These preparations include a mild abrasive, e.g., diatomaceous earth, pumice, in their formula. Mechanical action, together with the suspending and emulsifying action of the detergent, is sufficient for most cleansing purposes. The last traces of detergents are firmly adsorbed on the clean glass surface and are removed with difficulty. If rinsing is not thorough this adsorbed material will be a source of future contamination.

TABLE 11.9 Silicates in the Practice of Pharmacy

	Abrasive	Absorbent	Adsorbent	Alkalizer	Anticaking Agent	Binder	Carrier	Clarifier	Detergent	Diluent, Excipient, Filler	Distributing Agent	Emulsifier	Filter Aid	Gelling Agent	Lubricant	Moisture Control	Suspending Agent	White Pigment	Whitening Agent
Asbestos													×						
Attapulgite			×		×		×							×			×		
Bentonite		×						×	×			×		×			×		
Fuller's earth								×		×									
Kaolin			×					×					×						
Magnesium trisilicate												×					×		
Pumice	×										×		×						
Silica gel		×	×		×	×	×						×	×		×	×		
Siliceous earth	×	×			×			×		×			×						
Sodium silicate				×		×						×							
Talc							×			×	×		×		×			×	
Ultramarine																			×
Vermiculite							×			×						×			
Zeolite																×			

Chromic acid is the standard drastic cleansing aid. It is best prepared by slowly adding one and a half liters of concentrated sulfuric acid to a solution of 200 Gm. sodium dichromate in 100 ml. water, with constant stirring. Chromic acid will remove most adherent materials, especially if used hot. However, it has serious disadvantages. Because of its extremely corrosive character it must be used with extreme care, even the glass surface is not immune to attack. Not only are inanimate materials destroyed, e.g., clothing, towels, but contact with the skin will cause severe burns. Chromium(III), the reaction product, is adsorbed tenaceously on the active glass surfaces produced by this reagent. The last traces of this ion are removed only after long and careful rinsing. Even in traces undesirable reactions may be initiated.

Hot nitric acid is also an effective cleansing agent, especially for glassware having tightly adherent carbonaceous residues. Hot sulfuric acid with small amounts of concentrated nitric acid is especially effective for removal of carbonaceous deposits—it must be handled with care. The reactions may become violent and severe burns may result from contact. Aqua regia, concentrated nitric acid, concentrated hydrochloric acid, and water in a 1:3:1 ratio by volume, is also used.

A rarely used but effective glass cleansing agent is the hydrochloric acid-hydrogen peroxide reagent, prepared by mixing equal volumes of HCl, 6*M*, and H_2O_2, 6%. This mixture has both oxidizing and reducing properties. It is safer to use and easier to handle. It is particularly useful for removal of adherent inorganic compounds, however, greases, etc., are also destroyed. Only freshly prepared solutions are effective.

The silicates of importance in the practice of pharmacy are compiled in Table 11.9, together with their more important applications. Table 11.10 similarly summarizes the most important medicinal applications. The more important nonsilicate substances having analogous properties, and their uses, are given in Table 11.11.

TABLE 11.10 SILICATES USEFUL AS MEDICINALS

Substance	Antacid	Cataplasm	Dusting Powders	Intestinal Adsorbent	Moisture Absorbent	Protective
Bentonite						×
Fuller's earth					×	
Kaolin		×		×	×	×
Magnesium trisilicate	×					×
Siliceous earth		×	×		×	
Talc			×			×

TABLE 11.11 Substances Having Applications Similar to Silicate Minerals

Abrasives	Absorbents	Adsorbents	Anticaking Agents
Calcium perborate Calcium carbonate Magnesium carbonate Magnesium oxide	Aluminas Charcoals	Aluminas Cellulose Charcoal	Magnesium oxide Magnesium trisilicate
Diluents, Extenders Excipients, Fillers	**Filter Aids and Clarifying Agents**	**Lubricants**	**Suspending Agents Gelling Agents**
Tribasic calcium phosphate Lactose Magnesium oxide Magnesium carbonate Sucrose Silicon Dioxide (Colloidal)	Aluminas Cellulose Charcoal	Calcium stearate Magnesium stearate Sodium stearate	Aluminum monostearate Soaps
Whitening Agents	**White Pigments**		
Prussian blue	Titanium dioxide Zinc oxide		

STUDY AND SELF-APPRAISAL AIDS

IMPORTANT TERMS. Define or identify in a single concise sentence if possible.

Anomaly, asbestosis, binder, carbonization, carrier, cation hull, clarifying agent, clathrate, close packing, diluent, electrodialysis, elutriation, flocculation, fluidity, free diameter, glass, iliostomy, isomorphous, isomorphous replacement, isotropic, kinetic energy, lubricant, macrocation, micella, molecular sieve, molecular vibrational energy, octahedral hole, petrological, polarizability, polymorphous, pyrolysis, tetrahedral hole, thixotropy, unctuous, vitreous, water hull, whitening agent, zeta-potential.

NOMENCLATURE OF INTEREST. Identify each of the following.

Activated charcoal, amphibole, attapulgite, bone charcoal, brucite, china clay, chrysotile, expanded mica, feldspar, French chalk, halloysite, kaolinite, montmorillonite, mullite, orthoclase, permutite, pumice, pyroxene, talcum, vermiculite, zeolite, zeolitic water.

SELF-APPRAISAL QUESTIONS

1. How do cubic and hexagonal closest packing resemble one another? How do they differ?

2. What conditions must be met for one cation to isomorphously replace another in an oxygen close packed system?

3. Explain why certain physical properties of silicates e.g., refractive index, are more or less independent of the cations present.

4. Identify the two chain silicate structures. Why are they useful as filter aids? Which one is preferable?

5. Name the two substances used to cement the *A-A* faces of silicate sheets. Why is the binding stable?

6. List the different types of materials which serve as cementing material between *B-B* faces of silicate sheets. Identify the binding force involved with each.

7. Explain how a silicate anionic network with the ideal formula, $(SiO_2)_n$, achieves its negative charge. How does a brucite sheet, ideal formula, $Mg(OH)_2$, achieve a positive charge?

8. Assume a macroanionic silicate-brucite-silicate layer of a chlorite has a charge of $-10{,}000$. How many magnesium atoms of a brucite slab between such layers must be isomorphously replaced by aluminum atoms to give it the required positive charge?

9. Why is the term aluminosilicate descriptive of the clays?

10. Describe the binding between the successive aluminosilicate layers of montmorillonite. Which of its characteristic properties are dependent upon the state of this binding.

11. Explain why montmorillonite is able to exchange greater than theoretical equivalents of the very large amine salt cations.

12. Give the correct procedure for preparation of a bentonite suspension.

13. Explain the binding between successive aluminosilicate layers of the kaolin clays. Why are large crystals of kaolinite unavailable?

14. Give the medicinal uses of the kaolins. What problems may arise on ingestion?

15. Name the usual impurities present in native clays.

16. Why are native clays purified by physical rather than chemical procedures? Describe the purpose of each of the usual purification stages.

17. Why do thixotropic clays never exist in the potassium form?

18. Name those properties which determine whether a given compound is adsorbable on a given clay; in a zeolite.

19. Why must wood be pyrolyzed in the absence of air when manufacturing charcoal?

20. Explain why a coarse, hard charcoal is desirable for purification of gases.

21. Name the properties which make expanded mica valuable as a packing material for breakable containers of corrosive chemicals.

22. What is the relationship between aluminum-silicon ratios and the type of guest allowed in the chambers of a clathrate? Explain.

23. Describe the gross structural arrangement of the zeolites.

24. What are the advantages of a zeolite as an adsorption agent as compared to a silica gel? Disadvantages?

25. Explain why a molecule can pass through a zeolite aperture even though it has a larger diameter than the free diameter of the aperture.

26. Why are zeolites not used as cation exchangers for demineralization of water?

27. Name the three theories of the vitreous state and give the salient features of each.

28. Why do elevated temperatures increase the rate of attack of water on a glass surface? When is this important in pharmaceutical practice?

29. Why are certain precautions essential when a very dilute solution is in contact with glass? Explain.

30. List the commonly used cleansing agents for laboratory glassware. Give the advantages and disadvantages of each.

31. When and why is particle size important in using the following: talc, bentonite, activated charcoal.

32. Identify those properties of substances, positive and negative, important in each of the following applications: filter aid, clarifying agent, filler, abrasive, lubricant, suspending agent. Illustrate each with a suitable silicate.

33. Distinguish between each of the following pairs of substances: bentonite, montmorillonite; kaolinite, montmorillonite; molecular sieve, zeolite; zeolite, ultramarine; feldspar, zeolite; mica, chlorite; zeolite, cation exchange resin; aquamarine, ultramarine; glass, crystalline solid.

34. Name the substances discussed which are official in the current compendia. Give the official category of each.

PRACTICE IN EQUATION WRITING

35. Write equations for all chemical reactions taking place during the purification of native talc.

36. Write equations showing why silicates dissolve in both fused caustic potash and in hydrofluoric acid.

37. Write equations for the reactions taking place when crude charcoal, treated with sodium carbonate, is heated during the activation process. When sodium sulfate is used instead of sodium carbonate.

38. Show, by suitable equations, the reactions taking place during the *U.S.P. XXI* tests of glass.

39. Give a general equation for the cation exchange reactions of the hydrogen form of a zeolite.

40. Write equations for the reactions taking place during the following official limit tests: bentonite, gel formation; kaolin, carbonate, iron; purified siliceous earth, sulfate; pumice, acid solubles, iron.

PART III · INORGANIC SUBSTANCES AS MEDICINAL AGENTS

Inorganic compounds and preparations play significant roles as medicinal agents. For the most part useful compounds may be arranged into well-defined groups, based on the application or mechanism of action, e.g., trace minerals, antacids. The ability of a given compound to act in any one category actually depends on electron structure and atomic properties. Thus oxidizing germicides are members of the halogen family in their zero and positive oxidation states or closely related forms. On the other hand, inert gas type ions from Groups IA, IIA, and VIIB constitute the cations and anions of the fluid and electrolyte replenishers and physiological fluids. The periodic table of inorganic medicinals, shown as a front end paper, gives these and other major relationships.

All substances having similar applications as medicinals will be considered concurrently. This method allows better correlation between mechanism of action and the physical, chemical, and structural properties of importance than the usual presentation.

TWELVE · A PROLOGUE—SOME ESSENTIAL BACKGROUND CHEMISTRY

Prior to a discussion of therapeutically important inorganic compounds, a background of the descriptive chemistry of certain families of elements must be provided. Such background is a prerequisite to the elucidation of the mode of action of the different medicinal compounds and an appreciation of the chemical and physical problems which may arise during preparation of suitable dosage forms. Since the initial area of study will be those ions normally present in the body as electrolytes and body fluids, the subject matter for this introductory chapter is easily determined.

First, the alkaline earth family[1] will be reviewed. Calcium and magnesium ions are important as minerals, and the other members of the family have limited medicinal uses. Second, since the phosphate moiety is so intimately involved in the body's solutions as a buffer system, in metabolic processes, and in bone, a section on phosphorus chemistry follows. Particular emphasis will be placed on the phosphorus(V) compounds. The chapter will conclude with a review of the halogen family. The halogens are important not only as body electrolytes but also in several other categories.

THE ALKALINE EARTH FAMILY

General chemistry. Data pertinent to a discussion of the chemical and physical properties of this family of elements are given in Table 12.1. The valence electron structure of these elements is a pair of *s* electrons overlying an $8e^-$ inert gas shell; beryllium, however, has a helium subshell.

[1] Some authorities consider only calcium, strontium, and barium as the true members of this family. In this textbook magnesium and beryllium are included.

The general trend of properties follows that of the alkali metal family. Differences in degree arise mainly from two factors: smaller size and higher ionization potential. The greater nuclear charge and the closing of the *s* orbital in the alkaline earth atoms result in a smaller atomic radius. In forming ions, two electrons must be removed. The resulting double positive electrostatic field is much stronger than that of the monovalent alkalis; this external field strength is enhanced by the volume contraction resulting from interaction of the excess double nuclear charge and the inert gas electron structure.

TABLE 12.1 ATOMIC CHARACTERISTICS OF THE ALKALINE EARTHS

Element	Beryllium	Magnesium	Calcium	Strontium	Barium
Symbol	Be	Mg	Ca	Sr	Ba
Atomic weight	9.013	24.32	40.07	87.63	137.4
Atomic number	4	12	20	38	56
Valence electron configuration	$1s^2 2s^2$	$2s^2 2p^6 3s^2$	$3s^2 3p^6 4s^2$	$4s^2 4p^6 5s^2$	$5s^2 5p^6 6s^2$
Atomic radius (Å)	0.88	1.36	1.74	1.91	1.98
Ionic radius (Å), +2	0.31	0.65	0.99	1.13	1.35
Ionization potentials (ev.), I	9.3	7.6	6.1	5.7	5.2
II	18.2	15.0	11.9	11.0	9.95
Electrode potential (v.)	+1.70	+2.34	+2.87	+2.89	+2.90
Electronegativity	1.5	1.2	1.0	1.0	0.9

Inherent in the divalent character of the alkaline earths is a decreased ease of ion formation from the metallic atom. The ionization potential for a second electron is always appreciably larger than that of the first electron. This must be so since the removal of the second electron requires sufficient energy to overcome the unit positive charge of the nucleus resulting from the loss of the first valence electron. The energy requirement for ionization of the beryllium atom is so large that this element does not ordinarily exist in the ionic state. Despite the higher ionization potential for the second valence electron, these elements do not exist in stable monovalent compounds.

The alkaline earth metals are strongly electropositive, being second only to the alkali metals. Within the family the usual order exists; electropositive character increases with increasing atomic weight. The ions of these elements are extremely stable. The reducing environments required to restore the ions to the elemental state will not be met in ordinary pharmaceutical preparative practice.

Because of the closely held nonpolarizable inert gas shell, these ions do not complex readily. All members of the family form poorly defined ammonia complexes. Magnesium forms many double salts but few true complexes. The instability constants of those complexes which do form are comparatively large, e.g., the pK_{inst} for the calcium citrate complex, $[Ca(Ci)]^-$ is 3.17. Calcium and magnesium form EDTA complexes; their successful analytical determination using sodium edetate reagent probably depends on the added stabilization of the complexes by chelation.

The strong double positive electrostatic fields of these small cations do, however, interact sufficiently with polar solvents to cause solvation. As a result their salts are frequently soluble in organic solvents; even the strontium and barium halides (except fluorides) are moderately soluble in alcohol. It is this solvation which makes calcium chloride unsuitable as a drying agent for the preparation of absolute alcohol. Due to the highly favorable concentration gradient the calcium alcoholate forms preferentially to the hydrate.

All five cations form stable hydrates, the lower members being more highly hydrated than the higher; e.g., barium usually forms a dihydrate, whereas magnesium usually exists as a hexaqua complex. Because of the very weak acidic character of these hydrated cations (except beryllium), the pH of aqueous solutions of their salts depend for practical purposes on the base strength of the accompanying anion. The hydroxides are strong bases comparable to the alkali bases. Frequently the limited solubility of the hydroxides, especially calcium and magnesium, is incorrectly interpreted as weak base character.

Group differences, except for beryllium, are a matter of degree. Calcium, strontium, and barium show the greatest similarities in both chemical and physical properties. Because of the extremely small size of the beryllium(II) ion, property deviations are rather extreme. It forms only a few ionic compounds, e.g., the sulfates. Most simple beryllium compounds are covalent, usually *sp* hybrids. If electron-rich ligands are available, beryllium's covalency may be expanded to four through sp^3 hybridization. Being a bridge element, beryllium shows a closer resemblance to aluminum than to the members of its own family; e.g., beryllium hydroxide is amphoteric.

Unlike the alkali metals, these elements have many insoluble compounds. For this reason solubility incompatabilities must be kept in mind when working with solutions of alkaline earth cations. The degree of insolubility of a series of salts of a given anion is usually in the order of atomic weight. Unfortunately the order of solubility is not consistent. Of the more common anions, the sulfate, carbonate, and phosphate become less soluble with increasing atomic weight of the alkaline earth cation, whereas the hydroxide and fluoride become more soluble with increasing atomic weight. Thus magnesium salts, except for the hydroxide and fluoride, tend to be more soluble than those of the other members of the family. Acidic media increase the solubility of alkaline earth salts for the usual reasons

$$CaSO_4\downarrow + H_3O^+ = Ca^{2+} + HSO_4^- + H_2O$$

Sources and preparative techniques. The common naturally occurring minerals of this family are given in Table 12.2. The raw material used

TABLE 12.2 NATURALLY OCCURRING COMPOUNDS OF THE ALKALINE EARTHS

Element	Per Cent of Earth's Crust	Common Natural Minerals		Commercial Raw Materials
		Name	Formula	
Be	0.0006	beryl	$Be_3Al_2Si_6O_{18}$	beryl
Mg	1.93	magnesite	$MgCO_3$	
		dolomite	$[Mg, Ca]CO_3$	sea water
		carnallite	$KMgCl_3 \cdot 6H_2O$	
		many silicates	—	
Ca	3.4	aragonite		
		calcite		
		limestone	$CaCO_3$	$CaCO_3$
		chalk		
		marble		
		anhydrite	$CaSO_4$	
		gypsum	$CaSO_4 \cdot 2H_2O$	
Sr	0.02	strontianite	$SrCO_3$	$SrCO_3$
		celestite	$SrSO_4$	$SrSO_4$
Ba	0.04	barytes	$BaSO_4$	$BaSO_4$
		witherite	$BaCO_3$	

for the manufacture of the commercially available compounds is included. With a carbonate as the raw material preparation consists either of a preliminary calcination followed by neutralization with the appropriate acid, or of direct neutralization of the carbonate:

$$CaCO_3\downarrow = CaO\downarrow + CO_2\uparrow$$

$$CaO + 2(H_3O^+ + Cl^-) = Ca^{2+}Cl^-_2 + 3H_2O$$

or

$$CaCO_3 + 2(H_3O^+ + Cl^-) = Ca^{2+}Cl^-_2 + 3H_2O + CO_2\uparrow$$

The thermal stability of a given carbonate is directly proportional to the electropositivity of the cation. The temperature necessary to achieve the dissociation pressure of one atmosphere for the carbon dioxide is 25°,

540°, 900°, 1290°, and 1360°C. for the beryllium carbonate through barium carbonate series. If high temperatures are to be avoided, the calcination of barium carbonate[2] must be carried out in the presence of carbon:

$$BaCO_3 + C = BaO + 2CO\uparrow$$

With the sulfate as the raw material, the sulfate is first reduced to the sulfide by calcination with carbon:

$$BaSO_4\downarrow + 4C = BaS + 4CO\uparrow$$
$$BaS + 2(H_3O^+ + Cl^-) = Ba^{2+}Cl^-_2 + 2H_2O + H_2S\uparrow$$

This latter method has the advantage that the barium sulfide may be leached from the crude reaction mixture, thereby separating it from insoluble impurities.

Magnesium ion is now obtained chiefly from sea water by precipitation as the hydroxide with a suspension of calcined dolomite; all other salts present are soluble. Magnesium compounds may also be prepared from magnesite and those dolomites low in calcium. With dolomites, however, contamination by calcium compounds may be appreciable, depending upon the rigor of the preparative methods. For this reason the various compendia (*U.S.P.*, *B.P.*) include limit tests for calcium in addition to the usual purity tests.

Beryl, $Be_3Al_2Si_6O_{18}$, the only economically useful source of beryllium, must be fused and quenched in sulfuric acid. Under this treatment the element together with aluminum dissolves as the sulfate. The silicon precipitates as insoluble silicic acid. After filtration the solution is partially neutralized with ammonia. Most of the aluminum present may be crystallized as the sparingly soluble ammonium alum by addition of the requisite quantity of ammonium sulfate. Beryllium is recovered from the mother liquids as the insoluble hydroxide by adjusting the pH to 10–11 with ammonia. This compound is used as the raw material for all other beryllium compounds.

Properties of some common compounds. In the following discussion of the common compounds of this group those of beryllium have been omitted. Its compounds have properties approximating those of the corresponding aluminum compounds (page 490).

The oxides are used as the starting material for the preparation of most compounds of this family. A given oxide is most economically obtained

[2] This is strictly an economic requirement since the barium oxide is stable to very high temperatures. Costs increase directly with temperature, not only because of the added heat energy required, but also due to the greater cost of construction materials stable at the higher temperatures.

by calcination of the respective native carbonate; purity is a function of the nonvolatile matter present in the raw material. Pure oxides may be obtained by calcination of chemically precipitated carbonates or by burning the element in air. The latter method is not applicable to magnesium, since appreciable quantities of its nitride form on combustion in air. Magnesium hydroxide obtained from sea water may serve as a source of pure magnesium oxide. If it is first carefully washed to remove soluble salts, low-temperature calcination yields a reactive, pure magnesium oxide.

The oxides are heat stable. Except for magnesium oxide, they slake readily with the vigorous evolution of heat:

$$CaO + H_2O = Ca(OH)_2$$

Magnesium oxide will react with water only if prepared at low temperatures, a factor of importance in the manufacture of magnesia magma. Calcium hydroxide is one of the few compounds with a negative temperature coefficient of solubility; it is more soluble in cold water than in warm water. Its solubility is 1 in 780 at 15°C., 1 in 714 at 25°C., and 1 in 3000 at 80°C. Its aqueous suspension is known as milk of lime, its saturated solution as lime water. A saturated barium hydroxide solution is known as baryta water.

Being strong bases, hydroxide solutions must be protected from the carbon dioxide of the air; insoluble carbonates precipitate.

$$OH^- + CO_2 = HCO_3^-$$
$$OH^- + HCO_3^- = CO_3^{2-} + H_2O$$
$$M^{2+} + CO_3^{2-} = MCO_3\downarrow$$

Barium hydroxide solutions are extremely sensitive reagents for the detection of carbon dioxide. The three heavier hydroxides are more soluble in sucrose solutions than in pure water. This increased solubility is due to the formation of soluble complex sucrates.

All four elements react with oxygen to form peroxides. Barium and strontium peroxides are prepared by heating the oxide with oxygen under pressure. Octahydrates of the peroxides crystallize from solutions of the hydroxides on addition of hydrogen peroxide.

Magnesium sulfide is prepared by burning the metal in sulfur vapor; the others are prepared by roasting the sulfate with coke. All are water soluble and hydrolyze in aqueous media:

$$2MgS + 2H_2O = Mg(OH)_2\downarrow + Mg^{2+} + 2HS^-$$
$$Mg^{2+} + 2HS^- + 2H_2O \rightleftharpoons Mg(OH)_2 + 2H_2S\uparrow$$

The latter reaction is temperature dependent. High temperatures decrease the solubility of gaseous hydrogen sulfide and favor the forward reaction. The sulfides are important as intermediates in the commercial preparation of other alkaline earth compounds.

The oxo salts of these elements have several interesting aspects. Strontium and barium sulfates are anhydrous and sparingly soluble. Calcium sulfate is sufficiently soluble as the dihydrate to be responsible for the so-called permanent hardness of water. It also forms a second hydrate:

$$\underset{\text{(gypsum)}}{CaSO_4 \cdot 2H_2O} \underset{}{\overset{120\text{–}130^\circ}{\rightleftharpoons}} \underset{\text{(plaster of Paris)}}{(CaSO_4)_2 \cdot H_2O} \overset{140\text{–}200^\circ}{\rightleftharpoons} \underset{\text{(anhydrite)}}{CaSO_4}$$

Gypsum is a layer structure. The calcium and sulfate ions are arranged in separate layers, and the water molecules are arranged between the layers. Each water molecule is hydrogen bonded to two different sulfate groups; the negative oxygen end of each water molecule dipole interacts with the positive field of the calcium ion layer. Magnesium sulfate (and beryllium sulfate) is water soluble. It forms many different hydrates; 1, $1\frac{1}{4}$, $1\frac{1}{2}$, 4, 5, 6, 7 H_2O have been identified.

Addition of a soluble carbonate to a solution of a soluble magnesium salt yields one of two basic carbonates depending on the temperature of precipitation: $Mg(OH)_2(MgCO_3)_4 \cdot 5H_2O$ at lower temperatures and a tetrahydrate at higher temperatures. The pentahydrate is the more common commercially available form. A trihydrate of a normal magnesium carbonate precipitates when a carbon dioxide-saturated solution of magnesium sulfate is heated to 50°C. Calcium, strontium, and barium precipitate as anhydrous normal carbonates when a soluble carbonate is added to solutions of their soluble salts. The phosphates of these cations form complicated systems (see page 329).

The solubilities of some common salts are given in Table 12.3.

TABLE 12.3 SOLUBILITIES OF COMMON ALKALINE EARTH COMPOUNDS[a]

	Cl	F	OH	CO_3	SO_4
Mg	0.6	i	80,000	3,330	0.8
Ca	0.6	i	590	4,000	375
Sr	0.8	8,500	50	10,000	8,800
Ba	2.8	620	11.6	44,000	500,000

[a] The solubilities are in parts water dissolving one gram at room temperature or lower. Compounds existing in more than one hydration state have different solubilities depending upon the hydrate involved. Ordinarily the solubility of the highest hydrate is given.

Biochemistry. Only calcium and magnesium are normal constituents of living matter. Strontium can be tolerated. It deposits in bone in sites normally occupied by calcium ions. Apparently it cannot replace soluble calcium ions; laboratory animals on high strontium, low calcium diets developed typical syndromes of calcium deficiency. Strontium-90 is one of the principal fission products of uranium-235 and plutonium. Beryllium and barium are extremely toxic. Radium, which is the sixth member of this family, is both extremely toxic and dangerously radioactive.

General uses of compounds. MAGNESIUM: Magnesium ion has two pharmacological actions. It is a cathartic and acts as a central depressant. However, except for the sulfate, most magnesium compounds do not find employment in medicine or pharmaceutical practice for the pharmacological action of the magnesium ion. More often than not, the magnesium ion is present for chemical reasons. It may be acting as a carrier for the anion, the anion being the medicinally active portion of the compound: e.g., the bromide as a sedative, the salicylate as an intestinal antiseptic, the hydroxide as an antacid, the peroxide as an antiseptic powder. The magnesium ion may act as a carrier for a chemically important anion: e.g., carbonate, hydroxide, or oxide for neutralization reactions, sulfate as an antidote for barium poisoning. Finally, magnesium ion, together with the accompanying anion, may impart desirable physical properties to a compound: e.g., a mild abrasive quality to the carbonate, a quality useful in tooth powders, or the covering and adhering properties to the stearate, useful in baby dusting powders. The following magnesium compounds have been or are at present important in pharmaceutical practice: bromide, carbonate, chloride, glycerophosphate, hydroxide, hypophosphite, lactate, mandelate, oxide, peroxide, dibasic and tribasic phosphate, salicylate, silicate, stearate, and sulfate.

CALCIUM: Calcium compounds of interest in pharmaceutical chemistry may be classified much as those of magnesium. Certain uses depend upon the action of the calcium ion itself, e.g., calcium gluconate as a calcemic. As with magnesium compounds there are many calcium compounds in which the calcium ion is not present for its own medicinal effects. The chemical properties of the calcium ion may be important, e.g., the use of a soluble calcium salt as an antidote for oxalate poisoning. The calcium ion may serve as a carrier for a medicinally active anion, e.g., bromide, penicillin G. In other cases the calcium serves as an inert cation accompanying a chemically active anion, e.g., the hydroxide as a base, palmitate as a thickener. Finally, physical properties of the compound may be important, e.g., calcium carbonate as a mild abrasive in toothpastes. Numerous calcium salts are used in pharmaceutical practice. These include the acetate, acetysalicylate, bisulfite (and sulfite), bromide, chloride, carbonate, gluconate, glycerophosphate, hydroxide, hypophosphite, iodate, iodide, lactate, levulinate, mandelate, oxide, palmitate, pantothenate, penicillin G, perborate, permanganate, peroxide, different phosphates, salicylate, stearate, sulfate, sulfide and polysulfides, and thiosulfate. The major uses of calcium compounds include: calcemic, allergen, antacid, systemic and urinary acidifier, protective, mild abrasive, and base.

BERYLLIUM: Beryllium compounds have no application in pharmacy or medicine. The salts have a sweet taste. For this reason the element was at one time named glucinium in continental Europe. Beryllium is now internationally recognized as the proper name for this element.

The toxic character of beryllium is caused by its interference with the carbohydrate metabolic cycle; it prevents the breakdown of phosphorylated compounds. It has been shown that beryllium in concentrations as low as $10^{-6}M$ inhibit alkaline phosphatase. Contact dermatitis, nonhealing ulceration on the site of injury, and subcutaneous nodules are common manifestations of beryllium poisoning. Pulmonary granulomatous disease may appear in up to six years after exposure, often after short periods of exposure to low concentrations. Complete recovery is uncertain. The Argonne National Laboratory has proposed aurin tricarboxylic acid as an antidote.[3] It inactivates beryllium by forming a nontoxic complex.

Beryllium salts were the original phosphor activators of fluorescent lamps. Because of numerous cases of accidental poisoning due to breakage of these tubes, this use of beryllium compounds has been discontinued. The metal finds wide use in metallurgy as an alloying agent for copper; its action in these alloys is similar to that of carbon in steels. Beryllium is used in nuclear chemistry as a convenient source of neutrons. When mixed with a small quantity of an alpha emitter, e.g., radium, a nuclear reaction results.

$$^{9}_{4}\mathrm{Be} + ^{4}_{2}\mathrm{He} = ^{12}_{6}\mathrm{C} + ^{1}_{0}n$$

STRONTIUM: Strontium compounds are used in medicinals largely as anion carriers, e.g., bromide, salicylate. It has been assumed that a slower release of anions is achieved, but this phenomenon has not been substantiated. Recently the chloride has been suggested as a toothpaste ingredient. It is claimed to reduce the sensitivity of teeth to hot and cold. Strontium carbonate and strontium hydroxide find use in sugar refining. The nitrate, sulfate, carbonate, and chlorate find use in pyrotechnics for the production of red flames.

[3] *Chem. Eng. News*, **29**, 5109 (1951).

BARIUM: Soluble barium compounds are extremely toxic. The action resembles that of the digitalis group of substances, but is inferior because of coronary vasoconstriction. Barium ion is locally irritating, stimulating the smooth muscle. For this reason the ion has a laxative effect. Except for barium sulfate (page 587) and barium hydroxide lime (page 594), compounds of this element find no regular use in pharmacy and medicine. When other treatments have failed, barium chloride has been used as a heroic treatment in complete heart block, (dose, 0.03 Gm., fatal dose, 0.8–0.9 Gm.). The chloride is used in veterinary medicine as a violent rapid cathartic in horses. The sulfide has depilatory action. The chloride, oleate, and fluosilicate are used as rodenticides; the oleate for roach control, the fluosilicate as an insecticide. The chlorate and nitrate are used in pyrotechnics for production of green flames. Barium hydroxide is used in sugar refining and in the refining of vegetable and animal oils.

PHOSPHORUS CHEMISTRY

Phosphorus is a member of Group VB of the periodic table. A discussion of group properties *per se* will be postponed (see page 500). At this time a study of the specific chemistry of phosphorus, especially in its +5 oxidation state, will be undertaken.

The free element was at one time an official substance (*N.F. VII*). It was thought to act as a nerve stimulant and to increase bone formation. The usual dose was 0.6 mg. However, its value is highly questionable. Large doses are lethal. Copper sulfate is used as an antidote. Despite the availability of more effective materials, phosphorus is used as a rodent poison.

The ground state electronic structure of the phosphorus atom is

$$1s^2 2s^2 2p^6 3s^2 3p_x{}^1 3p_y{}^1 3p_z{}^1 (3d^0)$$

Its valence shell structure is similar to that of nitrogen. However, important differences in chemistry exist as compared to nitrogen because of a difference in the underlying closed shells ($1s^2$ only for nitrogen) and the availability of the $3d$ orbitals for hybrid formation in the case of phosphorus.

Phosphorus has two fundamental series of compounds. In the first, phosphine and its derivatives the phosphonium compounds, the atom acts as an electron acceptor to form σ bonds with three or four ligands. In so doing, it achieves a −3 oxidation state. In phosphine these bonds are nearly pure p^3 character, the molecule being pyramidal with a H—P—H angle of 93°. Phosphine is a much weaker base than ammonia but a much stronger reducing agent. Phosphonium compounds are sp^3 hybrids. They

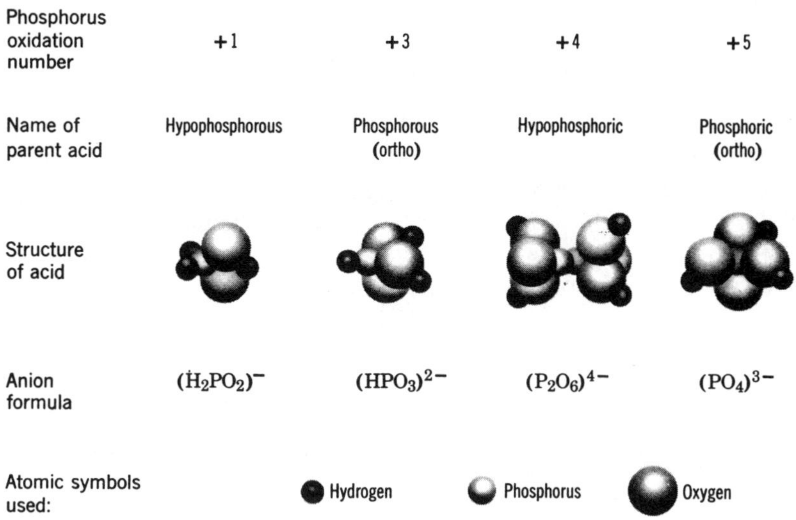

FIGURE 12.1 The positive oxidation states of phosphorus.

are less stable than the corresponding ammonium compounds. The quaternary phosphonium bases are highly ionized.

In the second group of compounds the phosphorus atom acts as an electron donor. The oxidation number of the phosphorus may be +1, +3, +4, and +5. The tetrahedral arrangement exists in all. The different members of this series of phosphorus compounds are illustrated in Fig. 12.1. From a mechanistic point of view the series may be considered to result from the successive substitution of oxygen atoms for the hydrogens of the phosphonium ion. The first substitution product, not shown in the figure, is phosphine oxide, PH_3O. It should be noted that the coordination number of the phosphorus atom remains constant at four throughout the series. Hydrogen atoms bonded directly to the central phosphorus atom are not acidic.

Hypophosphorous acid and the hypophosphites. The structure of this acid has not been conclusively established. Structural studies of its ammonium and magnesium salts have demonstrated the existence of a tetrahedral $(H_2PO_2)^-$ ion. X-ray studies do not directly locate the two hydrogen atoms, but indirect evidence indicates that they are bonded to the central phosphorus atom rather than the oxygen atoms.

The hypophosphites are prepared by the reaction of phosphorus with a strong base. Calcium hypophosphite is prepared by reacting a thin water suspension of calcium hydroxide and finely divided phosphorus at

50°C. until phosphine evolution ceases. This reaction should be conducted under a hood because of the poisonous nature of the phosphine by-product.

$$6OH^- + 2P_4 + 6H_2O = 6(H_2PO_2)^- + 2PH_3$$

The reaction mixture is then filtered to remove unreacted solids and the filtrate treated with enough carbon dioxide to just precipitate the excess dissolved calcium hydroxide. After refiltration calcium hypophosphite is obtained by crystallization or by precipitation with ethanol.

Pure hypophosphorous acid is prepared by treating one of its salts with an acid capable of precipitating the cation present as an insoluble salt:

$$(Ca^{2+} + 2H_2PO_2^-) + H_2(C_2O_4) = Ca(C_2O_4)\downarrow + 2H(H_2PO_2) \text{ (aqueous solution)}$$

$$[K^+H_2PO_2^-] + H_2(C_4H_4O_6) = H(H_2PO_2) + KHC_4H_4O_6\downarrow \text{ (hydroalcoholic solution)}$$

Pure hypophosphorous acid occurs as deliquescent crystals or as an oily liquid. Its melting point is 26.5°C. The acid is available commercially as aqueous solutions of various concentrations; that official in *N.F. XV* is 30%. It is a moderately strong acid, having a pK_a of 2 at 25°C.

The alkali hypophosphites are very soluble in water. The potassium salt is soluble in 9.0 parts of alcohol. The alkaline earth and heavy metal salts are moderately soluble—usually 1 Gm. of salt in about 6 parts of water.

Since the phosphorus atom of this acid is in an intermediate oxidation state, hypophosphorous acid and its salts are capable of acting as either oxidizing agents or reducing agents. Its oxidizing action is weak. The standard potential as an oxidizing agent is pH dependent:

$$H(H_2PO_2) + H_3O^+ + e^- = \tfrac{1}{4}P_4 + 3H_2O \qquad \mathscr{E}^0 = -0.51 \text{ v. (acid media)}$$

$$(H_2PO_2)^- + e^- = 2(OH)^- + \tfrac{1}{4}P_4 \qquad \mathscr{E}^0 = -2.05 \text{ v. (basic media)}$$

Hypophosphorous acid and its salts are powerful reducing agents. This reducing ability is, in fact, the property of importance. The standard potential as a reducing agent is also pH dependent:

$$H(H_2PO_2) + 3H_2O = H_3PO_3 + 2H_3O^+ + 2e^- \qquad \mathscr{E}^0 = 0.50 \text{ v. (acid media)}$$

$$(H_2PO_2)^- + 3(OH)^- = HPO_3^{2-} + 2H_2O + 2e^- \qquad \mathscr{E}^0 = 1.57 \text{ v. (basic media)}$$

The reducing action of these compounds on some inorganic ions is illustrated in Table 12.4. Because of their powerful reducing action, hypophosphites must not be triturated with oxidizing agents; explosive reactions result with, say, chlorates or nitrates.

As a result of the bonding of a positive hydrogen directly to the positive phosphorus atom, repulsive forces exist which render these compounds unstable. The autoxidation

$$2H(H_2PO_2) = PH_3 + H_3PO_4$$

is pH dependent, taking place at lower temperatures in basic media than

TABLE 12.4 Hypophosphite as a Reducing Agent

Ion	Reduced To
Bi^{3+}	Bi^0
Hg(II), Hg(I)	Hg^0
Fe(III)	Fe(II)
Cu(II)	Cu(I)
$(MnO_4)^-$ (acid)	Mn(II)
$(AsO_4)^{3-}$, $(AsO_3)^{3-}$	As^0
$(ClO_3)^-$, $(ClO_4)^-$	Cl^-

in acidic media. Finely divided metals of the transition group catalyze the decomposition.

If heated to about 300°C., the hypophosphites burst into flame because of the phosphine evolved. This phenomenon serves as a qualitative test for the hypophosphites since, except for the salts of organic acids, no common inorganic crystalline salts of importance in pharmaceutical chemistry will undergo combustion. The organic salts may be distinguished from the hypophosphites by the black carbon residue remaining after ignition.

The hypophosphite series of compounds is therapeutically inert. They are slightly toxic, but much less so than phosphorus or phosphine. Their use as stabilizing agents has been previously discussed (page 242). Their use in phosphorus tonics will be covered in Chapter 13.

The phosphoric acids. Phosphorus pentoxide, P_2O_5, is the parent anhydride of an intricate series of acids. It is a white solid formed by burning the element in a plentiful supply of air. It has a strong affinity for water, making it an excellent drying agent. In fact this affinity is so strong that it can serve as a dehydrating agent:

$$2HNO_3 + P_2O_5 = N_2O_5 + 2HPO_3$$

Unfortunately the mass of phosphorus pentoxide quickly becomes coated with a film of inactive glassy metaphosphoric acid and loses its effectiveness.

The origin of the phosphoric series of acids may be theoretically, if not practically, demonstrated by writing equations for reactions between the oxide and water. The various acids differ in degree of hydroxylation:

$$P_2O_5 + H_2O = 2HPO_3 \text{ (meta)}$$
$$P_2O_5 + 2H_2O = H_4P_2O_7 \text{ (di)}$$
$$P_2O_5 + 3H_2O = 2H_3PO_4 \text{ (ortho)}$$

The meta acid is usually assigned its empirical formula, HPO_3, but the existence of the monomer is extremely doubtful. It forms on the initial reaction of water with phosphorus pentoxide, but its aqueous solutions change slowly to the ortho acid at room temperatures. At elevated temperatures this change is rapid. Metaphosphates are ring and chain structures.

Diphosphoric acid (pyrophosphoric acid), $H_4P_2O_7$, is the first member of a series of polyphosphoric acids. The pK_a of 0.8 for its first ionization stage makes it a slightly stronger acid than the ortho acid. Each of the four hydrogens is replaceable; the dihydrogen and the normal salts are most common. The structure of the diphosphate ion has been shown to be a pair of phosphate tetrahedra linked through a common oxygen.

The chain polyphosphates and ring metaphosphates are useful chelating agents. Oxygen atoms on adjacent phosphate tetrahedra are used to form six-membered rings. Their tendency to form complexes decreases with increasing size of the complexed cation. The instability constants of the calcium complexes, $[CaP_2O_7]^{2-}$, $[CaHP_2O_7]^{-}$, $[CaP_3O_{10}]^{3-}$, and $[CaHP_3O_{10}]^{2-}$, expressed as negative logs, are 4.9, 2.3, c. 5.0, and 3.1, respectively.

The metaphosphates and diphosphates are used largely in laundry and other cleansing preparations. Potassium pyrophosphate, $K_4P_2O_7$, soluble in 0.5 parts water, is used as a builder synergist in liquid general purpose cleansers. The metaphosphate glass, sodium hexametaphosphate, is used as a water softener.

Orthophosphoric acid and the orthophosphates. Orthophosphoric acid is the only monomeric member of the phosphate series of acids. Pure phosphoric acid[4] is prepared by burning elemental phosphorus under conditions insuring complete oxidation and dissolving the resultant oxide in pure water. The syrupy commercial acid results. Other concentrations

[4] Technical grades of phosphoric acid are prepared by the reaction of calcium phosphate rock with sulfuric acid.

are obtained by dilution. Orthophosphoric acid is a colorless crystalline solid melting at 42.4°C. It is prepared by removing water from syrupy (glacial) phosphoric acid under reduced pressure. The arrangement around the central phosphorus atom is approximately tetrahedral; the HO—P—OH angles are 106°, the O—P—OH angles 112°. The single nonprotonated oxygen is associated with one of the adjacent hydroxyl groups through a hydrogen bond. This is a contributing factor to the extremely weak acidic character of the third proton of the acid.

The first ionization of phosphoric acid is moderately strong.

$$H_3PO_4 + H_2O = H_3O^+ + H_2PO_4^- \qquad pK_1 = 1.96 \tag{1}$$

The ionization of the second hydrogen is that of a weak acid.

$$H_2PO_4^- + H_2O = H_3O^+ + HPO_4^{2-} \qquad pK_2 = 7.12 \tag{2}$$

The monohydrogen phosphate ion is a strong base and undergoes reaction with water.

$$HPO_4^{2-} + H_2O \rightleftharpoons H_2PO_4^- + OH^- \tag{3}$$

When a soluble dihydrogen phosphate salt is dissolved in water, the net result of reactions 2 and 3 is such that the acid ionization is slightly predominant. The pH of such solutions will be about 6–7. On the other hand, the hydrolysis reaction 3 predominates in a solution of a soluble monohydrogen phosphate salt. Such solutions have a pH comparable to that of normal carbonates, about 10–11. The third ionization of phosphoric acid is that of an extremely weak acid.

$$HPO_4^{2-} + H_2O \leftarrow H_3O^+ + PO_4^{3-} \qquad pK_3 = 12.32 \tag{4}$$

The phosphate ion is too strong a base to exist in the presence of water. On solution of a soluble tertiary phosphate the reaction

$$PO_4^{3-} + H_2O \rightarrow HPO_4^{2-} + OH^- \tag{5}$$

proceeds to such an extent that the solution becomes very strongly alkaline.

Phosphoric acid would by ordinary standards be considered an oxidizing agent, the phosphorus central atom being in its highest positive oxidation state. However, both the acid and its anions are inert to all but the very strongest reducing agents. Since the materials ordinarily met in pharmaceutical practice are not strong enough reducers to be affected by phosphorus(V), redox incompatabilities may be considered nonexistent.

The phosphates present an interesting example of the relationship between the solubility of an electrolyte and the charges of its cation and anion. Essentially the dissolution of an ionic compound depends upon the ability of the solvent to overcome the electrostatic interactions existing between the positive and negative fields of the ions, the interactions responsible for the fixed crystalline arrangement of the solid state (page 190). The phosphate ion, with its formal -3 charge, is accompanied by a very intense field (see Figure 3.1, page 53). Exceptionally strong interactions with the positive cationic fields result, as evidenced by the paucity of soluble phosphates.

The charge of the cation is therefore very important in determining the solubility of a given phosphate. Monovalent cations, having the least intense positive field, should and do give rise to the most soluble group of phosphates. The alkali metal phosphates together with ammonium and a few ammonium derivative phosphates are the only soluble phosphates. Even here the degree of solubility is a function of field strength as moderated by ionic radius. Because of its extremely small radius, the lithium ion has a sufficiently intense field to react rather strongly with the field of the phosphate ion. As a result lithium phosphate is only slightly soluble in water (1 in 100). Water is unable to overcome the electrostatic attractions existing in the phosphate of divalent, trivalent, and tetravalent cations.

The solubility of the insoluble phosphates may be improved by any procedure reducing the high negative charge of the phosphate moiety. In practice this is accomplished most readily by controlling the pH of the solvent. Because of the strongly basic character of the phosphate ion, it will readily accept protons from the hydronium ion of acidic media,

$$PO_4^{3-} + H_3O^+ = HPO_4^{2-} + H_2O$$

As a result the equilibrium between the solid phosphate and its dissolved ions cannot be established. As quickly as a phosphate ion separates from the solid to maintain the equilibrium, it reacts with a hydronium ion, thereby being effectively removed from participation in the equilibrium. The solubility is shifted to that allowed by the K_{sp} of the monohydrogen salt in moderately acid solutions. In strong acid solutions the reaction

$$HPO_4^{2-} + H_3O^+ = H_2PO_4^- + H_2O$$

becomes extensive; in these solutions the solubility of the phosphate is that allowed by the K_{sp} of the dihydrogen phosphate salt. For trivalent cations the interactions between the cation and anion in the solid state are so strong that this latter condition is necessary before the salt will dissolve, e.g., $AlPO_4$, $FePO_4$. Phosphates of tetravalent cations, e.g., $Zr_3(PO_4)_4$, are insoluble in all aqueous media. The relationships discussed have been summarized in Table 12.5. The acetic acid-ammonium acetate buffer may be considered the moderately acidic medium; hydrochloric acid will give the strongly acid medium.

The preparation of the soluble phosphates is straightforward: neutralization of the carbonate or the hydroxide of the desired cation to the required end point, followed by filtration and crystallization. However, the preparation of insoluble phosphates by the usual precipitation techniques is frequently unsuccessful. Especially in alkaline media the stoichiometry of the compound obtained may not be that suggested by the reacting ratios used. Due to the presence of appreciable hydroxide ion in these solutions, insoluble hydroxy phosphates may precipitate. The nature of

TABLE 12.5 Solubility of Phosphates as a Function of Cation Charge

pH \ Cation Charge	+1	+2	+3	+4
≥7	soluble	insoluble	insoluble	insoluble
4-6	soluble	soluble	insoluble	insoluble
0–2	soluble	soluble	soluble	insoluble

these problems will be elucidated by a discussion of the preparation of the alkaline earth phosphates.

The preparation of the alkaline earth phosphates is dependent on several factors. The degree of ionization of phosphoric acid is important, and thus the pH must be controlled. The concentration of reactants and the method of mixing are also important. Temperature has a strong effect. Impurities present may change the nature of the precipitate, particularly those forming insoluble salts or chelating with the alkaline earth cation. In effect, anything which affects the ionization of the acid and the concentration of free cations will alter the nature of the phosphate obtained from the reaction mixture.

For some time there was uncertainty as to the existence of a true tricalcium phosphate in aqueous media. When crystallization of the compound is attempted, using highly purified solutions of the proper stoichiometry, octacalcium phosphate, $Ca_4H(PO_4)_3 \cdot 3H_2O$, is obtained. The compound has the *apatite* structure of most calcium phosphate minerals, and the bones and teeth. For this reason the apatite formula, $Ca_8H_4(OH)_2(PO_4)_6 \cdot 4H_2O$, rather than that commonly given is preferable. Tricalcium phosphate does, however, occur in nature as whitlockite, a mineral having 6–8 atom% magnesium ion substituting for calcium ion.

TABLE 12.6 PREPARATION OF THE COMMON CALCIUM PHOSPHATES

Compound	Reagents	pH	Temperature, °C.
$Ca(H_2PO_4)_2$	CaO + excess H_3PO_4	—	evaporate at 125–130
$Ca(H_2PO_4)_2 \cdot H_2O$	CaO + excess H_3PO_4	—	evaporate at 35–40
$CaHPO_4$	CaO + excess H_3PO_4 or,	—	100–110
	$Ca(NO_3)_2 + (NH_4)_2HPO_4$	3	83
$CaHPO_4 \cdot 2H_2O$	CaO + excess H_3PO_4, slowly ammoniated or	—	25
	ammoniated $CaCl_2$ added to $(NH_4)_2HPO_4$	5	30
β-$Ca_3(PO_4)_2$	$Ca(NO_3)_2 + Na_2HPO_4$ + 1% Mg^{++} or Mn^{++}	—	70
$Ca_8H_4(OH)_2(PO_4)_6 \cdot 4H_2O$	$CaHPO_4 \cdot 2H_2O$ + 0.5M sodium acetate	—	40

As a result of this observation it was found that impurities stabilize the tertiary structure under crystallization conditions which ordinarily produce the more common hydroxyapatite. In practice 0.1–1% magnesium ion, 1% manganese ion, or 10% iron(II) ion is effective in stabilizing the tertiary calcium phosphate. It has been postulated that only cations smaller than calcium can produce this stabilization and that the effect is one of seeding.

The conditions for preparation of the common calcium phosphate salts are given in Table 12.6. Calcium hydroxyapatites may be formed having various Ca:P molar ratios; the preparative procedures for some common ratios are given in Table 12.7. The higher phosphates may also be prepared in the solid state. Fusion of dicalcium diphosphate, ($Ca_2P_2O_7$), with calcium carbonate at 1000°C. yields the β-$Ca_3(PO_4)_2$. Fusion of calcium monohydrogen phosphate (anhydrous) with water at 300°C. and

TABLE 12.7 PREPARATION OF HYDROXYAPATITES

Ca:P Molar Ratio	Reagents	Temperature
1.41	dilute $CaCl_2$ + excess dilute Na_2HPO_4	25°C.
1.50	$Ca(OH)_2$ added to dilute H_3PO_4; to phenolphthalein end point	—
1.61	$Ca(OH)_2$ added to dilute H_3PO_4; to phenolphthalein end point	heat to boiling after neutralization
1.67	$Ca(OH)_2$ added to dilute H_3PO_4	neutralize at boiling temperature
1.75	freshly precipitated $Ca_3(PO_4)_2 + Ca(OH)_2$	—

under 1250 lb./in.[2] pressure yields the hydroxyapatite, octacalcium phosphate. A more thorough discussion of the phosphates of the entire alkaline earth family is available.[5]

The solubility behavior of these calcium compounds is important to the pharmaceutical chemist not only for help in the preparation of calcemics, mineral supplements, etc., but also for an understanding of the calcium-phosphate relationships in the body and for the prevention of solubility incompatabilities involving these ions. The order of solubility shows typical pH dependence. The solubility products are given in Table 12.8.

TABLE 12.8 SOLUBILITY PRODUCTS OF CALCIUM PHOSPHATES

Compound	$K_{sp}^{25°C.}$	$pK_{sp}^{25°C.}$
$Ca(H_2PO_4)_2$ and $Ca(H_2PO_4)_2 \cdot H_2O$	7.25×10^{-2}	1.14
$CaHPO_4$, and $CaHPO_4 \cdot 2H_2O$	2.5×10^{-7}	6.60
	(1.2×10^{-8} at 90°C.)	(7.91 at 90°C.)
$Ca_3(PO_4)_2$	1×10^{-26}	26.0
	(3×10^{-33} at 38°C.)	(32.5 at 38°C.)
$Ca_8H_4(OH)_2(PO_4)_6 \cdot 4H_2O$	1.3×10^{-47}	46.9
$Ca_{10}(OH)_2(PO_4)_6$	1×10^{-115}	115

Although ordinarily only the orthophosphates are used in pharmaceutical chemistry, the meta- and diphosphates find occasional employment. Tests useful in distinguishing between these three families of phosphates are given in Table 12.9.

The phosphates are used in pharmacy and medicine almost entirely for their acid-base properties. The insoluble tribasic phosphates are ordinarily used as antacids, but may also serve to maintain proper phosphate levels in the gastrointestinal tract. The soluble monobasic salts are valuable as urinary acidifiers. The phosphate buffer systems are frequently used to provide stable pH conditions for medicinal preparations.

Phosphoric acid itself is rarely used in a medicinal capacity.

Glycerophosphates. Glycerophosphoric acid (phosphoglyceric acid) is prepared by direct esterification of glycerine using phosphoric acid. The pure acid is difficult to prepare since conditions required for concentration result in decomposition. For this reason it is sold as a 25–50% aqueous solution. The commercial product is a mixture of two isomers. The α acid has the ester linkage at one of the primary hydroxyls of the glycerine. The β acid is esterified at the secondary hydroxyl of the glycerine. Since

[5] *Chem. Rev.* **61,** 436 (1961).

TABLE 12.9 DISTINGUISHING TESTS FOR THE MAJOR PHOSPHATE GROUPS

Reagent	Phosphate Groups[a]		
	ortho	di	meta
$AgNO_3$	yellow↓	white↓	white, gelatinous↓
$Zn(OAc)_2$	—	white↓	—
$(NH_4)_2MoO_4$ (in HNO_3)	yellow↓	—	—
$BaCl_2$			
acid medium	—	—	white↓
alkaline medium	white↓	white↓	white↓
$AlCl_3$	white↓ soluble in HOAc	white↓ insoluble in HOAc	white↓ insoluble in HOAc
$CoCl_3$ (in strong ammoniated solution)	—	red-yellow↓	—

[a] Downward arrow indicates formation of precipitate.

only one of the hydroxyl groups of the phosphoric acid is removed during esterification, glycerophosphoric acid is a dibasic acid.

The glycerophosphates (phosphoglycerates) are prepared by neutralization of glycerophosphoric acid with a suitable base. They may be precipitated by addition of ethanol to the reaction mixture. The sodium, calcium, iron(II), and manganese(II) salts were at one time official. The commercial salts are mixtures of the two isomers in varying states of hydration. They may be separated. The sodium α isomer is a colorless or yellowish syrupy liquid from which, on drying under reduced pressure, a white, very hygroscopic solid separates. The β isomer occurs as plates or scales. Aqueous solutions of glycerophosphates are subject to hydrolysis, especially if heated:

$$2Na^+ + C_3H_5(OH)_2OPO_3^{2-} + HOH = C_3H_5(OH)_3 + 2Na^+ + HPO_4^{2-}$$

The salts may be determined by titration with a strong acid to a methyl orange end point. One equivalent of acid is consumed per mole of salt.

$$2Na^+ + C_3H_5(OH)_2OPO_3^{2-} + H_3O^+ = Na^+ + C_3H_5(OH)_2OPO_2(OH)^- + Na^+ + H_2O$$

The glycerophosphates were at one time considered valuable sources of phosphate as dietary supplements. Their use in this application is discussed in Chapter 13.

HALOGENS

Except for hydrogen these physiologically important elements have the ground state electron structure, ns^2np^5. In certain of its chemical properties hydrogen acts as though it were the first member of the halogen group. For this reason it should be included in this family as well as the alkali family. The halogens are all very reactive. Thus all exist in nature as compounds rather than in the free state. Important structural properties are given in Table 12.10.

Oxidation states. Minus one is by far the most important oxidation number, both in naturally occurring compounds and in physiologically important compounds. This follows logically from the ground state

TABLE 12.10 STRUCTURAL PROPERTIES OF THE HALOGENS

Element	Hydrogen	Fluorine	Chlorine	Bromine	Iodine
Symbol	H	F	Cl	Br	I
Atomic weight	1.008	19.00	35.46	79.92	126.9
Atomic number	1	9	17	35	53
Electronic structure	$1s$	$2s^2 2p^5$	$3s^2 3p^5$	$4s^2 4p^5$	$5s^2 5p^5$
Covalent radius (Å)	0.30	0.72	0.99	1.14	1.33
Ionic radius (Å), −1	2.08	1.36	1.81	1.95	2.16
+7 (calc.)	—	0.07	0.26	0.39	0.50
Ionization potential (ev.)	13.60	17.42	13.01	11.84	10.44
Reduction potential (v.)	—	−2.85	−1.36	−1.07	−0.54
Electronegativity	2.1	4.0	3.0	2.8	2.5

electron structure and high electronegativity; ready acceptance of one additional electron gives a closed electron structure.

$$ns^2np^5 + e^- = [ns^2np^6]^-$$

Not only the strongly electropositive but also moderately electropositive metals will contribute valence electrons to a halogen atom, giving rise to ionic compounds. With weakly electropositive elements the ns^2np^6 state is achieved by formation of a σ bond, e.g., $BeCl_2$, $AlCl_3$, $HgCl_2$. In fact, the electronegativity of the halogens is so great that they form compounds with other rather strongly electronegative elements, e.g., NF_3, ICl.

Once the closed inert gas shell is achieved, one or more of the four pairs of nonbonding electrons of the reacted halogen are available for coordinate covalent bond formation. With electronegative elements the halogen atom assumes the role of the central atom of a complex, e.g., ClO_2, IBr_3. With the metallic elements the halogen acts as a ligand.

In the role of a central ion the halogen effectively acts as though it were giving up a share in one, two, three, or four of the electron pairs constituting its closed shell. As a result the central halogen unit assumes a formal positive charge or oxidation number. If one electron pair is shared the formal charge of the halogen becomes $+1$; if two pairs are shared, $+3$; with three shared pairs, $+5$; and if all four pairs are shared, $+7$. This accounts for the existence of molecules and ions such as IO_3^-, HBrO, ClF_3, ClO_4^-. However, except for the chlorine-oxygen series (page 119), compounds for each possible oxidation state of a given halogen with a given ligand are not known.

The maximum coordination number for the individual halogens has not been fully established. On the basis of periodic classification fluorine should have a maximum coordination number of four; chlorine, bromine, and iodine, six. However, no compound has been definitely characterized in which fluorine has achieved a maximum coordination number of four. Nor has a coordination number of six been realized for chlorine. It is doubtful that bromine can achieve a maximum coordination number of six. Only iodine is definitely known to achieve this maximum coordination number.

All halide ions, including the hydride ion, actively participate as ligands in complex formation. Most of these complexes form readily and all are at least moderately stable. Complexes involving the fluoride ion as a ligand are frequently the only complex in which the central ion is able to achieve its maximum coordination number.[6] Some of these complexes important in pharmaceutical chemistry are HgI_4^-, $AlCl_4^-$, $FeCl_2^+$.

Solubility behavior of halogen compounds. The solubility behavior of the simple and complex fluorides does not follow that of the other halides. All common fluorides are soluble, except calcium fluoride. The other halide salts are soluble, except those of Ag^+, Hg_2^{2+}, Pb^{2+}, Cu_2^{2+}, and Tl^+. For a given cation the solubility decreases with increasing atomic weight of the halogen. Thus silver iodide is less soluble than silver bromide, which in turn is less soluble than silver chloride. Silver fluoride is very soluble in water. The unusual behavior of fluoride is evidenced in Table 12.11.

The complex salts in which the halogens participate either as the central unit of the anion or as ligand are, in general, water soluble—again with the exception of some of the fluorine complexes. Fluoride complexes frequently behave quite differently from the expected; e.g., sodium fluosilicate is quite insoluble, whereas lead(II) fluosilicate is very soluble in water. The oxohalide anions are water soluble.

Acid-base behavior. The halogens display well-defined group acid-base behavior as a function of atomic weight. The simple halide anions are all bases, the order of base strength being

$$I^- < Br^- < Cl^- < F^- < H^-$$

[6] Hydroxyl ion frequently shares this distinction.

It follows that the hydrogen halides are acidic, having an inverse order of acid strength.

$$HI > HBr > HCl > H_2F_2 > H_2$$

In aqueous media the first three are among the strongest acids known. Hydrogen-bonded hydrogen fluoride is a moderately strong acid. Hydrogen is a very weak acid.

All hydrogen halides are gases under standard conditions. For this reason, though they are strong acids, they may be displaced from their salts if acidified, even by weaker nonvolatile acids.

TABLE 12.11 PROPERTIES OF SOME HALIDE SALTS

	F^-	Cl^-	Br^-	I^-
Solubility Ca^{2+} salt (Gm./L.)	0.016	427	588	676
Solubility Ag^+ salt (Gm./L.)	1.72×10^4	0.0154	0.00084	0.000028
Boiling point HX(°C.)	19.4	−85	−67	−35.5
Solubility HX in water (%, w/w)	35.3	42	49	57
Melting point Ca^{2+} salt (°C.)	1333	780	760	740

$$Na^+Cl^- + H_2SO_4 = HCl\uparrow + Na^+HSO_4^- \text{ (room temperature)}$$
$$2Na^+Cl^- + H_2SO_4 = 2HCl\uparrow + Na^+{}_2SO_4^{2-} \text{ (elevated temperature)}$$

Volatility of the hydrogen halides gives rise to a possible chemical incompatibility when working with halide salts or their solutions. Acidification can lead to loss of the gaseous acid from the solution.

The hydrogen halides are extremely soluble in water, their saturated aqueous solutions consisting of dissolved acid molecules and the ionization products. Since the dissolved gas exists in equilibrium with the free gas above the solution (Henry's law), the hydrogen halide molecules will escape to the atmosphere unless the containers are well stoppered.

Because chloride, bromide, and iodide are very weak bases, the extent of hydrolysis of their salts in aqueous media is a function of the accompanying cation. Sodium chloride solutions are approximately neutral, whereas dissolved ammonium chloride imparts as acidic reaction to the solution.

$$Na^+ + Cl^- + H_2O \leftrightharpoons \text{no appreciable reaction}$$
$$NH_4^+ + Cl^- + H_2O = NH_3 + H_3O^+ + Cl^-$$

Because of the volatility of the hydrogen halides (except for the fluorides), the common practice of preparing anhydrous salts by heating a hydrate cannot be used.

$$[Mg(H_2O)_6]^{2+}Cl^-{}_2 = MgO + 2HCl\uparrow + 5H_2O\uparrow$$

Redox properties. Although theoretically these elements may act as reducing agents, the halogens generally act as oxidizing agents. They arrange, strongest oxidizer first:

$$F_2 > Cl_2 > Br_2 > I_2$$

This order is as expected. Fluorine being the smallest member of the family should have the greatest electron affinity. Typical reactions are

$$F_2 + H_2O \rightharpoonup H_2F_2 + \tfrac{1}{2}O_2$$
$$Cl_2 + 2H_2O \rightleftharpoons HClO + H_3O^+ + Cl^-$$
$$I_2 + 2S_2O_3^{2-} \rightharpoonup 2I^- + S_4O_6^{2-}$$

The first reaction explains why, unlike chlorine and bromine, fluorine aqueous solutions cannot be prepared.

The halogens find wide use as oxidizing agents, both commercially and in the laboratory. Because of their oxidizing action, chlorine and iodine find use as germicides (page 449).

The halide ions are reducing agents; the order is

$$I^- > Br^- > Cl^- > F^-$$

For most purposes in pharmaceutical chemistry, this property is largely of nuisance value. The iodides and bromides, especially in solution, must be stabilized by the use of antoxidants. Even relatively mild oxidizing agents, e.g., oxygen of the air, cannot be tolerated in preparations including these salts. In pharmaceutical chemistry, for practical purposes, the chloride ion has no reducing properties and may be considered redox stable. The fluoride ion is redox stable except under the most severe oxidizing conditions.

The oxo anions of the halogens constitute another important aspect of the redox properties of these elements. The known oxo complexes, recorded as the acids, appear in Table 12.12 together with some of their properties. Unless otherwise indicated, a given acid is known in aqueous solution only. Fluorine, being more electronegative than oxygen, has no oxo anions.

Although the +1, +3, and +5 complexes may act as oxidizing or reducing agents, in most common applications these compounds, e.g., $HClO$, $KClO_3$, act as oxidizing agents. Perchloric and the periodic acids and their salts do not have medicinal applications. However, they are valuable analytical reagents. Hypoiodic acid may act as either an acid or a base:

$$HIO + H_2O = H_3O^+ + IO^-, \qquad pK_a = 2.5 \times 10^{-11}$$

and

$$IOH = I^+ + OH^-, \qquad pK_b = 3.2 \times 10^{-10}$$

The I^+ ion is stabilized by complexation with pyridine (py). Salts of the general formula $[I(py)_2]^+X^-$ are well established.

The chemistry of the halogens and their compounds has been summarized in Table 12.13.

Natural occurrence of halogen salts. Although fluorides are widely distributed, they are not abundant. Fluorspar, CaF_2, is the usual source of the element, although cryolite, Na_3AlF_4, also serves. Sodium chloride

TABLE 12.12 THE OXO ACIDS OF THE HALOGENS[a]

General formula	HXO	HXO_2	HXO_3	HXO_4
Halogen oxidation number	+1	+3	+5	+7
Chlorine compound	$HClO$	$HClO_2$	$HClO_3$	$HClO_4$ (oily liquid) $H_3O^+ClO_4^-$ (crystalline)
Bromine compound	$HBrO$	(?)	$HBrO_3$	—
Iodine compound	HIO	(?)	HIO_3(cryst.)	HIO_4, $H_4I_2O_9$, H_3IO_5, H_5IO_6 (cryst.)
pK_a	~8	~2	~−2	~−10
Redox character	reducer oxidizer	reducer oxidizer	reducer oxidizer	oxidizer

[a] The prefixes for the periodic acids, upper first, are meta, dimeso, meso, and para.

is available in abundant deposits as rock salt, or it is obtained by crystallization from salt brines. This compound serves as the most economical source of chlorine. Deposits of sylvine, KCl, also exist. Other than a silver bromide ore, no bromide deposits exist as such. Ordinarily this element must be obtained from sea water or salt lakes or as a by-product from the working of salt brines. The only commercial ore serving as a source of iodine is caliche, in which $NaIO_3$ occurs as a recoverable impurity. Iodine is also recovered from salt brines and seaweed.

Biochemistry. Fluoride is present to some degree in all bony structures. It occurs in comparatively high concentration in the enamel of the teeth. The free ion has no known metabolic role. In high concentrations it acts as a protoplasmic poison, inhibiting or diminishing cellular respiration.

Chloride ion is present in intracellular and extracellular body fluids. It is a constituent of all body excretions. Its prime function is to maintain

TABLE 12.13 Reactions of the Halogen Family

Typical Reaction	Limitations and Remarks
$2M + nX_2 = 2MX_n$ $Fe + I_2 = FeI_2$	React with virtually all metals (metallic closures corrode).
$X_2 + 2H_2O = H_3O^+ + X^- + HXO$ $Cl_2 + 2H_2O = H_3O^+ + Cl^- + HClO$	Except F_2 (see text).
$X_2 + 2MX' = X_2' + 2MX$ $Br_2 + 2K^+I^- = I_2 + 2K^+Br^-$	Halogen displacement series—a more electronegative halogen will displace a less electronegative halide from its salts.
$X_2 + X'^- = [X'X_2]^-$ $Br_2 + I^- = [IBr_2]^-$	Reaction especially prominent with iodides.
$M^{n+} + yX^- = [MX_y]^{(n-y)}$ $Hg^{2+} + 4I^- = [HgI_4]^{2-}$	All halide ions are strong ligands.
$M^+X^- + H_3O^+ = M^+ + HX\uparrow + H_2O$ $Na^+Cl^- + H_3O^+ = Na^+ + HCl\uparrow + H_2O$	Volatile hydrogen halides are displaced from their salts by non-volatile acids (oxidizing acids may not be used with Br^- and I^-).
$2M^+X^- + MnO_2 + 4H_3O^+ =$ $2M^+ + Mn^{2+} + X_2\uparrow + 6H_2O$ $2Na^+Cl^- + MnO_2 + 4H_3O^+ =$ $2Na^+ + Mn^{2+} + Cl_2\uparrow + 6H_2O$	Oxidizing agents displace the elemental halogens from their salts.
$M_yO_z + zX_2 + zC = yMX_{2z} + zCO\uparrow$ $Fe_2O_3 + 3Cl_2 + 3C = 2FeCl_3\uparrow + 3CO\uparrow$	Anhydrous halide salts may be prepared in this way at elevated temperatures.
$3X_2 + 2P = 2PX_3$ $3Br_2 + 2P = 2PBr_3$	Excess P must be used; As, Sb, and Bi react similarly.
$PX_3 + X_2 = PX_5$ $PCl_3 + Cl_2 = PCl_5$	Excess X_2 must be used, PI_5 does not form.

charge balance within the various body fluids. Together with sodium ion it is responsible for maintenance of the tonicity of the body fluids. Except in very high concentrations this ion is inert physiologically.

Bromide ion occurs in traces in all body fluids. Its particular role, if any, is unknown. Its physiological behavior is similar to that of chloride ion in that it may substitute for it to some extent in all tissues except nerve tissue. The medicinal use of this ion as a sedative is not a normal body function.

Iodide is an essential trace element. It is required as a structural element in the thyroid hormone, thyroxine.

Pharmaceutical and medicinal uses. Simple and complex fluorides, are used for the fluoridation of municipal water supplies. They also find use in dental therapy. Fluorine atoms are frequently substituted in organic molecules in order to modify their medicinal or pharmaceutical properties. The fluosilicates are used as insecticides, the sodium salt as a laundry sour.

Chloride is used in pharmacy largely in the preparation of electrolyte replenishers. Trace element preparations in solution form are often prepared using soluble chloride salts. Since chloride is physiologically inert, many insoluble basic organic medicinally valuable compounds are rendered soluble by converting them to hydrochloride salts (page 224). Sodium chloride is usually the compound of choice in adjusting the tonicity of pharmaceutical preparations (page 204). Hydrogen chloride is used in achlorhydria, and is a valuable acid.

Chlorine (as hypochlorite) has excellent disinfecting and germicidal properties. Chlorine dioxide is finding acceptance in laundry bleaches and may be useful as a disinfectant. Potassium chlorate has been used as an antiseptic.

Bromide is used medicinally as a sedative. It is a useful reagent in synthetic drug research. Although the oxobromine compounds, e.g., hypobromite, should have germicidal properties, their cost mitigates against such use since hypochlorite and iodine are comparatively inexpensive germicides.

Iodides find use as antigoiterogenics in cases of thyroid insufficiency. They are also used as expectorants. Elemental iodine is one of the most effective and useful germicides known. Its usefulness surpasses that of the mercurials. The element is also a valuable analytical reagent and finds wide use in organic and inorganic synthetic procedures.

The pseudohalogens or halogenoids. Several inorganic anions, e.g., CN^-, OCN^-, SCN^-, and N_3^-, remarkably resemble the halide anions. The similarity is especially marked in the cyanide ion; its properties are intermediate between the chloride and bromide ions. These similarities are best illustrated by comparing corresponding reactions of the halogens and pseudohalogens (see Table 12.14). The most striking difference is the very weak acidic character of the pseudohalogen hydrides as compared with the halogen hydrides: e.g., HCN, $pK_a = 8.9$; HCl, $pK_a \sim -10$.

Hydrogen cyanide is thought to be an equilibrium mixture of two tautomeric forms—hydrocyanic acid, HCN, and isohydrocyanic acid, HNC.

$$HCN \rightleftharpoons HNC$$

The respective ionizations may be written

$$HCN + H_2O = H_3O^+ + CN^- \qquad [:C\vdots\ddot{N}:]^-$$

$$HNC + H_2O = H_3O^+ + NC^- \qquad [:N\vdots\ddot{C}:]^-$$

Close examination of the two ionic structures, however, reveals that they are identical. The apparent difference is nothing more than a reverse order of writing. This is in accordance with the behavior of the cyanide salts in solution. However, in organic bonding, where permanent σ bonds form, either the carbon or the nitrogen atom may be bonded to the organic moiety, giving rise to nitriles or isonitriles, respectively.

TABLE 12.14 REACTIONS OF THE PSEUDOHALOGENS

Preparation of X_2:
$2Na^+CN^- + MnO_2 + 2H_2SO_4 \rightarrow (CN)_2\uparrow + Mn^{2+}SO_4^{2-} + Na^+_2SO_4^{2-} + 2H_2O$
$2Na^+Cl^- + MnO_2 + 2H_2SO_4 \rightarrow Cl_2\uparrow + Mn^{2+}SO_4^{2-} + Na^+_2SO_4^{2-} + 2H_2O$
Preparation of HX:
$Na^+CN^- + H_2SO_4 \rightarrow HCN\uparrow + Na^+HSO_4^-$
$Na^+Cl^- + H_2SO_4 \rightarrow HCl\uparrow + Na^+HSO_4^-$
Solubility behavior, AgX:
$CN^- + Ag^+ \rightarrow AgCN\downarrow$
$Cl^- + Ag^+ \rightarrow AgCl\downarrow$
Complex formation:
$AgCN\downarrow + (excess)CN^- \rightarrow [Ag(CN)_2]^-$
$AgCl\downarrow + (excess)Cl^- \rightarrow [AgCl_2]^-$
Formation of polyhalide ions:
$K^+CNS^- + (CNS)_2 \rightarrow K^+(CNS)_3^-$
$K^+I^- + I_2 \rightarrow K^+I_3^-$
Reaction with hydroxyl ion:
$(CN)_2 + 2OH^- \rightarrow CN^- + CNO^- + H_2O$
$Cl_2 + 2OH^- \rightarrow Cl^- + ClO^- + H_2O$
Addition to π bond:
$(SCN)_2 + C_2H_4 \rightarrow C_2H_4(SCN)_2$
$Cl_2 + C_2H_4 \rightarrow C_2H_4Cl_2$

Hydrogen cyanide, hydrocyanic acid (prussic acid), and the cyanides are poisons. Sodium thiosulfate injection is used as an antidote (page 586). The cyanides have no important applications in medicine. Mercury(II) cyanide, a linear covalent molecule, is used for sterilization of metallic surgical instruments (page 481). A 0.67% solution of potassium cyanide has been used on the conjunctiva to remove the olive-colored stains resulting from silver therapy.

Gaseous hydrogen cyanide is a useful fumigant. Because of the dangers involved, it should be used in this application only by trained experts. The gas is generated *in situ* using egg-like pellets of sodium cyanide and diluted sulfuric acid.

$$Na^+CN^- + H_3O^+ + HSO_4^- \rightharpoonup HCN\uparrow + HSO_4^- + Na^+$$

The fumigating action is very efficient, the gas penetrating all nooks, cracks, etc. Hydrogen cyanide is similarly generated in the gas chamber for administering the death penalty in certain states. Calcium cyanide is a useful fumigant and rodenticide. A dry portion of this salt is placed in the burrow of the animal, which is then sealed. Being a strong base, cyanide ion reacts with moisture,

$$Ca^{2+}(CN^-)_2 + 2H_2O = Ca^{2+}(OH)_2^- + 2HCN\uparrow$$

The cyanide group is a very valuable reagent in synthetic organic chemistry.

Cyanates are prepared by oxidation of either potassium or sodium cyanide with permanganate or lead(II) oxide. The potassium salt can be extracted from the reaction mass with dilute alcohol. Ammonium cyanate is of historical interest. By heating this salt, F. Wohler was able to show that there is no difference between the urea formed and that of organic origin.

The thiocyanate ion forms by direct reaction between sulfur and the cyanide ion.

$$CN^- + S = SCN^-$$

It is unstable in acid media.

$$SCN^- + H_3O^+ = HSCN + H_2O$$

followed by

$$HSCN = HCN + S\downarrow$$

On ingestion thiocyanate ion is distributed in the tissues much as chloride. It is quickly absorbed from the gastrointestinal tract and is excreted without change. Sodium and potassium thiocyanates have been used for the control of high blood pressure and until very recently were officially recognized in this category. However, due to dangerous side effects and poor tolerance by the aged, their use for this purpose has been gradually discontinued.

Thiocyanate ion has a goiterogenic effect. A. M. Chesney first noted the high incidence of goiter in rabbits maintained on cabbage. This was traced to the high thiocyanate levels of the cabbage plant. Iodine is not taken up as bound iodine in its presence. A dose (1–2 Gm.) of the potassium or the sodium thiocyanate, taken orally or intravenously, causes the thyroid cell membrane to lose its power of concentrating iodide. The favorable iodine ratio of 150:1 rapidly drops to 1:1. (The synthesis of diiodotyrosine is not blocked here, but is blocked by thiouracil.)

Thiocyanate is widely used as an analytical reagent, especially for iron(III). The exact nature of the highly colored blood-red reaction product has been a matter of dispute for some time. Several complex ions have been identified: $Fe(SCN)^{2+}$, $Fe(SCN)_2^+$, $Fe(SCN)_3^0$, $Fe(SCN)_4^-$.

Recently reported work seems to indicate that $Fe(SCN)_2^+$ is the major light-absorbing unit in thiocyanate solutions, 0.04*M* or higher.

To some degree the two complex hexacyanoferrate ions also act as pseudohalogens. The reaction products of these two ions with the two iron cations are of particular interest. The formulas usually assigned to the dark blue precipitates or colloids obtained by reacting iron(II) with hexacyanoferrate(III) and iron(III) with hexacyanoferrate(II), respectively, are based on reaction stoichiometry and have no experimental basis. These reactions are conveniently written

$$K^+{}_3[Fe(CN)_6]^{3-} + Fe^{2+} = \underset{\text{Turnbull's blue}}{KFe^{II}[Fe^{III}(CN)_6]} + 2K^+$$

$$K^+{}_4[Fe(CN)_6]^{4-} + Fe^{3+} = \underset{\text{Prussian blue or Berlin blue}}{KFe^{III}[Fe^{II}(CN)_6]} + 3K^+$$

Despite the difference in route of synthesis, evidence indicates that the two compounds are structurally identical. A super complex has been postulated. The cyanide ion apparently complexes two different iron cations, one at the carbon end, the other at the nitrogen end.

These compounds are used as whitening agents in laundry bluing and in food products, e.g., sugar, starch. Conceivably they could have a similar application in pharmaceuticals. They are also used as a dark blue pigment for paints and pencil leads. Ingestion of these leads by children may produce symptoms of cyanide poisoning.

STUDY AND SELF-APPRAISAL AIDS

Important Terms, Nomenclature of Interest, Self-Appraisal Questions, Practice in Equation Writing—these items will be worked into the corresponding sections of succeeding chapters. In this way the chemistry, which has been the predominant theme of the present chapter, will be integrated into the appropriate pharmaceutical applications.

THIRTEEN · MINERALS I: *fluid electrolytes and trace anions*

Minerals consist of those substances capable of providing the ions which are normal constituents of body fluids and support structures and which normally play a role in enzyme systems and metabolic processes. They are administered to maintain or restore normal levels of the specific ions of interest. The study of minerals will consider the normal role and concentration level of a given ion and a review of those compounds suitable for providing the ion; identification of preparations used for given medicinal needs; and a thorough investigation of the important physical and chemical properties in the formulation, preservation, and use of various compounds and preparations.

A critical examination of the ions present in this group of medicinal agents reveals two distinctly different ionic types. The first consists of chemically stable cations and anions which impart the necessary physical properties to the body fluids and the support structures, and appear in some important compounds. These ions are the subject of the present chapter. The second type consists of cations of variable oxidation numbers. This group of cations are the so-called trace elements, the individual cations having active roles in metabolic processes. They are the subject of Chapter 14.

IONIC STRUCTURE AND PHYSIOLOGICAL SUITABILITY OF ELECTROLYTES

Consideration of the cations, Na^+, K^+, Ca^{2+}, and Mg^{2+}, and anions, Cl^-, F^-, (I^-), CO_3^{2-}, PO_4^{3-}, important as fluid electrolytes and in bony structures, reveals one common characteristic—chemical stability. The simple cations and anions all have the chemically stable inert gas electron structure. The two complex anionic moieties have structures characterized by high chemical stability: a cationic central unit of high charge and small radius coordinated with a strongly electronegative anionic ligand. Both the carbonate and phosphate moieties are extremely stable, especially under the conditions existing in the body. The changes they undergo in their role as buffers do not affect the structure of the complex. Rather they merely involve the interaction of the strong negative electrostatic fields of these complexes with the positive field of protons to an extent necessary to maintain constant pH. Were it not for the mediation of the powerful enzyme carbonic anhydrase, the utility of the bicarbonate ion as a transport medium for carbon dioxide would be limited.

Because of the stability of the ionic species involved, chemical incompatibilities are for the most part nonexistent in fluid and electrolyte preparations. Only physical incompatibilities are common. Usually these are related to the insolubility of certain compounds as a function of pH, i.e., the possible precipitation of certain phosphates in alkaline media and the possible evolution of carbon dioxide (or hydrogen fluoride) in acidic media. In this group of medicinals it is more often the pharmacist's knowledge of the physiological functions and concentration levels of the substances in a given preparation that is important. He can use this knowledge, especially in formulation work, to minimize as much as possible untoward side effects of a given preparation. For this reason a modern interpretation of certain aspects of the biochemistry of inorganic ions will be stressed.

FLUID AND ELECTROLYTE REPLENISHERS

Normal levels and functions of electrolytes in body fluids. The human body is more than 60% water. This water serves in two different capacities. First, it acts as the body's heat sink; as such it is the principal factor in the maintenance of constant body temperature. Second, water acts as the body's solvent. As such it provides transport for nutrients, for metabolic wastes, and, of special interest here, for the electrolytes essential in maintaining pH, charge balance, tonicity, etc. The nature and the normal function of these electrolytes will be considered at this time.

Several different methods are used to express the concentration of a given substance in the body fluids. Gases are usually expressed in volume per cent (%, v/v). Substances, such as bicarbonate, that have an equivalent gas volume are frequently expressed similarly. Other dissolved substances, such as proteins, are usually expressed in grams per 100 ml. or in gram per cent (%, w/w). However, since both physical and chemical reactions take place on a particulate rather than a weight basis, milliequivalents per liter (me./L.) will usually be used in this textbook.

Ionic substances will not be treated on a molecular basis, e.g., $NaHCO_3$, KCl, as is so often done. This is inconsistent with present knowledge of the nature of electrolytes. In biological systems, as in ordinary aqueous systems, a given ion exists independently of other ions present irrespective of origin. In a solution including the ions of several different salts it is ambiguous to assign certain cations and anions to one another, even though they may have initially originated from that particular combination as a crystalline solute. Although interactions do exist between the electrostatic fields of the individual ions, the solutions under discussion are rather dilute with respect to any one ion. Thus the interactions are minimal. It is reasonable, however, that a few higher-charge ions, e.g., calcium or magnesium ions, associate electrostatically with anions to give rise to loosely bound lower-charge species, e.g., $CaCl^+$ or $MgCl^+$.

Chemically the actual number of each species of ion present in a given solution is not important, but the number of cations and anions must be such that charge balance is maintained. In biological systems, however, the relative abundance of each species of ion is important because of the specific roles each plays.

There are three unspecialized body fluids—the blood plasma (or serum), the interstitial fluid, and the intracellular fluid. Of these the blood plasma is most readily available for study and for adjustment through medical treatment. The intracellular fluids are not amenable to direct medical treatment.

The blood plasma is essentially a 0.9% salt solution containing 70 Gm. dissolved protein per liter. Under normal conditions the following ionic species are present: sodium, potassium, calcium, magnesium, chloride, bicarbonate, mono- and dihydrogen phosphate, sulfate, organic acid, and proteinate ions. The normal electrolyte composition of the blood plasma is found in Table 13.1. Except for the proteins these ions move freely into the interstitial spaces. Thus interstitial fluid has the same relative composition as the blood plasma, except for the protein component. The intracellular fluid is essentially similar to the interstitial fluid, except for the concentrations of sodium and potassium ions and for the existence of concentration gradients arising from the metabolic activity of the cell.

SODIUM ION: The sodium ion is the predominant cation of all extracellular fluids, supplying more than 90% of the cations present on an equivalent basis. It is present in all body secretions. The cells are able to maintain an effective negative concentration with respect to this ion. Although present to 142 me./L. in the extracellular fluids, only 35 me./L. are found in the intracellular fluid. Excess sodium ion is eliminated through the kidneys by glomerular secretion.

Acting through its osmotic effect in the extracellular fluids, sodium ion (together with chloride ion) regulates fluid distribution throughout the body. The ion must be present if water is to remain in the tissues. Efforts to alter the quantity and distribution of sodium ion are common medical procedures, e.g., the salt-free diets frequently recommended in the control of the cardiac edema of renal and hypoproteinema origin.

TABLE 13.1 CONCENTRATION OF ELECTROLYTES IN BLOOD PLASMA

Ions	Normal Range		Average Value
	Gm./L.	me./L.	me./L.
Cations			
Na^+	3.10–3.40	135–145	142
K^+	0.16–0.22	4.5–5.5	5
Ca^{2+} (total)	0.09–0.12	4.5–6.0	5
(ionic)	0.04–0.05	2.1–2.6	2.5
Mg^{2+}	0.02–0.04	1.5–3.0	3.0
Total cations			154
Anions			
Cl^-	3.5–3.75	98–105	102
HCO_3^- (adult)	1.50–1.90	25–31	28
(child)	1.20–1.60	20–26	23
$H_2PO_4^-$ and HPO_4^{2-}			
(adult)	0.06–0.15	1.2–3.0	2
(child)	0.12–0.21	2.5–4.5	3.5
SO_4^{2-}	0.02–0.07	0.3–1.5	1
$[\text{Proteinate}]^{n-}$	60–80	14.6–19.0	15
$[\text{Organic acid anions}]^{n-}$	—	6.0	6
Total anions (adult)			154

POTASSIUM ION: The potassium ion, together with magnesium ion, is the cation of the intracellular fluids. Although potassium ion easily passes through the cell wall in either direction, the cell is able to maintain a high positive concentration gradient of this ion with respect to that present in the surrounding interstitial fluid. The average cellular concentration is 115 me./L. vs. 5 me./L. for the extracellular fluids. Approximately 97% of the 1.6–2.1 Gm./Kg. of this ion present in the body is found within the cells.

Intestinal absorption and urinary excretion of potassium ion is rapid. Normally it is excreted almost exclusively by the kidneys where, unlike sodium ion, it is excreted by both the glomeruli and tubules. Certain hormones act to conserve potassium, e.g., testosterone and the growth hormone of the anterior pituitary. The 11-oxysteroids of the adrenals have the opposite effect.

Potassium ion is essential for growth, normal body function, and cardiac function. It has a role in protein biosynthesis, glycogen synthesis, and the phosphorylation of glucose. Deposition of glycogen in the liver is accompanied by deposition of potassium ion. It actively enters the cell during glucose metabolism and during acetylcholine synthesis. Associated with acetylcholine, it is essential for nerve impulse conduction and transmission.

Potassium deficiency appears at about the 3.5 me./L. level. It is manifested by general weakness, depression of the heart, and a low blood pressure. Since this ion is abundant in all foods, deficiencies are not common. Potassium deficiency may be produced by conditions which incur excessive loss of body fluids: excessive vomiting, loss of potassium-rich gastrointestinal secretions by severe diarrhea or chronic purging, hemorrhage, burns, and diabetic coma. It is frequently seen with patients maintained on diazides; for this reason a potassium salt is used concurrently. The ion is toxic in high concentrations. The effects are especially evident in patients with impaired cardiac function.

Potassium deficiencies are usually treated with potassium chloride, potassium gluconate and potassium bicarbonate (and acetate and citrate), usually in tablet form. Because some individuals find the taste objectionable or have gastrointestinal irritation, enteric coated potassium chloride tablets were introduced. However, severe ulcerations appeared in the small intestine of some patients so that this dosage form has been discontinued. The newer slow release forms utilize a wax matrix to slowly release the potassium salt in the intestine over several hours. Potassium Chloride Extended Release Tablets and Potassium Chloride Extended Release Capsules are official in *U.S.P. XXI.* Concentrated aqueous solutions, e.g., Potassium Chloride Oral Solution, *U.S.P. XXI,* are popular since they are easy to administer by adding to fruit juices, etc. Potassium Chloride and Potassium Gluconate are also available in elixir form (*U.S.P. XXI*).

CALCIUM ION: About 99% of the body's calcium ion is deposited in the bones and teeth. The remaining 1% is present in the body fluids, approximately equally divided between diffusible calcium and nondiffusible calcium. The diffusible calcium is largely ionic, but some is found in very loosely bound citrate-like complexes. The nondiffusible calcium is bound, chiefly to albumin, although a small amount is bound by the globulins.

Ionic calcium is the only physiologically active form of this element. Bone, though considered primarily as the structural support for the soft tissues of the body, also serves as a storage depot for this cation. Its calcium is available to the body and is drawn upon to maintain normal blood calcium levels during periods of low calcium intake. The protein-bound calcium of the blood probably serves as a secondary reservoir, becoming available locally during excessive loss or use of the ionic species.

Maintenance of a positive calcium balance by the body depends on dietary intake and the efficiency of absorption of calcium ion from the intestinal tract. However, because calcium forms insoluble compounds with many of the anions present in the food, efficient absorption of calcium is fraught with problems. Normally the phosphate ion is the principal interfering anion. Only calcium dihydrogen phosphate is sufficiently soluble to maintain the necessary levels for efficient absorption of the cation (see Table 12.8, page 331). Unfortunately this salt is stable only in acid media; in alkaline media the much less soluble mono-hydrogen phosphate or the highly insoluble tertiary phosphate are the stable forms.

Conditions in the stomach are sufficiently acid for the stable existence of free calcium ion in the presence of phosphate ion; but here absorption cannot take place. As the acid chyme discharged from the stomach moves through the duodenum, it is neutralized by the alkaline bile.

$$Ca^{2+} + H_2PO_4^- + OH^- = CaHPO_4\downarrow + H_2O$$

It is apparent that while calcium absorption may take place in the duodenum, solubility considerations act to prevent this uptake except during the relatively short period of time before the chyme is completely neutralized. Absorption in the remainder of the intestine is nil; the calcium has been precipitated from solution due to the alkalinity produced by the bile.

It is evident that high phosphate diets are unfavorable to efficient calcium absorption. Because of the common ion effect, excess phosphate depresses the solubility of the particular calcium phosphate stable at the given pH. Therapy involving use of strong bases, e.g., gastric antacids, decreases the efficiency of calcium absorption through its effect on the pH of the gastrointestinal tract. Faulty fat digestion due to high fat intake or an inadequate bile secretion interferes with calcium absorption through the precipitation of insoluble calcium soaps. Some plant foods, e.g., rhubarb, Swiss chard, spinach, beets, have high oxalate contents; calcium absorption is poor due to its precipitation as insoluble calcium oxalate. Proper calcium absorption also requires an adequate level of vitamin D, dietary or by supplement. This vitamin apparently controls the absorption of calcium ion. The parathyroid hormone is also involved in the calcium metabolism.

There is a very delicate balance between the two hydrogen phosphate ions, calcium ion, pH, and the solubility of the different calcium phosphate salts with respect to the equilibrium between the blood levels of these ions and their deposition in or solution from bone. The relationship between blood calcium and blood phosphate is an inverse one; when calcium is high, phosphate is low, and vice versa. High phosphate blood levels favor deposition of calcium in bone, whereas conditions reducing calcium blood levels or administration of salts having this effect (acidifying salts) favor release of calcium from bone.

Neither the chemistry involved in laying down the insoluble salts constituting the inorganic portion of bone nor the structure of these salts has been fully elucidated. It has been suggested that ions equivalent to $CaHPO_4 \cdot 2H_2O$ are first laid down as the insoluble octacalcium phosphate, $Ca_8H_4(OH)_2(PO_4)_6 \cdot 4H_2O$, and that this is slowly converted to a highly insoluble hydroxyapatite. During crystallization other anions, principally carbonate, citrate, and fluoride, are codeposited in the predominantly calcium-phosphate structure.

Structural studies indicate an arrangement similar to the mineral, chlorapatite. Chemical studies indicate similarity to podalite, a carbonato-apatite, with some carbonate being replaced by hydroxyl, chloride, citrate, and fluoride ion. Carbonate ion is deposited in the apatite structure despite a negative concentration gradient; the solution from which bone forms is unsaturated with respect to calcium carbonate. The carbonate ions present in bone do not have a uniform distribution and except for enamel can be preferentially removed. Citrate is present in an insoluble form in amounts approaching that of the carbonate. Enamel deposits are characterized by comparatively high fluoride levels.

The inorganic portion of the bone deposits through the mediation of bone cells, known as osteoblasts. It is laid down in a mold of collagen activated by a complex mucopolysaccharide. Since calcium, phosphate, and the other ions are being effectively concentrated from surrounding fluids, energy must be expended. This energy is obtained from enzyme systems involving high-energy phosphate bonds and possibly the citric and glycolic acid cycles. The release of calcium ion from bone involves large multinuclear bone cells known as osteoclasts. They release the enzyme phosphatase which actively causes solution of the inorganic material of bone. This dissolution is favored by a condition of acidosis in the body fluids.

Tribasic calcium phosphate is not present as such in bone. Examination of the data in Table 12.8 reveals that it is more soluble than the hydroxy-apatites. However, stones deposited in the bladder, kidney, and other parts of the body are essentially the normal tribasic compound. It is frequently present in dental calculi (the octacalcium compound is a minor constituent). Deposits of pure β-$Ca_3(PO_4)_2$ have been found in teeth.

An interesting example of isomorphous replacement exists in bone when certain foreign cations gain access to the body fluids. Strontium and lead(II) ions having ionic radii closely similar to that of calcium and having nearly identical solubility behavior will normally deposit in bone rather than being excreted. Painter's colic is a condition arising from the release of such deposited lead(II), for reasons not fully understood.

The excretion of calcium is largely through the mucosa of the small intestine. A comparatively small quantity (25–35%) is excreted in the urine as phosphate. Since excretion is continuous, a negative calcium balance can result if dietary intake is too low.

The story of calcium metabolism has been summarized in Fig. 13.1. It is important to remember that calcium metabolism is intimately

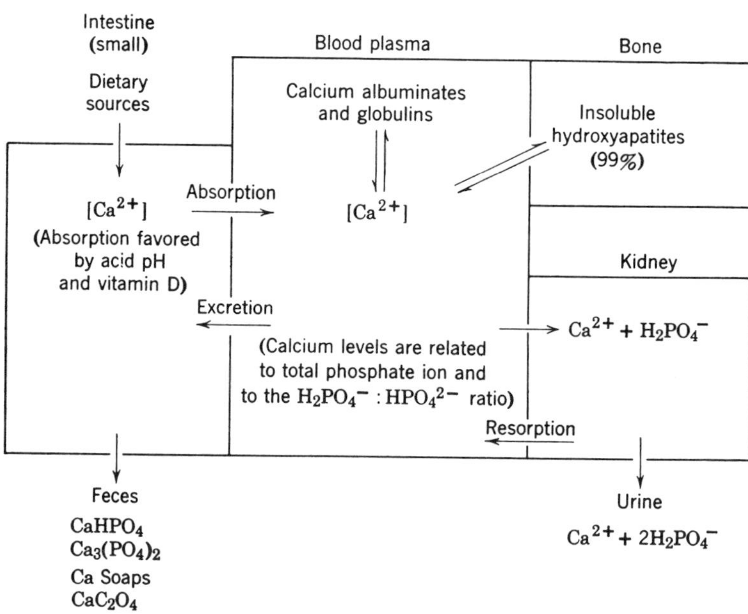

FIGURE 13.1 Calcium metabolism.

intertwined with that of the phosphate ions, vitamin D, and parathyroid hormone.

Calcium ion is indispensible for normal neuromuscular function. Low calcium levels lead to increased irritability of both nerve and muscle tissue. High calcium blood levels result in nerve sedation. Calcium ion is essential for proper cardiac action. The beat mechanism requires a delicate balance between potassium and calcium ion ratios; excess calcium ion results in systolic arrest (contraction) and excess potassium in diastolic arrest (relaxation). Calcium ion is essential for proper bone formation and for maintenance of proper membrane permeability. It is required for the coagulation of blood and for the coagulation of casein.

Hypocalcemia, a deficiency of ionized calcium in the blood stream, produces the syndrome known as tetany. A calcium-deficient diet or a vitamin D-deficient diet will result in the condition known as rickets in children and osteomalacia in adults. Calcium deficiencies are especially likely during periods of high calcium requirements, as in adolescent growth or pregnancy.

Hypercalcemia is a condition in which an abnormally high calcium ion level is present in the blood. It is characterized by anorexia, weight loss, constipation, bradycardia, arrhythmia, muscular pain, and kidney impairment. The effect of excess calcium on the heart is similar to that of digitalis; fatalities have resulted from simultaneous calcium and digitalis medication. Hypercalcemia occurs as a result of hyperparathyroidism, impaired kidney function, or very large doses of vitamin D or similar compounds.

MAGNESIUM ION: Magnesium ion is essential to life. It is present to about 0.012 me./Kg. The distribution is such that slightly over half occurs in the bone, most of the remainder being found in the cells. The blood plasma level is approximately 3 me./L. Because of the insolubility of its hydroxide, phosphates, and soaps, magnesium presents absorption problems virtually identical to those of calcium. However, vitamin D and the parathyroid hormone are not involved. Excretion is largely through the kidneys.

Magnesium ion appears to be indispensable for the smooth functioning of the neuromuscular system. It has a role in several enzyme systems. On intravenous or intramuscular injection it is a powerful anesthetic, acting similarly to chloroform. Its anesthetic value is questionable, however, since the useful anesthetic dose is too close to the lethal dose. Soluble magnesium salts (usually the sulfate) are used in depressive doses to relieve the tetany resulting from overdoses of oxytocin and convulsive states, and to relax striated muscle in eclampsia. Soluble magnesium salts and, to a lesser extent, insoluble salts have a saline laxative action (page 577). The sulfate is used as an antidote for poisonous cations having insoluble sulfates.

CHLORIDE ION: The chloride ion is the predominant anion of all body fluids. Two thirds of the anions present in the plasma are chloride ions. Chloride ions are absorbed readily throughout the intestinal tract. They are excreted in the urine and through the skin during active sweating. They are present in all body secretions.

Chloride ion is virtually without pharmacological action. It serves two functions of a physical nature in the body fluids. Together with sodium ion it is responsible for maintenance of osmotic balance between the different body fluids. Being able to pass freely through all membranes, it is responsible for maintenance of charge balance between intracellular and extracellular fluids. For example, in actively metabolizing tissues carbon dioxide passes into the red blood cells and is converted to bicarbonate anion; equivalent chloride moves out of the cell to maintain charge neutrality.

BICARBONATE ION: The bicarbonate ion is the second most abundant anion of the body fluids. On a particle basis there is one dissolved bicarbonate ion for every four chloride ions. Their functions are, however, entirely different. Bicarbonate serves the body in two capacities. First, it represents the predominant mode of transport for carbon dioxide from the tissues to the lungs. Second, together with carbonic acid, it gives rise to the most important buffer system of the body fluids. The other buffers are the $H_2PO_4^-/HPO_4^{2-}$ system, the proteins in the body fluids, and the hemoglobin and the oxyhemoglobin within the red blood cells.

Carbon dioxide transport takes place by several mechanisms. A small part, 5%, is transported as dissolved molecular carbon dioxide. The remaining 95% is transported as bicarbonate or equivalent forms. The equilibria involved are

$$CO_2(\text{gaseous}) \rightleftharpoons CO_2(\text{dissolved}) \qquad (1)$$

$$CO_2 + H_2O \rightleftharpoons H_2CO_3 \underset{-H_2O}{\overset{+H_2O}{\rightleftharpoons}} H_3O^+ + HCO_3^- \qquad (2)$$

$$CO_2 + OH^- \rightleftharpoons HCO_3^- \qquad (3)$$

$$CO_2 + RNH_2 \rightleftharpoons RNHCOOH \underset{-H_2O}{\overset{+H_2O}{\rightleftharpoons}} RNHCOO^- + H_3O^+ \qquad (4)$$

The reaction represented by Equation 1 is controlled by Henry's law. This law applies only to molecular carbon dioxide, not to reaction products. The equilibrium expressed in Equation 1 is especially important during ventilation of the blood in the lungs. The reactions in Equations 2 and 3, known as hydration reactions, account for the formation of bicarbonate ion. It is not known to what extent the carbon dioxide diffuses from actively metabolizing cells as such and to what extent bicarbonate forms within the cell and diffuses into the body fluids. Carbon dioxide diffusion through the cell walls is the generally favored process since it represents a conservation method; no charge imbalance results and a compensating chloride migration is not required. It is known, however, that the red blood cells actively transport large quantities of carbon dioxide from the cells to the lungs. Both bicarbonate formation and the carbamino reaction, Equation 4, take place actively within the red blood cells while they are in the vicinity of actively working cells.

Carbamino transport of carbon dioxide (up to 30% of the total) is ordinarily overlooked since the usual methods for determination of blood bicarbonate fail to distinguish between bicarbonate-bound carbon dioxide and carbamino-bound carbon dioxide. The reaction is very rapid. Both the α amino groups of the amino acids of proteins and the ϵ amino group of lysine residues react readily with carbon dioxide. The carbamino reaction is active within the red blood cells, the free amino groups of the hemoglobin are utilized.

Reactions 2, 3, and 4 represent the acidifying action of carbon dioxide. The reactions shown in Equations 2 and 4 result in formation of hydronium ion, whereas that represented in Equation 3 accomplishes the same effect by reducing hydroxyl ion levels. Although Equation 3 is rarely considered in standard works, it has been shown *in vitro* that up to 40% of carbon dioxide is converted to bicarbonate via this reaction at physiological temperatures and pH levels.

The normal H_2CO_3/HCO_3^- ratio is 1/20. From the relationship

$$pH = pK_a - \log \frac{[H_2CO_3]}{[HCO_3^-]}$$

it is seen that an increase of bicarbonate, whether systemic or oral, results in an increase in the pH of the body fluids. On the other hand, an increase in carbon dioxide levels with its resultant increase in the carbonic acid levels has an acidifying action in the body fluids. Both effects are counteracted by excretion. Carbon dioxide gas is actively excreted in the lungs. The kidneys act to counteract acidity by excretion of dihydrogen phosphate ion and excess basicity by excretion of bicarbonate ion.

The reformation of carbon dioxide from bicarbonate ion and carbamino ion and its active elimination into the alveolar sacs in the lungs represent the active interplay of several equilibria. Bicarbonate ion and oxygen actively diffuse into the red corpuscles, whereas carbon dioxide diffuses out of the red blood cells at rates many times greater than those allowed by the known reaction rates for dehydration of carbonic acid. Hemoglobin (HHb) is a weak acid. Oxygenation to oxyhemoglobin ($HHbO_2$) converts it into a stronger acid, which serves as a source of hydronium ions.

$$HHb + O_2 = HHbO_2$$

$$HHbO_2 + H_2O = H_3O^+ + HbO_2^-$$

At the same time carbonic acid molecules are actively synthesized by reactions such as

$$H_3O^+ + HCO_3^- = H_2CO_3 + H_2O$$

$$HCO_3^- + H_2O = H_2CO_3 + OH^-$$

$$(OH^- + H_3O^+ = H_2O)$$

Carbonic acid is dehydrated through enzymatic action involving carbonic anhydrase

$$H_2CO_3 = CO_2\uparrow + H_2O$$

It is only through the mediation of this enzyme that this dehydration reaction proceeds at a rate sufficient to account for the carbon dioxide eliminated at observed ventilation rates. Carbon dioxide is also reformed by a reversal of the carbamino reaction.

The buffer action of the carbonic acid-bicarbonate buffer is straightforward. Excess acid is neutralized by reactions such as

$$H_3O^+ + HCO_3^- = H_2CO_3$$

and

$$H_2PO_4^- + HCO_3^- = H_2CO_3 + HPO_4^{2-}$$

Excess base is accommodated by the reaction

$$OH^- + H_2CO_3 = H_2O + HCO_3^-$$

PHOSPHATE IONS: There are about 700 Gm. phosphorus present in the body of an average adult. Of this about 85% is deposited in the bone as phosphate. The remainder of the element is present in body fluids as the monohydrogen phosphate and dihydrogen phosphate ions and as various phosphorylated compounds in actively metabolizing cells.

Phosphate serves many roles in the body. Structurally it is bound in the complex hydroxyapatites of the bone. Theoretically this serves as a phosphate reservoir; this ion is very abundant in most foods so that deficiencies are virtually unknown. Because of slow absorption, ingestion of large quantities of a soluble phosphate has a mild saline laxative effect (page 580).

The $H_2PO_4^-/HPO_4^{2-}$ buffer system of the body fluids is important, but is subsidiary to the H_2CO_3/HCO_3^- system. The normal ratio of the dihydrogen to the monohydrogen phosphate is 1:4. Natural or medically induced acidosis involves a disturbance of the phosphate buffer system.

Phosphorylation is very important in metabolic processes. It is the process by which phosphate esters of various organic compounds are

TABLE 13.2 ELECTROLYTE CONCENTRATIONS OF SOME COMMON FLUID AND ELECTROLYTE REPLENISHERS

Injection[a b]	Component	Gm./L.	me. cation/L.	me. anion/L.
Sodium Chloride (Isotonic Saline)	0.9% NaCl	9.0	154	154
Hypotonic Saline	0.45% NaCl	4.5	77	77
Hypertonic Saline	3.0% NaCl	30.0	513	513
Hypertonic Saline	5.0% NaCl	50.0	855	855
Sodium Bicarbonate	4.2% $NaHCO_3$	42.0	500	500
Sodium Lactate	1.87% $Na(C_3H_5O_3)$	18.7	167	167
Potassium Chloride	15% KCl[d]	150	2000	2000
Ammonium Chloride	2.14% NH_4Cl	21.4	400	400
	26.8% NH_4Cl[d]	267.5	5000	5000
Calcium Chloride	10% $CaCl_2 \cdot 2H_2O$	100	1360	1360
Calcium Gluconate	10% $Ca(C_{12}H_{22}O_{14}) \cdot H_2O$	100	465	465
Calcium Levulinate	10% $Ca(C_{10}H_{14}O_6) \cdot 2H_2O$	100	655	655
Dextrose and Sodium Chloride	NaCl	4.5	77	77
	$C_6H_{12}O_6 \cdot H_2O$	25.0	--	--
		50.0	--	--
	NaCl	9.0	154	154
	$C_6H_{12}O_6 \cdot H_2O$	25.0	--	--
		50.0	--	--
		100.0	--	--
Ringer's[c]	NaCl	8.6	147	
	KCl	0.3	4	155
	$CaCl_2 \cdot 2H_2O$	0.33	4	
Lactated Ringer's[c]	$Na(C_3H_5O_3)$	3.1	130	28
	NaCl	6.0		
	KCl	0.3	4	110
	$CaCl_2 \cdot 2H_2O$	0.2	3	

[a]Generic monographs in *U.S.P.* permit various concentrations to be official as long as purity rubric statement is respected.

[b]*U.S.P.* requires that label state osmolar concentration in mOsm/L.

[c]Not a generic monograph. *U.S.P.* specifies concentrations.

[d]Concentrates. Desired dose obtained by dilution, e.g., with sterile water for injection.

formed. Combustion of glucose or its conversion to glycogen requires phosphorylation. Muscular energy requires utilization of high-energy phosphate ester bonds. After a preliminary phosphorylation several vitamins function as coenzymes.

Salts of interest. The salts commonly used as sources of the ions of the body fluids will be discussed at this time. Group physical and chemical properties of these compounds have for the most part been previously outlined. Discussion will be limited to reiteration of properties of special importance in the application under study.

SODIUM CHLORIDE: Being the primary source of both sodium and chloride ions, sodium chloride may possibly be considered the one most important compound in this field. It is administered as a source of the component ions by itself and in almost all fluid and electrolyte preparations. Its dosage forms include tablets, parenteral solutions, and physiological salt solutions for external use.

Commercial sodium chloride of ordinary purity is prepared by slow evaporation of water from saturated solutions. As crystallization proceeds, soluble impurities are concentrated in the mother liquors. Since small pockets of mother liquids are occluded in the growing salt crystals, successive crops from a given mother liquid will be less and less pure. The usual commercial grade is about 98–99% sodium chloride. Especially pure sodium chloride is prepared by passing hydrogen chloride gas into a saturated solution of the commercial salt. The crystals are separated by filtration or centrifugation. Adherent mother liquor is removed from the crystals by a careful rinse with purified water. The crystals are then heated to drive off any adhering hydrogen chloride and water. The recourse to the common ion effect for recrystallization of pure sodium chloride is necessary because of the negligible temperature coefficient of solubility of this salt. Cooling a hot solution does not change the solubility enough to achieve a worthwhile crop of crystals.

Pure sodium chloride is not hygroscopic; less costly commercial grades exhibit this undesirable property because of impurities, largely magnesium and calcium chlorides. "Free flowing salt" includes small amounts of tertiary calcium phosphate, magnesium carbonate, or starch to overcome this hygroscopic tendency.

Chemical and physical incompatibilities are virtually nonexistent when sodium chloride is used in fluid and electrolyte replenishers. Since the preparations are to be taken orally or by injection, solubility problems should not appear since the cations forming insoluble chlorides will not be used in such preparations. Redox problems will not be present since both ions are stable to ordinary redox conditions. Since hydrated sodium ion is a very weak acid and chloride ion is a very weak base, this salt exhibits no acid-base effects when in solution. However, an unusual solubility problem may appear in solutions involving both sodium bicarbonate and sodium chloride (or another soluble sodium salt). Sodium bicarbonate has a comparatively low solubility; 1 Gm. in 10 Gm. water at 25°C. High concentrations of more soluble sodium salts decrease this solubility sufficiently by the common ion effect of the added sodium ion to cause precipitation of some bicarbonate.

Sodium Chloride is official in *U.S.P. XXI* as the salt and in many different forms: inhalation, injection, bacteriostatic injection, irrigation, ophthalmic solution, tablets, tablets for solution, injections in

combination with dextrose, fructose, and mannitol, and tablets with dextrose. (See Table 13.2).

OTHER SODIUM SALTS: Sodium lactate, sodium acetate and sodium bicarbonate are used as sources of sodium ion when a systemic alkalizing action is also desirable. Other than their alkalizing action, neither anion has any physiological effect. Sodium lactate has two advantages over sodium bicarbonate. First, it can withstand sterilization temperatures. Second, its effect is achieved slowly and over a longer period of time since there is a time lag between administration and the appearance of its bicarbonate oxidation product.

Sodium lactate is prepared by the usual neutralization procedures. Because of the existence of the so-called lactic anhydride form in the commercial acid, the reaction mixture must be maintained at elevated temperatures for a period sufficient to insure complete hydrolysis of the ester linkage. The reaction mixture is concentrated by evaporation to give a 70–80% solution. The compound is commercially available either as this concentrate or as a crystalline salt. Solutions of sodium lactate are approximately neutral.

Having an asymmetric carbon atom, lactate ion exists in a D- and L- form. Commercial sodium lactate is the racemate. The metabolic disposition of the two isomers proceeds via different pathways. The L-(+)-lactate is oxidized to equivalent bicarbonate, whereas the D-(−)-lactate is converted to glycogen in the liver.

Sodium bicarbonate solutions are slightly alkaline. Owing to loss of carbon dioxide they become more alkaline on standing, continued agitation, or heating. The solutions show evidence of free carbon dioxide at 50°C.; loss is complete at 100°C. The increasing alkalinity is due to formation of normal carbonate and its reaction with water.

$$2HCO_3^- = H_2O + CO_2\uparrow + CO_3^{2-}$$

$$CO_3^{2-} + H_2O \rightleftharpoons HCO_3^- + OH^-$$

Sodium Lactate is official in *U.S.P. XXI* as the salt and injection. Sodium Bicarbonate is official as the salt, injection and tablet.

POTASSIUM SALTS: Potassium chloride and potassium bicarbonate are the usual sources of potassium ion for fluid and electrolyte preparations. The discussion under sodium bicarbonate is pertinent to potassium bicarbonate except for solubility, the potassium salt being about four times more soluble.

AMMONIUM CHLORIDE: Ammonium chloride (sal ammoniac, ammonium muriate, salmiac) serves as a source of chloride ion when an accompanying acidifying effect is desirable. The ammonium ion as a stronger acid than water imparts a pH of 4.5–6.0 to its solutions. It is used in tablet form and as an injection.

CALCIUM SALTS: Calcium salts find employment in fluid and electrolyte replenishers and as calcemics for treatment of hypocalcemic conditions. The usefulness of a particular calcium salt for a given method of administration depends upon its physical and physiological properties. Obviously only the soluble salts may be used for solutions intended for parenteral administration. The relative irritating effects of a particular salt will determine whether it is suitable for intramuscular or intravenous administration. Either soluble or insoluble compounds may be used orally; palatability and gastric irritation now become physiological factors.

Calcium Chloride (muriate of lime, $CaCl_2 \cdot 2H_2O$), Calcium Gluceptate (calcium α-glucoheptate), Calcium Gluconate, Calcium Lactate, and Calcium Levulinate are useful soluble salts cited in *U.S.P. XXI* together with suitable dosage forms. Insoluble dibasic calcium phosphate (monohydrogen phosphate) *(U.S.P. XXI)* is considered an excellent calcium source for oral administration. The other calcium phosphates are not suitable as calcium sources (page 361).

Calcium chloride is very soluble and is absorbed satisfactorily. However, it is locally so irritating and corrosive that it cannot be administered intramuscularly. Injection must always be intravenous, slowly and with great caution. It is irritating orally and likely to disturb the digestive process and thus hinder absorption. Because it interferes with the phosphate buffer system, calcium chloride is a systemic acidifier (page 430).

Four hydrates have been claimed for calcium chloride—$6H_2O$, $4H_2O$, $2H_2O$, and $1H_2O$. Of these the hexahydrate and dihydrate are best known. The anhydrous compound and all hydrates are deliquescent, rapidly dissolving in their own water of hydration on exposure to the atmosphere. The dihydrate is probably most convenient for use. The hexahydrate solidifies into a solid mass on exposure to moisture. These hydrates cannot be dehydrated by heating in the atmosphere; both hydrogen chloride and moisture are evolved:

$$[Ca(H_2O)_6]Cl_2 \rightleftharpoons Ca(OH)_2 + HCl\uparrow + 4H_2O\uparrow$$

However, dehydration will take place normally in an atmosphere of hydrogen chloride. Calcium chloride is an inexpensive by-product of the Solvay process for the manufacture of sodium carbonate.

The other soluble calcium salts employed in electrolyte replenishers or as calcemics represent attempts to alleviate or overcome the irritating and corrosive properties and the unpalatability of the chloride. Calcium lactate, $(CH_3CHOHCOO)_2Ca \cdot 5H_2O$, is less irritating than the chloride. It is prepared by reacting lactic acid with calcium carbonate, filtering, and crystallizing the salt from the resultant clear solution. The hydrate is slightly efflorescent and becomes anhydrous at 120°C. Aqueous solutions are sufficiently acid to cause slow effervescence when a carbonate is added. This salt is a satisfactory dosage form of calcium; it is commonly administered orally in tablet form.

Calcium gluconate is less irritating than either the chloride or the lactate; its palatability is superior. Calcium gluconate tablets are used extensively for diet supplementation. Its solutions are sufficiently nonirritating to allow both intramuscular and intravenous administration. Calcium Gluconate Injection may be stabilized with calcium D-saccharate. (*U.S.P. XXI* allows its addition in amounts not greater than 5% of the calcium present as gluconate. *U.S.P. XXI* allows the addition of a small amount of sodium hydroxide to adjust the pH). These solutions show improved stability if slightly alkaline (pH 6.0-8.2). The preparative method for calcium gluconate is similar to that used for calcium lactate. Calcium Gluceptate is also official as the injection.

Calcium levulinate is sufficiently nonirritating to be administered subcutaneously. Its higher solubility allows preparation of more concentrated solutions than the other calcium organic acid salts: 50% w/v solutions have been prepared, although the 10% solution is commonly used. This compound also has a higher calcium content (13%) than the gluconate (9%). It may be administered orally, intravenously, intramuscularly, and subcutaneously. Because muscular necrosis has been observed in children following injection of the levulinate and gluconate, the use of these salts should be restricted to adults.

Solubility problems are the most probable source of chemical incompatabilities in preparations including calcium salts. Here the role of pH must also be considered; e.g., while the carbonate is insoluble, at pH values permitting formation of the bicarbonate, appreciable quantities of calcium ion will enter solution. The hydrated calcium ion is so weak an acid that the pH of solutions of calcium salts will be largely determined by the strength of the anion as a base. Calcium ion is stable to all except the most powerful chemical reducing agents.

MAGNESIUM SALTS: While magnesium ion has many functions in the bodys' economy it is so common in foods that there is ordinarily no need to supplement it. Magnesium Gluconate and Magnesium Gluconate Tablets are official in *U.S.P. XXI.*

MULTICOMPONENT ELECTROLYTE REPLENISHERS. The Ringer's solutions are commonly used when concentrations of major electrolytes, approximating those of blood, are desirable. This solution creates little disturbance of the sodium-postassium ratio when administration in relatively large volumes is necessary. The anionic balance is, however, disturbed since only chloride is present. The partial replacement of chloride by lactate in time helps maintain the bicarbonate level. However, phosphate is uncompensated. Ringer's Injection, Lactated Ringer's Injection, and Ringer's Irrigation are cited in *U.S.P. XXI.* (See Table 13.2.)

Solutions are also available combining the nutrient function with the fluid and electrolyte function. Dextrose and Sodium Chloride Injection, Fructose and Sodium Chloride Injection, and Mannitol and Sodium Chloride Injection are cited in *U.S.P. XXI.*

Salt substitutes. A demand exists for salt substitutes because of the necessity of a salt-free diet in the treatment of certain conditions involving circulatory difficulties. Although the salty taste imparted to foods by table salt is generally considered desirable and pleasing to the palate, it is not essential. Compounds used in the preparation of such substitutes must meet two requirements: they must impart a suitable salt-like taste, and they must have no undesirable side effects. Not all salts possess the pleasant salty taste of sodium chloride; this eliminates many otherwise fairly innocuous compounds as salt substitutes. The second requirement narrows the field of suitable substances to an even greater extent.

Chloride is the most common anion choice. Organic anions such as the glutamate, formate, and citrate are also used, usually in combination with chloride. The organic anions are selected from those participating in the normal oxidative metabolic cycles; thus the only side effect introduced is a slight systemic alkalosis. Iodide is frequently present, but as a trace element to supply this essential mineral.

The choice of cations is difficult, since other than sodium most suitable cations have physiological and pharmacological actions of their own. Potassium and ammonium ions are the usual choice—largely on the basis of being "the lesser evil." The potassium ion imparts a diuretic effect to the preparation; ammonium ion imparts a systemic acidifying action. This acidifying effect of the ammonium ion may be partially counteracted by the alkalizing action of a suitable organic anion. Soluble calcium and magnesium salts of organic anions are sometimes used in small quantities.

As a member of the alkali family, lithium would seem an obviously suitable cation for salt substitutes. However, the most recent salt substitute including lithium was hastily withdrawn from the market following several fatalities among its users. Although, while this element should be able to substitute for sodium on the basis of family similarities, it is entirely probable that its bridge element character (page 208) leads to systemic problems. Examination of the syndromes reported in the cases of high lithium intake reveals several remarkable resemblances to those of a hypercalcemic condition. It is claimed that the untoward effects involved in the use of lithium salt substitutes were found only in patients with a blood serum lithium level of at least 1 me./L. in association with a depleted serum sodium ion level; restoration of normal sodium levels relieved the symptoms.

Glutamic acid is frequently an ingredient of salt substitutes, as are inert excipients such as tricalcium phosphate and starch. The starch also functions as a free flow agent. If necessary a drying agent may be present to counteract deliquescence.

As a sidelight, consider the salt shaker. It is an interesting instrument, plagued with plugged holes. In the past century the U.S. Patent Office has issued over 300 patents for devices designed to solve this problem, while 1000 patents have been issued on salt shakers in general. There is still room for improvement.

CALCEMICS AND TONICS

The body's requirement for calcium is dependent on age and special conditions. Except for periods of high demand (pregnancy, rickets) adult calcium requirements range between 0.2 and 1.0 Gm. daily. A well-balanced diet easily meets this requirement. Although calcium occurs in most foods, its level is subject to wide variation. Certain foods such as milk, milk products, and beans serve as good sources. Meats, except shell fish, are generally lower in calcium than foods of vegetable origin. Such vegetables as spinach and rhubarb are poor sources of calcium because of their relatively high oxalic acid content.

Dietary supplementation of calcium may become necessary during periods of rapid growth, pregnancy, and lactation or with individuals on calcium-deficient diets. Certain abnormal conditions such as tetany, rickets, and osteomalacia require calcium supplementation for their correction (page 351).

It is impossible to consider calcium requirements without a simultaneous consideration of phosphate because of their extremely close biochemical interrelationship. According to the best evidence phosphorus deficiencies are unlikely as phosphate is abundant in virtually all foods. When calcium supplements become necessary, nutritionists generally favor a Ca:P atomic ratio of 1:1. Calcium monohydrogen phosphate has this ratio. This accounts for its popularity in calcium supplements. Although the compound is insoluble, the following reaction takes place under the acid conditions prevailing in the stomach:

$$CaHPO_4\downarrow + H_3O^+ = Ca^{2+} + H_2PO_4^- + H_2O$$

This reaction is favored by the strong basic character of the monohydrogen phosphate ion. As a result the calcium ion is available for absorption in the duodenum. The dihydrogen phosphate is unsuitable as a calcemic because of its systemic acidifying action (page 429). The acid consumption of the tertiary salt

$$Ca_3(PO_4)_2\downarrow + 2H_3O^+ = 2CaHPO_4\downarrow + Ca^{2+} + 2H_2O$$

is such that the resulting increase in pH is unfavorable for calcium absorption. If calcium salts other than the phosphates are used, appropriate levels of soluble phosphate salts should be included to achieve the 1:1 atomic ratio.

Another school of thought holds that phosphate need not be considered and that calcium gluconate or lactate are preferable as dietary supplements. This group is of the opinion that calcium-phosphate combinations should be restricted to less severe forms of calcium deficiency. It has been shown that a calcium deficiency is responsible for leg cramps in pregnancy due to a preferential absorption of phosphate.

Calcium carbonate also is used as a calcemic, although its antacid properties may prove undesirable. The chloride is generally unsatisfactory as a calcemic because of gastric irritation.

Hypocalcemic conditions are frequently coupled with subminimal vitamin D intake. In temperate zones this is especially likely during the winter months. For this reason calcium supplements frequently include this vitamin; 600 U.S.P. units per gram are commonly added. Problems related to the stability of this vitamin are outside the province of this text.

Phosphorus tonics have been and continue to be popular despite the fact that it is highly unlikely that the human diet will be inadequate. Not only is phosphate very widespread in human foods, but it also occurs in highly utilizable forms. The lay public's inclination toward phosphorus supplementation probably persists on the same basis as the idea that fish are good brain food because of their high phosphorus content.

The hypophosphites and glycerophosphates are most commonly used as the phosphorus source in tonics. The hypophosphites have a higher percentage of phosphorus than other available phosphorus salts and were introduced as tonics for this reason. It has been shown that they pass through the body virtually unchanged. Sodium, potassium, calcium, iron(III), and manganese(II) hypophosphite were at one time official

(*N.F. X*). Compound Hypophosphites Syrup[2] is no longer recognized as an official preparation. Hypophosphorous acid and its sodium salt are useful antoxidants (page 241).

The glycerophosphates owe their popularity to the belief that organically bound phosphorus should be more available to nerve tissue. However, the ease with which the ester linkage involved cleaves renders it highly unlikely that it survives digestive action. Glycerophosphates are therefore equivalent to any corresponding quantity of inorganic phosphate. The claims put forward for the glycerophosphates have not been substantiated. The sodium, calcium, iron(III), and manganese(II) compounds and Compound Glycerophosphates Elixir are no longer recognized officially.

FLUORIDES AND DENTAL HEALTH

During the past decade or two it has become evident that there is a close relationship between fluoride intake during childhood and healthy teeth. It is now well established that the surest and best way to avoid fluoride deficiencies is through fluoridation of municipal water supplies at a level of 1 p.p.m. However, with the opposition to fluoridation sometimes reaching almost a point of hysteria, efforts to extend fluoridation have frequently been thwarted, despite support of the dental and medical professions. For this reason a particularly thorough study of this problem will be undertaken in this section. The *in vitro* and *in vivo* chemistry of the fluoride ion will be thoroughly discussed. It will be evident that fluoride is not the dangerously poisonous substance that some uninformed sources assume it to be. The chemistry of the teeth will then be discussed to show why fluoride is essential. Finally, fluoridation of water supplies, the dental application of fluorides, and home fluoride sources, e.g., tooth pastes, will be discussed.

Chemistry of fluorides. Two common types of fluoride salts are available. The first group is made up of the simple fluorides, e.g., NaF. Because it is a strong ligand, fluoride ion also appears in many complexes, e.g., fluosilicate ion (hexafluosilicate(IV), silicofluoride), $(SiF_6)^{2-}$, fluoborate ion (tetrafluoborate), $(BF_4)^-$. The amount of free fluoride present in solutions of salts of the complex ions named is negligible; the chemical properties of the two free complex acids and their salts are similar to those of sulfuric acid and the sulfates. In fact, a series of alums exist in which the fluoborate ion substitutes for the sulfate ion. In addition to

[2] Caution must be exercised when working with hypophosphites because of their strong reducing action. Do not triturate or heat hypophosphites with nitrates, chlorates, permanganates, or other oxidizing agents. Explosive reactions may result.

its inorganic salts fluorine is present in many organic compounds, in most of which it is innocuous. However, a few such as the fluoroacetates are extremely poisonous and have been employed as insecticides.

Perhaps the most striking property of the fluorides is the ability of hydrofluoric acid to dissolve glass. This is due to the volatility of the primary product, SiF_4. With silica representing the silicate component of glass, the reaction may be written

$$SiO_2 + 2H_2F_2 = 2H_2O + SiF_4$$

All too frequently this reaction is interpreted to mean that hydrogen fluoride is an exceptionally strong acid, which it is not. Hydrofluoric acid, fluosilicic acid, and fluoboric acid are all extremely corrosive to the skin, but no more so than other common acids such as sulfuric acid.

The chemical and physical properties of the fluoride ion are such that there are remarkably few possible incompatibilities. The small size of the ion and the strong electronegativity of the element render the ion resistant to all common oxidizing agents; no redox incompatibilities exist. Because of the small size of the ion its electrostatic field is intense. For this reason, the ion is a moderately strong base. Due to this fact and the volatility of hydrogen fluoride, the acid may be displaced from its salts by stronger acids.

Except for salts involving the high-charge cations of small radius, fluorides are soluble in water. Only one of the insoluble salts, calcium fluoride, is likely to form and thereby cause problems in pharmaceutical preparations. Accidental ingestion of large quantities of a soluble fluoride can reduce serum calcium ion levels due to formation of calcium fluoride. Calcium gluconate injection is used to counteract this effect.

Physiology of fluoride ion. The toxicity of this ion has been greatly exaggerated. The lethal dose of sodium fluoride ranges between 5–15 Gm. taken orally, depending on individual variation. On the other hand, as little as 0.05 Gm. of atropine may be fatal. Many areas normally use drinking waters which naturally contain 2–10 p.p.m. fluoride. Other than a marked reduction in the incidence of dental caries, the only physiologically recognizable effect of continued ingestion of these comparatively high fluoride levels is a cosmetic effect—mottling of the teeth. Up to 10 mg. of fluoride may be ingested daily without storage or untoward effects. Ingestion of 20–30 mg. daily may be tolerated over rather long periods without any chronic effects or visible changes in the bones or teeth. At this high level 90% of the ingested fluoride is eliminated immediately in the urine. In fact, aside from that deposited in the bones and teeth there is no accumulation of this element in the body. A recent report by an orthodontist concerned use of a 0.25% NaF mouth wash by school age children over a ten-year period without any signs of toxicity. Other studies over shorter time periods confirm this finding.

Fluoride occurs in bones and teeth to 0.5—1.5 me./100 Gm. Highest concentrations are found in the enamel of the teeth. Like bone, the teeth are probably a hydroxyapatite, although a carbonate apatite has also been proposed. The empirical structure approaches $3Ca_3(PO_4)_2 \cdot Ca(OH)_2$ in composition. However, F^-, Cl^-, and CO_3^{2-} substitute isomorphously for a portion of the hydroxyl ion. Because the chloride and carbonate concentrations of the fluids from which the teeth are laid down are constant, the quantity of these ions remains fairly constant in teeth. However, the fluoride content varies widely and has been shown to be related to the dietary fluoride level during the tooth's formative years. It is especially in the enamel that this is evident.

The enamel of the erupted teeth is the one body tissue essentially independent of body metabolism, except possibly for some ionic exchange reactions with the saliva. This becomes the more evident when it is seen that the organic matter of dry enamel is only 0.5–1.5% as compared to 20% in dry dentin and 30–40% in cementum and bone. Thus enamel is for practical purposes a pure inorganic crystalline salt. Unlike bone, it is not being continuously torn down and rebuilt.

Since the teeth are laid down in the gums and are completely formed by ten years of age (though not necessarily erupted), during fetal life and the first ten years of life, dietary fluoride levels are of extreme importance. Fluoride ion enters the tooth structure during its growth and is incorporated into its crystal lattice directly or by subsequent adsorption or ion exchange with hydroxyl ion already present in the lattice. Since the developing tooth (and before eruption the mature tooth) is continuously bathed in body fluids, it is likely that during the period before eruption fluoride is actively being incorporated into the enamel structure by the latter processes. This is supported by studies of the enamel layer. Based on special surface grinding techniques, it has been shown that the fluoride content of a given tooth's enamel layer is highest at the outer surface. The fluoride content drops to consistently lower levels in the deeper layers of the enamel. After eruption there may be a small continued exchange or adsorption of fluoride on the surface only.

The action of fluoride (especially in enamel structure and in general in the remainder of the tooth) is twofold. First, it reduces the solubility of the hydroxyapatite; the solubility of this substance is inversely related to its fluoride content. It is generally accepted that the erosion of the tooth structure, known as dental caries, is due to the solubility of phosphates in acid media. In the developing caries the acid is supplied by fermentation. At this stage of the developing caries fluoride plays a second role. As the tooth structure is eroded, fluoride ion is released into the solution within the developing cavity. Here the fluoride ion acts as a protoplasmic poison, interfering with the acid-forming enzyme of the bacteria involved. The higher the fluoride content of the enamel, the more effective the resistance of the tooth to dental caries.

In the study of the incidence of dental caries it was shown that residents of areas deficient in fluoride have a high incidence of carious teeth. The enamel of carious teeth is consistently lower in fluoride than normally sound teeth. These studies have established the optimum level of fluoride in drinking water within the range 0.7–1.2 p.p.m., with 1.0 p.p.m. as a good average.

Certain cities[3] have fluoridated water at the 1 p.p.m. level for many years. The experience in all has been a two thirds reduction in dental caries in the populace affected. Other cities with shorter histories of fluoridation are showing similar reductions.

The upper limit of the suggested fluoridation level is determined by the incidence of fluorosis (mottled teeth) in regions having natural water supplies with fluoride levels greater than 2 p.p.m. (There are actually natural water supplies in use with fluoride levels as high as 8 p.p.m.) Fluorosis is characterized by the appearance of chalky white patches on the enamel. These are due to an interference with the calcification enzyme system. The size of the areas is directly proportional to the fluoride content of water. The mottled appearance of these teeth is actually produced by preferential adsorption of dyes from foods on the surface of these areas. The stains become pronounced at 6 p.p.m. fluoride levels. However, the mottling is entirely a cosmetic problem since the teeth themselves are not affected in any way.

Administration of fluoride. The most effective way of providing fluoride to growing children is through fluoridation of municipal water supplies where necessary. This treatment, to the extent required to give the desirable 1 p.p.m. level, is inexpensive (7-14 cents per person per year). It insures available fluoride to all requiring it, irrespective of economic or educational level.

In communities with fluoride-deficient water supplies which do not practice fluoridation, the beneficial effects may be gained in the home by use of fluoride supplements. The following fluoride compounds and preparations are at present official in *U.S.P. XXI:* Sodium Fluoride, Sodium Fluoride Oral Solution, Sodium Fluoride Tablets, Sodium

[3] Newburgh, N.Y.; Brantford, Ont.; Grand Rapids, Mich.

Fluoride and Phosphoric Acid Gel, Sodium Fluoride and Phosphoric Acid Topical Solutions, Sodium Monofluorophosphate, Stannous Fluoride.

Home fluoridation is easily achieved by adding the proper quantities of fluoride to the water, fruit juice, or milk consumed by the children. Milk is especially recommended since it is a common food for pregnant women, infants, and children. Theoretically this procedure makes fluoridation available to all, but for many reasons (disinterest on the part of the parents, for one) too many children actually fail to receive the necessary fluoride supplement. The amount of fluoride naturally present in the municipal water must be taken into consideration in any home fluoridation techniques.

For convenience in home administration, fluoride concentrates may be prepared in tablet form or as solutions. A convenient tablet has 2.21 mg. sodium fluoride per tablet, with sodium chloride as an inert filler, 94.49 mg. per tablet. One of these tablets dissolved in one liter (or one quart) of water or other liquid gives a solution 1 p.p.m. in fluoride ion. Alternatively, a solution may be prepared by dissolving 0.132 Gm. sodium fluoride in 240 ml. purified water. One teaspoonful of this solution added to a liter (or quart) of water, fruit juice, or milk provides 1 p.p.m. fluoride ion. Lozenges have become available with a fluoride concentration comparable to that of the tablets just cited.

Where fluoridation has not been practiced, topical application of solutions of fluoride to the surface of the erupted teeth is of value. A 2% solution of sodium fluoride is widely accepted for this treatment. It is swabbed on the *dry* teeth four times over a ten-day period, the treatment being repeated several times a year. With children some surface ion exchange is possible during these treatments. Fluoride ion enters the surface layers of the enamel and becomes part of the enamel structure, possibly by isomorphous replacement of hydroxyl ions. It has been suggested that a deposit of calcium fluoride might also build up on the surface. The effectiveness of the procedure is attested by a 40% reduction of caries when topical application is practiced, in contrast to control groups without topical treatment. Results obtained with dental patients over fourteen years of age are conflicting. Topical application of fluoride is superfluous if the local water supplies have adequate fluoride levels.

Sodium fluoride was the original source of fluoride ion for topical application. Controlled laboratory tests indicated that tin(II) fluoride (stannous fluoride), lead(II) fluoride (plumbous fluoride), and sodium fluosilicate are also effective in reducing dental caries. However, after clinical testing it was found that only the tin compound gave consistent results. A freshly prepared 8% solution of tin(II) fluoride is applied to the cleaned dry teeth. The teeth are kept moist with this solution for about four minutes by repeated application. The mouth must not be rinsed, and no food or drink taken for thirty minutes after this application is completed. A single application is required every six to twelve months. It is claimed that the tin(II) fluoride treatment is more effective than the sodium fluoride treatment, even though fewer visits to the dentist are required.

It is thought that unlike the sodium ion of sodium fluoride, the tin(II) ion participates in the protective action of tin(II) fluoride. There is reason to believe that the tin(II) cation may act by isomorphous replacement of calcium in the surface of hydroxyapatite structure.

From the preparative point of view the use of tin(II) fluoride is fraught with problems. Exposure to air rapidly causes oxidation of the tin(II) to tin(IV), in which form it is ineffectual.

$$[Sn(H_2O)_6]^{2+} + \tfrac{1}{2}O_2 = [Sn(H_2O)_4(OH)_2]^{2+} + H_2O$$

Both hydrated tin(II) and hydrated tin(IV) ions are strong acids compared to solvent water molecules.

$$[Sn(H_2O)_6]^{2+} + 2H_2O = [Sn(H_2O)_4(OH)_2]\downarrow + 2H_3O^+$$
$$[Sn(H_2O)_4(OH)_2]^{2+} + 2H_2O = [Sn(H_2O)_2(OH)_4]\downarrow + 2H_3O^+$$

Tin(II) hydroxide will begin precipitating at a pH of 3, a pH usually considered sufficiently acid to prevent such reaction. Since the free tin(II) ion is the active form, dental protective action is lost by these solutions on oxidation and/or precipitation.

Because of these unsatisfactory properties of tin(II) ion, stock solutions cannot be prepared and maintained until needed. A fresh solution of tin(II) fluoride must be prepared at the time of application. For expeditious preparation of fresh solutions it is most convenient for the dentist to have the pharmacist weigh out 0.8 Gm. portions of the compound and place them in No. 2 hard gelatin capsules. Both the stock bottle of tin(II) fluoride and the bottle containing the capsules must be kept tightly sealed to keep out moisture. Surface hydrolysis accompanied by surface oxidation of the tin(II) ions will take place in moist atmospheres. When needed, the dentist can prepare a fresh 8% solution by emptying the contents of one capsule into 10 ml. purified water. Tin(II) fluoride has an unpleasant astringent taste; however, masking with flavoring agents must not be attempted because of the reactivity of the metallic ion.

Tin(II) fluoride in a "stabilized" form is available in fluoridated tooth pastes[4] also. The A.D.A. (American Dental Association) was sufficiently

[4] U.S. Patent 2839448 to Hager and Reiss, assigned to Proctor and Gamble, lists the following as a suitable formulation: tin(II) fluoride, 0.4%; calcium pyrophosphate, 42%; glycerin, 25%, sodium carboxymethylcellulose, 0.8–1.5%; water, flavoring, sweetening and sudsing agents. A compatible container is recommended.

impressed with the results reported following controlled use of a fluoridated tooth paste to allow its seal of approval on the product. However, the A.D.A. cautions that the tooth paste is no cure-all and is no substitute for water fluoridation.

IODIDE AS AN ANTIGOITEROGENIC

Chemical and physical properties of iodine. Iodine is a blue-black solid laid down in a platy crystalline structure. The element has a high vapor pressure; this is easily shown by placing a few crystals in the bottom of a clear glass bottle and noting the development of a light violet-pink color in the air space of the bottle.

Elemental iodine is soluble in a wide range of solvents; the color of the resulting solutions is a function of the solvent's basicity.[5] Neutral solvents such as hydrocarbons, chloro- and bromohydrocarbons, and carbon disulfide give violet solutions. In these solutions, as in the vapor state, the iodine exists as independent diatomic molecules. With very weakly basic solvents, (e.g., benzene) red or orange solutions result, and interactions are weak. With strongly basic solvents (e.g., water, alcohols, ethers, ketones, organic acids, nitriles, nitrogen bases, and some sulfur compounds) brown solutions result. Here strong interactions exist between the solvent molecules and the iodine molecules. With very strong bases (e.g., hydroxyl ion) the forces are sufficiently strong to cause complete solvolysis, and colorless solutions result.

$$OH^- + I_2 = HOI + I^-$$

The field interactions existing between the iodine molecules and the solvent molecules can be shown by subjecting iodine solutions to conditions known to increase or decrease the effectiveness or permanence of such interactions. Brown solutions become violet when heated and revert to a brown color on cooling. Similarly, violet solutions become brown on cooling in a mixture of solid carbon dioxide and ether.

Precise molecular weight determinations have shown that the iodine is present in the diatomic or equivalent state in all solvents investigated. In general it is more soluble in solvents giving brown solutions. The heat of solution is higher for brown than for violet solutions. Iodine in the brown solutions appears to be more reactive.

Iodine is slightly soluble in water, 0.016% at 0°C. and 0.096% at 60°C., giving rise to light brown solutions. It dissolves readily in an aqueous solution of iodides or other halides to give brown solutions.

[5] The Lewis definitions are necessary for this interpretation. The strength of a base is directly proportional to its efficiency as an electron donor.

The iodides have colors ranging from the colorless potassium salt to the light yellow silver salt. The following iodides are insoluble in aqueous media: Ag^+, Hg^{2+}, Pb^{2+}, Hg_2^{2+}, Cu_2^{2+}, and Bi^{3+}. Excess iodide will redissolve the mercury(II) and bismuth(III) compounds to give complex ions of the type $[MI_4]^{-n}$. Many alkaloids form insoluble iodides, e.g., strychnine iodide. Many iodides are deliquescent.

Iodine is a mild oxidizing agent. It is also a weak reducing agent, but requires a strong oxidizing agent to bring out this latent property. It is unlikely that iodine as a reducing agent will be encountered in ordinary pharmaceutical practice. But iodides are reducing agents. Oxygen of the air slowly liberates elemental iodine from iodide solutions or causes a slow discoloration of crystals, especially in moist atmospheres. This necessitates the use of antoxidants or other stabilizing agents with almost all preparations including iodide as a constituent. Hypophosphorous acid and sucrose are commonly used as antoxidants in solutions, e.g., hydriodic acid, iron(II) iodide syrup. Hypophosphites and sodium carbonate are frequently included with solid iodides for the same purpose, e.g., ammonium hypophosphite in ammonium iodide, sodium carbonate in potassium iodide. Sodium stearate-coated crystalline potassium iodide is also available.

The interesting property of the iodide ion for the pharmaceutical chemist is its tendency to form polyiodide ions. All common pharmaceutical iodide-iodine solutions, e.g., Lugol's Solution, are polyiodide preparations. (The other halide ions exhibit the same property but to a lesser degree.) The general equation for the reaction is

$$I^- + nI_2 = I^-_{(2n+1)}$$

Doubly charged ions are also possible:

$$2I^- + nI_2 = I^{2-}_{(2n+2)}$$

The instability constants of some of the complexes which have been identified in solutions are given in Table 13.3. Evidence for the existence of the double negative charge is based on a decrease in light absorption (at 2880 Å) resulting from adding iodide to a dilute solution of I_3^-. The existence of these doubly charged negative ions is interesting since in solid state work only such ions as I_3^-, I_5^-, I_7^-, and I_9^- exist. Solid hydrates, $KI_3 \cdot H_2O$, $KI_3 \cdot 2H_2O$, $KI_7 \cdot H_2O$, have also been prepared.

If a triiodide solution is made alkaline, there is an extremely rapid change to a pale yellow color. The fading rate is proportional to the degree of alkalinity. The pale yellow color is probably due to one or more of I_3^-, IO^-, I_2OH^-, and I_2O^{2-}. I_2OH^- is the hydroxy analog of I_3^-.

The ability of iodide ion to react in this manner is not restricted to

TABLE 13.3 INSTABILITY CONSTANTS OF POLYIODIDE IONS AT 25°C. AND AT CONSTANT IONIC STRENGTH OF 2.00

Ion	K_{inst}	pK_{inst}
I_3^-	1.34×10^{-3}	2.87
I_4^{2-}	7.3×10^{-3}	2.14
I_6^{2-}	5.2×10^{-6}	5.28
I_8^{2-}	6.2×10^{-9}	8.21
I_9^{3-}	1.3×10^{-9}	8.89

reaction with iodine. Interhalogen compounds $[ClICl]^-$, $[BrICl]^-$, $[ICl_4]^-$ have been identified.

Physiology. Iodine is present in all cells of the body. The total body content is about 50.0 mg., of which one fifth is concentrated in the thyroid gland. The blood serum level is 0.4–1.0 me./L. Irrespective of the form of the iodine used for oral administration, absorption from the intestinal tract occurs as the simple iodide ion. Absorption is rapid; the administered iodine may be detected in the thyroid gland and the saliva within minutes after ingestion. It is believed that the sodium salt is absorbed more efficiently than other forms. Oral administration may be accompanied by gastric discomfort. The absorbed iodine is taken up by the thyroid gland and probably stored in follicle cells as inorganic iodide until required for incorporation into the organic form. Large doses of iodine are fatal, the lethal dose being 2–3 Gm. Iodine is largely eliminated via the urine.

Endemic goiter is usually associated with a dietary deficiency of iodine. Thyroxin, the iodine-containing thyroid hormone, is actively concerned with the level of body metabolism. Hypothyroidism, which produces cretinism in the infant and young and myxedema in the adult, is characterized by an abnormally low basal metabolic rate. Hyperthyroidism is characterized by a very high basal metabolism. There is apparently no danger of creating a hyperthyroid state by continuous ingestion of relatively large quantities of iodine.

Iodide also has an expectorant action. It has no germicidal properties.

Pharmaceutical problems. Since iodine is volatile, any dispensed solutions must have tight, well-fitting closures. Due to the oxidizing action of this element metallic closures will be attacked:

$$Fe + I_2 = Fe^{2+} + 2I^-$$

$$2Fe^{2+} + \tfrac{1}{2}O_2 \overset{\text{(air)}}{=} 2Fe^{3+} + O^{2-}$$

etc. Cork is also oxidized. For short periods rubber stoppers may serve, but soon they too are oxidized. Only glass and certain oxidation-resistant plastic closures are suitable for long storage periods.

Stabilization of the iodide ion is always required. Fortunately, lack of a stabilizing agent or its exhaustion is quickly evidenced by coloration (light yellow to brown depending on the extent of oxidation of iodide). Colored solutions or salts should not be dispensed.

Solubility incompatabilities are unlikely in antigoiterogenic preparations. (They may appear, however, in the preparation of mineral supplements containing iodide.)

Compounds and solutions used as antigoiterogenic agents. Potassium and sodium iodide are the usual source of iodide ion. The iodides are used either in tablet form or as a solution. Potassium Iodide Oral Solution, *U.S.P. XXI,* is a saturated solution of potassium iodide in purified water (approximately 6 me./ml). On standing crystals of potassium iodide may separate. Since iodide ion is subject to air oxidation to the element, if the solution is not to be used within a short time 0.5g. per liter sodium thiosulfate should be added. At one time Hydriodic Acid Syrup was officially cited as a source of iodide.

The free element is also used, but because of its corrosive nature it must be diluted by dispensing it in solution. Strong Iodine Solution (Lugol's Solution) is the preparation commonly dispensed if elemental iodine is desired. Being prepared from I_2, 50 Gm., KI, 100 Gm., and made up to one liter, it is essentially a solution of polyiodides in excess iodide. Since the element will dissolve more readily in concentrated KI, Lugol's Solution should be prepared by first dissolving the potassium iodide in a small amount of water and then dissolving the iodine in this solution. Efficient stirring is required since the heavy iodine crystals tend to settle out. For small volumes a mortar and pestle provide the necessary grinding action to reduce the size of the iodine crystals, thereby hastening solution. When the iodine has completely dissolved, the concentrate is diluted to the required final volume.

Other iodine-iodide solutions are listed in the compendia, but are intended for use as germicides rather than as antigoiterogenics. These include Iodine Tincture, Strong Iodine Tincture, and Iodine Solution. If the proper dosage and dilution are used, there is no reason why these solutions cannot also serve as antigoiterogenics.

As a prophylactic against iodine deficiencies "iodized salt" is available. These salts usually include 0.01–0.02% sodium iodide (1–2 part/10,000).

STUDY AND SELF-APPRAISAL AIDS

IMPORTANT TERMS. Define or identify in a single concise sentence if possible.

Acidosis, alkaline earth, alkalosis, astringent, calcemic, caries, carbonic anhydrase, chalcogen, dental prophylactic, electrolyte, electrolyte replenisher, excipient, fluid replenisher, halide, halogenoid, hydroxyapatite, hypertonic solution, hypocalcemia, hypothyroidism, pseudohalogen, tonic, volume per cent.

NOMENCLATURE OF INTEREST. Identify each of the following.

Ammonium muriate, anhydrite, baryta water, calcite, calcium diphosphate, chalk, chlorine dioxide, dolomite, α-glycerophosphoric acid, limestone, Lugol's Solution, marble, muriate of lime, octacalcium phosphate, potassium triiodide, Ringer's Solution, sal ammoniac, sodium hexafluosilicate(IV), sodium monofluorophosphate, sodium silicofluoride, stannous fluoride.

SELF-APPRAISAL QUESTIONS

1. Why is there an inverse relationship between the atomic radius and ionization potentials of the different alkaline earth atoms? How is ionization potential of indirect importance in the pharmaceutical and medicinal uses of the elements of this family?

2. Within a series of alkaline earth compounds with a given anion, etc., the characteristic properties of the series are modified in degree as a result of the differing cationic radii. Discuss this radius effect on the property named for each of the following series of alkaline earth compounds: solubility of the sulfates, extent of hydration of the cations, acid-base behavior of the hydroxides, resistance of the carbonates to thermal degradation.

3. Calcium chloride is an inexpensive by-product of the Solvay process for soda ash. However, it is not used as a starting material for synthesis of other calcium compounds. Why?

4. Explain on the basis of the properties of the ions involved why it is impossible to prepare anhydrous calcium chloride by simply heating one of its hydrates.

5. Why is it advantageous to use calcium carbonate rather than calcium hydroxide as the starting material for the preparation of calcium lactate?

6. Distinguish between calcination and roasting.

7. Give reasons why lithium ion shows certain similarities to the

alkaline earths, especially Mg^{2+} and Ca^{2+}.

8. Name some common nonmedicinal uses of barium ion. What property of this ion is the basis for the usual treatment for its accidental ingestion?

9. List the common oxidation states of the phosphorus atom. Which states exist in compounds useful in pharmacy and medicine? Illustrate.

10. Given the formulas Na_3PO_4, Na_3PO_3, Na_2HPO_3, $NaPO_3$, Na_2HPO_2, $K_2H_2P_2O_7$. Indicate those which represent the empirical or molecular formula of an actual compound and give their correct chemical, and any other commonly used names.

11. The compound $Na_2HPO_4 \cdot H_2O$ was formerly cited in the compendia as Sodium Phosphate. Give the correct chemical name based on modern rules of nomenclature. Give three other unambiguous names for this compound. Why was the officially cited name actually equivocal?

12. Discuss the effect of the following on the solubility of a given phosphate salt: cationic charge, anionic charge, pH.

13. Differentiate between octacalcium phosphate and tricalcium phosphate. What are the conditions for formation of each? Is there any good reason why calcium and phosphate ions deposit in the bone as the former rather than the latter?

14. What are the roles of bone? Name the ions which are normally present. Which are the major components? Explain the effect of pH on the deposition or dissolution of bone.

15. Ingested lead(II) ion is isomorphously deposited in the bone. Thus the body effectively "neutralizes" the poisonous character of this ion. Taking a clue from the behavior of calcium ion deposited in the bone, postulate the conditions under which lead(II) ion may be released from the bone as the free ion.

16. Name those substances or classes of substances which may interfere with the absorption of calcium ion from the gastrointestinal tract. Explain the chemical basis of the interference in each case, with equations if possible.

17. Name the substances which are intimately involved in calcium absorption and utilization.

18. Explain why the halogens form monovalent anions so easily. Give the effect of the different structural factors such as ionic radius on the stability of these ions.

19. Why should hydrogen be considered a member of both the alkali and halogen families?

20. Given the acid series $HClO_3$, $HBrO_3$, HIO_3, predict the acid-base and redox properties of these acids as a family. Indicate the gradation in properties due to changing halogens. The compounds HFO_3 and $HAtO_3$ are not included. Explain.

21. The ionic radius of chlorine in the -1 oxidation state is 1.81 Å, and its covalent radius in the $+7$ oxidation state is 0.26 Å. Explain why this difference arises. How does this affect the comparative properties of the respective chlorine compounds?

22. The native chlorides are the usual natural sources of sodium and potassium ions. Explain why purification procedures differ.

23. When iodine dissolves in a solvent, e.g., water or diethyl ether, the solvent acts as a Lewis base. However, the basic properties are weak, and the resultant coordinate covalent bond is not a strong bond. Write some possible electron structures for the interaction products between water and the iodine molecule. (Do not disregard the possibility of *d*-level occupancy in the iodine atom.)

24. Explain on a structural basis why the iodide ion is susceptible to air oxidation, whereas fluoride ion is not.

25. Fluoride ion reduces the solubility of the enamel of the teeth. Considered in conjunction with the other anions present is it possible to explain this effect?

26. Why are the silicofluorides rather than NaF or SnF_2 preferred for fluoridation of water supplies?

27. A municipal water supply has 0.3 mg. fluoride ion per liter. Is this a satisfactory level? Show calculations.

28. How many mL. of hydrofluosilicic acid must be added per 1000 gallons of natural water having 0.2 p.p.m. fluoride to bring the concentration to the desired 1.0 p.p.m. level? (H_2SiF_6, 30% solution, density, 1.3 Gm./mL.) (9.83 mL)

29. Identify the pseudohalogens. How do they resemble the halogens? How do they differ?

30. Why do the two hydration reactions for carbon dioxide have the same effect on the pH of the blood, a fact not immediately apparent from the equations for the reactions?

31. Identify the carbamino reaction.

32. Identify those physical properties of water which make it so valuable as the body's heat sink. Explain how these properties accomplish this function.

33. Why is it preferable to consider specific ions, e.g., Na^+, rather than compounds of that ion, e.g., $NaHCO_3$, when discussing body fluids?

34. Why is it expedient to use milliequivalents rather than grams in expressing the concentration of a given ion in the body fluids?

35. List the specific roles played by each of the following ions in the body's normal economy: Na^+, K^+, Ca^{2+}, Mg^{2+}, F^-, Cl^-, I^-, HCO_3^-, HPO_4^{2-}, and $H_2PO_4^-$. Under what conditions might supplementation be required?

36. In addition to the ions listed in the preceding question, Fe^{2+}, Cu^{2+}, Zn^{2+}, Co^{2+} are important in the mineral balance of the body. Classify the ions listed in questions 35 and 36 as to type. Write the electron structure of each ion. Is there any correlation between the electron structure of an ion and the normal physiological function of the ions?

37. Why is the chloride ion chemically ideal for its roles in the body fluids?

38. Why is the sodium ion chemically ideal for its roles in the body fluids? Can the high hydration state of this ion at least partially explain the ease with which the cells reject sodium but maintain a positive potassium concentration gradient with respect to the extracellular fluids?

39. Explain the stability of the phosphate structure. Why is this complex so useful as an aid in maintaining a constant pH in the body fluids?

40. Explain why the bicarbonate ion, as opposed to carbonic acid and dissolved carbon dioxide, is such an ideal transport form for the carbon dioxide formed during the oxidative process of the cell.

41. Identify those compounds, solutions, and injections, studied in this chapter which are official in the current compendia.

42. What is the purpose of each ion present in the following fluid and electrolyte replenishers: Ringer's Solution, Sodium Lactate Injection, Dextrose and Sodium Bicarbonate Injection, Ammonium Chloride Injection.

43. Give specific reasons for using the following parenteral solutions: 5% dextrose, potassium chloride, calcium gluconate, ammonium chloride.

44. Give the advantages of sodium lactate over sodium bicarbonate as an alkalizing agent in fluid and electrolyte replenishers.

45. What are the advantages in using Ringer's Injection rather than isotonic saline?

46. How do the requirements for an injection differ from those for the same solution if it is intended only for topical use?

47. By means of calculations show whether Lactated Ringer's Solution is hypo-, iso-, or hypertonic. (Use 0.18 as the freezing point depression of 1% sodium lactate.) (hypotonic)

48. Calculate the change in pH taking place if the normal 1:20 ratio of carbonic acid to bicarbonate is changed by a 1% increase in bicarbonate level.

49. An experimental electrolyte replenisher was prepared using 500 me. bicarbonate ion and 100 me. chloride ion per liter, both as sodium salts. Will this solution show any solubility incompatibilities?

50. Calculate the pH of a solution of the sodium salts of dihydrogen phosphate ion, 50 me./L., and monohydrogen phosphate, 100 me./L. If 10 ml. of 0.01*N* hydrochloric acid is added to 500 ml. of this buffer, what is the resulting change in pH? (pH original solution = 7.51) (pH after addition of acid = 7.50).

51. A salt substitute includes ammonium lactate, potassium chloride, calcium tertiary phosphate, and starch. What is the rationale, if any, for including each member of the combination?

52. Why is calcium chloride generally unsatisfactory as a calcemic? What lines have generally been followed in attempting to overcome these deficiencies?

53. Explain why calcium monohydrogen phosphate is a useful calcemic, whereas the dihydrogen and the tribasic salts are not suitable in this use. What is the present official name of this compound? The former name?

54. Name the three methods in use for administration of essential fluoride ion. Compare their relative effectiveness.

55. Why is it inadvisable to have the "fluoridized" equivalent of "iodized salt?"

56. Name the phosphorus compounds commonly used in tonics. What is the rationale advanced for the use of each? What is the value of phosphorus tonics? Explain.

PRACTICE IN EQUATION WRITING

57. Write equations to show why: (*a*) fluoride salts are unstable in acidic media; (*b*) ammonium salts impart an acidic reaction to their solutions; (*c*) $(PO_4)^{3-}$ cannot exist in aqueous media; (*d*) glycerophosphates are not absorbed as such; (*e*) an aqueous solution of fluorine cannot be prepared; (*f*) gastric antacids interfere with calcium absorption.

58. Barium hydroxide solution is an excellent test solution for carbon

dioxide. Write the equation for a positive test for this gas.

59. Write equations demonstrating conditions leading to possible chemical or physical incompatibilities for each of the following ions in preparations of the type discussed in this chapter: F^-, Ca^{2+}, Na^+, HCO_3^-, HPO_4^{2-}, Sn^{2+}, I^-.

60. Write equations showing how hypophosphorous acid acts as an antoxidant for hydrogen iodide solutions.

61. Write the equations for a simple test for the hypophosphites.

62. Write the equation(s) showing the usual method used for the preparation of the following: (*a*) calcium gluconate, (*b*) potassium hypophosphite (elemental phosphorus as one raw material), (*c*) precipitated calcium carbonate, (*d*) barium chloride (starting with native barium sulfate), (*e*) hydrogen cyanide, (*f*) chlorine gas, (*g*) hydrofluosilicic acid, (*h*) tin(II) fluoride, (*i*) Lugol's Solution, (*j*) magnesium sulfate (from sea water).

63. Write the equations for the reactions used for the quantitative determination of the following: (*a*) calcium ion in calcium gluconate tablets, (*b*) sodium chloride in an aqueous solution thereof, (*c*) potassium dihydrogen phosphate, (*d*) sodium bicarbonate, (*e*) magnesium chloride, (*f*) sodium fluoride, (*g*) iodine in Lugol's Solution, (*h*) potassium iodide.

FOURTEEN · MINERALS II: *essential trace elements*

This group of inorganic ions is made up of those minerals present in the body in trace quantities. For the most part they play active roles in the metabolic processes, usually in enzyme systems. The elements considered members of this category are iron, cobalt, copper, zinc, manganese, magnesium, and molybdenum. Except for magnesium (discussed previously under electrolytes), all proven essential trace elements are members of the transition element group. For this reason a preliminary discussion of the properties of this family of elements is necessary.

Consideration of the medicinal applications of these elements may be divided into two categories: hematinics and mineral supplements. Iron is usually considered separately from the others because of a better knowledge of its roles and a proven need for iron supplementation under certain conditions. The remaining trace elements are considered under the mineral supplement heading.

THE TRANSITION ELEMENTS AND THEIR COMPOUNDS

Electronically the ground state of these elements is characterized by two incomplete valence shells overlying an inert gas structure. The electron structure of the typical element may be written

$$[\text{Inert gas core}](n - 1)\, d^y ns^x$$

where x may be 0, 1, or 2, and y may have values of 1–10. Groups IIIA through VIII have the typical transition structures, featuring an incomplete d level in the ground state. Group IB, the coinage elements—copper, silver, and gold—has the structure $(n - 1)\, d^{10} ns^1$. Even though the d shell is completely occupied in the ground state, it is unstable under

TABLE 14.1 ATOMIC CHARACTERISTICS OF SELECTED TRANSITION ELEMENTS[a]

Element	Titanium	Chromium	Manganese	Iron	Cobalt	Copper	Zinc	Zirconium	Molybdenum	Tantalum
Symbol	Ti	Cr	Mn	Fe	Co	Cu	Zn	Zr	Mo	Ta
Atomic weight	47.9	52.0	54.9	55.9	58.9	63.5	65.4	91.2	96.0	181.0
Atomic number	22	24	25	26	27	29	30	40	42	73
Group	IVA	VIA	VIIA	VIII	VIII	IB	IIB	IVA	VIA	VA
Valence electron configuration	$3d^2 4s^2$	$3d^5 4s^1$	$3d^5 4s^2$	$3d^6 4s^2$	$3d^7 4s^2$	$3d^{10} 4s^1$	$3d^{10} 4s^2$	$4d^2 5s^2$	$4d^5 5s^1$	$5d^3 6s^2$
Atomic radius (Å)	1.32	1.17	1.17	1.16	1.16	1.17	1.25	1.45	1.29	1.34
Ionic radius[b] (Å)		0.64(III)	0.91(II)	0.83(II)	0.82(II)	0.96(I)	0.74(II)			
	0.64(IV)	0.52(VI)	0.62(III)	0.67(III)	0.65(III)			0.74(IV)	0.68(IV)	0.73(V)
Ionization potential (ev.), I	6.83	6.76	7.43	7.90	7.86	7.72	9.39	6.95	7.18	7.0
II	13.63	16.49	15.46	16.16	17.3	20.29	17.89	13.97	15.72	—
Reduction potential	−0.95	−0.71	−1.05	−0.44	−0.28	+0.35	−0.76	−1.53	−0.2	−0.7
(v.)	Ti, TiO^{2+}	Cr, Cr^{3+}	Mn, Mn^{2+}	Fe, Fe^{2+}	Co, Co^{2+}	Cu, Cu^{2+}	Zn, Zn^{2+}	Zr, ZrO^{2+}	Mo, Mo^{3+}	Ta, Ta^{5+}
Electronegativity	1.32	1.56	1.60	1.64	1.70	1.75	1.66	1.22	1.30	1.33

[a] Some elements of this group with applications other than mineral supplements have been included.
[b] Oxidation states are given in Roman numerals.

oxidizing conditions; one or two electrons of this shell may be removed. Group IIB, the volatile metals—zinc, cadmium, and mercury—has the structure $(n - 1)\ d^{10}ns^2$. The *d* shell is stable in this group; these elements show none of the typical transition element properties. The actual electron structures of some transition elements, together with other structural properties of interest, are given in Table 14.1.

Physical properties. The transition elements, except for the volatile metals, have two outstanding physical properties. First, the atoms of all elements of the group are paramagnetic, a property due to unpaired electron spins. This property reaches its acme in the Group VIII triad of the first long period—the ferromagnetic elements, iron, cobalt, and nickel. Paramagnetism has been extremely valuable in the elucidation of the electron structure in these elements and their compounds, and in the development of the modern theory of complexation.

The second distinguishing physical property of these elements is their ability to impart color to their compounds. With only a few exceptions, colored inorganic compounds include one or more transition elements.[1] Color is due to the relative ease with which the valence electrons of the incompletely filled *d* levels of these elements may achieve an excited state. Since the wavelength of light is inversely related to its energy, the color of these compounds is characteristic of the energies of the permitted electron transitions. The typical elements do not impart color to their compounds because of the comparatively high energies required to achieve excitation of their *s* and *p* valence electrons.

Chemical properties. The ease of electron excitation in the transition element family results in exceptionally involved chemistry, characterized especially by extremely variable valence. For a given transition element the permitted oxidation states are all possible numbers from zero to the group number of the element. However, not all will necessarily appear; e.g., manganese appears in the oxidation states 0, 2, 3, 4, 6, 7; titanium, 0, 2, 3, 4; and iron, 0, 2, 3, 4, 6. Only the coinage elements commonly exhibit the +1 oxidation state. Because of the instability of their ground state *d* shell, Group IA elements also differ from the general behavior in that they may achieve oxidation numbers (and ionic valences) in excess of their group number. The oxidation states of these elements are 0, 1, 2, and 3.

---

[1] Compounds in which an element is present in more than one oxidation state are also colored. A few illustrations are $\mathit{Cs}_3\mathrm{O}$, dark green, $\mathrm{Na}_2\mathit{S}_5$ orange, $\mathrm{K}\mathit{I}_3$, orange-brown The element responsible for the color is italicized. If the element involved is also a transition element, very intense colors result, e.g., $\mathrm{KFe^{II}Fe^{III}(CN)_6}$, intense blue-black. Many covalent compounds not satisfying either of these requirements for color are colored, e.g., CdS, HgI_2, ICl.

The chemical behavior of a given transition element may be related to its oxidation state. In general the transition elements display typical metallic properties in their lowest oxidation states. When in the +1 or +2 state, the resultant compounds are generally ionic. The hydroxides, though insoluble, act only as bases. Complex ions form readily with these low-valenced cations, although their instability constants are usually large compared to those in which the same element exists in a higher oxidation state.

In their highest oxidation states the transition elements exhibit nonmetallic properties. They never appear as simple (or even hydrated) cations, but always as the small, highly charged central units of complex anions. These complex ions are frequently very strong oxidizing agents. The hydroxides are always acidic in character (see Fig. 4.3 and Table 4.10). In their highest oxidation states these elements resemble the nonmetals of the corresponding B subgroup rather closely. Thus the chromate ion is isomorphous with sulfate; permanganate is isomorphous with perchlorate.[2]

The intermediate oxidation states of a transition element generally exhibit both metallic and nonmetallic behavior; i.e., the hydroxides are amphoteric. If the element involved has a sufficiently large radius, trivalent positive cations may exist independently. However, in intermediate states the elements will more commonly exist as complex ions.

Because of the similarity in external electron structure in passing from one transition element to the next, these elements show well-defined horizontal resemblances, especially in the lowest oxidation states. However, as oxidation numbers are increased, these similarities become less pronounced. Examples of these horizontal similarities of properties are: manganese(II) and iron(II), silver(I) and palladium(I).

Biochemistry and medicinal applications. Except for their role as trace minerals, these elements have no other known role in normal biological processes. Except for the use of iron, manganese, cobalt, copper, zinc, and molybdenum compounds in mineral supplements, medicinal applications have been based on either the corrosive and germicidal properties

[2] This permanganate-perchlorate pair may be used as a very good demonstration of solid solutions and their properties. Solid solutions of their potassium salts may be prepared by crystallization from saturated solutions including the two salts. The crystals forming will include permanganate and perchlorate ions in ratios corresponding to their ratio in the parent solution. The color of the crystals, ranging from a very faint or pale pink to the characteristic dark purple-black of permanganate, depends upon the ratio of the two salts in the mother liquid solution. The components of solid solutions cannot be separated by fractional crystallization; the new crop of crystals forming will always have the same composition as the crystals used to prepare the mother solution from which recrystallization takes place.

(e.g., silver nitrate, iron(III) chloride, potassium permanganate) or on the protective action in topical preparations (e.g., zinc oxide, titanium (IV) oxide. Platinum, a member of the third triad of Group VIII, is used as cis-diamminedichloroplatinum(II) in cancer therapy. Metallic tantalum and silver have been used to replace bone, etc. Some compounds which have been used as medicinals, but find no application today, are uranyl nitrate (*U.S.P. IX)*, tantalum pentoxide, the vanadates and the oxide, hydroxide and carbonate of zirconium.

IRON AND THE HEMATINICS

Chemical properties of interest. The iron halides, sulfates, nitrates, and acetates are soluble. However, the resulting solutions are complicated systems. Free simple iron ions in either oxidation state do not exist in aqueous media. They occur either as hydrated cations or in association with the anions present to give complex ions of varying charge and composition, depending upon the concentration, both absolute and relative, of the cations and anions. In hydrochloric acid solutions of iorn(III) chloride, $FeCl^{2+}$, $FeCl_2^+$, $FeCl_3$, and $FeCl_4^-$ have been identified.

The common insoluble iron compounds become soluble in acid media. Some typical solubilizations are

$$FeO + 2H_3O^+ + 3H_2O = [Fe(H_2O)_6]^{2+}$$
$$FePO_4 + 2H_3O^+ + 4H_2O = [Fe(H_2O)_6]^{3+} + H_2PO_4^-$$
$$FeCO_3 + H_3O^+ + 5H_2O = [Fe(H_2O)_6]^{2+} + HCO_3^-$$

Generally, an iron(III) compound will be less soluble in water or at a given pH than the corresponding iron(II) compound. As iron is more metallic in the +2 oxidation state, iron(II) hydroxide is a stronger base than the iron(III) compound.

Because of the comparatively large acid ionization constants of the hydrated iron cations, solutions of soluble iron salts, especially those of iron(III), are subject to hydrolysis. The pK_a of the hexaquairon(II) ion and the hexaquairon(III) ion are 5.3 and 2.2, respectively. Until recently it was assumed that these ions undergo hydrolysis via the reactions expressed in either Equation 1 or 2.

$$[Fe(H_2O)_6]^{3+} + H_2O = [Fe(H_2O)_5(OH)]^{2+} + H_3O^+ \quad (1)$$
$$[Fe(H_2O)_6]^{3+} + OH^- = [Fe(H_2O)_5(OH)]^{2+} + H_2O \quad (2)$$

Recent studies have shown that ionization of the aqua iron cations, and similar ions of other transition elements, may proceed via a dimerization process.

$$2[Fe(H_2O)_6]^{3+} = [(H_2O)_4Fe(OH)_2Fe(H_2O)_4]^{4+} + 2H_3O^+$$

Evidence indicates that this latter reaction is most important in 0.001*M* and more concentrated solutions. The proposed structure consists of two octahedra, each with an iron central atom, joined on a common edge. The positions at the ends of this common edge are occupied by the two hydroxyl ions. The unshared corners of each octahedron are occupied by water molecules. The structure is illustrated in Fig. 14.1.

Depending on the concentration of the iron salt and on the pH of the medium, ionization may proceed to further stages, ultimately resulting in the precipitation of the insoluble hydroxides and hydrated oxides. Conventionally these reactions may be written

$$[Fe(H_2O)_5(OH)]^{2+} + H_2O = [Fe(H_2O)_4(OH)_2]^+ + H_3O^+$$

etc. It is more likely, however, that the polymerization process proceeds to further stages; e.g.,

$$2[Fe_2(H_2O)_8(OH)_2]^{4+} = [Fe_4(H_2O)_{12}(OH)_6]^{2+} + 2H_3O^+$$

This process is more probable since it accounts for the ultimate precipitation by a mechanism easily building up macro particles too large to remain in solution (see Fig. 14.1).

Iron(II) may act as an oxidizing or reducing agent, but the latter action proceeds much more easily. Oxidation of iron(II) to iron(III) even by atmospheric oxygen is one of the continuing problems of the pharmaceutical chemist. Aqueous solutions of the iron(II) ion are so unstable that

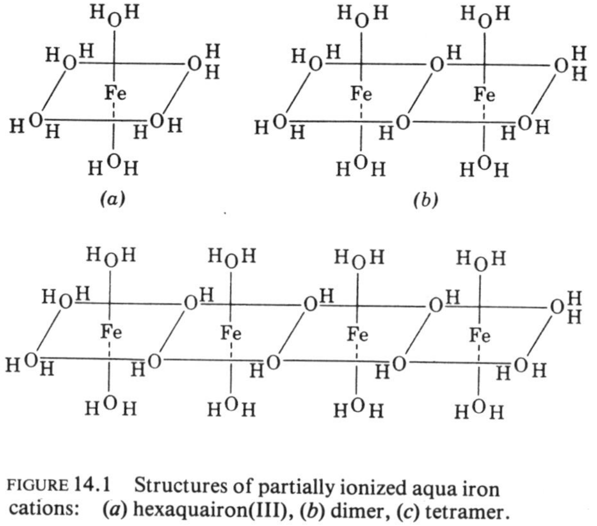

FIGURE 14.1 Structures of partially ionized aqua iron cations: (*a*) hexaquairon(III), (*b*) dimer, (*c*) tetramer.

unless protected from oxygen by inclusion of a suitable antoxidant (e.g., sucrose in Ferrous Iodide Syrup) they cannot be dispensed. The reactions taking place may be written

$$2[Fe(H_2O)_6]^{2+} + \tfrac{1}{2}O_2 + 2H_3O^+ = 2[Fe(H_2O)_6]^{3+} + 3H_2O$$

or

$$2[Fe(H_2O)_6]^{2+} + \tfrac{1}{2}O_2 + H_2O = 2OH^- + 2[Fe(H_2O)_6]^{3+}$$

depending on the pH of the medium. Oxidation is followed by varying degrees of hydrolysis of the hexaquoiron(III) ion, again depending on the pH. Crystalline iron(II) salts, e.g., iron(II) sulfate, must be stored in containers protected by good closures, especially in moist atmospheres. Otherwise surface oxidation takes place.

Iron(III) is an oxidizing agent only.[3] In the presence of reducing agents it reacts easily; e.g.,

$$Fe^{3+} + I^- = Fe^{2+} + \tfrac{1}{2}I_2$$

and

$$Fe^{3+} + Fe^0 = 2Fe^{2+}$$

The latter equation demonstrates why solutions of iron(III) salts cannot be stored or reacted in common metallic vessels or pumped through iron piping. Metallic closures on bottles containing solutions of this ion will be corroded.

In common with the other transition elements, iron in any oxidation state complexes with a very wide range of ligands to give exceptionally stable complexes. Iron(III) employed in medicinals is most frequently in the form of one of its soluble complexes. The two "ferric" ammonium citrates, "ferric" citrochloride, "ferric" phosphate, "ferric" pyrophosphate, "ferric" ammonium tartrate, "ferric" potassium tartrate are examples. Some of these complexes are so stable that they fail to respond to the usual tests for iron(III) ion in aqueous solution. Iron(II) also complexes readily, although generally the binding is not as firm. Some interesting reversals of stability are found on complexation. Although ordinarily iron(II) salts are less stable in aqueous solution, hexacyanoferrate(III) solutions are less stable than those of hexacyanoferrate(II).

Thus the pharmaceutical chemist's problems in working with the two iron cations center around the instability of the +2 oxidation state to even weak oxidizing agents and the strong tendency for hydrolysis of the

[3] Iron in higher oxidation states exists in the crystalline state. Its compounds are very powerful oxidizing agents incapable of existence in aqueous media. They are prepared in fused bases.

salts of iron(III). Solubility incompatabilities are probable for soluble salts in neutral and alkaline pH ranges. Acid media will cause solution of the insoluble compounds.

Gallium(III) is very similar to iron(III). It has substituted for iron in *in vitro* experiments.

Biochemistry. Iron is an essential constituent of the structure of hemoglobin and in certain redox enzymes, e.g., the cytochromes, cytochrome oxidase, catalase, and the peroxidases. The human body has between 2.0–2.5 me. iron/Kg. body weight. About 60–70% of this iron is in the hemoglobin. The balance is distributed between the iron stores (about 20%) and the iron-based enzymes of the cells.

Iron metabolism is virtually a closed system. Unlike sodium, potassium, and vitamin C, nearly all of which are absorbed from the intestinal tract and any excess excreted, iron absorption is regulated according to the needs of the body. Normally intake exactly balances losses; 1–2 mg. are absorbed and excreted daily. In cases of iron deficiency, however, 5–15 mg. may be absorbed daily over short periods.

Because of relative insolubility of its hydroxides and phosphates, iron can be absorbed only from acidic media; the optimum pH for absorption is approximately 5. Insignificant quantities may be absorbed in the stomach and jejunum. Major absorption takes place in the duodenum. The availability of iron for absorption is mediated by the acidity of the gastric juice and reactions taking place while in the stomach. Much of the iron in foods is in the iron(III) state and in insoluble forms. The acid medium of the stomach converts these insoluble forms into soluble iron(III) compounds. Reducing agents normally present in foods, e.g., ascorbic acid, glutathione, now reduce the iron to the +2 oxidation state. The hexahydrate of iron(II), being a weaker acid than that of iron(III), is less subject to hydrolytic influences; in addition iron(II) hydroxide is more soluble.

Unless administered at mealtime, hematinics do not have the advantage of the reducing action of foods to convert and/or maintain their iron in the favorable +2 oxidation state. For this reason ascorbic acid is frequently administered with hematinics. In addition to its reducing action, ascorbic acid probably aids in reducing the pH of the intestinal fluids, thus further increasing the availability of iron.

When needed, iron passes through the intestinal mucosa as iron(II). In the mucosal cell it is oxidized to the +3 state; colloidal particles of iron(III) hydroxide form. These particles combine with the protein apoferritin (m.w. 465,000) to give ferritin—the stable, nonionic, nontoxic storage form of iron. The iron is not bonded chemically to the protein, but is found in its interstitial spaces as hydroxide particles in amounts up to 23% dry weight. When the ferritin content of a mucosal cell reaches a certain level, absorption of iron ceases.

When the blood iron falls below its normal level of about 0.05 me./L., the intestinal mucosal cells must release iron(II) to the blood. The ferritin gives up its iron(III) which is reduced by cell enzymes to iron(II). The iron(II) passes into the blood plasma where it is immediately reoxidized. Two iron(III) ions combine with the protein siderophilin (m.w. 90,000) to give transferrin—the circulating transport form of iron. The iron level of the blood plasma is normally only about one third of saturation. If saturation is temporarily achieved through injection of an iron salt, iron is rapidly eliminated to re-establish the normal level.

The circulating iron store is available to the liver, spleen, bone marrow, and cells, whenever the ferritin stores of these organs fall below normal. The chain of events during this transfer is identical to that described for transfer of iron to and from the intestinal mucosal cells. Figure 14.2 summarizes the different steps taking place during storage and transport of iron in the body.

Gastric action

$$Fe(OH)_3 + 3H_3O^+ + 2Cl^- = FeCl_2^+ + 6H_2O$$

$$FeCl_2^+ + \text{(reducers in foods)} = Fe^{2+} + 2Cl^- + \text{oxidation products}$$

Intestinal action

Lumen | Mucosal cells

Fe^{2+} → $2Fe^{2+} + \frac{1}{2}O_2 = 2Fe^{3+} + (O^{2-})$

$Fe^{3+} + 3OH^- = Fe(OH)_3$

$Fe(OH)_3$ + apoferritin = ferritin

Blood plasma transport

Mucosal cells | Blood plasma | Liver

Reversal of the reactions directly above.

Fe^{2+} → $2Fe^{2+} + \frac{1}{2}O_2 = 2Fe^{3+} + (O^{2-})$

$2Fe^{3+}$ + siderophilin = transferrin

Transferrin = siderophilin + $2Fe^{3+}$

$2Fe^{3+}$ + reducer = $2Fe^{2+}$ → Reaction identical to that in the mucosal cells, forming ferritin.

FIGURE 14.2 Absorption and transport of iron.

Medicinal applications. Iron compounds have two important uses: as a mineral dietary supplement, especially in iron deficiency conditions associated with secondary anemias; as an astringent. The hematinic response is best when administration is in the form of iron(II); however, iron(III) compounds are also used. The astringent action is restricted to iron(III) compounds and will be discussed in a more appropriate section (page 496). Insoluble iron oxides are used as pigments in topical preparations. This cosmetic application will also be discussed (page 536).

Pharmacology. Iron compounds are toxic. The LD_{50} varies from 11–300 mg. iron/Kg. body weight. The soluble ionic salts are most toxic. The more highly complexed forms are less toxic, but the availability of the iron is also reduced.

Children are especially susceptible to an overdose. Several accidental deaths have been reported. The etiology is vomiting, lethargy, coma, death. Iron preparations, especially chocolate or candy-coated tablets, must be kept out of the reach of children.

Conditions requiring hematinics. Because of conservation of the body's supply of iron, dyshemopoiesis,[4] iron deficiency anemia, may arise only infrequently. Anemias rarely develop because of a lack of iron in the food intake. When the condition does exist, it is usually because of extraordinary requirements for iron, e.g., periods of rapid growth or hemorrhagic conditions, or because of conditions resulting in impairment of iron absorption, e.g., achlorhydria and sprue.

In the absence of bleeding only 1–2 mg. of iron are lost daily and must be replaced from that in the diet. Children and menstruating, pregnant, or lactating women require higher levels of dietary intake. Women lose 20–40 mg. iron in addition to the normal loss during a menstrual period. The requirements during pregnancy are about twice the normal, during lactation about 1.5 times normal.

Excessive bleeding may deplete iron stores. Hemorrhagic anemias may be acute or chronic. Once acute hemorrhage has been controlled, a varying degree of anemia may persist for some time. Patients having a history of chronic bleeding prior to an acute episode may require restoration of iron reserves by administration of a suitable iron supplement.

Failure of response to oral administration of iron may be due to intolerance arising from gastrointestinal irritation, inability to absorb iron from the digestive tract, or a sensitivity to iron. In such cases intravenous administration of iron may be necessary. Usually 25 mg. of iron (calcu-

[4] Dyshemopoiesis develops when inadequate amounts of iron are available for the development of fully hemoglobinized red blood corpuscles. Relatively normal numbers of red blood cells develop, but these cells are *hypochromic*, that is, deficient in hemoglobin.

lated as the metal) is required for each 1% deficiency in hemoglobin. The total dose should not be administered at one time, but should be broken up into individual doses not exceeding 50 mg. as the initial dose and 100–150 mg. daily thereafter until the total requirement has been supplied.

Iron therapy is of no value in other anemias, e.g., anemias due to folic acid or vitamin B_{12} deficiencies, hemolytic anemias, etc.

While iron-deficiency anemia is well known, the opposite condition, iron overload, is less well known because it is comparatively rare. It is usually more serious and frequently fatal. Iron overload exists as a hereditary disease, hemochromatosis, and as secondary hemochromatosis in which diet plays an important role. Accumulations of 20-30 grams of excess iron have been found in victims.

A recent review of iron metabolism by Emory is available.[5]

Compounds and preparations used as hematinics. This presentation is arranged in groups, based on the oxidative state of the iron.

ELEMENTAL IRON: Reduced iron (iron by hydrogen, Quevenne's iron, *ferrum reductum*) was at one time popular as a hematinic and is still available. The hydronium ion of the stomach oxidizes the element to the +2 state.

$$Fe^0 + 2H_3O^+ = Fe^{2+} + H_2\uparrow + 2H_2O$$

If pure and in a finely divided state elemental iron is a satisfactory hematinic. Iron filings, the usual chemical source of iron, are unsatisfactory on both counts. A satisfactory product must pass a No. 100 standard sieve (*N.F. X*), since a large surface area is required for reaction at an adequate rate. Because of its stability on storage, reduced iron is useful in food and feed fortification. Obviously this hematinic is of no value to patients with achlorhydria, or if administered with a gastric antacid.

Three methods are available for the preparation of reduced iron: thermal decomposition of pentacarbonyl iron,

$$Fe(CO)_5 = Fe + 5CO\uparrow$$

reduction of iron(III) oxide (ferric oxide, hematite) by hydrogen

$$Fe_2O_3 + 3H_2 = 2Fe + 3H_2O\uparrow$$

and cathodic reduction of a solution of an iron(II) salt under conditions producing a powder rather than massive crystals. Due to a slow surface oxidation of the particles iron oxides are usually present in reduced iron

---

[5]American Scientist, *70,* 626-32 (1982).

samples. However, this presents no problem since the oxide is also soluble in the acid gastric juice.

IRON(II) COMPOUNDS AND PREPARATIONS: Iron(II) salts are absorbed most efficiently and are therefore the most prevalent form of iron in hematinics. The sulfate, gluconate, lactate, succinate, fumarate, and ascorbate are most common.

Iron(II) sulfate (ferrous sulfate, iron sulfate, green vitriol, iron vitriol, iron protosulfate) (technical product, copperas) is the most efficient, cheapest, and most popular hematinic. It is available either as the heptahydrate, $FeSO_4 \cdot 7H_2O$, or as the exsiccated salt. The heptahydrate occurs as pale bluish green crystals or granules. It melts at 64°C. It dehydrates in stages (a tetrahydrate and monohydrate exist), becoming completely anhydrous at about 300°C. At higher temperatures the salt decomposes.

$$FeSO_4 = FeO + SO_3\uparrow$$

It is quite soluble in water; 1 Gm. dissolves in 1.5 ml. at room temperature and in 0.5 ml. at boiling temperature. It is insoluble in ethanol. The heptahydrate is subject to two reactions on exposure to air. Because it is efflorescent in dry air, the crystals gradually become coated with a white film of partly dehydrated salt. In moist air the surface of the salt rapidly oxidizes, acquiring a film of yellow-brown basic iron(III) sulfate. In order to improve the stability of the crystals manufacture is controlled to give a water content slightly less than theoretical. Thus a drying action is built into the product, making the moisture unavailable for the redox reaction. Freshly opened bottles may assay on the high side by as much as 1% because of this manufacturing control.

Dried iron(II) sulfate is a greyish white powder consisting primarily of the monohydrate. Varying quantities of the tetrahydrate are also present; the commercial product averages about $2H_2O$. The exsiccated salt is slowly soluble in water. Its solubility, on a $FeSO_4$ basis, is ultimately the same as that of the heptahydrate. Due to some hydrolysis the solution is usually slightly cloudy. The exsiccated form has better stability on atmospheric exposure than the heptahydrate, but storage in well-closed containers is necessary.

Because of its better solubility behavior, the heptahydrate is preferred for the preparation of solutions, syrups, and elixirs. Because of its better stability, the exsiccated form is used for the preparation of tablets, capsules, etc. Tablets are usually coated to hide the iron taste. In order to improve its palatability the compound often is administered in the form of an elixir or syrup. Iron(II) sulfate is frequently used as an iron source in multivitamin and mineral preparations.

Iron(II) sulfate is also used for the preparation of a number of pharmaceutical products in which the active hematinic compound is an iron(II) carbonate. Iron(II) sulfate is reacted with either sodium or potassium carbonate or bicarbonate under suitable conditions. The freshly precipitated product is white or nearly so. Limited contact with air causes a slight green tinge. Excessive contact with air and moisture results in the appearance of orange-brown iron(III) compounds. Preparations of this type are Ferrous Carbonate Pills (Chalybeate Pills, Blaud's Pills, Ferruginous Pills), Ferrous Carbonate Mass (Vallet's Mass), and Saccharated Ferrous Carbonate. Although these products are no longer popular as hematinics, their preparation illustrates some interesting chemistry.

Iron(II) iodide (ferrous iodide, iron protoiodide) is another iron(II) hematinic employing an inorganic anion. The chemistry of the preparation of this compound has been previously discussed (page 85). Concentrates are commercially available. It is usually used as Ferrous Iodide Syrup, Hypophosphorous acid and sucrose are present as protectives.

Iron(II) chloride (ferrous chloride) is sometimes used as a hematinic in elixirs. The crystalline salt (tetrahydrate) is deliquescent in moist air and efflorescent in dry air.

Other iron(II) salts used as hematinics represent attempts to reduce the irritating quality and ferrugenous taste of the iron(II) salts by incorporation of various organic anions. The most successful is probably the gluconate. It is claimed to cause less gastric irritation on oral administration, usually as tablets or in elixir form. Intramuscular administration of the gluconate is said to be less painful than similar administration of other iron(II) salts. The fumarate is probably as effective as other oral forms, but an occasional patient shows the usual iron reactions. It is widely used in multiple mineral preparations. The lactate and ascorbate are also used.

In working with any of these compounds, the pharmacist must anticipate the usual incompatibilities of the iron(II) cation arising from ease of oxidation, insolubility and extreme ease of oxidation in alkaline media, ease of complexation with such ions as citrate, tartrate.

The officially cited iron(II) compounds at this time are: Ferrous Fumarate and Tablets, Ferrous Gluconate and Capsules, Tablets and Elixir, Ferrous Sulfate and Oral Solution, Syrup and Tablets, and Dried Ferrous Sulfate (Exsiccated).

IRON(III) COMPOUNDS AND PREPARATIONS: Despite poor absorption properties, iron(III) compounds continue in use as hematinics. The stability of the strongly complexed forms in which iron(III) is administered further reduces the availability of the iron. However, complexation is necessary to reduce the astringency and local irritating action of the iron(III) ion. Because of the solubility of these complexes, it is possible to administer iron(III) hematinics as solutions.

Iron(III) is available as a citrate complex in the form of Ferric Ammonium Citrate, Green Ferric Ammonium Citrate, Ferric Citrochloride Tincture, Soluble Ferric Phosphate, and Ferrocholinate. All must be stored in light-resistant containers and protected from heat.

$$[\text{Fe(III) citrate complex}] + h\nu = Fe(OH)_3\downarrow + Fe^{2+} + CO_2\uparrow$$

The ammonium citrate complexes are prepared by reaction of freshly precipitated iron(III) hydroxide with citric acid. A slight stoichiometric excess of ammonia is added, giving the solution a neutral, slightly acid, or basic reaction depending upon the degree of excess. The preparative procedure for the green form uses more citric acid and less ammonia, giving the resultant solution an acid reaction. Ferric Ammonium Citrate is free of astringency and local irritating action. It is available as thin garnet red scales or plates or as a brownish yellow powder. It may have a slight ammoniacal odor and a saline, mildly ferruginous taste. Fifty per cent solutions are usually used as the dosage form, but must be well diluted before administration and taken through a straw to prevent discoloration of the teeth. The compound may also be administered in capsules. Green Ferric Ammonium Citrate may be used for intramuscular injection, but is painful. For this reason a local anesthetic should be employed.

Ferrocholinate is a chelate formed by reaction of freshly precipitated iron(III) hydroxide with choline dihydrogen citrate in a 1:1.25 molar ratio. Since one carboxy group of the citrate ion is utilized for esterification of the choline, the iron is probably chelated to the remaining two. This complex is more soluble than iron(II) sulfate or gluconate. Clinical testing seems to indicate reduced toxicity and less gastrointestinal irritation. There is no evidence for better oral efficacy, however.

Tartrate is also used as a complexing agent for iron(III) ion. Both Ferric Ammonium Tartrate and Ferric Potassium Tartrate have been used as hematinics and were at one time official.

Ferric Chloride Tincture (Iron Tincture) is sometimes used as a hematinic, although primarily used as an astringent. It is a yellow acidic solution with a very astringent taste. Since the Ferric Chloride Solution used to prepare the tincture is aqueous iron(III) chloride stabilized with hydrochloric acid, the iron is present as the tetrachlorferrate(III) ion:

$$FeCl_3 + H_3O^+ + Cl^- = H_3O^+ + [FeCl_4^-]$$

This complex is less stable than the citrate or tartrate complex. This decreased stability accounts for the higher residual astringency of Ferric Chloride Tincture.

Iron Dextran Injection and Iron Sorbitol Injection are injectable hematinics. The former is a colloidal solution of iron(3+) hydroxide in a partially hydrolyzed dextran. The latter is a complex of iron with sorbitol and citric acid. Both are relatively nonirritating when used intramuscularly, providing a well-tolerated parenteral therapy for iron deficiency anemia. These preparations are intended only when parenteral administration seems preferable because intolerance or gastrointestinal conditions preclude oral administration. (U.S.P. XXI)

Some insoluble iron(III) compounds also are used as hematinics. The pyrophosphate is widely used in proprietary hematinics and is found in numerous mineral supplements. Saccharated Ferric Oxide was at one time official (*N.F. VII*). Both compounds are rendered soluble by the acid medium of the stomach.

MINERAL SUPPLEMENTS

Trace minerals play a vital role in the nutritional field. The human body requires a relatively large supply of calcium, phosphorus, sodium, potassium, and chloride, and traces of copper, iron, zinc, cobalt, magnesium, manganese, molybdenum, iodine, and fluorine. Recent work has shown arsenic, chromium, nickel, silicon, selenium and vanadium are also essential for higher animals. Since the body cannot synthesize any of these elements they must be supplied in the diet. Whether or not a given mineral deficiency exists may depend upon the presence or lack of that element in the soils upon which the food plants of an area are grown.[7] One of the best known examples is the high incidence of endemic goiter in persons living in iodine-deficient areas.

Dietary deficiencies in certain minerals may, however, develop in other ways. Poor selection of the foods making up an individual's diet, either because of personal taste or selection favoring highly refined foods, may lead to long-standing deficiencies. Refining invariably seems to remove the essential trace elements. Vegetarian diets may result in deficiencies, e.g., zinc. While zinc is abundant in most plants it is not easily freed for absorption by the human digestive process. Poor absorption of trace elements from food of an ordinarily adequate diet may result from the nature of the diet, e.g., excessive precipitation of calcium by presence of high oxalate foods; medication, e.g., adsorption of trace minerals, etc.,

[7] Plants, like animals, develop trace element deficiencies, each with its typical syndromes. New Jersey soils used for onion production require copper. Danish soils are cobalt deficient, resulting in low cobalt levels in the plants upon which animals of this agricultural nation graze. During World War II the Danes found it necessary to bring in this element from Sweden, via the underground.

on gels such as aluminum hydroxide; or disease, e.g., sprue, characterized by poor iron absorption. Deficiencies may also arise temporarily during periods of high requirements, e.g., pregnancy, lactation, rapid growth in children, excessive loss of fluids in hemorrhage, diarrhea. Assimilation of trace elements appears to become poorer with age. A review of trace elements in human nutrition, including numerous references, was recently published.[8]

Roles of trace minerals. The transition elements and magnesium appear to play vital roles in various enzyme systems; they are the complexly bound central unit of protein structures involved in enzymatic reactions of all types. These protein prosthetic groups possess no activity in the absence of the metallic ion usually associated with them. All redox enzymes involve a transition element, e.g., iron, copper, easily undergoing reversible oxidation-reduction. In other cases the complexing ability, e.g., zinc, apparently plays a role in forming the temporary bonds essential for progress of definite enzymatic reactions.

Recently the mode of action of specific therapeutics is being shown to involve trace elements. It is believed that the activity of certain antibiotics, e.g., the tetracyclines, depends upon their complexation of certain trace elements, making these elements unavailable to the bacterium, fungus, etc. The tetracyclines apparently can complex trace elements in two or more ways. With divalent elements such as copper, nickel, or zinc, each tetracycline molecule acts as a tridentate ligand bridging to the ion through hydroxyl groups at C-12, 12*a*, and C-3, and through the dimethylamine group at C-4 (see Fig. 14.3). Calcium and magnesium may be chelated at the β-diketo system between C-11 and C-12. Foye[9] and co-workers have found that the antibacterial, analgesic, antipyretic, and fungicidal effects of salicylic acid derivatives are related to the magnitude both of the ionization constants and the instability constants of complexes with certain metal ions. Antibacterial activity requires a high instability constant; analgesic activity requires constants about equal to those of the parent acid; antifungal activity requires low instability constants.

It is possible that aging involves increased mineralization of tissues; it has been observed, for example, that calcium increases particularly in those organs rich in connective tissue. Some believe that the build-up of certain elements (e.g., copper) is an important factor in aging. There is

---

[8] Dairy Council Digest, *53,* No. 1, Jan.-Feb. 1982, 1SS-0011-5568.

[9] W.O. Foye and J.G. Turcotte, "Stabilization of Metal Complexes of Nuclear Substituted Salicyclic Acids—Correlation with Biological Effects," *J. Pharm. Sci.,* **51,** 329-332 (1962).

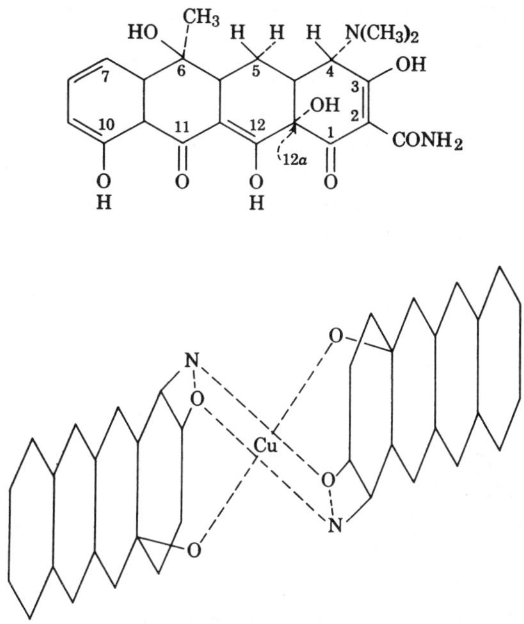

FIGURE 14.3 Complexation of metal ions by tetracycline.

also a belief extant that cancer in the aged may be related to high accumulation of trace elements. The cancer picture is two-sided, however. It has recently been shown that certain metal complexes are more active anticarcinogenics than the ligands themselves (Pt and Pd complexes of 6-mercaptopurine).

Recently it has been postulated on good evidence that some of the effects of exposure to γ radiation are due to interference with the metal moiety of certain enzymes. Radiation increases the peroxide concentration in the cells. This peroxide may oxidize the copper(I) of cytochrome oxidase to copper(II). In this form the enzyme is inactive.

Roles of individual trace minerals. It should be noted that the essential trace minerals appear in the first row transition series, except molybdenum, a second row transition element and silicon, fluorine and iodine. Except for silicon and fluorine all have easily polarized valence shells and thus in turn can polarize other reacting moieties. All, except for zinc and fluorine, can exist in several oxidation states and are thus able to act as "electron sinks" in physiological redox reactions.

IRON: Iron has already been discussed under hematinics. In addition to hemoglobin it is the central atom of the cytochromes, involved in the oxidation processes of the cell, homogentisic acid oxidase and pyrocatechase, and 3-hydroxyanthranilic oxidase. Tryptophan pyrrolase utilizes an iron-porphyrin cofactor.

COPPER: This element is widely distributed in foods. Deficiencies are highly unlikely except in infants on an exclusive milk diet. It is a beneficial adjunct in the treatment of nutritional anemias. It functions in the mobilization of stored iron.

Copper occurs in ceruloplasmin, a mammalian serum protein. Human red blood corpuscles include 100 μg. copper per 100 ml. packed cells. Erythrocuprein holds 40–70% of this copper. It has been separated into two fractions clearly distinct from hemoglobin. One of these fractions is associated with the carbonic anhydrase fraction. The element also occurs as the oxygen-carrying respiratory pigment, hemocyanin, of the cephalopods. It occurs in the plant enzymes ascorbic acid oxidase and polyphenol oxidase.

ZINC: This element is present in all naturally growing materials. It is widely distributed in foods. Numerous zinc activated enzymes are known. These include many dehydrogenases. It is an essential component of carbonic anhydrase. Dehydropeptidase, phosphatase, and urease appear to contain the element. It also increases the activity of a number of enzymes as a cofactor, e.g., arginase, histidine deaminase. It is present to a particularly high concentration in the pancreas. It is essential in plant nutrition. Many areas have a soil deficiency of this element.

MAGNESIUM: This element has been discussed previously under fluid and electrolyte replenishers. It is essential to the enzyme system for fatty acid oxidation. It is involved in ATP (adenosine triphosphate) formation. Inorganic phosphatase becomes more heat resistant in the presence of magnesium ion.

COBALT: This element is the central feature of vitamin B_{12}, or cobalamin, which functions as a coenzyme in protein synthesis. It is essential for hematopoiesis. The anemia of chronic infection (etiology unknown) responds favorably to 300 mg. of the element orally per day.

MANGANESE: This element was thought to have a hematopoietic function. It is essential in the enzyme system involved in fatty acid synthesis. It is involved in ATP, polysaccharide, glycoprotein and cholesterol synthesis. This element acts as the activator of several hydroxylases and decarboxylases. It is essential in plant nutrition.

MOLYBDENUM: Low levels of this element markedly inhibit the phosphatases (molybdo-polyphosphate?). It is a constituent of various oxidases, e.g., xanthine oxidase, aldehyde oxidase. There is some evidence that it potentiates iron absorption, especially during pregnancy.

SELENIUM: This element is the essential constituent of the glutathione peroxidase of the red blood cells. Like other trace elements it is toxic in larger quantities. It first appeared in biochemical literature because of its toxicity.

VANADIUM, NICKEL: On the basis of latest evidence these elements are probably essential for mammals but no exact biological function has been established.

IODINE, FLUORINE: These trace elements have been discussed in chapter thirteen.

The better understanding of the importance and roles of trace elements has led to a recognition of the necessity to include these elements in total parenteral nutrition (TPN). They are now classified as micronutrients. The compendia, *U.S.P. XXI,* now includes monographs for Manganese Chloride, $MnCl_2 \cdot 4H_2O$, Manganese Sulfate, $MnSO_4 \cdot H_2O$, Chromic Chloride, $CrCl_3 \cdot 6H_2O$, Ammonium Molybdate, $(NH_4)_6Mo_7O_{24} \cdot 4H_2O$, and Cupric Chloride, $CuCl_2 \cdot 2H_2O$. In addition Cupric Sulfate, $CuSO_4 \cdot 5H_2O$, Zinc Sulfate, $ZnSO_4 \cdot 7H_2O$, and Zinc Chloride, $ZnCl_2$, were already officially cited in earlier compendia in other categories, as was cobalt, as Cyanocobalamin. All compounds named above also have a monograph as injections, e.g., Manganese Chloride Injection.

Considerations in preparation of mineral supplements. The most important way to prevent or overcome mineral deficiencies is to eat wisely those foods generally available. If supplementation is deemed necessary or desirable, ordinarily on sound medical advice, the particular needs are to be considered. Conditions which may require mineral supplementation include pregnancy and inadequate diet. The regimen for control of obesity, especially if over an extended period, should provide for minerals lost due to withdrawal of foods ordinarily supplying such minerals. There is evidence that absorption of the required minerals becomes difficult with age.

In the selection or formulation of a mineral supplement, the question should not be *How many*? but *How much*? Too often the quantities included in proprietary mineral supplements are insignificant. Witness the mineral supplement capsule with 5 mg. potassium. A normal daily diet supplies about 2500 mg. of this element; one third of a lima bean will better supply the 5 mg. Elements for which a human need has not been established are often included. A discussion of these problems in respect to vitamin-mineral supplements is available.[10]

In formulation the amount of each ingredient should be geared to

[10] Roger J. Williams, "Nutritional Insurance with Vitamins," *Chem. and Eng. News,* March 9, 1959, pp. 86–89.

recommended dietary allowances (R.D.A.).[11] For some of the trace minerals no such information is available, or at best the information is an educated guess. It should be kept in mind that many of the substances involved are toxic in too high concentration.

The choice of the actual salts to be used evolves around the stability of the compound in the medium selected, incompatibilities which may arise involving other substances present, and the actual dosage form to be used. If a solution, say, a syrup or elixir, is to be prepared, only soluble salts can be used. Solutions are especially convenient since they may be added dropwise to liquids normally in the diet, e.g., fruit juice. Stabilization may be necessary for some salts to prevent hydrolysis or oxidation. Gluconates, sulfates, and chlorides are usually suitable. On the other hand, a loose powder (a convenient form for mineral supplements for pets) must have only compounds which are not deliquescent, not efflorescent, and otherwise stable to air oxidation, etc. Usually insoluble salts such as the oxides, hydroxides, phosphates, and carbonates serve well for such preparations. However, in order to be available for absorption the insoluble compound must be soluble in the acid medium of the stomach. Capsules and tablets may have either soluble or insoluble compounds. If moisture stability problems exist, the tablet may be coated.

High bulk is frequently a problem in formulation of mineral supplements, especially if certain calcium salts such as the carbonates are used as the source of this element. For this reason the calcium-phosphorus content of supplements is frequently below R.D.A. For the same reason inert fillers are not necessary in these preparations in the powder, capsule, or tablet form.

Because of the practice of combining vitamin and mineral supplements into a single package, possible incompatabilities due to these combinations must be taken into consideration. Many of the vitamins, e.g., ascorbic acid, are catalytically decomposed in the presence of even small quantities of transition elements.

Possible side effects arising from the anions employed as carriers for the cations must be considered for possible physiological effects. Oxide, hydroxide, carbonate, and monohydrogen phosphate have an antacid effect on solution in the stomach. Many organic anions such as citrate, acetate, and lactate have a similar systemic effect on oxidation. Sulfates in large quantities may have a laxative effect.

In rural areas mineral supplementation for animal feeds is important. Generally the same salts are used as for human supplements. The F.D.A.

---

[11] *Recommended Dietary Allowances,* Food and Nutrition Board, Nat'l Res. Council, Ninth Revised Edition, Nat'l Acad. of Sci., Washington, D.C. (1980).

TABLE 14.2 COMPOUNDS USED IN MINERAL SUPPLEMENTS

Element	Soluble Compounds	Insoluble Compounds
Calcium + phosphorus	Ca(glycerophosphate), $Ca(H_2PO_2)_2$	$CaHPO_4 \cdot 2H_2O$
Calcium	$CaCl_2 \cdot 2H_2O$, Ca(lactate)$_2 \cdot 5H_2O$, Ca(gluconate)$_2 \cdot H_2O$, Ca(ascorbate)$_2$	$CaCO_3$, $CaC_4H_4O_4 \cdot 3H_2O$ (succinate)
Magnesium	$MgCl_2 \cdot 6H_2O$, $Mg(OAc)_2 \cdot 4H_2O$ Mg(glycinate)$_2$, Mg(gluconate)$_2$, Mg(ascorbate)$_2$	MgO, $Mg(OH)_2$, $MgCO_3$
Iron(0)	—	Reduced iron
Iron(II)	$FeSO_4 \cdot 7H_2O$ or exsiccated, Fe(gluconate)$_2 \cdot 2H_2O$, Fe(fumarate)	$FeCO_3$
Iron(III)	$FeCl_3 \cdot 6H_2O$, Ferric ammonium citrate	Fe_2O_3, $Fe_4(P_2O_7)_3 \cdot 9H_2O$
Copper	$CuSO_4 \cdot 5H_2O$, Cu(gluconate)$\cdot H_2O$, Cu(citrate), $CuCl_2 \cdot 2H_2O$, $Cu(OAc)_2 \cdot 4H_2O$	CuO, $Cu(OH)_2$, $CuCO_3$
Manganese	$MnSO_4$, Mn(glycerophosphate), Mn(gluconate)$_2$, $MnCl_2 \cdot 4H_2O$, MnNa(citrate), $MnSO_4 \cdot H_2O$	MnO_2, $MnCO_3$, $MnHPO_4 \cdot 3H_2O$
Cobalt	$CoSO_4 \cdot 7H_2O$, $CoCl_2 \cdot 6H_2O$, $Co(OAc)_2 \cdot 4H_2O$, Co(gluconate)$_2$, Vitamin B_{12}	$CoCO_3$, CoO
Zinc	$ZnSO_4 \cdot 7H_2O$ (or $1H_2O$), $ZnCl_2$, $Zn(OAc)_2 \cdot 2H_2O$ Zn(gluconate)$_2$, Zn(lactate)$_2 \cdot 3H_2O$, Zn(glucoheptonate)$_2$	ZnO, $ZnCO_3$
Chromium	$CrCl_3 \cdot 6H_2O$	
Molybdenum	$(NH_4)_6Mo_7O_{24} \cdot 4H_2O$, $Na_2MoO_4 \cdot 2H_2O$	MoO_3
Boron	H_3BO_3, $Na_2B_4O_7 \cdot 10H_2O$	—
Potassium	K_2SO_4, KCl, (KI)	—
Iodine	KI, $Ca(IO_3)_2 \cdot 6H_2O$, KIO_3	Cu_2I_2
Fluoride	NaF —	CaF_2

(Food and Drug Administration) sets the amounts of each mineral which may be used. Cobalt is restricted as a feed supplement for cattle, sheep, and goats, and zinc is restricted to swine and poultry; experiments show these elements of value only in the animals specified.

Table 14.2 charts the compounds used as sources of the various minerals included in supplements. They have been divided into soluble and insoluble categories. When calcium and phosphorus are included, the 1:1 ratio discussed under calcium should be observed.

STUDY AND SELF-APPRAISAL AIDS

SPECIAL TERMS. Define or identify in a single concise sentence if possible.

Achlorhydria, chronic, dimer, exsiccated, hematopoiesis, mineral supplement, paramagnetism, scaled salt, secondary anemia, solid solution, trace element.

NOMENCLATURE OF INTEREST. Identify each of the following.

Catalase, cobalamin, cytochrome, ferric chloride, ferric glycerophosphate, ferritin, ferrum, ferrous gluconate, ferrous sulfate, green vitriol, hematite, iron perchloride, iron protochloride, iron sesquisulfate, iron trichloride.

Basham's Mixture, Blaud's Pills, Chalybeate Pills, Ferric Chloride Tincture, Ferric Citrochloride Tincture, Ferrous Carbonate Mass, Ferrous Carbonate Pills, Ferrous Iodide Syrup, Ferrous Sulfate Elixir, Iron Tincture, Soluble Ferric Phosphate, Vallet's Mass.

SELF-APPRAISAL QUESTIONS

1. The valence electron structure of the transition elements is given as $(n - 1)\, d^y ns^x$. Give the value of x and y for the following elements and indicate the number of unpaired electrons in each subgroup: chromium, cobalt, copper, iron, manganese, and zinc.

2. Which of the elements in question 1 are paramagnetic? Arrange them in order, according to the magnitude of this property. Why is this property of importance to the inorganic and theoretical chemist?

3. Name the trace elements that are not transition elements.

4. Give the general chemical properties of the transition elements. Why are these properties important when acting as trace elements?

5. Given sodium ion as an osmotic regulator, calcium ion as the major cation of the bone, and iron(II) as the active center of hemoglobin. Explain why the individual electronic structures (and resultant properties) of each makes it ideal for the particular application given.

6. Which of the following compounds should exhibit color? CuI, CrS, $ZnCrO_4$, Tl_2SO_4, VCl_3, Mn_3O_4, RbI_3.

7. Cobalt(II) chloride (hexahydrate) has a light red or rose color. Which of the three structural entities of the compound is responsible for the color? What colors of the spectrum are absorbed when white light falls on this compound? Explain.

8. Although cobalt(II) chloride as the hexahydrate has a light red color, the anhydrous compound is blue. Explain. Use an electronic basis if possible.

9. Give the properties of the iron(II) ion which lead to stability problems in its compounds and preparations.

10. Explain why iron(II) complexes are generally less stable than iron(III) complexes.

11. Why should bicarbonate be more suitable for preparation of pure iron(II) carbonate than a normal carbonate?

12. Is exsiccated ferrous sulfate anhydrous? Explain.

13. Why is the heptahydrate of iron(II) sulfate preferred for preparation of aqueous solutions of the compound?

14. Explain on a theoretical basis why iron(II) ion will be absorbed more readily from the gastrointestinal tract than iron(III) ion.

15. Why have iron(II) salts of certain organic acids, e.g., gluconic acid, been introduced as hematinics?

16. Why are iron(II) compounds favored over iron(III) compounds for use as hematinics? Give all aspects—chemical, physiological, etc.

17. Name the most efficient, most economical hematinic. Why is it so often administered orally as an elixir, syrup, or coated tablet? How is the elixir prepared?

18. What is the common structural form of iron(III) compounds used as hematinics? Why is this type of compound required?

19. Why does no precipitate form when ammonia water is added to Soluble Ferric Phosphate, whereas iron(III) hydroxide precipitates when a strong base such as NaOH is added? Explain on a sound theoretical basis.

20. What is the determining factor as to whether a given iron(III) hematinic preparation will give the usual positive test with a thiocyanate?

21. Distinguish between Ferric Ammonium Citrate and Green Ferric Ammonium Citrate.

22. Reduced iron (electrolytic, 325 mesh) is used to fortify certain foods. Give several reasons why this form is preferable to one of the iron(II) or iron(III) salts.

23. Why is particle size important when reduced iron is to be used as a hematinic? Varying amounts of iron oxides are present in this preparation; why are there no problems when reduced iron is used as a hematinic?

24. Why are trace element deficiencies sometimes present in otherwise healthy individuals on an apparently adequate diet? Give some other reasons why trace element deficiencies may develop.

25. What are the usual types of roles of the trace elements? How does this correlate with the electron structure and electrostatic properties?

26. List compounds which would be suitable for a mineral supplement which is to supply the following ions in an unprotected aqueous solution: calcium, iron, manganese, copper, iodide, and phosphate. (Use compounds listed in Table 14.2.)

27. What criteria should be used in the evaluation of a given mineral supplement?

28. Name those compounds and preparations studied in this chapter which are official in the current compendia.

PRACTICE IN EQUATION WRITING

29. A mineral supplement includes potassium iodide and iron(III) chloride (hexahydrate). Write the equation for a reaction which will take place on dissolving this preparation.

30. Write the equation for the reactions taking place if a pharmacist inadvertently uses a metallic closure on a bottle of Ferric Chloride Solution.

31. Write the equation for the antoxidant action of ascorbic acid when included in an iron(II) hematinic.

32. Write equations for all possible equilibria existing in an aqueous solution containing both iron(II) and iron(III) chlorides.

33. A dilute sodium hydroxide solution is added to the solution of question 32, with constant stirring. Write the equation for the initial reaction taking place on this addition.

34. Write the equation(s) for the reaction(s) taking place during the preparation of each of the following: iron(III) pyrophosphate, ferrous gluconate, iron(III) tartrate, Ferric Chloride Solution, Soluble Ferric Phosphate, cobalt(II) carbonate, manganese(II) acetate, iron(III) hydroxide.

35. Write the equation(s) for the reaction(s) used for the following analytical procedures: (*a*) a common qualitative test for iron(II) ion; (*b*) two common qualitative tests for iron(III) ion; (*c*) a qualitative test for cobalt(II) ion; (*d*) a quantitative procedure for the iron of a soluble iron compound; (*e*) a quantitative evaluation of copper(II) in copper(II) sulfate.

FIFTEEN · MEDICINAL AGENTS FOR CONTROL OF pH

When the pH of a body fluid deviates from normal, the condition is usually described in terms of acidity. Hypoacidic and hyperacidic describe the pH drift, higher or lower than normal, respectively. There are two aspects to each condition—gastric (and intestinal) and systemic (and urinary); they may be interrelated.

Gastric hypoacidity is evidenced in the condition known as achlorhydria. Gastric hyperacidity, real and imagined, is widespread if the plethora of gastric antacids is taken as a measure. Like vitamin and mineral supplementation, conditions of hyperacidity are subject to much self-diagnosis and self-medication, unfortunately aided and abetted by suggestive advertising aimed at a sometimes gullible public.

Systemic hypoacidity and hyperacidity are known as alkalosis and acidosis, respectively. The body's defense mechanisms against these conditions include urinary excretion of bicarbonate for the former and dihydrogen phosphate for the latter. As a result the pH of the urine will also drift from normal.

Even though a distinction is not always made during the following discussion, there are two aspects to the use of pH-controlling medicinals. A pharmaceutical of this type may be used to correct an abnormal pH condition, e.g., hydrochloric acid for treatment of achlorhydria, or it may be used to artificially induce a drift from normal pH levels, e.g., the administration of sodium dihydrogen phosphate with methenamine for the treatment of urinary tract infections.

THERAPEUTICS FOR ACHLORHYDRIA

Logically the treatment of choice for this condition should be diluted hydrochloric acid. However, direct oral administration is impossible owing to the highly acidic nature of this acid, and the resultant severe irritation of the mucous membranes and dissolving action on the enamel of the teeth. Administration of hydrochloric acid must be by stomach tube or by gelatin capsules. Although the capsule dosage form should eliminate the irritation and make administration possible at any place, leakage problems make this dosage form quite unsatisfactory. Phosphoric acid, a milder acid, has been used, but the same problems persist.

The most successful mode of administration of acid has been as amino acid hydrochlorides. Glutamic acid hydrochloride is most widely used; glycine hydrochloride is also used. The problems of administration are greatly decreased with the amine hydrochlorides. Being crystalline solids, they do not irritate the mucous membranes or injure the teeth. They can be prepared in safe, transportable dosage forms. The chief disadvantage is the hygroscopic nature of these salts. Larger oral doses are occasionally accompanied by nausea and vomiting. (Sodium glutamate is used to impart a meat flavor to foods.)

On solution in the gastric fluids the amino acid hydrochlorides react with water.

$$\begin{array}{c} H \\ | \\ RCNH_3^+ \\ | \\ COOH \end{array} + H_2O \rightleftharpoons \begin{array}{c} H \\ | \\ RC{-}NH_2 \\ | \\ COOH \end{array} + H_3O^+ \rightleftharpoons \begin{array}{c} H \\ | \\ RC{-}NH_3^+ \\ | \\ COO^- \end{array} + H_3O^+$$

In this way the oxonium ion, essential for gastric digestive action, is regenerated. However, Table 15.1 reveals that the two amino acid salts used in this application are slightly weaker acids than phosphoric acid, and

TABLE 15.1 pK_a VALUES FOR GASTRIC ACIDIFIERS

Acid	Group	pK_a
HCl	—	−7
H_3PO_4	First ionization	1.3
	Second ionization	7.2
$[C_2H_3O_2NH_3^+]Cl^-$	First ionization (ammonium)	2.4
	Second ionization (carboxyl)	9.8
	First ionization (ammonium)	2.1
$[C_5H_7O_4NH_3^+]Cl^-$	Second ionization (first carboxyl)	4.1
	Third ionization (second carboxyl)	9.5

much weaker than hydrochloric acid. Thus correspondingly larger doses would be required to produce the normal pH of the gastric juice, especially when compared with hydrochloric acid.

In practice achievement of the normal acidity of the stomach fluids is not attempted; only enough of the acidifier is administered to provide relief. The usual doses are 1–2 ml. 10% HCl in 25–50 volumes of water, or 0.3 Gm. of glutamic acid hydrochloride (or 5 ml. 6% solution). In order to stimulate secretion of acid by the gastric glands at least 10 ml. of the 10% HCl must be administered. The usual doses for phosphoric acid or glycine hydrochloride are given in Table 15.2.

The normal pH of the gastric juice is 0.9–1.5. In order to calculate the effective pH produced by administration of the usual doses of these substances, certain assumptions must be made. The volume of fluid will be considered to be one liter, and no buffering or salt effects should exist due to the presence of other substances. It is further assumed that the hydrochloric acid is 100% ionized. The results of such calculations are shown in Table 15.2. Glutamic acid has an advantage in this application

TABLE 15.2 TREATMENT OF ACHLORHYDRIA—DOSES AND CALCULATED pH EFFECTS

Acid	Dose Range	Calculated	
		Dose Basis	pH
10% HCl	1–2 ml.	1 ml.	2.6
		2 ml.	2.3
	10 ml.	10 ml.	1.6
10% H_3PO_4	1–4 ml.	1 ml.	2.6
		4 ml.	2.3
$[C_2H_3O_2NH_3^+]Cl^-$	0.2 Gm.	0.2 Gm.	2.6
$[C_5H_7O_4NH_3^+]Cl^-$	0.3 Gm.	0.3 Gm.	2.4

in that the pK_a of the first carboxyl group is sufficiently large to allow a small additional ionization (calculated pH, 2.44 without carboxyl ionization contribution; 2.40 with carboxyl).

GASTRIC ANTACIDS AND ANTACID ACTION

Several new factors with respect to gastric antacids have brought changes within recent years. The approval of cimetidine, and recently ranitidine, introduced a new approach for the treatment of gastric dis-

orders such as ulcers by controlling the release of gastric acid. However, since these drugs are available by prescription only, self-medication with OTC antacids has not been affected.

Over-the-counter drugs such as antacids have a history of non-regulation by the government. Only lack of toxicity had to be assured. Recently the FDA became concerned with effectiveness. The antacids were the first group of products considered.

Regulations have been issued[1] which must be met if an OTC oral antacid is to be recognized as safe and effective. In addition to listing approved antacids, maximum daily doses are included for some. The standards also cover the inclusion of non-antacid laxatives, analgesics, and the antiflatulent simethicone.

Sodium ion content of antacids is also receiving attention because of its implication in water retention and hypertension. Sodium ion may be present as the carrier cation but more often it remains as a by-product of the manufacturing process. It is difficult to remove sodium ion from precipitates and gels. The FDA requires a sodium warning at 5 me. or more per dose. It sets the maximum daily dose as 200 me. for individuals less than 60 years and 100 me. for those 60 or older.

The chemistry of antacid action. The chemistry of antacid action is simple—the reduction of oxonium ion concentration by reaction with a base stronger than water.

$$H_3O^+ + B_1^- = HB_1 + H_2O$$

The extent to which the reaction proceeds is a function of the relative strengths of water as a base and the antacid as a base, but the effect is an increase in pH. The product of antacid action is a weak acid.

There is disagreement on the pH to which the acidity should be reduced; a value of 3.5–4.0 is generally given. There is, however, agreement that neutralization carrying much beyond this pH range is undesirable, both for gastric digestion and because of the possibility of *acid rebound.* As the stomach contents become less acid, the gastric glands tend to increase hydrochloric acid secretion in an effort to restore a more normal pH. If this occurs, the desired effect of the gastric antacid is counteracted. It is believed that the greater the pH shift resulting from antacid action, the greater the chances of acid rebound. The appearance and extent of this phenomenon are dependent on the nature of the antacid and its reaction products, and on the individual variation of the patient.[2]

---

[1]Federal Register, *39,* No. 108, 19862-83, June 4, 1974.

[2]J. Pereiva-Lima and F. Hollander claim that a review of experimental and clinical reports together with the hypothesis offered show no agreement and give no conclusive evidence for the existence of "acid rebound." *Gastroenterology,* 37, 145 (1959).

If ingested in excess, certain antacids have a continuing action; pH shifts far beyond the desirable range result. Sodium bicarbonate is a classical example of an antacid having such action. The antacid reaction may be written

$$H_3O^+ + HCO_3^- = H_2CO_3 + H_2O$$

followed by

$$H_2CO_3 = H_2O + CO_2\uparrow\text{(burps!)}$$

Being soluble, any excess unneutralized bicarbonate undergoes further reaction.

$$HCO_3^- + H_2O = H_2CO_3 + OH^-$$

In pure water this latter reaction proceeds until a pH of 8.0 or even slightly higher is achieved. In the presence of other materials, as in the stomach fluids, it is unlikely to proceed to this extent except possibly on the ingestion of a very large excess of bicarbonate. Rapid neutralizations to give pH values around 7 are common with this antacid.

Other antacids, usually insoluble, do not exhibit this continuing neutralization. Aluminum hydroxide is generally considered an ideal example; it is soluble only in acidic (or very strongly basic) media. Neutralization proceeds in a series of steps, depending on the amount of antacid ingested and the pH.

$$[Al(H_2O)_3(OH)_3]^0 + H_3O^+ = [Al(H_2O)_4(OH)_2]^+ + H_2O$$
$$[Al(H_2O)_4(OH)_2]^+ + H_3O^+ = [Al(H_2O)_5(OH)]^{2+} + H_2O$$
$$[Al(H_2O)_5(OH)]^{2+} + H_3O^+ = [Al(H_2O)_6]^{3+} + H_2O$$

As the acidity of the gastric juice decreases on consumption of oxonium ions, the rate of dissolution and reaction of the base with oxonium ion also decreases. Neutralization proceeds at a decreasing rate until a pH is reached at which the antacid is insoluble, or its solution is so slow that for practical purposes reaction has stopped.

As the reaction proceeds, a back reaction becomes increasingly prominent. The aquoaluminum cations are weak acids; e.g., hexaquaaluminum cation has a pK_a of 4.9. Their ionization has an acidifying effect, especially after active dissolution of the base has stopped. This interplay of reactions establishes an equilibrium pH of about 4.0, irrespective of original dose. The undissolved excess aluminum hydroxide has virtually no effect on this equilibrium pH..

Systematic vs. nonsystemic antacid action. If an antacid's effects are limited to the gastrointestinal tract, it is called a *nonsystemic antacid.* Compounds having such limited action are generally insoluble substances: the carbonates, phosphates, hydroxides, oxides, or silicates of cations such as aluminum, magnesium, calcium, and bismuth(III). However, excessive use of some of these compounds may result in systemic effects.

An antacid which affects the pH not only of the gastrointestinal tract but also that of all body fluids is called a *systemic antacid.* These antacids act by increasing the normal bicarbonate levels of the blood and other body fluids. The action is through the change of the [base]/[acid] ratio of the carbonate buffer.

$$\mathrm{pH} = 6.4 + \log \frac{[\mathrm{HCO_3}^-]}{[\mathrm{H_2CO_3}]}$$

A general systemic alkalosis results depending upon the size and frequency of the antacid dose. The normal bicarbonate level is slowly restored by urinary excretion. As a result, systemic antacids are also urinary alkalizers.

$$\mathrm{HCO_3}^- + \mathrm{H_2O} = \mathrm{H_2CO_3} + \mathrm{OH}^-$$

The reaction is not extensive but is sufficient to cause a definite alkaline shift in the urine.

All true systemic antacids are water soluble; they are the bicarbonate, acetate, or citrate of sodium or potassium. The organic anions have a delayed systemic action. Initially they exert a gastric antacid action by the usual reaction

$$\mathrm{H_3O^+} + B_1^- = HB_1 + \mathrm{H_2O}$$

Since the intestinal mucosa is permeable to both the neutralized and any excess unneutralized antacid, both forms find their way into the blood stream, probably because of favorable concentration gradients. Acetic and citric acids and their ions, being constituents of the normal carbohydrate oxidation mechanism, are oxidized to an equivalent quantity of the bicarbonate. Thus, as with direct ingestion, the blood levels of the bicarbonate ion increase, and alkalosis results.

In order that the difference in the physiological effects of systemic and nonsystemic antacids be better understood, a series of schematic metabolic balance diagrams are shown in Figs. 15.1, 15.2, and 15.3. To simplify the mathematics of presentation certain assumptions are necessary. It is assumed that the carbon dioxide level of the blood remains relatively constant. If this is true, all pH changes may be referred to the bicarbonate member of the carbonic acid-bicarbonate buffer system. The normal bicarbonate level is assumed to be 20 me., and the level of sodium ion and chloride ion are each assigned a value of 100 me. Figure 15.1 shows the acid-base balance existing during normal digestive action. Figures 15.2 and 15.3 show the manner in which this normal pattern is changed by ingestion of a 10 me. dose of a nonsystemic and systemic antacid, respec-

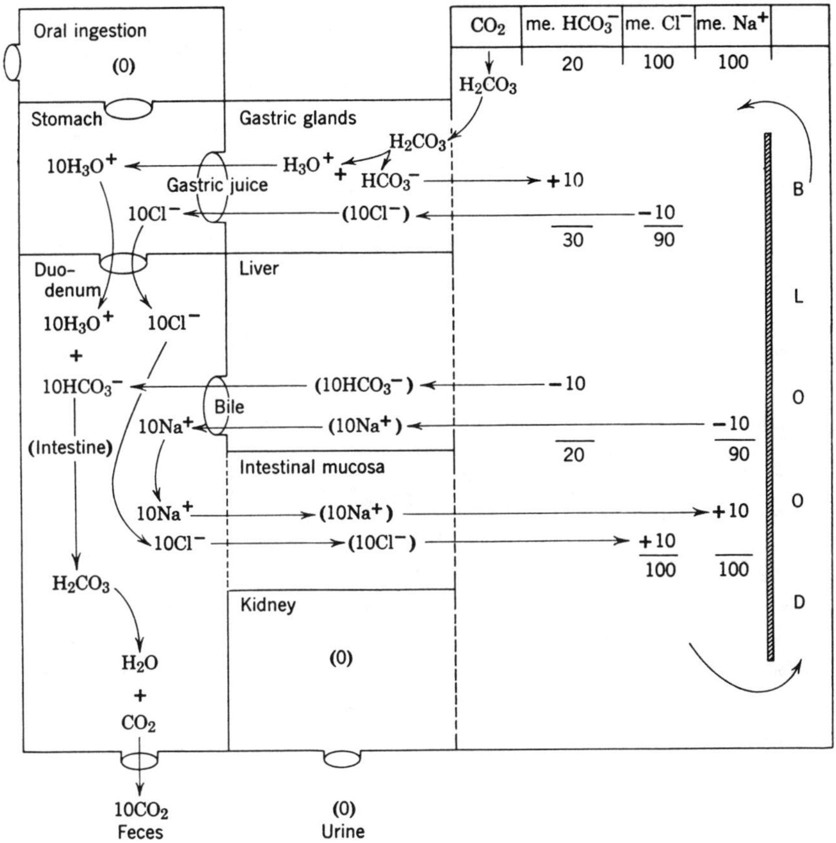

FIGURE 15.1 Normal gastrointestinal systemic acid-base relationships. The numbers appearing before a given ion or compound represent the number of milliequivalents of that substance present.

tively. Magnesium hydroxide and sodium bicarbonate were arbitrarily chosen as representative of their respective antacid type.

Comparison of the three figures shows that with a nonsystemic antacid the reaction product is either water, which has no effect on the bodys' pH, or an insoluble substance excreted with the feces. Systemic antacids always result in a temporary alkalosis because of the equivalent resorption of the bicarbonate normally excreted in the bile and neutralized by hydronium ion from the stomach. This resorbed bicarbonate must be re-eliminated via the slower urinary route. Since a considerable time lapse occurs between ingestion of a systemic antacid and final excretion of the excess bicarbonate in the urine, a continuing alkalosis results in the body fluids and in the urine.

Compounds suitable as antacids. The compounds in common use as gastric antacids are listed in Table 15.3. Soluble and insoluble compounds are clearly indicated in order to facilitate later discussion of systemic alkalosis and the urinary alkalizers. A few of the infrequently used antacids, e.g., bismuth subgallate, do not appear in this table, but will be discussed briefly later.

Although the anion's ability to act as a base is mainly responsible for antacid activity of a compound, the cation cannot be ignored. In effect it controls or moderates the antacid activity of the anion, usually through

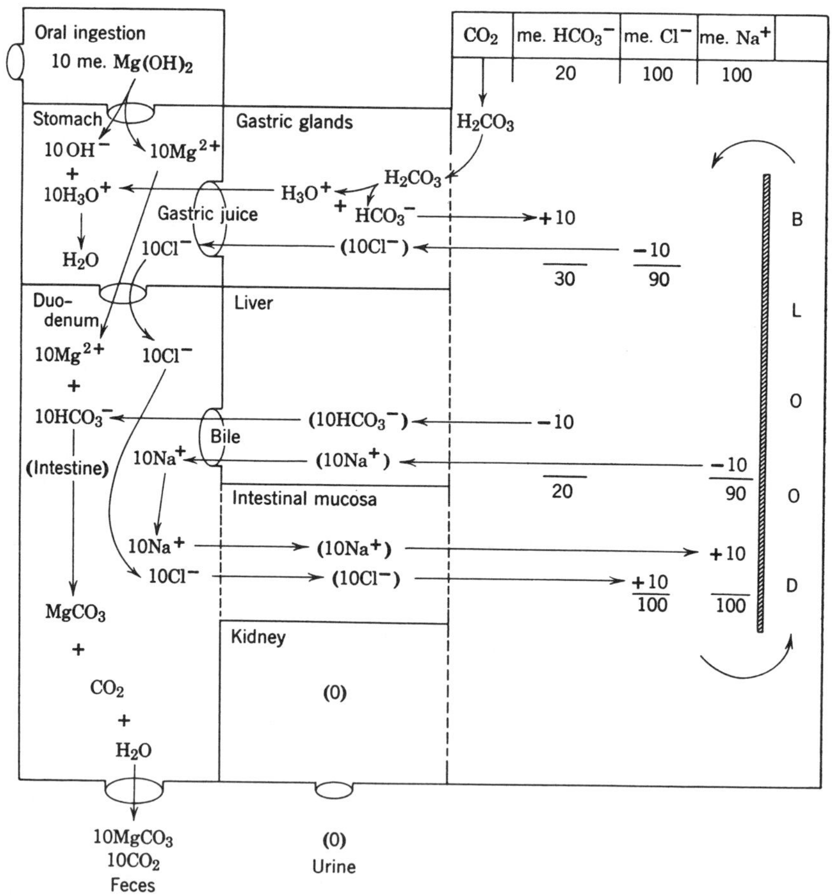

FIGURE 15.2 Effect of nonsystemic antacids on body acid-base balance. The numbers appearing before a given ion or compound represent the number of milliequivalents of that substance present.

control of the solubility of the compound. For example, ostensibly the hydroxyl ion should have excellent antacid properties since it is a strong base. However, in practice its usefulness is limited to a few specific compounds whose solubility is controlled by proper selection of the cation. Sodium hydroxide is too soluble to be a useful antacid; very high hydroxyl ion concentrations of its solutions render it excessively corrosive. However, both magnesium and aluminum hydroxides are useful antacids. The K_{sp} of each of these compounds is such that the free hydroxyl ion concentrations allowed are too low for appearance of the corrosive properties ordinarily associated with hydroxyl ion. The compounds are sufficiently soluble, however, to act as excellent antacids. It is evident

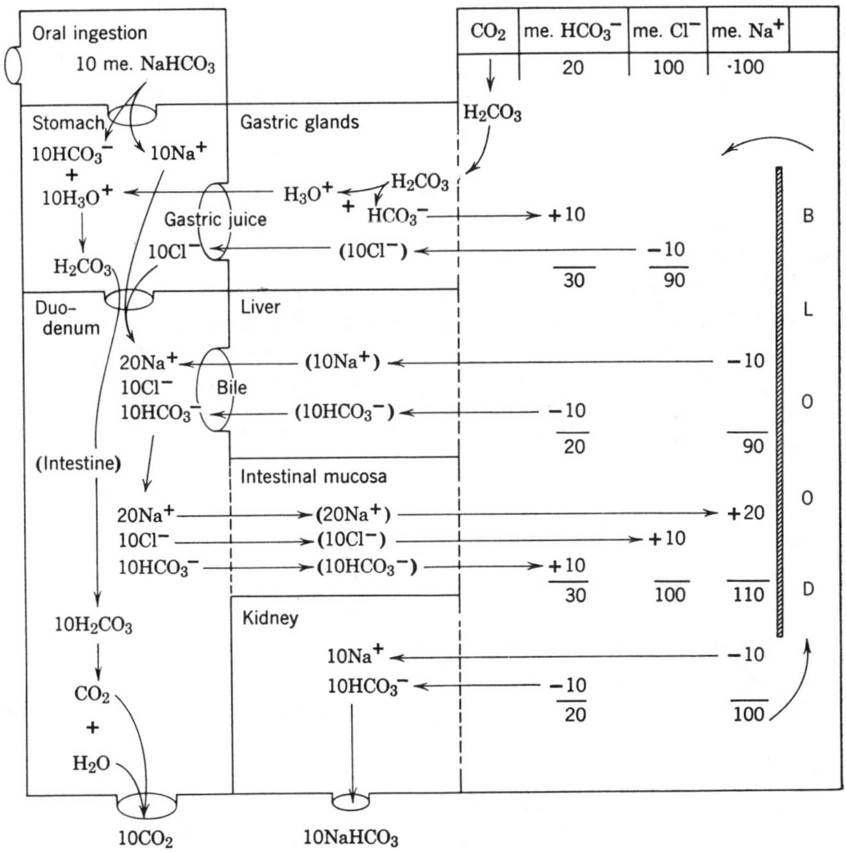

FIGURE 15.3 Effect of systemic antacids on body acid-base balance. The numbers appearing before a given ion or compound represent the number of milliequivalents of that substance present.

from this discussion that consideration of a given antacid has two aspects: properties due to its anion and properties due to its cation.

In the discussions of antacid anions and cations which follow, the compounds named after a given anion or cation are FDA approved unless enclosed in parentheses or otherwise indicated. In several instances variations of a given compound have been approved, e.g., aluminum hydroxide (or as aluminum hydroxide-hexitol stabilized polymer, aluminum hydroxide-magnesium carbonate codried gel, aluminum hydroxide-magnesium trisilicate codried gel, aluminum hydroxide sucrose power hydrate). Such variations are not discussed separately; the chemistry, etc. involved is essentially that of the pure compounds(s) present. The approved antacids and simethicone are also cited in *U.S.P. XXI.*

Anions as bases for antacids: THE ACETATES: pK_b—9.28; $NaOAc{\cdot}3H_2O$, KOAc. Being a strong base, the acetate ion is a good antacid.

$$OAc^- + H_3O^+ = HOAc + H_2O$$

Little back ionization results since the acetic acid formed is a weak acid. Acetate ion and acetic acid are absorbed readily from the intestinal tract, with no metabolic problems occurring. Both are oxidized to an equivalent quantity of bicarbonate in the normal carbohydrate oxidation system. Because of its absorption and conversion to bicarbonate, acetate always acts as a systemic antacid. For this reason it is rarely used as a gastric antacid; rather it is administered when systemic alkalosis is required or a systemic acidosis is to be counteracted. Both salts, particularly the potassium, have a diuretic effect.

THE CITRATES: pK_b values—Ci^{3-}, 8.70; HCi^{2-}, 9.26; H_2Ci^-, 10.94; $Na_3C_6H_5O_7{\cdot}2H_2O$, $K_3C_6H_5O_7{\cdot}H_2O$. Being a somewhat stronger base than acetate, citrate (Ci) ion is an effective antacid. Like acetate, it is absorbed from the intestinal tract; in ordinary quantities it is oxidized to equivalent bicarbonate. In larger doses citrate has a secondary or "salt" action, which increases the flow of various secretions, e.g., urine. Because of systemic involvement, citrate, like acetate, is seldom employed as a gastric antacid. When used for its gastric antacid action, it is commonly present to protect a simultaneously administered medicinal, especially one which may be decomposed by the acid medium of the stomach, e.g., penicillin. Palatability is superior to that of acetate. The maximum daily dose limit for citrates, as citric acid or a salt, is given by the FDA as 8 grams.

THE BICARBONATES: pK_b—HCO_3^-, 7.63; $NaHCO_3$, $KHCO_3$. Bicarbonate ion is an effective antacid and is widely used in the home and in proprietary preparations.

TABLE 15.3 GASTRIC ANTACIDS AND THEIR USUAL DOSES[a]

	Anions → / Cations ↓	Weak Bases			Strong Bases				
		OAc^-	$[C_6H_5O_7]^{3-}$	HCO_3^-	CO_3^{2-}	O^{2-}	OH^-	PO_4^{3-}	Misc.
Soluble	Na^+[b]	1.5 Gm.	1.0 Gm.	2.0 Gm.	—	—	—	—	—
	K^+	1.0 Gm.	1.0 Gm.	1.0 Gm.	—	—	—	—	—
Insoluble	Mg^{2+}	—	—	—	0.6 Gm.	0.25 Gm.	0.3 Gm. 5.0 ml.	1.0 Gm.	1–4 Gm.[c]
	Ca^{2+}	—	—	—	1–2 Gm.	—	—	1.0 Gm.	—
	Al^{3+}	—	—	—	8.0 ml.[d]	—	0.3–0.5 Gm. 8.0 ml.	15.0 ml.	0.5 Gm.[e]
	BiO^+	—	—	—	1.0 Gm.	—	—	—	5.0 ml.[f]

[a] If a magma is available, the dose is given in ml.
[b] Injections often include sodium lactate for its systemic alkalizing action (page 357).
[c] Magnesium trisilicate.
[d] Essentially $[Al(OH)]CO_3$.
[e] $[Al(OH)_2]$ glycinate.
[f] Bismuth magma, a mixture of the hydroxide and carbonate.

$$HCO_3^- + H_3O^+ = H_2CO_3 + H_2O$$
$$H_2CO_3 = H_2O + CO_2\uparrow$$

It has two disadvantages. First, its systemic action may be undesirable, especially on repeated self medication. Second, the evolution of carbon dioxide can result in discomfort because of undue distention of the stomach and intestine. The gas pressure is especially dangerous to patients with ulcerations and may lead to perforations. The FDA gives the maximum daily dose as 200 me. bicarbonate (or carbonate) for persons up to 60 years and 100 me. for those 60 years and older.

Bicarbonate should never be used as an antidote for accidental ingestion of strong acids. The carbon dioxide evolution can cause perforation of the stomach, and as a result lead to peritonitis.

THE CARBONATES: pK_b—CO_3^{2-}, 3.75; $CaCO_3$, $MgCO_3$, $Al(OH)CO_3$, $(BiO)_2CO_3$. Carbonate ion is a very strong base—too strong for use as an antacid if in soluble form. For this reason its sodium and potassium salts are never used as antacids. By the use of insoluble carbonates the concentration of available dissolved carbonate is controlled by the K_{sp} of the compound. In all compounds used this is so low that solution can take place only as long as oxonium ion is present to consume the small concentration of carbonate ion allowed by the K_{sp}.

$$CaCO_3(\text{solid}) \rightleftharpoons Ca^{2+} + CO_3^{2-}(\text{dissolved})$$
$$CO_3^{2+} + H_3O^+ = HCO_3^- + H_2O$$
$$HCO_3^- + H_3O^+ = H_2CO_3 + H_2O$$
$$H_2CO_3 = H_2O + CO_2\uparrow$$

Upon consumption of the gastric oxonium ion to levels allowed by the K_b and K_{sp}, reaction stops. For this reason insoluble carbonate antacids do not impart an excess alkalinity to the intestinal contents.

As with the bicarbonates, the evolution of carbon dioxide gas is disadvantageous. Unlike the bicarbonates, carbonates are nonsystemic antacids if used in moderation.

THE HYDROXIDES AND OXIDES: pK_b—(−2); $Al(OH)_3$, $Mg(OH)_2$, MgO, magnesium trisilicate, bismuthyl compounds, dihydroxyaluminum glycinate. For reasons already discussed the soluble hydroxides and oxides are never used as antacids.

The excellent antacid properties of these two bases depend upon the K_{sp}-controlled release of hydroxyl ion from an insoluble compound. Solution proceeds as long as oxonium ion is available to react with dissolved hydroxyl ion.

$$Mg(OH)_2(\text{solid}) \rightleftharpoons Mg(OH)_2(\text{dissolved})$$

$$Mg(OH)_2 = Mg(OH)^+ + OH^-$$
$$Mg(OH)^+ = Mg^{2+} + OH^-$$
$$OH^- + H_3O^+ = H_2O$$

or oxide

$$MgO + H_3O^+ = Mg(OH)^+ + H_2O$$

Water as the reaction product imposes no problems. Since reaction stops when the equilibrium hydroxyl ion concentration allowed by the K_{sp} and K_w is reached, the antacid action of this group of compounds is self-limiting. The dissolved cations resulting from antacid action react with bicarbonate present in the bile to form insoluble carbonates in the alkaline medium of the intestine. They are eliminated in the feces. Used in moderation, this group of antacids show no systemic affects.

These antacids should be protected from the carbon dioxide of the atmosphere. If exposed in a humid atmosphere or in magma form, they react to form basic carbonates.

Tribasic magnesium silicate belongs to the hydroxide-oxide group in its antacid action. It is essentially a mixture of hydrated magnesium oxides and silicic acid rather than an actual compound, even though its stoichiometric composition approaches that of a disilicate (page 271). The antacid action is due to the insoluble magnesium hydroxide and hydrated magnesium oxide present. The silicic acid, having its full quota of protons, cannot participate in the antacid activity except possibly to a very limited extent by adsorption of oxonium ions.

Bismuth compounds, bismuth carbonate, bismuth subcarbonate (bismuth cream), bismuth subgallate, bismuth subnitrate, represent a special case of the oxo and hydroxo groups as antacids: in addition to anion action (carbonate and hydroxide) the cation can contribute a secondary action. This may be represented by

$$BiO^+ + H_3O^+ = [Bi(OH)]^{2+} + H_2O$$

These compounds are nonsystemic.

THE PHOSPHATES: pK_b—PO_4^{3-}, 1.7; $AlPO_4$, $Ca_3(PO_4)_2$, ($Mg_3(PO_4)_2 \cdot 5H_2O$). Tertiary phosphate ion is a powerful base. If present in a soluble form, each ion is the equivalent of a hydroxyl ion.

$$PO_4^{3-} + H_2O = OH^- + HPO_4^{2-}$$

Because of this, as with the carbonates, hydroxides, and oxides, only insoluble tertiary phosphates are suitable as antacids. Here again the K_{sp} of the insoluble phosphate determines the extent to which the reaction may proceed:

$$Mg_3(PO_4)_2(\text{solid}) \rightleftharpoons Mg_3(PO_4)_2(\text{dissolved})$$

$$Mg_3(PO_4)_2(\text{dissolved}) + 18H_2O = 3[Mg(H_2O)_6]^{2+} + 2PO_4^{3-}$$

$$PO_4^{3-} + H_3O^+ = HPO_4^{2-} + H_2O$$

or

$$PO_4^{3-} + H_2O = HPO_4^{2-} + OH^-$$

and

$$H_3O^+ + OH^- = 2H_2O$$

Since the cations are eliminated in the feces as tertiary (or secondary) phosphates, these compounds have no systemic effects unless used excessively.

The FDA has maximum daily dose limits for the two approved tertiary phosphate antacids: aluminum phosphate, 8 grams, calcium phosphate, 24 grams. The monobasic and dibasic calcium phosphates are given maximum daily dose limits of 2 grams.

Cation involvement in antacid therapy. SODIUM AND POTASSIUM COMPOUNDS: $NaHCO_3$, $KHCO_3$. Only the bicarbonates of these cations are commonly used as gastric antacids. The normal carbonates are too **alkaline for this use. The acetates and citrates are ordinarily not employed as gastric antacids, but rather for their systemic alkalizing action. The maximum daily dose for sodium is given on page 406. If a product contains more than 25 me. potassium in the maximum daily dose the label must provide a warning for those with kidney disease.**

Since the bicarbonates of these cations are most commonly used as the dry powder or in tablet or capsule form, no incompatibility problems will ordinarily appear. Both are heat labile; decomposition begins at 50°C. and is complete at 100°C.

$$2HCO_3^- = H_2O + CO_2\uparrow + CO_3^{2-}$$

In solution some incompatibilities may result. Although potassium bicarbonate is quite soluble (in 2.8 parts of water at room temperature), sodium bicarbonate (baking soda) is one of the less soluble sodium salts (10 parts water at room temperature). For this reason highly concentrated solutions of the sodium bicarbonate cannot be prepared. In fact if other more soluble sodium salts are present, the common ion effect results in a decreased solubility of bicarbonate and resultant precipitation. The following prescription demonstrates this incompatibility:

Rx

Sod. bicarb.	10.0 Gm.

Sod. chloride	3.5 Gm.
Spirit $CHCl_3$	2.5 cc.
Anise water qs	120 cc.

Acid-base incompatibilities may result when sodium or potassium bicarbonate, acetate, or citrate are used in solution. Since the anion must be a strong base in order to act as an antacid, reaction with the weakly acidic water molecule will impart a basic reaction to the solution. Bicarbonate, acetate, and citrate solutions have a pH of 8 or slightly higher. This means that most alkaloid salts and other amine salts will separate as the free base. Many inorganic cations, e.g., iron(III), will also precipitate.

ALUMINUM COMPOUNDS: Aluminum hydroxide, basic aluminum carbonate, dihydroxyaluminum aminoacetate, and aluminum phosphate. The over-all mode of action of a given aluminum-based antacid is a function of the anion(s) present. With the hydroxide the antacid action is associated with the six-coordinate octahedral aluminum complex. Solution of aluminum hydroxide is limited by its K_{sp}.[3] The mechanism of solution may be written

$$\text{(Aluminum hydroxide gel)} + 3H_2O \rightleftharpoons [Al(H_2O)_3(OH)_3]^0$$

Aluminum hydroxide continues to dissolve since in an acid medium this equilibrium cannot be established. As the molecules separate from the solid and go into solution, they are converted into an aluminum cation by reaction with the hydrochloric acid present in the gastric juice.

$$[Al(H_2O)_3(OH)_3]^0 + H_3O^+ = [Al(H_2O)_4(OH)_2]^+ + H_2O \qquad (1)$$

This is followed by the reaction

$$[Al(H_2O)_4(OH)_2]^+ + H_3O^+ = [Al(H_2O)_5(OH)]^{2+} + H_2O \qquad (2)$$

Still further neutralization may take place:

$$[Al(H_2O)_5(OH)]^{2+} + H_3O^+ = [Al(H_2O)_6]^{3+} + H_2O \qquad (3)$$

With a pK_a of 4.85 for the hexaquaaluminum cation, it is unlikely that reaction 3 proceeds to completion. Assuming antacid action ceases at or slightly before a pH of 4 is reached, the approximate ratio of the hexaqua and pentaqua forms may be determined by assuming the existence of a buffer involving this conjugate pair (which indeed it is).

[3] The K_{sp} at body temperature is not available. However, it must be somewhat larger than that at 25°C., 3.7×10^{-15}.

$$4.0 = 4.85 + \log \frac{\{[Al(H_2O)_5(OH)]^{2+}\}}{\{[Al(H_2O)_6]^{3+}\}}$$

$$\frac{[Al(H_2O)_5(OH)]^{2+}}{[Al(H_2O)_6]^{3+}} = 0.14$$

On the basis of this calculation it is possible to say that at final equilibrium 10–15% of the complex aluminum cations are present in the pentaquo form. At this point solution of the solid also ceases. Since the dissolved aluminum hydroxide molecules are no longer neutralized as they separate from the suspended solid, their concentration cannot exceed that permitted by the K_{sp} of this compound. The equilibrium concentration is so low that they have no effect upon the final pH.

Reactions 2 and 3 are responsible for at least half of the antacid action of the basic aluminum carbonate and the dihydroxyaluminum aminoacetate. Reaction 1 is not involved in antacid action by these two compounds; it took place prematurely during their preparation. The anions present, carbonate and aminoacetate, respectively, provide additional neutralizing capacity. Both carbonic acid and the bicarbonate ion are weaker acids than the hexaquaaluminum ion (pK_a—6.4 and 10.2 compared to 4.85). Thus at equilibrium the reaction

$$CO_3^{2-} + H_3O^+ = HCO_3^- + H_2O \qquad (4)$$

and

$$HCO_3^- + H_3O^+ = H_2CO_3 + H_2O \qquad (5)$$

will have gone to practical completion before reaction 3 can take place to any appreciable extent. Similarly the carboxyl moiety of the aminoacetate ion is a strong base; the pK_a of its conjugate acid is 9.78. This anion too will be protonated before reaction 3 can take place.

$$NH_2CH_2COO^- + H_3O^+ = NH_2CH_2COOH + H_2O \qquad (6)$$

Theoretically the reaction

$$NH_2CH_2COOH + H_3O^+ = [NH_3CH_2COOH]^+ + H_2O \qquad (7)$$

may also take place. However, the pK_a of the resultant substituted ammonium ion is 2.4. Consequently it is a much stronger acid than the hexaquaaluminum ion. It can exist in appreciable concentrations only at pH levels much lower than achieved by use of this antacid.

Aluminum phosphate alone does not involve the six-coordinate aluminum ion in its antacid action. Its antacid activity is due strictly to the tertiary phosphate ion. The pK_a of its conjugate acid, 12.3, and that of the dihydrogen phosphate ion, 7.2, are the driving force involved. Being a 3-3 type of electrostatic compound, aluminum phosphate is an extremely

insoluble substance (page 328). It is insoluble in acetic acid solutions, but is soluble in strong mineral acids. To the small extent allowed by its K_{sp}, it will dissolve to give hexaquaaluminum ions,

$$Al^{3+} + 6H_2O = [Al(H_2O)_6]^{3+}$$

and tertiary phosphate ions. This anion is an extremely strong base; the reaction

$$PO_4^{3-} + H_3O^+ = HPO_4^{2-} + H_2O$$

proceeds to completion. In so doing, the solubility equilibrium is disturbed and more of the insoluble antacid compound dissolves. In addition the monohydrogen phosphate ion may react.

$$HPO_4^{2-} + H_3O^+ = H_2PO_4^- + H_2O$$

This cycle of reactions continues until the pH of the gastric contents is increased to levels at which aluminum phosphate is no longer sufficiently soluble to provide the ions necessary for the antacid action.

The dissolved six-coordinate aluminum ions which are the reaction products of all aluminum-based antacids impose certain characteristics upon their behavior. Since these cations are very poorly absorbed, the aluminum-based compounds have little or no systemic antacid action. However, these cations do impart an astringent and antiseptic action to the intestinal fluids. It is likely that this astringent effect interferes with water excretion from the blood vessels in the mucosa of the large intestine. Since this water is necessary to maintain a soft stool, aluminum antacids have a constipating action.[4] If aluminum-based antacids are used in excess and over long periods, this constipation can be quite obstinate.

Aluminum antacids (except $AlPO_4$) interfere with phosphate absorption from foods.

$$[Al(H_2O)_5(OH)]^{2+} + HPO_4^{2-} = AlPO_4 + 6H_2O$$

The alkaline medium of the small intestine favors the precipitation of this insoluble salt. However, because of the abundance of phosphate in foods, the effect will become serious only on continuing use of aluminum antacids. It may be overcome by insuring the presence of at least 750 mg. phosphorus in the diet daily.

Insoluble alumina gels, particularly the hydroxide, are good adsorption agents. If used continuously over long periods, deficiencies of trace food substances may develop. Concomitant medication with, say, antibiotics,

[4] Soluble aluminum compounds, e.g., aluminum chloride, have been administered to provide a constipating action in the control of severe diarrhea.

may prove valueless owing to adsorption on the gel. It has been found that aluminum antacids interfere with tetracycline absorption. The FDA requires an explanatory warning to this effect.

Aluminum hydroxide, available as tablets, magma, and exsiccated powder, is one of the best nonsystemic antacids. It is used widely, particularly for treatment of peptic ulcer. Since neutralizing action is controlled by its insolubility, there is no acid rebound. Because of the presence of the unreacted solid phase, its neutralizing action is available over a long period of time. It has an inhibiting action on pepsin even under acid conditions. The gelatinous quality of the unreacted solid gives the preparation some protective action.

Aluminum hydroxide is frequently administered as an adjunct with medications. It has the effect of relieving the distress frequently attending administration of salicylates, corticosteroids, and aminophylline. Simultaneous administration of aluminum hydroxide is also used to control absorption rates of drugs; e.g., the absorption of atropine sulfate is slowed to give a longer lasting action, and triple sulfonamides may be maintained in the blood at higher concentrations over longer periods.

Aluminum hydroxide is probably the most satisfactory of the aluminum antacids. For most applications the phosphate, carbonate, and dihydroxyaminoacetate offer no particular advantages. The neutralizing power of the phosphate is about one half that of the same quantity of aluminum hydroxide. The phosphate gives as good results with peptic ulcer, if in sufficient doses. As pointed out previously, it cannot interfere with phosphate absorption even if large antacid doses continuing over a long period are required. Aluminum subcarbonate, like the hydroxide, may interfere with phosphate absorption. It is claimed to bind up to one third more phosphate. For this reason it is recommended as an adjunct in management of the low phosphate diets frequently prescribed for patients susceptible to urinary phosphatic calculi. Examination of the possible formulas fails to show any sound stoichiometric basis for the claim of greater phosphate precipitating action.

Dihydroxyaluminum aminoacetate appears to have an antacid action similar to that of aluminum hydroxide. It has a lower percentage of aluminum than the hydroxide, but this is offset by the larger doses used; the usual doses of the two preparations are almost identical on the basis of aluminum equivalents present.

The preparation of the aluminum antacids involves some interesting applications of chemical theory and practice. Aluminum hydroxide is prepared by adding a hot aqueous solution of potassium alum to a hot solution of sodium carbonate, with good stirring. The order of mixing is important; if reversed, sulfate ion is included in the precipitate as an alkali metal salt which is difficult to remove by washing techniques. The essential reactions taking place are

$$3CO_3^{2-} + 3H_2O = 3HCO_3^- + 3OH^-$$
$$[Al(H_2O)_6]^{3+} + 3OH^- = [Al(H_2O)_3(OH)_3]\downarrow + 3H_2O$$
$$(2HCO_3^- = CO_3^{2-} + H_2O + CO_2\uparrow)$$

Beyond maintaining the necessary electrical neutrality in the mother liquor, the potassium, sodium, and sulfate ions play no part in the reactions. Upon completion of the reaction they must be removed by washing with hot water. Boiling water cannot be used as a wash; the precipitated hydroxide redissolves, possibly due to peptization of the insoluble particles.

Precipitation of the basic carbonate requires a careful control of pH since it is more soluble than the hydroxide at higher pH levels. A lower pH is assured by adding sufficient sodium bicarbonate to the sodium carbonate solution to repress the degree of reaction with water.

$$CO_3^{2-} + H_2O \rightleftharpoons HCO_3^- + OH^-$$

The same conditions may be obtained by saturating the carbonate solution with carbon dioxide. As a solution of soluble aluminum salt is added to this base solution, the hexaquoaluminum cation is partially neutralized.

$$[Al(H_2O)_6]^{3+} + OH^- = [Al(H_2O)_5(OH)]^{2+} + H_2O$$

and

$$[Al(H_2O)_5(OH)]^{2+} + OH^- = [Al(H_2O)_4(OH)_2]^+ + H_2O$$

Being reacted under conditions maintaining mildly alkaline pH levels, the basic carbonates will precipitate before formation of the trihydroxoaluminum compound.

$$[Al(H_2O)_5(OH)]^{2+} + CO_3^{2-} = [Al(H_2O)_5(OH)]CO_3\downarrow$$
$$2[Al(H_2O)_4(OH)_2]^+ + CO_3^{2-} = [Al(H_2O)_4(OH)_2]_2CO_3\downarrow$$

It is probable that the latter compound should predominate, although lower temperatures and lower precipitating pH's would favor the former. Basic carbonate is unstable; it loses carbon dioxide to the atmosphere if stored in improperly sealed containers.

Aluminum phosphate is prepared by adding a solution of sodium monohydrogen phosphate to a solution of aluminum chloride.

$$[Al(H_2O)_6]^{3+} + HPO_4^{2-} = AlPO_4\downarrow + H_3O^+ + 5H_2O$$

The reaction is driven to completion by the extreme insolubility of the 3-3 type of ionic compound. The phosphate ion allowed by the equilibrium,

$$AlPO_4 + 6H_2O \rightleftharpoons [Al(H_2O)_6]^{3+} + PO_4^{3-}$$

is less than that allowed by the equilibrium,

$$HPO_4^{2-} + H_2O \rightleftharpoons H_3O^+ + PO_4^{3-}$$

Upon completion of the reaction the pH is adjusted to 6.8–7.4 with dilute ammonia solution. The soluble by-product ions are removed by washing.

The aminoacetate is prepared by homogeneous precipitation involving a nonaqueous medium. A solution of aluminum isopropoxide in isopropyl alcohol is added to an aqueous solution of glycine with good mixing. A series of concurrent reactions take place. Their nature may be represented as follows (iPrOH = isopropyl alcohol):

$$NH_2CH_2COOH + H_2O = NH_2CH_2COO^- + H_3O^+$$

$$Al(iPrO)_3 + 6H_2O = [Al(H_2O)_3(OH)_3] + 3iPrOH$$

$$[Al(H_2O)_3(OH)_3] + H_3O^+ = [Al(H_2O)_4(OH)_2]^+ + H_2O$$

$$[Al(H_2O)_4(OH)_2]^+ + NH_2CH_2COO^- = [Al(H_2O)_4(OH)_2^+][NH_2CH_2COO^-]\downarrow$$

The insoluble compound is washed with water to remove the excess glycine and the isopropyl alcohol.

MAGNESIUM COMPOUNDS: MgO, $Mg(OH)_2$, $MgCO_3$, $(Mg_3(PO_4)_2)$, $Mg(NH_2CH_2COO)_2$ and magnesium trisilicate. Magnesium hydroxide is probably the most popular. It is available as magnesium magma (milk of magnesia, cream of magnesia), tablets, etc. Like aluminum hydroxide, it has a high neutralizing power and a lasting action. Magnesium oxide is also an effective antacid. The carbonate is about equivalent to the hydroxide in neutralizing power, but has the disadvantage of causing distress owing to gas formation. The phosphate has the advantage of noninterference with phosphate absorption and is an effective, long-lasting antacid. The trisilicate is less potent in neutralizing power than the other magnesium antacids, but has added value because of its protective action and its adsorptive action.

Unlike the aluminum antacids, the magnesium antacids owe their activity strictly to the anion. The hydrated magnesium ion, while it exists, does not play the role of its aluminum counterpart. Basic magnesium salts do exist, however.

Owing to the soluble magnesium ion set free during antacid action, all magnesium antacids have a laxative effect.

$$MgO + H_3O^+ = [Mg(OH)]^+ + H_2O$$

$$[Mg(OH)]^+ + H_3O^+ = Mg^{2+} + 2H_2O$$

Although not shown as such, these dissolved magnesium ions probably exist in the form of octahedral aqua complexes. The hydroxide, oxide, and carbonate are used for their antacid effect or for their cathartic action. The cathartic dose is considerably larger: e.g., magnesia magma, 15 mL. vs. 4 mL.; magnesium oxide, 4 Gm. vs. 250 mg.

Although 15-30% of the magnesium of a given dose of a magnesium antacid is absorbed, normal individuals can tolerate absorption of up to 150 me. However, individuals with impaired renal function may develop hypermagnesemia toxicity at these levels. For the latter reason the FDA limits the maximum daily dose to 50 me. of magnesium.

Magnesium hydroxide is prepared by hydration of magnesium oxide.

$$MgO + H_2O = Mg(OH)_2$$

Formerly it was prepared by adding a solution of sodium hydroxide to a boiling solution of magnesium sulfate. However, owing to difficulty in removal of the sodium sulfate by-product, this method has been gradually replaced by the hydration procedure. Magnesia magma, when prepared by hydration of magnesium oxide, has an unpleasant alkaline taste. For this reason one tenth of 1% citric acid is included to reduce the alkalinity and give the slight salty taste of the product manufactured by double displacement.

$$H_3Ci + (OH)^- = H_2Ci^- + H_2O$$

Milk of magnesia is light stable. The use of blue bottles is intended only to impart a more pleasing appearance to the chalky product.

Magnesium carbonate is a product whose actual analytical stoichiometry is a function of the mode and conditions of preparation. The product prepared by precipitation from a solution of a soluble magnesium salt by sodium carbonate will vary in composition and physical structure depending upon concentration of the reacting media and the temperature of reaction. The degree of dissociation of the reactants involved is inversely proportional to concentration and directly proportional to the temperature.

Light magnesium carbonate is prepared by reacting dilute solutions of each of the reactants at low temperatures. The reactions involved are

$$CO_3^{2-} + H_2O \rightleftharpoons HCO_3^- + OH^-$$

$$[Mg^{2+} + SO_4^{2-}] \rightleftharpoons Mg^{2+} + SO_4^{2-}$$

$$Mg^{2+} + OH^- = [Mg(OH)]^+$$

$$2[Mg(OH)]^+ + CO_3^{2-} = [Mg(OH)]_2CO_3\downarrow$$

$$Mg^{2+} + CO_3^{2-} = MgCO_3\downarrow$$

On completion of the reaction the precipitate must be washed thoroughly to remove the bicarbonate, sulfate, and sodium ions. The washed precipitate on drying in air yields a light friable powder. If 0.5*M* magnesium sulfate and 1.2*M* sodium carbonate are reacted at 55°C., the product has a composition which may be represented as $Mg(OH)_2(MgCO_3)_4$.

Heavy magnesium carbonate precipitates if hot concentrated solutions are reacted. Because of the higher temperature the equilibrium

$$CO_3^{2-} + H_2O \rightleftharpoons OH^- + HCO_3^-$$

is displaced further to the right to give very alkaline solutions. The higher temperature also forces the reaction

$$2HCO_3^- = CO_3^{2-} + H_2O + CO_2\uparrow$$

The precipitate forms according to the following reactions:

$$Mg^{2+} + CO_3^{2-} = MgCO_3\downarrow$$
$$Mg^{2+} + OH^- = Mg(OH)^+$$
$$Mg(OH)^+ + OH^- = Mg(OH)_2\downarrow$$
$$2Mg(OH)^+ + CO_3^{2-} = (MgOH)_2CO_3\downarrow$$

Washing and drying produce a dense product.

Magnesium carbonate preparations must be carefully dried at comparatively low temperatures; both the hydroxide and carbonate decompose to the oxide at relatively low calcination temperatures.

$$Mg(OH)_2 = MgO + H_2O\uparrow$$
$$MgCO_3 = MgO + CO_2\uparrow$$
$$(MgOH)_2CO_3 = 2MgO + H_2O + CO_2\uparrow$$

Magnesium oxide is classified as either "light" or "heavy." The volume of a 5-Gm. sample is used to determine the classification; 10–20 ml. for the heavy oxide and 40–50 ml. for the light. Magnesium oxide can be prepared in a range of densities depending upon the carbonate used and the temperature of calcination. Light carbonates calcined at red heat give light oxides. Use of higher temperatures results in more compact structures. Both forms of the oxide absorb carbon dioxide and moisture from the atmosphere. The light forms, because of greater surface area, are more reactive.

Tribasic magnesium phosphate, $Mg_3(PO_4)_2{\cdot}5H_2O$, is prepared by reaction of magnesium sulfate and tertiary sodium phosphate. The tertiary sodium salt is necessary to provide the alkaline medium required for precipitation of the normal phosphate. If monohydrogen phosphate is used, magnesium monohydrogen phosphate precipitates. This latter compound is unsatisfactory as an antacid because of its solubility (320 parts of water) and the inferior base strength of the monohydrogen phosphate ion.

The antacid action of magnesium trisilicate has been discussed (page 415). This hydrated gelatinous silicic acid has good adhering properties. This gives the product a protective action, coating craters of ulcers, etc. Like aluminum hydroxide, silicic acid is a good adsorptive agent; it is able to remove pepsin, some alkaloids, bacterial toxins, putrefaction amines, and food poisons from the intestinal tract. It has been presumed that silicic acid is not absorbed from the intestinal tract. However, a recent x-ray study of urinary calculi indicates the presence of silica in stones removed from five patients on this antacid over a period of several years.

CALCIUM COMPOUNDS: Calcium carbonate, tribasic calcium phosphate. Calcium compounds are excellent antacids, effective in the management of peptic ulcer and other conditions of gastric hyperacidity. The soluble calcium ion released as a result of the antacid action is not astringent and has no laxative action. Calcium antacids have a limiting action and do not impart an excessive alkaline reaction to the gastrointestinal contents. Large doses lead to a systemic effect, but this is counteracted by simultaneous administration of sodium chloride. In large doses calcium antacids have a constipating effect. Compound Chalk Powder is intended for treatment of diarrhea associated with intestinal acidity. The two approved calcium antacids are given maximum daily doses by the FDA: calcium carbonate, 160 me. calcium ion (8g. calcium carbonate), calcium phosphate (tertiary), 160 me. calcium ion (24 g. calcium phosphate).

The conditions for preparation of the calcium phosphates have been previously outlined (page 330). The product recognized for antacid use is prepared by reaction of calcium chloride, sodium monohydrogen phosphate, and ammonia at boiling temperatures. The composition of this product approximates $Ca(OH)_2 \cdot [Ca_3(PO_4)_2]_3$. It has the controlled antacid action of all tertiary phosphates. Despite the insolubility of this compound, calcium ion is absorbed from the intestinal tract when this compound is ingested, giving it some value as a calcemic. It has the advantage over the other calcium antacid, calcium carbonate, in that there is no gas evolution.

Calcium carbonate is recognized in two forms; the native compound and the product manufactured by precipitation. Precipitated calcium carbonate (precipitated chalk) is prepared by the reaction of calcium chloride, the most economical soluble calcium salt, and sodium carbonate.

It has an antacid capacity almost as good as that of aluminum hydroxide and is less expensive. It has an unpleasant "chalky" taste, however. Precipitated calcium carbonate may have some protective and adsorptive properties.

Prepared chalk (drop chalk, *creta preparata*) is produced by elutriation of finely powdered native chalk. (The heavier, coarser by-product of this process is known as whiting.) Prepared chalk is slightly less reactive than precipitated calcium carbonate, largely because of a more compact crystal structure resulting in a smaller surface area per unit weight. This form of calcium carbonate is used for preparation of Chalk Mixture, described in *N.F. XI*[5]

Calcium carbonate will dissolve in weak acids, e.g., acetic, ammonium chloride, with the evolution of carbon dioxide. The purity rubrics for the two forms allow for the possibility of greater impurities in the native form.

BISMUTH COMPOUNDS: Bismuth subcarbonate, bismuth subgallate, bismuth magma. These compounds largely owe their antacid action to their anion, although the bismuthyl ion does make a contribution to the action (page 504). Because of their adherent properties these bismuth compounds also have protective action. The small amount of dissolved bismuthyl ion exerts a mild antiseptic action. Bismuth antacids thus have a combined antacid, antiseptic, and protective action. There is no evidence that bismuth compounds are constipating. The compounds listed are used in treatment of hyperacidity, diarrhea, etc. Hydrogen sulfide, formed by decomposition of proteins, reacts with bismuth ion in the intestinal tract to give very dark brown to black insoluble bismuth sulfide. This is responsible for the black-appearing stools during bismuth antacid therapy.

Bismuth magma (milk of bismuth, bismuth cream) is an aqueous suspension of a mixture of basic bismuth carbonate and bismuth hydroxide. It is prepared by reacting an acidified solution of bismuthyl nitrate with a solution of ammonium carbonate in diluted ammonia.

$$NH_3 + H_2O = NH_4^+ + OH^-$$
$$(NH_4)_2CO_3 = 2NH_4^+ + CO_3^{2-}$$
$$BiO^+ + OH^- = BiO(OH)\downarrow$$
$$2BiO^+ + CO_3^{2-} = (BiO)_2CO_3\downarrow$$

Antacid preparations. Although some antacids may be used singly (e.g., aluminum hydroxide, sodium bicarbonate), the majority of antacid

---

[5] *N.F. XI*, p. 80.

preparations includes two or more substances with antacid activity. Earlier combinations include those by B. W. Sippy; two of his antacids—sodium bicarbonate and calcium carbonate, and sodium bicarbonate and magnesium oxide—are official in *N.F. XI.*

Antacid preparations, according to FDA rules and regulations, will consist of one or more active ingredients from the approved list. Each active ingredient must contribute at least 25% of the acid neutralizing capacity of the product (within any applicable maximum daily dose limits). Each minimum dose of the product must have at least 5 me. of acid neutralizing power. Under prescribed conditions the minimum dose, in contact with 5.0 me. of hydrochloric acid for 10 minutes must achieve a pH of 3.5 or greater. The lableing of the product must include appropriate warnings.

The most common combination is that of an aluminum compound, usually the hydroxide, with one or more of the magnesium compounds. This assures the excellent nonsystemic antacid action of aluminum and magnesium compounds; at the same time the constipating action of the aluminum ion is counteracted by the laxative effect of the magnesium ion. Preparations made up of the hydroxides of these cations are probably the most popular combination at this time.

In addition, substances whose special properties are desirable may appear in antacid preparations. The most common action and the compounds used are: antiseptic action, provided by bismuthyl compounds; adsorbant action, provided by kaolin, attapulgite, or charcoal; and sedatives.

Antacid preparations are available in virtually all dosage forms: capsules, powders, tablets, wafers, suspensions. The suspension form is especially popular for the hydroxy compounds. The pH of these suspensions is adjusted to 6.5–7.5. Because of the hydroxyl group present, these magmas absorb carbon dioxide on exposure to air. As with all suspensions, they must be protected from freezing temperatures.

SYSTEMIC ALKALOSIS AND ALKALINE URINES—PRODUCTION AND CONTROL

A hypoacidic condition of the body fluids involves an upset of the normal carbonic acid-bicarbonate buffer system. Examination of the equation

$$pH = pK_a + \log \frac{[HCO_3^-]}{[H_2CO_3]}$$

reveals that an alkaline pH shift will result either by an increase in bicarbonate levels or a decrease in carbonic acid levels (or dissolved carbon dioxide levels). Hyperventilation results in reduction of the carbon dioxide levels. Ingestion of bicarbonate or bicarbonate-producing substances, e.g., lactate, acetate, results in increased bicarbonate levels. The latter pathway is obviously the method of choice for medically induced systemic alkalosis. Urinary alkalosis results as the kidneys slowly restore the normal pH to the blood by excretion of the excess bicarbonate; at normal bicarbonate levels very little is excreted in the urine. The over-all balance is shown in Fig. 15.3.

The soluble antacids given in Table 15.3 are used to produce systemic alkalosis. If the injection route is to be used, sodium lactate is usually employed (page 357). Sodium bicarbonate is most commonly used in urology. Sodium citrate is the choice if protracted periods of alkalization are necessary, as in the prevention of uric acid and cystine stones in the urinary tract; both of these substances are soluble in an alkaline urine. However, a low calcium diet is also required since an alkaline urine may cause precipitation of calcium salts. Potassium citrate is used as a diuretic in sulfonamide therapy to prevent crystaluria. Potassium bicarbonate has been suggested as an alternate for sodium bicarbonate for patients on the verge of or in frank cardiac failure. Sodium ion causes pulmonary and systemic edema in such cases; potassium avoids these conditions and even has a diuretic effect.

SYSTEMIC ACIDOSIS AND ACID URINES—PRODUCTION AND CONTROL

Re-examination of the equation for buffer action reveals that systemic acidosis may result either from a decrease in bicarbonate levels or an increase in the carbonic acid levels of the blood and body fluids. Bicarbonate levels are unlikely to fall; very little if any of this ion is excreted at normal blood levels (27 me./L.) Hypoventilation can produce increased carbon dioxide levels, but conditions of acidosis are more commonly caused by failure in the oxidative cycle, allowing accumulation of organic acids, e.g., lactic acid.

Acidosis is produced artificially for the treatment of a number of conditions: systemic alkalosis, calcium tetany, certain cystitis, reduction or prevention of urinary calcium calculi, solubilization of lead deposits. It is also employed to produce a pH favorable for chemotherapy of the urinary tract, e.g., mandelic acid or penicillin therapy.

Hyperacidic conditions are largely counteracted by an increase in the dihydrogen phosphate level of phosphate buffer system of the body fluids.

$$HPO_4^{2-} + H_3O^+ = H_2PO_4^- + H_2O$$

This buffer is better suited for control of excess acid since it is a conservation system. The carbonate buffer system is intimately bound to carbon dioxide levels through hydration-dehydration reactions of carbonic acid.

$$HCO_3^- + H_3O^+ = H_2CO_3 + H_2O$$

$$H_2CO_3 = H_2O + CO_2$$

Except under the mediation of carbonic anhydrase in the red blood cells, these equilibria are not particularly mobile.

The kidneys play a major role in the elimination of excess acid, both that which is medicinally induced and that which comes from faulty metabolism. This elimination is accomplished by excretion of dihydrogen phosphate together with equivalent cation (usually sodium ion) in the urine. The kidney tubules are also able to deaminate amino acids

$$R\text{—}\underset{\underset{NH_2}{|}}{\overset{\overset{H}{|}}{C}}\text{—}COOH + \tfrac{1}{2}O_2 = R\text{—}\overset{\overset{O}{\|}}{C}\text{—}COOH + NH_3,$$

and use the ammonia formed to eliminate excess protons as ammonium salts.

$$NH_3 + H_3O^+ = NH_4^+ + H_2O$$

Urinary acidification results from either excretion product.

The compounds used to medicinally induce systemic and/or an acid urine are sodium dihydrogen phosphate, ammonium chloride, calcium chloride, and, very rarely, phosphoric acid. Phosphoric acid has been administered for treatment of cystitis and, over long periods, for treatment of chronic lead poisoning, but sodium dihydrogen phosphate is superior. This latter compound is the drug of choice in the treatment of calcium-containing calculi. It acts by depressing serum calcium levels and increasing urinary biphosphate.

Orally administered dihydrogen phosphate is absorbed readily from the intestinal tract. It immediately changes the normal one to four $[H_2PO_4^-]/[HPO_4^{2-}]$ ratio of the blood and body fluids to give more acid ratios. Ammonium chloride (ammonium muriate) is a more effective acidifier than biphosphate. It is most frequently used in chemotherapy of the urinary tract. However, extended use produces bone demineralization and increased urinary calcium phosphate. The action of ammonium chloride as an acidifier depends upon the relative acid strengths of the NH_4^+-NH_3 conjugate system and the dihydrogen-monohydrogen phosphate conjugate system. Since monohydrogen phosphate is the stronger base, the reaction

$$NH_4^+ + HPO_4^{2-} \rightleftharpoons NH_3 + H_2PO_4^-$$

takes place in the body fluids on ammonium ion absorption. The neutral ammonia formed has no effect on the pH of the fluids. It is converted to urea in the liver. Thus, the effect of ingestion of an ammonium salt is an equivalent increase of the blood dihydrogen phosphate level.

Calcium chloride, though not used frequently, also has a strong systemic and urinary acidifying effect. It acts in the intestinal tract as follows:

$$3Ca^{2+} + 6Cl^- + 2HPO_4^{2-} + 2H_2O = Ca_3(PO_4)_2\downarrow + 2H_3O^+ + 6Cl^-$$

The equivalent of the acid released appears in the blood as an increase in the dihydrogen phosphate ion. Acidosis is achieved through disturbance of the phosphate ratio.

A regimen consisting of alternate alkalinization and acidification has been used for the treatment of urinary infections. This is accomplished by a period of treatment with oral sodium bicarbonate followed by a period of treatment with oral monobasic sodium phosphate. The cycle is repeated several times.

STUDY AND SELF-APPRAISAL AIDS

IMPORTANT TERMS. Define or identify in a single concise sentence if possible.

Acidosis, alkalosis, antacid, calculi, heat labile, hypoacidity, hyperacidity, nonsystemic antacid, systemic antacid.

NOMENCLATURE OF INTEREST. Identify each of the following.

α-aminoglutaric acid, baking powder, basic aluminum carbonate, bismuth subgallate, bismuthyl carbonate, chalk, cream of magnesia, dihydroxyaluminum glycinate, diluted hydrochloric acid, drop chalk, glycine hydrochloride, light magnesia, limestone, magnesia magma, marble, milk of bismuth, prepared chalk, sal ammoniac, sal soda, soda ash, whiting.

SELF-APPRAISAL QUESTIONS

1. Considered from the modern point of view, a Brønsted base is essential for antacid activity. What are the limitations which determine whether or not a given base is suitable for antacid use?

2. The hydrated sodium, potassium, calcium, and magnesium ions show no antacid activity in those of their compounds useful as antacids. Why? What is their role?

3. Unlike the cations named in question 2, aluminum and bismuthyl ions actively participate in the antacid activity of most of their useful antacid compounds. Explain.

4. In many cases antacid activity may actually be considered a buffering action. Explain.

5. Explain on a chemical basis why soluble bicarbonates may be employed as antacids but soluble normal carbonates are unsatisfactory.

6. List the inherent problems in the use of each of the following compounds as antacids: sodium bicarbonate, aluminum hydroxide, magnesium phosphate, calcium carbonate.

7. Name and discuss each of the valuable ancillary properties of the following antacids: bismuthyl carbonate, aluminum hydroxide, magnesium silicate.

8. Theoretically tertiary sodium phosphate, lithium bicarbonate, calcium hydroxide, and sodium tetraborate have anions suitable for antacid activity. Name the practical considerations which make each of these respective compounds unsuitable as antacids.

9. Iron(III) hydroxide has a K_{sp} and properties very similar to those of aluminum hydroxide. The two cations in soluble form are also very similar. Give possible explanations for the failure of the iron compound to find application as an antacid.

10. Magnesium carbonate is assigned the following formulas, depending upon author: $Mg(OH)_2{\cdot}(MgCO_3)_4$, $[Mg(OH)]_2CO_3{\cdot}(MgCO_3)_3$, and $MgO{\cdot}(MgCO_3)_4{\cdot}H_2O$. Explain the existence of so many presumably correct formulas for the same substance.

11. What is the practical method used to distinguish between the "light" and "heavy" forms, respectively, of magnesium carbonate and magnesium oxide? Name the physical property basic to this criterion.

12. Name two incompatibilities caused by sodium bicarbonate. Discuss fully the chemistry involved.

13. Compare the stability of crystalline $Na_2CO_3{\cdot}H_2O$ in dry air at 100°C. and the stability of its aqueous solution at boiling temperatures. Discuss.

14. Magnesium magma prepared by the so-called "double decomposition" process has a salty taste, whereas the raw product resulting from the "hydration" method of synthesis has an unpleasant alkaline taste. Explain. How is the alkaline taste overcome? Explain why this treatment is effective.

15. Describe the procedure for preparation of both precipitated chalk and prepared chalk. Explain the purpose of all steps. How do the

chemical and physical properties of the two forms differ? Why should precipitated chalk have fewer impurities?

16. Why must pharmaceutical preparations in magma form be protected from freezing temperatures?

17. Explain why $AlPO_4$ precipitates when a solution of a soluble aluminum salt is added to a solution of sodium monohydrogen phosphate, whereas $Ca(OH)_2 \cdot [Ca_3(PO_4)_2]_3$ precipitates when a soluble calcium salt solution is added to an ammonia solution of sodium monohydrogen phosphate.

18. Explain why the addition of a solution of a soluble aluminum salt to a solution of sodium carbonate yields a precipitate of aluminum hydroxide, whereas when sodium bicarbonate is present in the carbonate solution the basic carbonate of aluminum precipitates instead.

19. Why is it necessary to wash the insoluble antacids prepared by precipitation methods? Explain.

20. Why is an acidified solution of bismuth(III) nitrate rather than bismuth subnitrate used in the preparation of bismuth subcarbonate?

21. Although some of the assumptions may not be completely valid, a qualitative comparison of the neutralizing action of the usual dose of common antacids may be made by means of calculations. Assume that the gastric fluids to be neutralized are equivalent to 250 ml. of hydrochloric acid solutions having a pH of 0.7. To simplify calculations assume complete neutralization of the added base rather than establishment of an equilibrium based on the pK_b of the base. The antacids to be compared and their usual doses are: aluminum hydroxide, 0.3 Gm.; calcium carbonate, 1.0 Gm.; magnesium tertiary phosphate, 1.0 Gm.; and sodium bicarbonate, 2.0 Gm. Calculate the ΔpH produced in each case. ($NaHCO_3$—pH = 0.98)

22. Referring to problem 21, change the initial pH to 1.0 and recalculate the ΔpH.

23. Discuss reasons for the opening statement of problem 21.

24. Discuss the chemical basis involved in the "Acid Neutralizing Test" for Aluminum Hydroxide Gel, *U.S.P. XXI*[7]

25. What is the normal pH range of the gastric juice? Why do achlorhydria treatments rarely attempt to re-establish a gastric pH within this range?

[7] *U.S.P. XXI*, p. 1192.

26. The end carboxyl group of glutamic acid hydrochloride makes a small contribution to the increased gastric acidity whereas the same group of the corresponding glycine compound makes no contribution. Explain.

27. Indicate the property(s) of diluted hydrochloric acid which make it difficult to administer in treatment of achlorhydria. How may the problems imposed be overcome?

28. Name the compounds usually used for oral administration in treatment of systemic alkalosis. Name the compound usually used for injections. Explain the mechanism of action of each compound.

29. Name the compounds which are administered orally for treatment of systemic acidosis or to induce a systemic alkalosis. Name the compounds usually used for administration by injection. Explain the mechanism of action of each compound.

30. Why does urinary pH closely follow any pH shift of the blood and body fluids?

31. Name those compounds and preparations discussed in this chapter which are official in the current compendia.

32. How does the basis or philosophy of acid-base applications discussed in this chapter differ from that in Chapter 7?

PRACTICE IN EQUATION WRITING

33. Write all equations needed to demonstrate the antacid activity of each of the following: dihydroxyaluminum aminoacetate, magnesium "trisilicate," potassium bicarbonate, sodium citrate, tribasic calcium phosphate, milk of bismuth.

34. Write equations for all chemical reactions involved in the preparation of each of the following: tribasic calcium phosphate, magnesia magma, aluminum subcarbonate, light magnesium oxide, bismuth magma.

35. Write equations demonstrating the acidifying action of glutamic acid hydrochloride.

36. Show why dihydrogen phosphate and ammonium ion, respectively, cause an acidic shift in the pH of the urine.

37. Show why higher than normal bicarbonate levels cause a basic pH shift in the urine.

38. Show by means of equations why an alkaline urine favors formation of calcium phosphate stones in the urinary tract. Show why an acid urine favors their resolution.

39. Write equations for all chemical reactions taking place during the assay, by official methods, of each of the following: aluminum hydroxide,

ammonium chloride, bismuth subcarbonate, tribasic calcium phosphate, magnesium carbonate, sodium bicarbonate.

40. By means of equations show why the aluminum based antacids are nonsystemic.

SIXTEEN · GERMICIDES AND RELATED SUBSTANCES I: *general background and the oxidizers*

GENERAL BACKGROUND

Definitions. A successful presentation of the material in Chapters 16, 17, and 18 requires a clear understanding of certain terms pertinent to the general area. In the following paragraphs these terms and the definitions will be given.

Germicide[1] denotes substances lethal to microorganisms. The more descriptive terms *bactericide*, *fungicide*, *algaecide*, *trypanocide* designate the specific activity of a given germicide. The suffix *-cide*, is reserved for substances capable of creating a sterile medium. In the general background discussion immediately following the terms germicide and germicidal action will be used for convenience only. However, in the succeeding discussions of specific compounds strict adherence to the definitions will be followed.

Other substances are not lethal to microorganisms but are able to maintain sterile or aseptic conditions. The suffix *-stat* designates substances having this action. Thus, a *bacteriostat* prevents the growth of bacteria, a *fungistat* prevents the growth of fungi, etc. The spores of restive organisms may be present and start growth upon removal of the inhibiting substance. "Stat" action may be considered preservative action.

A *disinfectant* is a chemical substance lethal to disease germs and other harmful microorganisms. The term is widely used; it may refer to treatment of equipment or instruments, water, rooms, food preservation, etc. *Sterilization* is the process by which inanimate objects are rendered aseptic.

[1] It has been suggested that another term, *antiseptic*, be dropped from usage because of its ambiguity. However, the U.S. Government recognizes antiseptic as the equivalent of germicide, since this association exists in the minds of the general public.

Certain terms have been developed in pharmacy and medicine which denote a germicidal preparation intended for use on or in specific parts of the body, e.g., gargles, mouth washes, internal antiseptics. Germicides may also be classified as internal and external depending on whether they are intended for use within the body or on local areas of its external surface. Substances suitable for internal use may be classified as gastrointestinal, e.g., bismuthyl nitrate, or systemic, e.g., inorganic arsenicals.

Mechanisms of action. In developing mechanisms for the chemical action of this group of medicinal agents it is enlightening to first consider those substances devoid of germicidal activity. This group consists almost entirely of those ions comprising the electrolyte component of the body fluids: Na^+, K^+, Ca^{2+}, Mg^{2+}, Cl^-, HCO_3^-, $H_2PO_4^-$, HPO_4^{2-}, and a few other physiologically important ions, e.g., I^-. These inert gas ions and stable complex structures are characterized by their weak electrostatic fields. Since they can exist in the body without interaction with the body proteins, they obviously have no effect on microorganisms. (In very high concentrations a destructive osmotic effect is possible, e.g., common salt as a preservative.) Other weak field ions, e.g., Li^+, Sr^{2+}, are similarly without germicidal action.

Examination of the remaining elements and their compounds, with a certain amount of anticipation of modes of action, suggests that the cations of the metals should be considered separately from the nonmetals and their anions. Metallic cations with germicidal action are strong field cations: silver(I), copper(II), and iron(III) are transition elements; zinc and mercury(II) have the $18e^-$ structure; aluminum, though having the inert gas structure, has a high charge and small ionic radius. Because of the high polarizability of the sulfur atoms and, to a lesser extent, of the nitrogen and even the oxygen atoms present in protein structures, strong electrostatic interactions are set up with the cations. These interactions result in protein precipitation.

This action is for the most part nonspecific. Protein precipitants do not distinguish between the proteins of the host and those of the parasite. Of the metallic elements only the alkali and alkaline earth (except barium) cations cannot act as protein precipitants. Oxonium ion also acts as a protein precipitant.

The nonmetallic compounds exhibiting germicidal action are generally those structures which are redox unstable, e.g., ClO^-, I_2, H_2O_2. Proteins having strong reducing groups, e.g., sulfhydryl, are susceptible to redox reactions. Germicidal activity by oxidation of labile protein groups results in reduction of the germicide to the comparatively stable ionic forms, Cl^-, I^-, O^{2-}, respectively, for the above substances.

The oxidizing germicides consist primarily of three groups of compounds: the peroxides, chlorine and its oxo anions, and iodine. There is a sprinkling of closely related compounds, e.g., permanganate.

Although protein precipitation and destructive oxidation are the principal modes of germicidal action by inorganics, a few, e.g., the arsenicides, appear to inhibit specific enzyme systems by complexation. Arsenic and antimony are strongly complexed with protein sulfhydryl groups as ligands. The affected groups are thus rendered unavailable for their usual activities.

Related medicinal actions. The extent of protein precipitation or oxidation is closely related to the concentration of the metallic cation or nonmetal compound or anion. Germicidal action results when the concentration is such that reaction is restricted largely to the parasite cell, the concentration being at levels too low to cause appreciable damage to the host's tissues. With *astringent* (*styptic*), *irritant*, and *corrosive* (*escharotic, caustic*) actions, the concentrations are higher; destruction of a parasite is no longer the object of application. Rather, varying degrees of reaction with the body's proteins is desired; each of the terms (in the order named) represents a successively greater degree of reaction.

If a mild, comparatively nonirritating action is produced, astringent action is achieved. This degree of reaction is characteristic of aluminum compounds as deodorants, and iron(III) chloride solutions for staunching the flow of blood from minor wounds. If a stronger reaction with tissue is produced, the action is said to be irritant. This degree of reaction is believed to stimulate growth and repair in slowly healing lesions. When destruction of a tissue is desired, corrosive action is attained. The removal of warts with toughened silver nitrate illustrates such corrosive action.

The relationship between the concentration of a protein precipitant or oxidant, and the extent of interaction with tissues is demonstrated by several substances, as shown for zinc ion in Table 16.1. By causing a variation in the degree of ionization, the anion present affects the extent of reaction of one salt as compared to another.

Control of precipitant and oxidative action. Since the degree of attack on the host's tissues will be related to the concentration of free active ion, various methods have been devised to minimize untoward or excessive action. With soluble compounds concentrations must be established for effective germicidal action, but with minimal irritant action. The concentrations used will differ depending upon the sensitivity of the area of application, e.g., delicate eye tissues vs. skin.

An old but effective tool for concentration control is the use of insoluble compounds of the desired ion. Here the solubility product of the compound selected is the concentration determinant. Within limits, the common ion effect can be used to modify the concentration. Many applications of this technique exist, e.g., calomel ointment, silver chloride suspension, zinc oxide paste.

Complexation techniques provide the pharmacist with still another tool for effective control of the concentration level of the active ion. The irritating effects of mercury(II) solutions may be reduced by complexation with halide ions, e.g., $[HgCl_4]^{2-}$, $[HgI_4]^{2-}$.

TABLE 16.1 RELATIONSHIP BETWEEN CONCENTRATION AND PROTEIN PRECIPITANT ACTION

Salt	Concentration	Action
$Zn(OAc)_2$	0.1–4.0%	Mild antiseptic to astringent (used on mucous membranes and skin)
$ZnSO_4$	0.2%	Ophthalmic astringent
$ZnCl_2$	0.2–1.5%	Astringent to styptic (for topical application)
$ZnCl_2$	(stick-100%)	Corrosive (caustic)

Research based on complexation techniques should reveal new applications for control of the undesirable irritant properties of otherwise excellent inorganic germicides and astringents. The complexes investigated must be such that the instability constant allows an optimum concentration of free active ion. This concentration level may be further controlled by the common ion effect of excess ligand. Obviously a ligand should have no complicating physiological or pharmacological properties of its own.

Selection of suitable substances. Most inorganic substances possess some degree of germicidal action, but still few find employment in this medicinal application. Other factors and properties of the compounds must be considered.

Speed and duration of action are important: ethanol is immediately effective, but being volatile it has no staying power. On the other hand, mild silver protein releases its silver ions slowly. Owing to precipitation of insoluble silver compounds within the wound, its action is prolonged.

Specificity and effectiveness are significant. Although most inorganics are nonspecific, e.g., iodine, some are highly specific in their action; e.g., copper(II) is an algaecide, antimony(III) is an antischistosomal. As to effectiveness, iodine is a germicide whereas boric acid is a bacteriostat.

The stability of the preparation is important: Tincture of Iodine is a valuable germicide, but the volatility of both the solvent and the iodine creates a problem. Chemical storage stability is also important. Hypochlorite, hydrogen peroxide, etc. slowly decompose on standing. If a germicide is present in an injectable as a preservative, its heat stability is important if the product is to be autoclaved.

There are also nonmedicinal and nonpharmaceutical factors in the determination of suitable germicidal agents. Economics should be considered. Frequently overlooked are odor, color, and staining properties of a compound. Iron(III) salts possess excellent astringent properties, but are seldom used because of the laundry problems created by discoloration of clothing. Owing to the acidity of their solutions, simple aluminum salts cause chemical deterioration of clothing.

Prologue. An ideal germicide would destroy all forms of infectious agents without being toxic or irritating to the human cells or inducing sensitivity; it would penetrate tissue; its solution would be stable, noncorrosive, and inexpensive.

As a result of the widespread introduction of antibiotics some of the compounds to be discussed find limited application at this time. They do, however, remain a second line of defense, e.g., bismuth-arsenical regimen for antibiotic-sensitive syphilitics. Although the systemic use of inorganics is limited, they offer advantages when used externally. Iodine has still to meet its first "iodine-fast" microbe; very few individuals are sensitive to this substance. There is a growing body of opinion against the indiscriminate use of antibiotics so frequently practiced at this time.

OXIDIZING GERMICIDES—PEROXIDES

Metal peroxides. A peroxide may be defined as a substance which includes a peroxo (peroxy) group, $[O_2]^{2-}$, in its structure. Peroxide salts yield hydrogen peroxide on acidification. This qualification eliminates compounds with apparent peroxide-like formulae, e.g., NO_2, PbO_2, MnO_2. These substances are normal dioxides, having two simple oxide ions, O^{2-}, in their structures.

True peroxides are formed by all alkali metals and all alkaline earths except beryllium. Sodium peroxide, the most common, is prepared by heating sodium metal in oxygen at 300°C.

$$2Na + O_2 = Na_2O_2\ (+Na_2O)$$

The pale yellow commercial product is about 10% Na_2O_2. The lithium compound, $Li_2O_2 \cdot H_2O_2 \cdot H_2O$, is prepared by adding alcoholic hydrogen peroxide to solid lithium hydroxide.

$$2LiOH + 2H_2O_2 \xrightarrow{\text{ethanol}} Li_2O_2 \cdot H_2O_2 \cdot H_2O\downarrow + H_2O$$

Ethanol acts to coordinate the excess water. Anhydrous lithium peroxide, Li_2O_2, may be prepared by careful dehydration of the solvate. Other metal peroxides are formed by controlled oxidation of the metal either by passing the calculated quantity of air over the metal at 300°C., or bubbling the calculated quantity of air through a liquid ammonia solution of the metal at −50°C.

Barium peroxide is the most stable alkaline earth peroxide. It may be prepared by heating barium oxide in air at 400°C.

$$BaO + \tfrac{1}{2}O_2 \rightarrow BaO_2$$

Barium, strontium, and calcium form peroxide octahydrates.

$$M(OH)_2(aq) + H_2O_2 \rightarrow MO_2{\cdot}8H_2O\downarrow$$

These compounds may be dehydrated. Slightly alkaline hydrogen peroxide added to a solution of magnesium sulfate yields $MgO{\cdot}MgO_2{\cdot}xH_2O$. Zinc, cadmium, and mercury(II) form peroxohydrates, $MO_x{\cdot}H_2O_2$. Recently zinc peroxide, ZnO_2, and cadmium peroxide, CdO_2, have been prepared and characterized.

In addition to the normal peroxides all alkali metals except lithium form superoxides by either of two reactions:

$$M_2O_2 + O_2 \xrightarrow[500°C.]{pressure} MO_2$$

$$(excess)\ O_2 + M(liq.\ NH_3\ solution) \xrightarrow{-78°C.} MO_2$$

To satisfy electron balance requirements these compounds were formerly assigned the formula M_2O_4. Recent evidence places an unpaired electron in these compounds; the formula, MO_2, is consistent with this fact.

Potassium superoxide is used as a high altitude oxygen source. In the presence of a trace of copper(II) chloride it reacts with CO_2, yielding oxygen.

$$2KO_2 + CO_2 = K_2CO_3 + 1\tfrac{1}{2}O_2\uparrow$$

One gram yields 236 cc. of oxygen at S.T.P. (standard temperature and pressure). Oxygen is also released by reaction with ice water.

$$2KO_2 + H_2O \rightarrow 2KOH + 1\tfrac{1}{2}O_2\uparrow$$

Care must be exercised in working with alkali metals encrusted with a peroxide film, especially potassium. Cutting through this film with a knife drags along sufficient peroxide to set off an explosive reaction with metallic potassium.

$$2K + K_2O_2 \rightarrow 2K_2O$$

It is wiser to destroy heavily encrusted potassium than to attempt to clean it of the oxide film.

Peroxy acids. Peroxy acids and their salts are also of interest. Three types exist: permono, perdi, and per acids. The permono acids have the general formula $HOOA(O)_n$, where n may be 0, 1, 2, or 3. The acids are prepared from hydrogen peroxide:

$$H_2O + 2H_2O_2 + P_2O_5 \rightarrow 2H_3[PO_3(O_2)]$$
$$H_2O_2 + ClSO_3H \rightarrow H_2[SO_3(O_2)] + HCl\uparrow$$
$$H_2O_2 + SO_3 \rightarrow H_2[SO_3(O_2)]$$

The permonoborates and permonocarbonates are prepared by action of hydrogen peroxide on the salts. The exact constitution of these products is unknown. Sodium perborate, usually represented as $NaBO_2 \cdot H_2O_2 \cdot 3H_2O$, or $NaBO_3 \cdot 4H_2O$, is prepared by reaction of hydrogen peroxide with a cold solution of borax. Reaction of 100 volume hydrogen peroxide with KBO_2 yields $KBO_3 \cdot H_2O$ or $KBO_2 \cdot H_2O_2$. $LiBO_2$ reacts similarly. The addition of hydrogen peroxide to a solution of sodium carbonate yields $(Na_2CO_3)_2 \cdot 3H_2O_2$ or $(Na_2CO_4)_2 \cdot 2H_2O \cdot H_2O_2$. Compounds represented by compositions $LiBO_4 \cdot H_2O$, $KBO_5 \cdot H_2O$, $KHCO_4$, $Li_2CO_4 \cdot H_2O$, $Na_2CO_4 \cdot H_2O$ have been prepared. This series of salts is distinguished from true peroxides by the ease with which hydrogen peroxide may be extracted with ether. This indicates peroxohydrates.

Many organic acids form permono acids by direct action between the acid and hydrogen peroxide. The most common are peroxoformic, peroxoacetic, peroxopropionic, peroxobutyric, and peroxobenzoic. (The organic chemist usually omits the oxo portion of the prefix.) They are valuable oxidizing agents in synthetic work.

The perdi acids have the general structure A-OO-A. Three acids have been established: perdisulfuric, $H_2S_2O_8$, perdiphosphoric, $H_4P_2O_8$, and perdicarbonic, $H_2C_2O_6$. They are prepared by electrolysis of the concentrated normal acid or its potassium or ammonium salt. The potassium perdicarbonate is prepared by electrolysis of very concentrated potassium carbonate at −15°C. It exists as pale blue crystals.

Perdisulfuric acid forms colorless crystals melting at 65°C. These crystals are stable when dry. This compound is the basis for the commercial preparation of hydrogen peroxide. Well-defined alkali, ammonium, barium, and lead(II) salts have been prepared. (The permonosulfate salts have not been established.)

The final group, the per acids, have the general formula, $H_nA(O_2)_n$. The red perchromic acid salts are best known. The potassium salt, K_4CrO_8, is prepared by adding H_2O_2 to a cooled, slightly alkaline potassium chromate solution. It is a tetrahedral arrangement of four peroxo groups about the chromium(VI) central ion. A second series of blue perchromate salts exists, but the formula usually given, $MCrO_5$, has not been established. These salts form when H_2O_2 is added to a dichromate dissolved in dilute sulfuric acid, and may be extracted by ether. The blue color in ether solution is probably due to the formation of a coordinate bond between the CrO_5 group and the oxygen of the ether, $Et_2O{:}CrO_5$.

Hydrogen peroxide. PRODUCTION AND PROPERTIES: At one time this compound was prepared almost entirely by the reaction

$$BaO_2 + H_2SO_4\ (\text{dil.}) \rightarrow BaSO_4\downarrow + H_2O_2$$

Today this method is used only when a market exists for the by-product barium compounds and for a dilute peroxide solution.

The present method of choice involves the preparation of peroxodisulfate by electrolysis of concentrated sulfuric acid or of a very concentrated solution of ammonium sulfate, the reaction taking place in a cell with a separate cathode and anode compartment. The essential reactions, using a sulfuric acid electrolyte, are

$$2H_2SO_4 \rightleftharpoons H_3SO_4^+ + HSO_4^-$$
$$2H_3SO_4^+ + 2e^- \rightarrow H_2\uparrow + 2H_2SO_4\ (\text{at cathode})$$
$$2HSO_4^- \rightarrow H_2S_2O_8 + 2e^-\ (\text{at anode})$$

The anode reaction, which is the heart of the process, is better understood by examining the reaction on a structural basis, as shown in Fig. 16.1. After electrolysis the anolyte is treated with water.

$$H_2S_2O_8 + 2H_2O \rightarrow 2H_2SO_4 + H_2O_2$$

The hydrogen peroxide formed is separated by distillation under reduced pressure. The sulfuric acid is recycled to the electrolysis cell.

The material balance and raw material used are a production chemist's dream. The economics are simple. A concentrated peroxide is produced with water as the only raw material. Sulfuric acid is almost entirely recovered. The hydrogen by-product may be sold or used as fuel. However, the energy economics depend on a cheap source of electric power. Manufacture, thus is concentrated near sources of hydroelectric power.

Pure hydrogen peroxide is stable. However, during manufacture contamination with traces of transition elements, e.g., iron, chromium, takes place. Transition metal ions and hydroxyl ion cause catalytic decomposition.

$$2H_2O_2 \rightarrow 2H_2O + O_2\uparrow.$$

But the compound is stable to the light metals. For this reason hydrogen peroxide is shipped in volume in aluminum tank cars, and small volumes are frequently sold in aluminum containers.

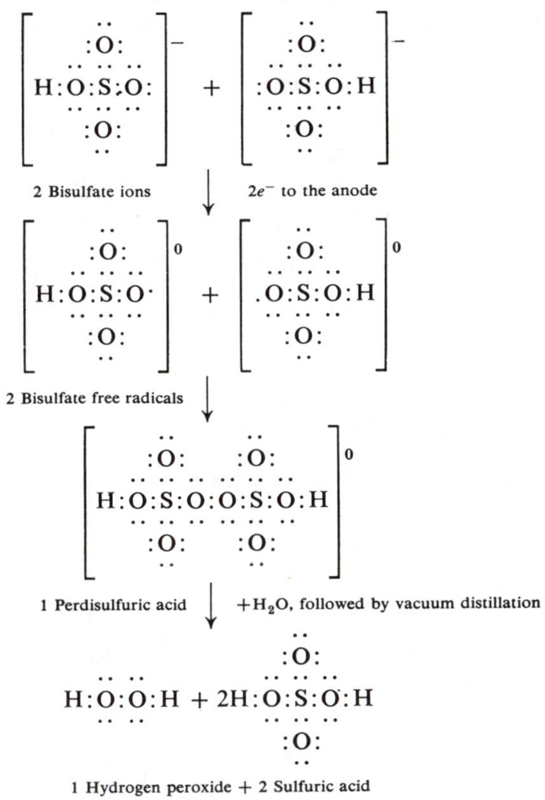

FIGURE 16.1 Preparation of hydrogen peroxide by electrolysis.

Stabilization of hydrogen peroxide is essential. The pH is adjusted to just acid with phosphoric acid. The transition element cations are removed either by passing the solution through a suitable adsorption column (hydrous stannic oxide, silica) or they are inactivated by chelation. Chelating agents include sodium pyrophosphate, 8-hydroxyquinoline, acetanilide, phenacetin, and quinine sulfate.

The product of the electrolysis process is a 25–35% solution. Fractional distillation of this solution under reduced pressure yields the commercially available 70% and 90% peroxide. Dilute peroxide, 3% and 6%, is prepared by dilution of the regular commercial product. The *U.S.P.* recognizes Hydrogen Peroxide Concentrate, the 30% solution, and Hydrogen Peroxide Topical Solution, the 3% solution.

Hydrogen peroxide concentration is also expressed in volume strength. The volume strength of a solution is equal to the volume of oxygen gas released from one volume of peroxide solution. A 10 volume peroxide is equivalent to the 3% solution, and a 100 volume peroxide is equivalent to the 30% solution.

The physical properties of 90% hydrogen peroxide are given in Table 16.2. The substance is miscible in water, ethanol, and diethylether. It is soluble in glycerol, recently introduced as a solvent for peroxide in certain medicinal applications. Peroxide can be extracted from water with ether. Heavy metal salts are insoluble.

$$H_2O_2 + Ba^{2+} + 2H_2O = BaO_2\downarrow + 2H_3O^+$$

Cations undergo solvation with hydrogen peroxide as readily as with water. Many of the so-called peroxo salts, e.g., perborate, are probably

TABLE 16.2 PHYSICAL PROPERTIES OF 90% HYDROGEN PEROXIDE

Color	Clear, colorless
Odor	None
Specific gravity	1.396 at 20°F.
Viscosity	0.0130 poise at 18°C.
Freezing point	−11°C.
Boiling point	140°C. (with decomposition)
Refractive index	1.3998 n_D 20°C.
Dielectric constant	95 at 0°C.

peroxohydrates. Organic substances solvate readily; one of the most useful is urea peroxide. It is prepared by the reaction

$$NH_2CONH_2 + H_2O_2 = [CO(NH_2)_2]\cdot H_2O_2$$

Hydrogen peroxide is a very weak acid.

$$H_2O_2 + H_2O \rightarrow H_3O^+ + HO_2^- \qquad K_a^1 = 2.4 \times 10^{-12}$$

$$HO_2^- + H_2O \rightarrow H_3O^+ + O_2^{2-} \qquad K_a^2 = 1.55 \times 10^{-12}$$

The pK_a values are 11.62 and 11.81, respectively. The pH of a dilute solution of pure hydrogen peroxide is 5.8.

Hydrogen peroxide may serve either as an oxidizing agent or a reducing agent. Figure 16.2 diagrams this dual behavior. Its action as a reducing agent is always identified by evolution of gaseous oxygen.

USES: The principal use of hydrogen peroxide is as an oxidizer, particularly as a bleaching agent. Substances of animal origin such as wool, hair, and furs are easily harmed by other bleaching agents, e.g., hypochlorite, sulfur dioxide. Peroxide is most effective as a bleach in a slightly alkaline medium. Hydrogen peroxide is made alkaline and the strongly alkaline sodium peroxide is made mildly alkaline by adding sulfuric acid.

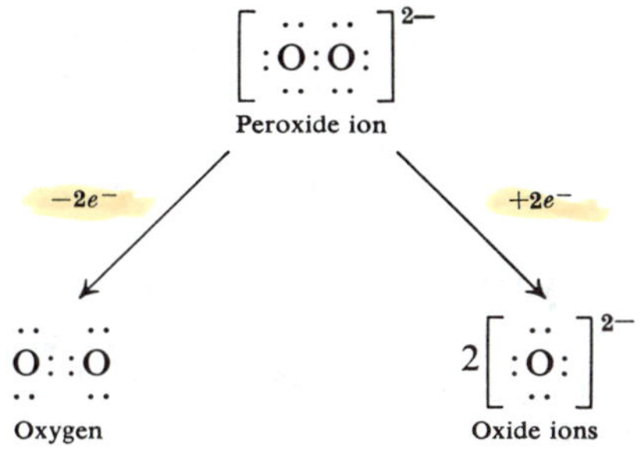

e.g.,

$$Ag_2O + H_2O_2 \rightarrow 2Ag^0 + H_2O + O_2\uparrow$$

e.g.,

$$\underset{\text{(black)}}{PbS} + 4H_2O_2 \rightarrow \underset{\text{(white)}}{PbSO_4\downarrow} + 4H_2O$$

FIGURE 16.2 Redox behavior of peroxide.

Sodium perborate has found wide use as a dry laundry bleach; it provides its own alkalizing action.

The gas-forming properties of hydrogen peroxide are valuable in producing lightweight, porous products. The gas can be produced without heat, a valuable attribute. It has been studied as a replacement for leavening agent in doughs, but the bread and cake produced is reported to have such a fine texture that a drying sensation is present in the mouth because of high water absorptive power.

Except for sodium peroxide, hydrogen peroxide is the starting material for most peroxide compounds. It is especially valuable for hydroxylation or epoxidation of unsaturated hydrocarbon bonds. In these applications the active agent is peroxyacetic acid, produced *in situ* by direct reaction of hydrogen peroxide with acetic acid. Peroxide is useful for the oxidation of sulfur, nitrogen, aldehydes, ketones, quinine, and iodide.

The use of hydrogen peroxide as a nonspecific germicide accounts for only a small percentage of production. Hydrogen peroxide (hydrogen dioxide, peroxide) (3%) destroys most pathogenic microorganisms, e.g., typhoid baccilli, anthrax, staphylococci. It inhibits the growth of all anaerobes. It also acts as a wound cleanser and deodorant. It is usually applied diluted with one part of water. The cleansing action is provided by the ebullition of oxygen gas released by the catalase of the exposed tissues. This effervescent action is valuable for the removal of dead organic matter from areas where mechanical removal is difficult. The liberation of oxygen gas dictates against its use in body cavities or abscesses from which the gas cannot escape. Because of the catalase involvement, effervescence is more vigorous on open wounds than on the unbroken skin.

Because of the breakdown to gaseous oxygen, the germicidal action of peroxide is transient. Rapid inactivation necessitates frequent reapplication. The high surface tension of the compound limits its penetrating power. Activity is increased by increased temperature and the presence of catalytic salts.

Dilute solutions of hydrogen peroxide are nontoxic, but are irritating if taken internally. Hydrogen peroxide produces a characteristic reaction on the skin: a stinging sensation accompanied by the formation of transient, somewhat hard, white spots. The 30% solution is corrosive.

Hydrogen peroxide is available as a 1½% solution in anhydrous glycerol, stabilized with 8-hydroxyquinoline. Urea peroxide, a stable 1:1 crystalline compound, is also available, usually as a 4-10% solution in anhydrous glycerol. Both solutions are preferred to aqueous hydrogen peroxide for treatment of ear and mouth infections. A 4% glycerol solution is also used as an antifungal.

Hydrogen peroxide has been proposed as a chemical antidote for cyanide and phosphorus poisoning. The following reactions take place readily:

$$CN^- + H_2O_2 \rightarrow CNO^- + H_2O$$

$$2P + 5H_2O_2 = 2H_2PO_4^- + 2H_3O^+$$

The products, cyanate ion and phosphate, are innocuous.

Hydrogen peroxide, 6%, (20 volume) has a cosmetic application as the active bleaching agent of hair bleaches and the prebleach for hair dyes. The bleaching action consists of the oxidation of the melanin pigments of the hair. In order to obtain conditions for optimum activity it is necessary to use an alkalizing agent in conjunction with the peroxide. This alkalinity, together with oxidation by peroxide, has a destructive side-effect on the protein bonds of the hair. The protein is reconstituted, as well as possible, after bleaching by means of an acid rinse. However, repeated bleaching affects the structure of the hair, making it brittle. Hydrogen peroxide is the only bleach safe for hair; other chemical bleaches are too harsh and destructive.

In Britain and in Europe hydrogen peroxide (and urea peroxide) are used as a preservative for milk and cream. The residual hydrogen peroxide must be destroyed by heat or chemicals to prevent taste impairment. Urea peroxide, 0.1%, is said to be an effective preservative for as long as 72 hours.

Miscellaneous peroxo germicides. MEDICINAL ZINC PEROXIDE: This preparation is an effective anti-infective agent. It is a mixture of zinc oxide, zinc hydroxide, zinc peroxide, zinc carbonate, and the hydroxohydrates of zinc oxide. Its chemical properties are the sum of those of insoluble zinc compounds (page 529) and hydrogen peroxide. Since the peroxo group is the distinguishing chemical feature, analysis consists of a determination of the volume of oxygen evolved under stated conditions. As with hydrogen peroxide, zinc peroxide is effective against microaerophilic organisms; the stricter an organism's oxygen requirements, the more effective the agent is as a germicide. Zinc peroxide is used topically as a suspension (usually 40%), paste, or dusting powder. The dusting powder has useful deodorizing properties. It is also used in sterilization procedures since heat activates oxygen evolution.

OTHER PEROXIDES: Sodium, magnesium, calcium, and strontium peroxides have found application in medicine and dentistry. Sodium peroxide (sodium dioxide, sodium superoxide) is freely soluble in water, reacting

$$Na_2O + H_2O \rightarrow 2Na^+ + 2OH^-$$
$$Na_2O_2 + H_2O \rightarrow 2Na^+ + OH^- + O_2H^-$$

The solutions are strongly alkaline. This high alkalinity limits their usefulness in medicine and hastens their decomposition.

In dermatological practice sodium peroxide is used in treating acne and for removal of comedones (blackheads). It has also been used in a $2\frac{1}{2}$–7% soap for acne and rosacea pustulosa. In dentistry it is used to bleach discolored teeth and to disinfect root canals.

Magnesium peroxide is insoluble in ordinary solvents. It is used in dusting powders and dentifrices. Calcium peroxide has been used in treatment of gastrointestinal disorders, as an antacid and antiferment. Strontium peroxide is used in dusting powders and ointments.

SODIUM PERBORATE: Sodium perborate of pharmacy is usually assigned the formula, $NaBO_3{\cdot}4H_2O$. It may be a true peroxo compound, a peroxohydrate, or, more likely, a combination of both. Its chemistry is the sum of borate chemistry (page 483) and peroxide chemistry. The salt is stable in dry, cold air. Oxygen is given off if stored under warm and/or moist conditions. Hydrogen peroxide, sodium, borate, and hydroxyl ions are present in its solutions. Being an alkaline solution, it shows the typical peroxide instability—solutions should be freshly prepared and used immediately. The compound must be stored in well-sealed containers.

Sodium perborate is used for its oxidant and anti-infective action in mouth washes for treatment of Vincent's Angina (Trench Mouth). Because of the poisonous nature of the borate ion, excessive swallowing must be avoided. Sodium perborate is used in dentifrices. Aromatic Sodium Perborate includes peppermint oil and sodium saccharin.

Sodium perborate is popular as a laundry bleach, both separately and as a constituent of soaps.

OTHER COMPOUNDS: The true constitution of calcium and magnesium perborates is also unknown, but the formulas are usually written $M(BO_3)_2 \cdot 7H_2O$. The calcium compound is used in the same applications as the sodium compound. Both magnesium and calcium perborate are used in tooth powders. Their solutions are faintly alkaline and show the usual peroxide instability.

Sodium percarbonate, usually assigned the formula, $2Na_2CO_3 \cdot 3H_2O_2$, has not found pharmaceutical application. It is used in powdered laundry bleaches.

Peroxide incompatibilities. Peroxides are unstable in moist, hot air and in alkaline solution. The presence of most heavy metal ions is also deleterious. Because of their inherent instability and the ease of oxidation or reduction, peroxides are rarely used in combination with other ingredients, except in the dry state with inert fillers.

Ozone. The triatomic ozone molecule is generally thought to be the only allotrope of oxygen, although there is some spectroscopic evidence for the existence of oxozone, O_4, at low temperatures. Ozone formation is exothermic and probably via the reaction

$$O_2 \rightarrow 2O \xrightarrow{+O_2} O_3$$

on passing a silent electrical discharge through oxygen at atmospheric pressure[2] or irradiating oxygen with ultraviolet light (1750 and 2500 Å). It may also be prepared by electrolyzing concentrated aqueous perchloric acid between a lead cathode and a platinum anode at −50°C. Ozone appears at the anode.

Ozone has a characteristic pungent odor best identified as the odor around certain operating electric motors, mercury arcs, and electronic equipment. Gaseous ozone is a deeper blue than oxygen. It liquefies to a dark blue liquid at −112°C., and freezes to a dark purple solid at −193°C.

Structural studies suggest an angular molecule with a bonding angle of 116.8°. The atoms are σ bonded, with a delocalized π bond extending over all three atoms. Since the bonding angle approaches 120°, the bonds may be considered sp^2 hybrids, with the end atoms having two lone pairs and the central oxygen one.

Ozone is one of the most powerful oxidizing agents known. Its oxidizing potential in acid medium is exceeded only by fluorine, atomic oxygen, and fluorine monoxide, F_2O. It oxidizes silver to the silver(II) (argentic) state, and reacts with olefins to form ozonides (valuable for

---

[2] At low pressures, e.g., 1 mm. mercury, atomic oxygen forms.

location of double bonds). It is metastable under all conditions. Reconversion to molecular oxygen is catalyzed by many substances, e.g., chlorine, manganese(IV) oxide, platinum.

In concentrations greater than 100 p.p.m. ozone acts as a respiratory irritant. At concentrations less than 100 p.p.m. it is used to sterilize and deodorize air, especially in crowded and confined areas. The use of ultraviolet lamps in sterile areas is partially based on a continuing formation of ozone. It is also used for sterilization of swimming pools, etc. The generators for these applications are either a silent electrical discharge or an ultraviolet lamp. To be effective the u.v. spectrum must be rich in very short wavelengths in or near the so-called vacuum u.v.

OXIDIZING GERMICIDES—HALOGENS AND THEIR OXO COMPOUNDS

The halogens are in a zero or a positive oxidation state when used as germicides. Their germicidal action is accompanied by the reduction of the halogen atom to the negative oxidation state. For practical purposes, the compounds used are restricted to elemental chlorine and iodine and their oxo compounds. Bromine and the oxobromates should be active germicides, but their cost and the commercial availability of suitable chlorine compounds and elemental iodine are probably responsible for their nonutilization. Fluorides, e.g., fluosilicates, have been used in dusting powders, but only to a limited extent. Zinc fluosilicate solution is useful as a rinse after washing of diapers. It acts both as a bactericide and urease inhibitor to control diaper rash due to ammonia formation.

Chlorine and its oxo compounds. Although sodium chloride solutions are sometimes used as wet compresses, soaks, and baths (e.g., 0.5% for antieczematic action), elemental chlorine and the hypochlorites are of greatest importance. The chlorates (and permanganates) are also used. With the development of suitable methods of preparation and handling, chlorine dioxide and the chlorites are finding increasing application as powerful oxidizing agents in chemical applications (bleaches). The possibility of pharmaceutical suitability should be investigated.

SODIUM HYPOCHLORITE: $NaClO \cdot 6H_2O$. This compound is a white, deliquescent, crystalline solid melting at 18°C. It is prepared by passing chlorine into a freshly precipitated aqueous suspension of mercury(II) oxide.

$$2Cl_2 + 2HgO + H_2O = HgO \cdot HgCl_2\downarrow + 2HClO$$
$$2HClO + 2NaOH = 2NaClO + 2H_2O$$

(The corresponding compounds of the other halogens may be prepared by substituting the desired halogen for chlorine.) The sodium hypochlorite of commerce is, however, prepared by electrolysis of brine. The cathode reaction may be written

$$2H_3O^+ + 2e^- = H_2\uparrow + 2H_2O$$

Since two oxonium ions are consumed, in effect two hydroxyl ions remain in the solution. The anode reaction may be written

$$2Cl^- = Cl_2 + 2e^-$$

Vigorous agitation prevents the escape of the gaseous chlorine, assuring its complete solution and reaction to hypochlorite.

$$Cl_2 + 2OH^- \rightarrow ClO^- + Cl^- + H_2O$$

The chloride ion by-product is available for reoxidation at the anode. The sodium ions of the brine take no part in the chemical reactions.

The ostensible end result of these reactions should be a complete conversion of sodium chloride and water into sodium hypochlorite and hydrogen. However, the reaction cannot be carried to completion. As the reaction proceeds, the efficiency of conversion becomes so low as to become prohibitive economically. For this reason the sodium hypochlorite solutions of commerce have varying but appreciable quantities of sodium chloride as an impurity. In addition, in order to improve the stability of the product the pH is adjusted to 10 or greater. Sodium ferrate, formed at the steel anode during the electrolysis, imparts a yellow color to the crude product. If this color is objectionable, it is destroyed by addition of sucrose to the solution. The finished solution has 4–6% hypochlorite as the sodium salt. Its germicidal action is improved by reducing its pH before use.

$$ClO^- + H_3O^+ = HClO$$

Labarraque's Solution is a diluted form, prepared by 1:1 dilution of the approximately 5% sodium hypochlorite solution.[3]

HYPOCHLOROUS ACID: This is a much more effective oxidizing agent than hypochlorite ion or chlorine. Its formation in chlorine water, the other aspect of the chlorine(I) system of germicidal agents, is favored by mildly acidic conditions. Oxonium ions act to repress the ionization of the acid molecules.

$$Cl_2 + 2H_2O \rightleftharpoons HClO + H_3O^+ + Cl^-$$

$$HClO + H_2O \leftrightharpoons H_3O^+ + ClO^- \qquad pK_a = 7.44$$

[3] A. G. Labarraque first chlorinated lime in 1829. He recognized the strong germicidal properties of hypochlorite and subsequently applied it to sanitation engineering.

Hydroxyl ion, formed during hydrolysis of the hypochlorite ion, is partially neutralized at the same time.

$$ClO^- + H_2O \rightleftharpoons HClO + OH^- \quad (0.05\% \text{ in } 1N \text{ at } 25°C.)$$

The effect of pH on $[HClO]/[ClO^-]$ ratios is shown in Fig. 16.3.

The hypochlorous acid molecule is unstable; this accounts for its effective oxidizing action which in turn is responsible for the excellent germicidal and disinfecting action. The instability reflects its positive heat of formation[4] at 25°C.

$$\tfrac{1}{2}H_2 + \tfrac{1}{2}Cl_2 + \tfrac{1}{2}O_2 + (aq) = HClO \qquad \Delta H_f = +25.8 \text{ kcal./mole}$$

Due to this inherent instability the pure compound has never been isolated. It is known only in solution, with a maximum known concentration in water of 5%. Although dilute solutions are colorless, the concentrated solutions are a pale golden yellow and have a chlorine-like odor. Dilute solutions will distill, but efforts at concentration by distillation are thwarted by the heat labile nature of the molecules. The over-all decomposition reaction may be summed:

$$2HClO + 2H_2O = 2H_3O^+ + 2Cl^- + O_2\uparrow$$

Ultraviolet light and heavy metal cations cause the same decomposition.

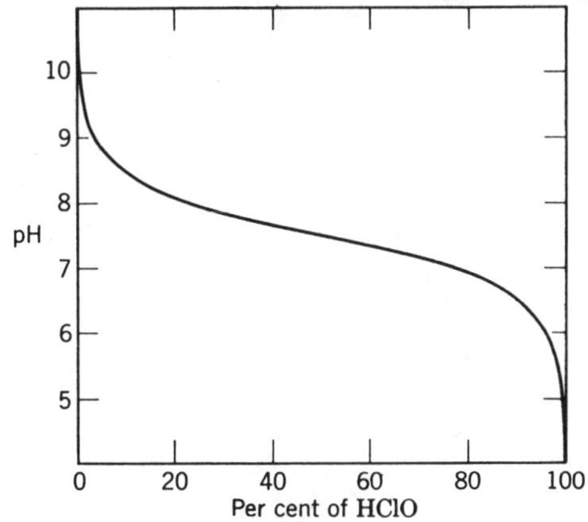

FIGURE 16.3 Hypochlorous acid: hypochlorite ratios as a function of pH.

[4] The energy absorbed or evolved when one gram-molecular weight of a compound is synthesized from its elements in their standard states. A positive heat of formation indicates that energy was expended during the synthesis; a negative value indicates the evolution of heat during synthesis. Compounds with positive values of this constant tend to be unstable.

Hypochlorous acid is also unstable in strong mineral acid solution (except nitric acid).

$$2HClO + 2Cl^- + 2H_3O^+ = 4H_2O + 2Cl_2\uparrow$$

This reaction is valuable as a qualitative test for hypochlorites. The evolution of chlorine on acidification is copious even with relatively dilute solutions. (Chlorine(I) compounds will release elemental iodine from iodides, another property valuable for identification of the hypochlorite moiety.)

The characteristic odor of hypochlorous acid solutions must not be confused with the odor of chlorine. The odor is caused by dichlorine monoxide, the gaseous acid anhydride of this acid.

$$2HClO \rightleftharpoons Cl_2O\uparrow + H_2O$$

HYPOCHLORITES AS GERMICIDES: The chlorine(I) series of compounds are excellent disinfecting agents with a strong deodorizing action. In neutral and slightly acid media they are effective germicides. However, unadjusted solutions of hypochlorites are unsuitable for application to wounds; their high pH is too destructive of protein. The solutions are relatively nonselective toward microorganisms. Pathogens of colon-typhoid and many pathogenic spores are sensitive to chlorine(I) compounds. Filterable viruses are inactivated. *Mycobacterium tuberculosis* resists destruction.

The effectiveness of these compounds is rapidly reduced in the presence of organic matter, probably because of an affinity of chlorine for the protein molecule. Chlorine reacts with the hydrogen of amino groups (page 455). Some chlorination probably takes place. The compounds cannot be used on colored materials because of their bleaching action.

Sodium hypochlorite solutions may be rendered suitable for use on wounds by adjusting the pH downward. However, reduction to the acid pH of maximum germicidal activity is inadvisable because of the concomitant decrease in the stability of the system. Since storage for variable periods of time is necessary, it is more practical to stop neutralization at the slightly alkaline range where stability is good and corrosiveness is greatly reduced. Diluted Sodium Hypochlorite (Modified Dakin's Solution) achieves a pH of approximately 8 by adjustment with sodium bicarbonate. The chemistry of this reaction has been discussed previously (page 184). However, the hydroxyl ion concentration at pH 8 is still sufficient to have a dissolving action on blood clots. This solution must not be used for injections.

OTHER HYPOCHLORITES: Recently lithium hypochlorite has become available commercially. It offers the convenience of a crystalline material.

The commercial product is about 30% LiClO, 30% NaCl, 20% [Na, K] SO_4 and small percentages of LiCl, Li_2CO_3, and LiOH. It dissolves readily, the pH of the resulting solution being about 10. It would seem to offer storage and handling convenience, if adaptable to pharmaceutical needs.

Chlorinated lime (bleaching powder, "calcium hypochlorite," chloride of lime) is one of the most effective and least expensive chemical disinfectants known. The composition of the compound has not been established, its name calcium hypochlorite notwithstanding. It is prepared by passing chlorine gas over moist slaked lime. Calcium hydroxide, hypochlorite, and chloride ions are identifiable in the product. (True calcium hypochlorite, $Ca(ClO)_2$, has been prepared.)

A 2% solution of bleaching powder is suitable for disinfecting urine, feces, and sputum (1–2 hours contact) and for washing down walls, etc. However, its powerful bleaching action must be considered when disinfecting colored materials. One to two grams per liter are effective for sterilization of water. The treated water should be exposed to air and allowed to settle before use. Chlorinated lime is used to decontaminate chemicals of the mustard gas type. It is also used as a laundry bleach.

CHLORATES: In addition to the chlorine(I) compounds a few chlorine(V) compounds and some related compounds of other elements have found medicinal application. Potassium chlorate and, less frequently, sodium chlorate have been used as oxidative anti-infectives (*N.F. X.*) Chlorate is irritating to the gastrointestinal tract and kidneys. Owing to hemolysis of the red blood cells, the compounds are toxic internally; 15 Gm. have been fatal to adults. Chlorate ion is not decomposed *in vivo*; it is slowly excreted unchanged in the urine. Within limits there is a cumulative effect. Chlorates should be used only over short periods.

Chlorate ion is an extremely feeble local antiseptic. The saturated solution of the potassium salt is without germicidal power. The idea that oxygen is released and acts as a germicide is a misconception. The over-all reaction

$$2KClO_3 = 2KCl + 3O_2\uparrow$$

takes place only at temperatures very much higher than physiological.

Potassium chlorate has been used as a local astringent cleansing solution for stomatitis, pharyngitis, and vaginitis. It has been used orally in gargles (to treat Vincent's Angina) and in mouth washes (1–4.5%). Tablets dissolved in the mouth are used for sore throat and stomatitis. (Sodium chlorate is used as a weed killer.)

Sodium chlorate is very soluble, potassium chlorate only slightly soluble: 1.0 Gm. $KClO_3$ dissolves in 16.5 ml. cold water, 1.8 ml. boiling; 1.0 Gm.

$NaClO_3$ dissolves in 1.0 ml. cold water, 0.5 boiling. The solutions of both salts are neutral. The substances are strong oxidizing agents. Violent to explosive reactions can take place if heated or triturated with organic substances, e.g., tannic acid, sucrose, charcoal, and inorganic reducing agents, e.g., sulfur, sulfide, phosphorus, hypophosphite, and iron. The chlorate ion will oxidize iodine to iodate.

PERMANGANATES: Since transition elements in their highest oxidation states resemble the typical elements of their group, manganese(V) and(VII) compounds resemble the corresponding chlorine compounds rather closely. This similarity extends to the use of the oxomanganese anions as oxidizing germicides. Potassium permanganate is official in the *U.S.P.* and has a wide range of medical applications for its germicidal, astringent, and oxidative properties. It also has an excellent deodorizing action. Dilute solutions are used for their powerful cleansing action in wounds and on mucous membranes.

Permanganate is useful in the 1:100 to 1:10,000 range of concentration. The solutions are bacteriocidal even in the lowest range. The 1:5000 or more dilute solutions are ordinarily nonirritating. Concentrations of about 1:5000 are used as an antieczematous agent, as wet compresses, and hand and foot soaks. This concentration is also used as a gastric lavage to destroy poisons such as morphine, strychnine, aconitine, and picrotoxin; it has no value for cocaine, phosphorus, or atropine. Dressings saturated with 1:500 to 1:300 solutions are used as wet compresses and wet dressings. For baths 1:40,000 to 1:60,000 solutions are used. For convenience in preparing these solutions potassium permanganate is available in 60, 120, 200, and 300 mg. tablets. Care must be exercised to insure complete solution, or particles of crystal will cause local chemical burns.

Since solutions of potassium permanganate are approximately neutral, manganese dioxide[5] is the product of its reduction.

$$2MnO_4^- + [R{-}SH] = 2MnO_2\downarrow + [R{-}SO_3^-] + OH^-$$

The manganese dioxide is deposited on the skin on topical use; a temporary brown discoloration results. Clothing and linens coming in contact with permanganate are badly stained or even ruined.

Care must be exercised when permanganate is in contact with reducing agents. Dangerously violent reactions may take place with solutions; trituration with such solids as those cited in the chlorate discussion will be explosive.

[5] In strong base manganese(VII) is reduced to manganese(IV); in acid media manganese(II) cation is the reduction product.

Manganese dioxide (*U.S.P. IX*) was at one time thought to possess a stimulating action on uterine mucosa. It was used for a time as a treatment for amenorrhea.

CHLORAMINES: The active structural feature of these compounds is the chloramine radical; the organic moiety merely serves as a carrier. The two principal compounds, chloramine (sodium *p*-toluenesulfonchloramide, tosylchloramine sodium, chloramine-T) and halazone (*p*-sulfonedichloramidobenzoic acid, *p*-sulfobenzoic acid dichloramide), slowly release hypochlorous acid by reaction with water.

$$[R{-}NCl_2] + 2H_2O = [R{-}NH_2] + 2HClO$$

The action of this group of compounds is similar to that of chlorine water and the hypochlorites. However, the disinfecting power is inferior. Because of the slow release of hypochlorous acid, these compounds have a longer lasting action than a simple hypochlorous acid or hypochlorite solution. Being less alkaline, they have less solvent action on necrotic tissue. Chloramine solutions are more stable than hypochlorite of similar strength. Because the hypochlorous acid is released only slowly, inactivation by organic matter is not as apparent. Halazone is available in 4–8 mg. tablets using sodium chloride as a base. A sufficient quantity of sodium carbonate or tetraborate is present in each tablet to provide a slightly alkaline medium. The use of these compounds for emergency sterilization of water has been discussed (page 151).

The chloramines form by reaction of chlorine water (or hypochlorite solution) with a dilute solution of ammonia or the desired amine.

$$NH_3 + HOCl = NH_2Cl + H_2O \qquad pH \sim 9$$

This reaction takes place even in the dilute media encountered in the control of microorganisms in potable water. In the presence of ammonia the superior action of chlorine and its oxo compounds is available only after chlorination of the ammonia molecule. The degree of chlorine substitution is a function of pH, as shown in Table 16.3. The parent compound

TABLE 16.3 EFFECT OF pH ON CHLORAMINE FORMATION

pH	Compound Formed	Reaction	*K*
<3	NCl_3	$H_3O^+ + 3NHCl_2 \rightleftharpoons 2NCl_3 + NH_4^+ + H_2O$	10^4
3–5	$NHCl_2$	$H_3O^+ + 2NH_2Cl \rightleftharpoons NHCl_2 + NH_4^+ + H_2O$	10^6
>8	NH_2Cl	$OCl^- + NH_3 \rightleftharpoons NH_2Cl + OH^-$	6.3×10^2

chloramine is a very weak acid.

$$NH_2Cl + H_2O \rightleftharpoons H_3O^+ + NHCl^- \qquad pK_a \sim 15$$

Iodine as a germicide. The element iodine is one of the best, if not the best, of the germicides. It is one of the most efficient chemical bactericides in current use. Its germicidal action does not vary greatly for the vegetative forms; it is effective against spores. It is one of the old established and still indispensable antifungals (together with Ammoniated Mercury Ointment and Calamine Lotion with Resorcinol). Iodine is also a virucide; for example, herpes simplex and rabies are affected. Iodine is effective over a wide pH range. Its action is rapid (10 p.p.m. are effective in 10 sec.). As a tincture the iodine and ethanol potentiate each others' action.

Germicidal applications are restricted almost entirely to elemental iodine. The iodide ion has little or no germicidal activity; it is used to solubilize the iodine as a polyiodide (page 370). The various iodine solutions in current use are listed in Table 16.4.

Iodine is difficulty soluble in dilute iodide solutions, but readily soluble in concentrated iodide. Thus, medicinal iodine solutions should be prepared by first dissolving the iodide in a minimum (25–50 ml.) of water or diluted alcohol. The iodine is then dissolved in this iodide concentrate. A mortar and pestle of proper size are useful tools in effecting the solution of iodine in the iodide concentrate. If necessary, additional solvent water (or diluted alcohol) may be added. After the iodine is completely dissolved, the solution should be transferred to a volumetric flask and made up to volume. Metallic equipment cannot be used in the preparation of iodine solutions or ointments since the element attacks most metals. Glass, hard rubber, and porcelain glaze are the only suitable materials for work involving iodine preparations. Metallic closures are similarly unsuitable; even rubber and cork are slowly attacked. Highly resistant plastic or glass closures should be used.

Tinctures were formerly solutions of elemental iodine only in alcohol. However, today the preference is for the use of NaI or KI as stabilizers. The vapor pressure of the iodine is thereby reduced. Sodium iodide is considered to have a more favorable action on tissue cells. Sodium ion being a normal constituent of body fluids, it is less likely to disturb natural equilibria.

Stronger iodine solutions have an irritant action. In fact, the one disadvantage of iodine solutions is the irritating action on the tissues of wounds, even in dilute solutions. However, the effectiveness of iodine preparations as germicides more than warrants their preferential use over the "painless" preparations having only bacteriostatic properties, e.g., organic mercurials.

All iodine preparations are chemically stable in glass containers, whether clear, blue, or amber. The solutions tend to become more concentrated

TABLE 16.4 IODINE SOLUTIONS

Solution	Components			Category
	Iodine (Gm.)	Iodide (Gm.)	Solvent[a] (mL.)	
Iodine Tincture *U.S.P. XXI*	20	24 NaI	500 mL. ethanol + Purified water	Anti-infective
Topical Iodine Solution[b] *U.S.P. XXI*	20	24 NaI	Purified water	Anti-infective
Strong Iodine Tincture *U.S.P. XXI*	70	50 KI	50 mL. Purified water + ethanol	Anti-infective irritant
Strong Iodine Solution[c] *U.S.P. XXI*	50	100 KI	Purified water	Iodine source Anti-infective
Iodine Solution[d] *(B.P.)*	100	60 KI	Purified water	Local irritant counterirritant

[a]All preparations on final one liter basis.
[b]Iso-osmotic.
[c]Compound Iodine Solution, Lugol's Solution, Schiller's Solution.
[d]Too strong for use as germicide on raw surface; intended for use on whole skin.

(and more irritant) in household use because of evaporation of solvent through improper closure.

IODINE COMPLEXES: Much research effort has been expended in attempts seeking to minimize the irritation and volatility of iodine without loss of its extraordinary germicidal activity. These efforts have produced the povidone, polyvinylpyrrolidone (1-vinyl-2-pyrrolidinone polymer, 2-pyrrolidinone-1-ethenyl-homopolymer, PVP) complexes and the polyiodides of organic cations.

Povidone-Iodine (PVP-iodine, PVP-I) is a water-soluble, chemically stable complex of polyvinylpyrrolidone and iodine. The oral toxicity of the iodine is reduced, permitting internal use of larger amounts. When used on open wounds, the "sting" is slight or eliminated. However, the product has all the typical germicidal and other useful properties of ordinary iodine solutions. The iodine is absorbable on ingestion.

When iodine is mixed with povidone, there is some chemical reaction between the two, possibly oxidation of unsaturated end groups by the iodine. Approximately one third of the initially added iodine appears as iodide ion as a result of this reaction. A very small quantity of the iodine becomes organically linked with the povidone. All the remaining iodine is available as such and titratable with thiosulfate.

The bond between povidone and the available iodine is not fully understood. But a rather strong bond must be present since the vapor pressure of the iodine is reduced virtually to zero. The iodine is not extractable by organic solvents such as carbon tetrachloride. Yet this iodine is available therapeutically and undergoes all of its characteristic chemical reactions.

Structurally it is thought in the presence of iodine the povidone assumes a helical arrangement; the iodine atoms line up inside the helix. The structure is very similar to that of the reaction product between starch and iodine. It has been postulated one iodine atom is bound per monomer unit of povidone. Povidone will bind iodine to the extent of approximately one part iodine to two to three parts of povidone. Above this ratio the vapor pressure of iodine rises sharply and iodine can be extracted with organic solvents. The most common ratio is that achieved by reacting 15 Gm. iodine with 85 Gm. povidone. Approximately 10 Gm. of iodine are available; 5 Gm. are converted to iodide during the reaction.

Povidone is much more soluble than iodine. Uncomplexed, elemental iodine will dissolve to about 0.034% in water at 25°C. A 1% aqueous povidone solution will dissolve iodine until a 0.58% iodine solution is attained. With larger amounts of povidone, solutions approaching 10% available iodine may be prepared. Comparative data show that povidone-iodine solutions are as stable or slightly more so than Lugol's Solution. The stability of Iodine Tincture is poor by comparison.

Povidone-iodine powder may be prepared by direct reaction between povidone powder and crystalline iodine. Alternatively, individual solutions may be mixed and the water evaporated. The powder is stable and may be redissolved in water to prepare the desired concentration. Povidone and Povidone-Iodine are official in *U.S.P. XXI.* The present official povidone-iodine preparations are given in Table 16.5.

A variety of other iodine complexes have been prepared, all claiming reduced irritation, etc. Some of these are given in Table 16.6. All are water soluble.

Although the cationics have some inherent antiseptic action, the antibacterial action of their iodine salts is due almost entirely to the iodine. The wetting action of the cationics facilitates contact of the iodine with the surface areas. The iodine is slowly released to the skin and mucous membranes.

The nonionic iodine complexes are more stable in solution; the vapor pressure of the iodine is lowered. They do not stain and are nonirritating

TABLE 16.5 Povidone-Iodine Preparations (*U.S.P. XXI*)

Preparation	Composition	pH	Category
Povidone-Iodine Topical Aerosol Solution	as labelled	<6	Anti-infective
Povidone-Iodine Topical Solution	as labelled	<6	Anti-infective
Povidone-Iodine Cleansing Solution	as labelled	1.5-6.5	Anti-infective
Povidone-Iodine Ointment	as labelled	1.5-6.5	Anti-infective

to the skin and eyes. They possess germicidal activity as long as they have visible color. The typical complex of this kind is prepared by dissolving iodine in the surfactant at 50°C. or higher. About 20% of the iodine in the resulting complex is unavailable; i.e., it is inactive as a germicide. These nonionic formulations are active only in acidic media. A typical preparation will include 30% complex (about 3% available iodine), hydrogen chloride 0.2%, and inerts, 69.8%.

The nonionics have been widely applied in sterilizing and sanitizing techniques. A sanitizing rinse requires less surfactant than a product intended to clean as well as act as a germicide. In formulations of this type three ingredients, iodine, surfactant, and acid, must be at proper levels. The ratio of surfactant to iodine is determined by the detergency needed. The ratio of available iodine to acid fixes the allowable dilution in use. Surfactants, in addition to the iodine complex, may be included to control properties such as storage stability, foaming tendency, solubility, viscosity.

The sanitizing rate is an important factor in these formulations since it determines the contact time required for effective sterilization. As an example of the importance of the sanitizing rate, consider the usual dishwashing techniques in restaurants and taverns; contact time in the sanitizing rinse is 5–15 seconds. Iodine sanitizing solutions of the non-ionic type are generally as effective in 10 seconds as in one minute, with iodine concentrations of 10 p.p.m. or greater. Experimental testing indicates a more effective action than either chlorine or cationics.

The miscellaneous compounds shown in Table 16.5 are claimed to have better solubility, greater stability, and lessened irritation than the simple iodine solutions. The germicidal properties are unimpaired. Tetra-

TABLE 16.6 Iodine Complexes

Cationic Surface Active Salts	
$(CH_3)_4N^+I_3^-$	Tetramethyl ammonium triiodide
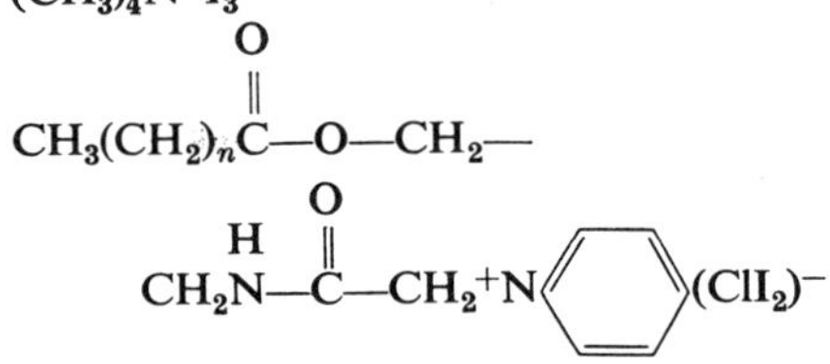	Benzene undecoylium chloride-iodine
(n = 8–14, preferably 10 or 11)	
Nonionic Surface Active Complexes	
	Polyethoxopolypropoxypoly-ethoxyethanol iodine complex
	Nonylphenoxypolyethoxyethanol iodine complex
Miscellaneous Salts	
$(NH_2CH_2COOH)_4KI_3$	Tetraglycine triiodide
$(NH_2CH_2COOH)_4HI \cdot 1\frac{1}{2}I_2$	Tetraglycine hydroperiodide
$Al[CO(NH_2)_2]_6(NO_3)_2I_3$	Aluminum hexaurea dinitrate triiodide
$Al[CO(NH_2)_2]_6(SO_4)I_3$	Aluminum hexaurea sulfate triiodide

glycerine hydroperiodide contains 39.5–42.6% active iodine and is used as a water purifier.

MISCELLANEOUS IODINE COMPOUNDS: Iodine trichloride is a powerful germicide. It is a yellow, brownish red, crystalline material soluble in water, ethanol, and diethylether. The 5% solution is fairly stable if protected from light. The 1% solutions are used as disinfectants.

Potassium iodide, 7.5% solution, has been used for eye drops. It is probable that traces of iodine formed by air oxidation are present. This iodine is tightly bound in triiodide ion. The solution would be relatively nonirritating, but possesses the germicidal properties of a dilute iodine solution.

Potassium tetraiodomercurate(II) (potassium mercuric iodide) is used as a germicide. It will be discussed in the section dealing with mercury germicides (page 481).

STUDY AND SELF-APPRAISAL AIDS

IMPORTANT TERMS. Define or identify each of the following in a single concise sentence if possible.

Acne, algaecide, antiseptic, astringent, bactericide, bleach, caustic, comedo, corrosive, deodorant, disinfectant, escharotic, fungistat, gastric lavage, germicide, heat of formation, herpes simplex, iodism, irritant, necrotic, pathogen, spore, sterilization, styptic, surface active agent, surfactant, virucide, wet compress.

NOMENCLATURE OF INTEREST. Identify each of the following.

Aluminum hexaurea sulfate triiodide, Aromatic Sodium Perborate, bleaching powder, calcium hypochlorite, chloramine, chlorinated lime, chlorine dioxide, chlorine monoxide, chlorine water, halazone, hypochlorous acid, Iodine Tincture, Labarraque's Solution, lithium chlorite, Lugol's Solution, manganese(IV) oxide, ozone, peroxoacetic acid, perchloric acid, peroxodisulfuric acid, potassium chlorate, potassium tetraiodomercurate(II), potassium tetraoxomanganate(VII), sodium dioxide, sodium monoxochlorate(I), sodium perborate, Strong Iodine Solution, tetraglycine triiodide, zinc peroxide.

SELF-APPRAISAL QUESTIONS

1. Explain the relationship which exists between the electrostatic field of a cation and its ability to act as a germicide. Which cation types have suitable fields?

2. Explain the relationship between oxidation number of a nonmetal, either in the elemental or a positive oxidation state, and its ability to act as a germicide. Give some illustrating compounds.

3. Explain why a single substance can be a germicide, astringent, irritant, and corrosive.

4. Name the three methods by which the pharmacist can control the effective concentration of a compound to give only a germicidal action. Illustrate each with an actual case.

5. What are the requirements of an ideal germicide? Select a substance you believe closely approaches the ideal. Give the reasons underlying your choice.

6. Compare the electronic structures of H_2O, H_2O_2, O_2, H_2, H_3O^+, H^+, OH^-, $[O]^{2-}$, $[O_2]^{2-}$, and $[O_2H]^-$. How do these structures influence the chemical properties of each? Explain your answer fully.

7. Give the important chemical and physical properties of hydrogen peroxide. How are these properties important in the medicinal use of this compound?

8. Explain why each of the following has a stabilizing action on hydrogen peroxide: phosphoric acid, sodium pyrophosphate, hydrous silicic acid, acetanilide, cation exchange resin.

9. Give and relate the methods for expressing hydrogen peroxide concentrations.

10. Compare the action and uses of 3%, 6%, and 30% hydrogen peroxide.

11. Distinguish between true peroxides, e.g., Na_2O_2, superoxides, e.g., KO_2, peroxohydrates, e.g., $Na_2CO_4 \cdot H_2O_2 \cdot 2H_2O$, and "peroxide-like" compounds, e.g., MnO_2.

12. Calculate the theoretical volume of oxygen measured at S.T.P. released from: 1.42 Gm. potassium superoxide, 100 ml. 3% hydrogen peroxide, 250 ml. 20 volume hydrogen peroxide. How would these volumes change if the gas was measured at 27°C. and 665 mm. Hg pressure?

13. Study the assay for Medicinal Zinc Peroxide (*U.S.P. XVII*) and explain why it is used rather than a gravimetric (or volumetric) determination based on the zinc. Would a permanganate titration be suitable? Explain.

14. How is ozone prepared?

15. Give the oxidation numbers common to the halogen family. Indicate those which appear in the halogens of pharmaceutical interest and give several illustrations for each category.

16. Give the electronic structure of hypochlorous acid. On the basis of structure and the atomic properties of its constituents explain: its weakness as an acid; its strong oxidizing action; its weak reducing action; the solubility of its sodium salt; its instability to heat, u.v., and strong mineral acids.

17. Why is sodium hypochlorite unsuitable for use on wounds but suitable as a disinfectant for inanimate objects? Name a property of the hypochlorite ion which may give problems on certain inanimate objects.

18. Compare the germicidal action of solutions of the following: chlorine, hypochlorous acid, potassium hypochlorite, and chloramine. Give advantages and disadvantages of each.

19. Why is sodium iodide sometimes preferred over potassium iodide in preparation of the polyiodide solutions? Why are the polyiodides used rather than strictly aqueous solutions?

20. Name and give the formulas of those polyiodides which have been identified. What is the advantage of polyiodide over an alcoholic solution of iodine?

21. Would other halide ions form polyhalide-like anions with iodine?

22. Give uses and advantages and disadvantages of each of the following: bleaching powder, ozone, povidone-iodine, hydrogen peroxide,

potassium permanganate.

23. Indicate those of the compounds and preparations discussed which are official in the current compendia. Which have cautions? Why is it considered important to have the caution in each of these cases?

24. Discuss the reaction(s) taking place during the preparation of providone-iodine.

PRACTICE IN EQUATION WRITING

25. Using R—SH as the formula of a protein, write equations for the germicidal action of: hypochlorite ion, iodine, hydrogen peroxide, silver ion, and arsenic(III).

26. Write the equations showing the effect of changing pH on an aqueous chlorine solution.

27. Write the equation(s) for the present commercial method for preparation of: hydrogen peroxide, sodium hypochlorite solution, iodine, bleaching powder, sodium perborate, and sodium peroxide.

28. Write the equation(s) for preparation of each of the following: Lugol's Solution, Strong Tincture of Iodine, potassium tetraiodomercurate(II), Modified Dakin's Solution.

29. Write the equations for the reactions taking place between each of the following: ozone and silver(I) oxide; sulfuric acid and potassium hypochlorite; hydrogen peroxide and a solution of sulfur dioxide in water; sodium peroxide and potassium permanganate; heat and hypochlorous acid; sodium chloride and iodine; iodine and a steel closure; permanganate ion and elemental phosphorus; sodium monoxochlorate(I) and water.

30. Write equations for a qualitative test for each of the following: a hypochlorite, hydrogen peroxide, a permanganate, chlorine, iodine, an iodide, a triiodide, a metallic peroxide.

31. Write equations for any possible untoward reactions which will take place if the following prescriptions are filled.

a. Rx	Sod. Iodide	gr x	**b.** Rx	Pot. chlorate	0.50
	Sol. hydrogen peroxide	℥ viii		Tannic acid	0.25
				Saccharose	0.25

c. Rx	Ferrous Iodide Syrup	8.0
	Pot. Iodide	8.0
	Iodine Tincture	10.0
	Aromatic Elixir	50.0

SEVENTEEN · GERMICIDES AND RELATED SUBSTANCES II: *protein precipitants and complexers*

This chapter surveys substances with one common property: the ability to inactivate proteins by precipitation or complexation. The compounds to be studied are those of elements which for the most part arrange into three major groups associated within well-defined areas of the periodic table. Most protein precipitant germicides and corrosives are found in Groups IB and IIB. The astringents are amphoteric elements located on, or to either side of, the diagonal line dividing the metals from the nonmetals. The inorganic compounds useful for control of parasitic infestations are usually those of elements of the arsenic subgroup. Because of this correlation, except for a brief review of the germicidal properties of acids and bases, presentation is based on families of elements.

ACIDS AND BASES

Earliest recorded history recognizes the germicidal action of acids and the corrosive action of bases. Over the centuries vinegar and sour wine have been used to dress wounds and for the preservation of foods. The inhibiting action of acids for microorganisms is attested by sour foods such as sauerkraut, pickles, and meats preserved with sour milk.

Inorganic acids. Diluted hydrochloric acid is an active germicide. A 0.5% solution is an efficient skin germicide. A 1:20,000 solution is used as a surgical antiseptic.

Nitric acid (concentrated and dilutions down to 10%) has been used as a caustic on warts, chancres, etc. It causes a characteristic yellow stain on contact with the skin. Ingestion causes a burning sensation; carbonated alkali should not be used as an antidote.

The pure concentrated acid is colorless. However, on exposure to light and air nitrogen oxides form, giving the solution a yellow to dark brown color. The color disappears on dilution and has no deleterious effect.

$$N_2O_4 + H_2O = HNO_3 + HNO_2$$

The concentrated acid undergoes self-ionization.

$$2HNO_3 \rightleftharpoons H_2NO_3^+ + NO_3^-$$

This reaction is catalyzed by concentrated sulfuric acid. Disproportionation also takes place

$$2HNO_3 \rightleftharpoons N_2O_5 + H_2O$$

followed by

$$N_2O_5 \rightleftharpoons NO_2^+ + NO_3^-$$

The nitronium ion, NO_2^+, gives rise to nitronium salts. (Nitrous acid gives rise to the nitrosonium ion, NO^+, and its salts.) Dilution restores the normal ionization pattern.

Nitric acid is a strong acid and strong oxidizing agent. Nitrogen oxides are essential for the latter action. This acid is used in its oxidizing capacity for preparation of Ferric Subsulfate Solution and Mercuric Nitrate Ointment.

Sulfur dioxide, sulfurous acid, and the sulfites find application in pharmaceutical chemistry largely as antoxidants (page 239). Dilute sulfurous acid solutions ($SO_2 + H_2O$) are used for treatment of parasitic skin diseases. Very dilute solutions are used as gargles.

Chromium(VI) oxide (Chromium Trioxide *N.F. X.*, chromic acid, chromic oxide) is a powerful caustic both in crystalline form and in concentrated solution. The 20% solution is used as a caustic for removal of growths, e.g., warts. A 5% solution is used as an antiseptic astringent wash for leukorrhea, ozena, hyperkidrosis (perspiring feet). The 2–3% solutions are used as astringents. Solutions in the range 0.05–0.025% have been used as urethral injections. Potassium dichromate has been used in treatment of gastric ulcers, 0.10–0.16 grain.

Owing to their local irritating effect and systemic action, chromic acid and its salts are powerful poisons. External burns result in slow healing ulcerations. Inhaled dust containing these compounds is irritating to the nasal mucosa, causing inflammation and ulceration.

Chromic acid is prepared by treating potassium dichromate (or chromate) with concentrated sulfuric acid:

$$2CrO_4^{2-} + 3H_2SO_4 \rightarrow Cr_2O_7^{2-} + 2SO_4^{2-} + H_2SO_4 \cdot H_2O$$

$$Cr_2O_7^{2-} + H_2SO_4 \rightarrow H_2Cr_2O_7 + SO_4^{2-}$$

$$H_2Cr_2O_7 \rightarrow H_2O + 2CrO_3\downarrow$$

$$H_2SO_4 + H_2O \rightarrow H_2SO_4 \cdot H_2O$$

The precipitated chromium(VI) oxide is filtered off, washed with concentrated nitric acid, and dried in a current of air. Because of the powerful oxidizing action, care must be exercised when it is in contact with reducing agents; explosive reactions may result.

Chromyl chloride, CrO_2Cl_2, the acid chloride of chromic acid, is a mobile red liquid, boiling at 116°C. It is nonpolar and miscible in all proportions with organic solvents, e.g., carbon tetrachloride and carbon disulfide. It is used as an oxidizing and chlorinating agent.

Organic acids: Acetic acid inhibits both gram negative and gram positive pathogens in as low as 0.33% solution. A 0.5% solution is a bacteriostat for most common bacteria and is used as a cleansing agent and moist wet compress for infected burns and other skin conditions. A 1% solution is bacteriocidal to many microorganisms. Alkali burns may be treated with a 3% (or lower) solution used with copious quantities of water.

In higher concentrations acetic acid is an active astringent and is often used for this action in treatment of skin diseases. Glacial acetic acid formerly used as a caustic for corns, warts, etc., has been replaced by the more efficient trichloracetic acid. Glacial acetic acid is deliquescent.

Glacial Acetic Acid is cited officially in *U.S.P. XXI* as is Acetic Acid Irrigation. The latter has a concentration of 250 mg. acetic acid per 100 ml. Acetic Acid (36-37%) is official in *N.F. XVI* as is Diluted Acetic Acid (6%). Common household vinegar is about 5% acetic acid.

Several other organic acids and their salts have uses related to germicidal activity; e.g., benzoic, salicylic, propionic, and mandelic acids.

Bases as caustics. The strong soluble bases are destructive of tissue. The compendia include cautions concerning their escharotic action. Although not in common use for such purposes, potassium hydroxide and sodium hydroxide have been used as cauterizing agents. Sodium hydroxide is used as a caustic to soften epidermis. Lime water has astringent action.

GROUP IB—THE COINAGE METALS

General group properties. Copper and gold (and slightly tinted bismuth) are the only colored metals. Copper, silver, and gold are the softest, most malleable and ductile of the metals, and are the best conductors of heat and electricity.

The coinage metals are relatively inert chemically. Because of their historical association with the reactive metals of Group IA this has not always been appreciated. The underlying electron shells of the two different subgroups involved impart entirely different reactivities even though both have the ns^1 external valence structure. The coinage metals

TABLE 17.1 ATOMIC CHARACTERISTICS OF THE COINAGE AND THE VOLATILE METALS

Element	Copper	Silver	Gold	Zinc	Cadmium	Mercury
Symbol	Cu	Ag	Au	Zn	Cd	Hg
Atomic weight	63.54	107.9	197.0	65.38	112.4	200.6
Atomic number	29	47	79	30	48	80
Valence electron structure	$3d^{10}4s^1$	$4d^{10}5s^1$	$5d^{10}6s^1$	$3d^{10}4s^2$	$4d^{10}5s^2$	$5d^{10}6s^2$
Atomic Radius (Å)	1.17	1.34	1.34	1.25	1.41	1.44
Ionic radius (Å) I	0.96	1.26	1.37	—	—	—
II	—	—	—	0.74	0.97	1.10
Ionization potential (ev.), I	7.72	7.87	9.22	9.39	8.99	10.43
II	—	—	—	17.89	16.84	18.65
Reduction potential (v.), M/M^+	+0.522	+0.80	+1.68	—	—	(+0.80 I)
M/M^{2+}	+0.345	+1.39	(+1.42 III)	−0.763	−0.40	+0.854
Electronegativity	1.8	1.4	1.4	1.7	1.5	1.4

are transition elements and should be compared with their horizontal neighbors rather than the alkali metals.

The important structural characteristics of these elements are found in Table 17.1. The ground state elements have the structure $(n-1)\,d^{10}ns^1$. Though a closed shell, the underlying group of ten *d* electrons is unstable in this family. The three elements appear in the +1 and +2, and with gold the +3 oxidation states. The +2 and +3 compounds are colored. The ions are oxidizing agents; a given element in the +2 state is a stronger oxidizing agent than when in the +1 state. All can undergo autoxidation when in the +1 state; e.g.,

$$2Cu^+ = Cu^0 + Cu^{2+} \qquad K = 1.2 \times 10^6$$

Because of the underlying $18e^-$ subgroup, the complexes of copper, silver, and gold are comparatively stable. In the +1 oxidation state a coordination number of two is common, e.g., $Na^+[CuCl_2]^-$. In the +2 oxidation state four coordinate square planar complexes are common (page 74). Some six-coordinate copper complexes with a distorted octahedral structure are known. Simple gold(III) salts are very unstable, especially at higher temperatures. The name "gold chloride" is given to the stable tetrachloroaurate(III) compound, $HAuCl_4{\cdot}4H_2O$, rather than the unstable gold(III) chloride, $AuCl_3$.

Most of the common coinage metal salts are soluble in aqueous media. Except for fluorides, the solubility of a given +1 salt follows that of the corresponding silver salt. (Cu_2F_2 is insoluble, AgF is very soluble). The salts are all light sensitive, those of copper only slightly so. Copper(II) iodide is unstable in aqueous media, disproportionating

$$2Cu^{2+}I^-_2 \rightarrow Cu_2I_2\downarrow + I_2$$

Copper and its compounds. Soluble copper salts give evidence of gastroenteritis when taken orally. Vomiting is usually so prompt that systemic poisoning does not occur, absorption being possible only with a corroded stomach. Soluble copper salts have an astringent action. Both soluble and insoluble copper salts have antifungal action. Copper(II) Chlroide, ($CuCl_2{\cdot}2H_2O$) and Copper(II) Sulfate, ($CuSO_4{\cdot}5H_2O$) are cited in *U.S.P. XXI.*

COPPER(II) SULFATE: Cupric sulfate, blue vitriol, bluestone, Roman vitriol, Salzburg vitrol, is toxic to microorganisms, especially protozoa, fungi, and algae. It is used as an algaecide in water treatment, usually 1:1,000,000. This concentration is so low that humans and fish are unaffected. The compound finds little employment as a medicinal agent. It has irritant, astringent, and antiseptic properties. It is not a reliable bactericide, though 0.5-1% solutions have been used as collyria, and urethral injections. A 1-2% solution is effective as a fungicide.

Copper(II) sulfate finds wider employment in nongermicidal applications, as a hematopoeitic (page 396), emetic (page 581), and poison antidote (page 585). It is the active component of Fehling's and Benedict's Solutions, the classical test solutions for reducing sugars. White, anhydrous copper(II) sulfate is used as a test for moisture in nonpolar organic solvents; if water is present the blue color of the pentahydrate is restored.

COPPER(II) CITRATE: Cupric Citrate (*U.S.P. XIII*); $Cu_2C_6H_4O_7{\cdot}2\frac{1}{2}H_2O$. This compound is somewhat less irritating. An 8% ointment has been used as an astringent antiseptic. It is prepared by reaction of hot aqueous copper(II) sulfate with sodium citrate. Two chelate rings per copper(II) ion form, the citrate ion acting as a tetradentate ligand.

OTHER COPPER SALTS: Copper(II) undecylenate, fluosilicate, and acetate are fungicides. Tetramminecopper(II) sulfate has been used for peronospera. Copper(II) oxide has been used for control of tapeworms. Copper(II) phenolsulfonate (hexahydrate) has been used as an anti-infective, styptic, and hemostatic.

An electrolytically prepared mixture of very finely divided copper and copper(I) oxide has been found effective as a barnacle-preventative paint on ships' hulls. Copper compounds are widely used as insecticides. These include Paris green (Vienna green, Pairot green, Schweinfurth green), $2Cu(AsO_2)_2{\cdot}Cu(OAc)_2$; Scheele's green (Swedish green), $CuHAsO_3$, copper(II) borate and copper(I) chloride. Copper(I) chloride is again as toxic as copper(II) chloride.

Silver and its compounds. The oligodynamic property of silver is interesting. Water in contact with silver metal dissolves the metal in levels lethal to suspensions of viable bacteria. Consequently, water distilled through a silver condenser is sterile for long periods of time. The lower effective concentration is 1:20,000,000.

Only silver(I) compounds are of interest in medicine. Solutions of the soluble salts are germicidal, astringent, irritant, or corrosive, depending on the concentration used. The insoluble compounds exhibit only germicidal action; the solubility products do not allow enough dissolved silver ion for the appearance of the other properties.

Silver ion precipitates both the albumin and the chloride ion of the tissue fluids. Trapped within the wound, the precipitates maintain a continuing germicidal action by slow release of the silver ions. The precipitating action also localizes the irritating effect of the silver ion within the area under treatment.

Despite the ability of silver to serve well in all four categories, silver salts are used infrequently for several cogent reasons. They are too poisonous for general use; 10 Gm. silver nitrate are toxic. Cosmetic problems result from their use; finely divided, black silver metal is deposited in or on the tissues because of photosensitivity. Irremediable discoloration of the skin or mucous membranes (argyria) may result from continued use. Finally, silver salts are "costic"; the price of silver compounds is comparatively high.

Silver compounds have properties leading to several types of incompatibilities. Because of light sensitivity, silver compounds and preparations must be stored and dispensed in dark bottles—only freshly prepared solutions or suspensions may be dispensed. Even mild reducing agents will deposit metallic silver as grey to black particles. Volatile oils, sugars, tannins, hypophosphites, iron(II), among others cause reduction.

SILVER NITRATE: Lunar caustic (*U.S.P. XXI*). Despite a distinctive preantibiotic record, this compound finds little employment today. In dilute solution it has excellent germicidal action. The concentrations used depend upon the degree of astringency or irritation desired or tolerated. Allergic reactions are highly unlikely. Excessive application can be neutralized by an application of sodium chloride solution.

Concentrations as low as 0.003% possess anti-infective action; 0.01% has been used for urethritis. Germicidal action is good at the 0.025% level; 0.1 to 0.2% solutions are irritating to the skin. Concentrations as high as 10% are used, e.g., 4–5% solutions as a swab in tonsilitis and diphtheria. All solutions are slightly acid. As the solid or in very concentrated solution, silver nitrate has a powerful escharotic action. Penetration is limited by the formation of a very dense coating of silver albuminate; the deposit is initially white but it gradually darkens owing to the action of light.

SILVER NITRATE OPHTHALMIC SOLUTION *(U.S.P. XXI)*: this compound is intended for routine instillation in infants' eyes at birth for prophylaxis of ophthalmic neonatorum.[1] It is a 1% solution which may be buffered at a pH of 4.5–6.0. It is available in wax vials containing a volume sufficient for an instillation.

TOUGHENED SILVER NITRATE *(U.S.P. XXI)*: Fused silver nitrate, lunar caustic, molded silver nitrate, silver nitrate pencils. This compound is silver nitrate rendered less brittle by the controlled inclusion of suitable additives. It is produced by fusing silver nitrate with either sodium chloride or concentrated hydrochloric acid to give a product with 5% silver chloride. It is also prepared by including 5% potassium nitrate. A product prepared with 1.4% sodium chloride plus 4% (w/w) hydrogen chloride has been reported to be less brittle than either. Toughened silver

[1] Such instillation is required by law in many states. Sulfonamides and penicillins are also effective.

nitrate is used as a caustic.

AMMONIACAL SILVER NITRATE SOLUTION *(N.F. XII)*: Howe's Solution. This is a solution of diamminesilver(I) cation in excess ammonia. It is prepared by adding strong ammonia solution, slowly and with constant stirring, to an aqueous solution of silver nitrate until the initial precipitate of silver hydroxide redissolves.

This solution is used in dermatological practice because of its ability to actively penetrate keratin. Mixed with formaldehyde or eugenol just before use, it is used as a dental protective, e.g., in a newly drilled cavity. A deposit of finely divided silver metal forms.

$$2[Ag(NH_3)_2]^+ + HCHO + OH^- = 2Ag^0\downarrow + 2NH_4^+ + HCOO^- + 2NH_3$$

TABLE 17.2 COMPARISON OF SILVER PROTEINS

	Mild	Strong
Protein	Serum albumin, casein, or alkaline gelatine	Partially hydrolyzed peptone and albumose
Total silver	19–23%	7.5–8.5%
Ag^0, Ag_2O, Ag proteinate	high	low
Ag^+, free	low[a]	high
Germicidal action	good	good
Irritation	negligible	irritating
Concentrations of aqueous suspension employed	1–20%[b]	0.25–0.5%

[a] The heated reaction mixture is allowed to react until a soluble chloride no longer precipitates silver chloride.
[b] May be stabilized with 1% EDTA.

OTHER SOLUBLE SILVER SALTS: Silver lactate (monohydrate) is used as a 0.05–1.0% solution in applications similar to those of the nitrates. A 1–2% solution of silver picrate is used for its germicidal action for control of *trichomonas vaginitis*, *monilia albicans*, and for urethritis. This compound is also used as an astringent. Sodium dithiosulfatoargentate(I), $Na_3[Ag(S_2O_3)_2]$, has a local action similar to silver nitrate. Irritant action is less than for corresponding concentrations of silver nitrate. Potassium dicyanoargentate(I) (silver potassium cyanide) has been used as a bactericide and as an antiseptic wash (0.5%). Silver arsphenamine (exact composition unknown) has been used as an antisyphilitic.

INSOLUBLE SILVER COMPOUNDS: These compounds are useful germicides and are generally nonirritating. Their action depends upon a slow release of silver ion from the insoluble compound.

Mild Silver Protein (*N.F. XIII)* and Strong Silver Protein are compared in Table 17.2. Both are used as antibacterials. Despite a lower total silver content, strong silver protein has a higher level of active free silver ion. Its solutions have an irritant action. Mild silver protein solutions were used in the following applications: rhinitis, tonsillitis, 0.5-10%; conjunctivitis, 2–10%; urethritis, acute, 0.25–1%, chronic, 10%; cervicovaginitis, 2–10%. Only freshly prepared suspensions should be dispensed, in amber bottles.

Colloidal Silver Iodide (*N.F. X*) is a mild local nonirritating, non-staining anti-infective. The concentrate is prepared by reacting an aqueous solution of silver nitrate with potassium iodide in the presence of gelatine. The by-product potassium nitrate is removed by dialysis. The dialyzed suspension is evaporated to dryness.

The suspension is reconstituted by addition of water before dispensing. Hot water must be used for concentrations greater than 25%. The reconstituted suspensions tend to become unstable after about one week.

Colloidal Silver Chloride (*N.F. IX*) is also used as a germicide, as a 9–11% suspension. It is stabilized with sucrose.

Silver oxide (*U.S.P. IX*) was formerly used for chorea, epilepsy, and venereal sores. A 5% ointment was used for the latter. ("Hopcalite," a mixture of colloidal manganese(IV) and copper(II) oxides, activated with silver and cobalt oxides, is used to detect carbon monoxide.)

REMOVAL OF SILVER STAINS. Contact between soluble silver salts and their solutions and clothing or the skin results in the formation of finely divided black particles of metallic silver. Light and reducing substances are responsible. The spots will gradually wear off the skin. If it is desirable to remove the stains it is recommended[2] that the affected area be painted with tincture of iodine. When dry, the treated area is painted with a sodium thiosulfate solution and rinsed. Several repetitions of the procedure may be required. Silver stains may be removed from clothing by judicious use of potassium cyanide solution. However, this should be done only by individuals with an appreciation for the safe handling of this dangerously poisonous substance.

Gold and its compounds. **A 2% solution of gold chloride has been used as an ophthalmic preparation. Except for the radioactive gold preparations, the only current use of gold compounds is as antirheumatics (page 573).**

GROUP IIB—THE VOLATILE METALS

General properties. The structural characteristics of these elements have been summarized in Table 17.1. The outstanding difference between

[2] U.S.D.A. Bulletin No. 1474.

these elements and those of Group IB is the stability of the $18e^-$ subshell. The ns^2 electrons only are available for reaction, either by formation of divalent cations, e.g., $Zn(NO_3)_2$, or by sharing, e.g., $HgCl_2$. Mercury(II) (mercuric) compounds exhibit a high percentage of covalent bonding, except for its salts with anions resistant to polarization. Mercury(II) chloride dissociates slightly.

$$HgCl_2 \rightarrow HgCl^+ + Cl^-$$

Mercury can exist in the +1 oxidation state. This state is achieved by formation of a rare metal-to-metal σ molecular bond, giving rise to diatomic, divalent mercury(I) (mercurous) ion, Hg_2^{2+}.

Compound stability is related to cationic size. Zinc compounds are generally stable to reducing conditions met in pharmaceutical situations. On the other hand, mercury, in either oxidation state, is easily reduced to the metallic state. Cadmium is intermediate to zinc and mercury in its oxidizing power.

The gradation of metallic properties as a function of atomic weight is nicely demonstrated in Group IIB. Zinc, the lightest member of the family, is amphoteric.

$$Zn + 2H_3O^+ = Zn^{2+} + H_2\uparrow + 2H_2O$$
$$Zn(OH)_2 + 2H_3O^+ = Zn^{2+} + 4H_2O$$
$$H_2ZnO_2 + 2OH^- = [Zn(OH)_4]^{2-}$$

Mercury is a true metal. Its appearance in negative ions is strictly a complexation phenomenon, e.g., $[HgCl_4]^{2-}$. Cadmium is intermediate; it exhibits slight amphoteric character.

The three metals form reasonably stable complexes. Except for the mercury(I) halides and mercury(II) iodide, the halides of this family are soluble.

Mercury and cadimum have no known normal role in the living organism. Except for the small quantities of zinc salts in mineral supplements, and mercury(I) chloride the compounds of the volatile metals are used only in external applications involving protein precipitant action. Zinc compounds, except in large quantities, are relatively nontoxic; 45.0 Gm. have proven fatal. Cadmium is considered to be more toxic. Soluble mercury compounds are highly toxic.

Zinc and its compounds. Depending on concentration, zinc ion exhibits a mild germicidal, astringent, or corrosive action. The action of insoluble zinc compounds in protective and topical preparations, to be discussed in Chapter 18, is probably due to a slow release of zinc ion as allowed by the solubility product and enhanced by the acid mantle of the skin.

All zinc salts show some degree of hydrolysis.

$$Zn^{2+} + 2H_2O \rightarrow [Zn(OH)]^+ + H_3O^+$$

At ordinary concentrations the reaction does not proceed sufficiently to result in precipitation of the hydroxide. However, the problem of surface hydrolysis and the subsequent accumulation of a layer of oxo salts and basic carbonate does exist. Its control has been previously discussed (page 234).

ZINC SULFATE: White vitriol, zinc vitriol; $ZnSO_4 \cdot 7H_2O$. This compound is the most stable of the soluble zinc salts. It is efflorescent. The heptahydrate tends to cake on exposure to moisture; the anhydrous salt and the monohydrate do not. Its solutions have a pH of about 5.

The principal use of this compound is in 0.1–1.0% solution as an ophthalmic astringent. Because of its acid reaction, these collyria should be buffered to a pH of about 6.5 with borate or phosphate, or to 7.0 with citrate. A 0.25–1.50% solution is used topically as an astringent. The compound is also used as a styptic. Compound Zinc Sulfate Powder (*N.F. XI*) was intended for use as a local antibacterial. The emetic action of zinc sulfate will be discussed later (page 581).

ZINC CHLORIDE: Butter of zinc; $ZnCl_2$. This compound is used as an antibacterial and astringent in mouth washes. It is also used in deodorant preparations. The solid in stick form is used for its caustic action on malignant growths and gangrenous sores.

Zinc chloride has many important nonmedicinal uses. It is used in fireproofing wood, is a constituent of embalming fluids, and is one of the compounds used to activate wood charcoal. It has a dehydrating action useful in synthetic procedures.

This compound is deliquescent. Its aqueous solutions have a pH of about 4. Freshly prepared solutions are usually cloudy because of basic carbonate; a small amount of hydrochloric acid brings about resolution. Because of oxychloride formation, very dilute solutions may hydrolyze sufficiently to become clouded. The compound is reasonably soluble in polar organic solvents: 1 Gm. dissolves in 1.3 mL. ethanol, 2.0 mL. glycerine, vs. 0.5 mL. water, and 0.25 mL. 2% HCl. Because of its deliquescence and carbon dioxide absorption, the compound must be stored in tight containers.

OTHER ZINC SALTS: Zinc iodide (*N.F. VIII*) has astringent properties. It is the active constituent of Talbot's Solution (for treatment of gingivitis). Zinc permanganate, $Zn(MnO_4)_2 \cdot 6H_2O$, possesses the astringent properties of zinc and the disinfecting properties of the permanganate ion. Zinc silicofluoride (zinc fluosilicate), $ZnSiF_6$, is used for prevention of diaper rash (page 449). Its widest application is as a laundry sour.

Zinc phenolsulfonate (zinc sulfocarbolate), $Zn(C_6H_4(OH)SO_3)_2 \cdot 8H_2O$, is used as an astringent. It was formerly used as an intestinal antiseptic. It is used in insecticide formulations. The compound is efflorescent. It is normally colorless but may become pink on exposure to air. Its solutions are acid.

Zinc caprylate, zinc undecylenate, and zinc propionate are used as fungicides. The latter finds special application in adhesive tapes. Zinc acetate (*N.F. VIII*), $Zn(OAc)_2 \cdot 2H_2O$, is used in 0.1–4.0% solution as an astringent and mild antibacterial for the mucous membranes and skin. Its solutions are neutral or slightly acid.

Cadmium and its compounds. Soluble cadmium compounds are active astringents; in concentrated solution they are irritants. Cadmium compounds have a strong tendency toward selfcomplexation when in solution.

$$2CdCl_2 \rightleftharpoons CdCl^+ + CdCl_3^-$$

Cadmium sulfate is a local astringent. It has been used in 0.5–2% in solution or as an ointment for eye infections and gonorrhea. This salt is efflorescent. Cadmium salicylate (monohydrate) has been used as an antiseptic. Cadmium iodide has been used externally for chronic arthritis. Cadmium sulfide has been introduced in treatment of seborrheic dermatitis of the scalp (page 543). Cadmium chloride has emetic properties (page 581). Currently, cadmium oxide is used in veterinary medicine for control of round worms in pigs.

Mercury and its compounds. At one time mercury compounds found rather wide application in pharmacy and medicine. Its uses in the home, industry, and agriculture were many, e.g., laxation, mold growth control in paints, scab control on seed potatoes. However, because of the extreme toxicity of mercury and its compounds and their pervasiveness in the environment, their use has been effectively banned by the EPA. For this reason, except for some very interesting chemistry, much of the material of this section is of a historical nature. The only surviving official mercury products cited in *U.S.P. XXI* are Ammoniated Mercury, Ammoniated Mercury Ointment, and Ammoniated Mercury Opthalmic Ointment.

Mercury found pharmaceutical application in all of its oxidation states. Mercury(I) and mercury(II) compounds are strong oxidizing agents; even mild reducing agents cause decomposition. Because the standard electrode potentials are so nearly equal, both yield elemental mercury on reduction.

Autoxidation of mercury(I) compounds is common. The equilibrium

$$[Hg_2]^{2+} \rightleftharpoons Hg^0\downarrow + Hg^{2+}$$

favors the mercury(I) ion by a factor of 83. However, since mercury(II) compounds usually have a much higher percentage of covalent character and form stable complexes with most common anions, the equilibrium tends to shift in favor of the mercury(II) compound involved. The usual qualitative test for mercury is an autoxidation in alkaline medium.

$$\underset{\text{(white)}}{Hg_2Cl_2\downarrow} + 2NH_3 \rightarrow \underset{\text{(white)}}{HgNH_2Cl\downarrow} + \underset{\text{(black)}}{Hg^0\downarrow} + NH_4^+ + Cl^-$$

When preparing solutions of mercury salts, especially mercury(II), solvolysis may be a problem. Even the nitrates give cloudy solutions unless excess nitric acid is present.

$$Hg^{2+} + 2H_2O + 2NO_3^- \rightarrow [Hg(OH)]^+(NO_3)^-\downarrow + H_3O^+ + NO_3^-$$

Ammonolysis proceeds via one of two routes. If mercury(II) is added to a hot solution of ammonium chloride in ammonia water an insoluble ammine results.

$$HgCl_2 + 2NH_3 \rightarrow Hg(NH_3)_2Cl_2\downarrow$$

If ammonia water reacts with mercury(II) chloride, ammoniated mercury forms.

$$HgCl_2 + 2NH_3 \rightarrow HgNH_2Cl\downarrow + NH_4^+ + Cl^-$$

Mercury(II) complexes readily with a wide range of ligands. The complexes are stable, as attested by the instability constants of the tetrachloro-, tetraiodo-, and tetracyanomercurate(II) ions, 8.3×10^{-16}, 5.3×10^{-31}, and 3×10^{-42}, respectively.

Many mercury compounds are light sensitive and must be properly protected. The presence of the mercury atoms give its salts a high density, characterized by heaviness when lifted.

Mercury compounds have a nonspecific protein precipitation action; they are, therefore, irritating. Most are bacteriostats. They cannot be relied on to kill spores. They are rapidly inactivated in the presence of serum and other proteins. Organic mercury compounds are less irritating, but the germicidal activity of their tinctures is frequently due to the alcohol present. As a result organic mercurials are overrated. They are employed as preservatives.

Mercury compounds are extremely toxic owing to the inactivation of essential sulfhydryl groups of essential enzymes. A powerful chelating agent for metals, dimercaptrol (2,3-dimercapto-1-propanol, BAL, British Anti

Lewisite) is usually injected as an antidote. Recently *N*-acetyl-DL penicillamine, a dimethyl analog of cysteine, has been proposed as an antidote. This compound is said to be effective orally.

Because mercury compounds are extremely toxic, control of the effective concentration of the free mercury ion is of utmost importance. Insoluble compounds or complexes are used almost without exception because of automatic control of free ion. The use of elemental mercury depends upon an unfavorable oxidation potential.

$$Hg \rightarrow Hg^{2+} + 2e^- \qquad \mathscr{E}^0 = -0.85 \text{ volts}$$

Mercury and its compounds have been superseded in many of their former useful applications, e.g., antisyphilitics. However, many important uses remain, especially as antifungals and disinfecting agents. The diuretic action and cathartic action of mercury compounds will be discussed (page 581, 578).

MERCURY: Quicksilver, hydrargyrum. Mercury was used for preparation of Mild Mercurial Ointment (diluted mercurial ointment, blue ointment), a local parasiticide, used for treatment of pediculosis pubis. Although it is incapable of killing the mites, they are controlled. Strong Mercurial Ointment (Ointment of Mercury) was primarily intended as a concentrate to be used for preparation of Mild Mercurial Ointment. It was at one time used in the treatment of late syphilis.

Particle size is very important in determining the activity of the mercurial ointments. The surface area available for solution and subsequent oxidation of the metallic mercury atoms is directly related to particle size. If the particles are too large the rate of solution will be insufficient to maintain the minimum effective concentration of mercury ions. It has been found that mercury particles no longer visible under a lens with a magnifying power of 10× maintain suitable levels of mercuric ion. Ultrasonic generators, if available, should be useful in preparing these mercury dispersions.

A so-called "water-soluble mercury" is prepared by suspending the metal (10–20%) in cetyl alcohol (70%) together with sodium lauryl sulfate, sodium dioctyl sulfosuccinate, glycerine, and water.

Mercury dissolves in water to give a $3.0 \times 10^{-7}M$ solution at 25°C. It amalgamates with many metals; care must be exercised to prevent contact with metallic surfaces during preparation of mercury ointments, etc. Lead amalgam is very brittle; lead traps in sinks disintegrate rapidly if exposed to spilled mercury or soluble mercury salts.

Metallic mercury has an appreciable vapor pressure. These vapors are highly toxic and should not be inhaled. The vapor absorbs u.v. light strongly. A striking demonstration of the existence of these invisible vapors, even at room temperature, is obtained by focusing a u.v. lamp onto a fluorescent screen and placing an open dish of mercury between at a level below the light beam. The rising mercury vapor casts a shadow on the screen. Warming the mercury slightly **(DANGER!)** results in complete absorption of the u.v. beam.

MERCURY(II) OXIDE: (Mercuric oxide, mercury oxide). This compound is available in two different colors. If prepared by precipitation

$$[Hg^{2+}2Cl^-] + 2[Na^+OH^-] = Hg(OH)_2\downarrow + 2Na^+ + 2Cl^-$$
$$Hg(OH)_2\downarrow = HgO\downarrow + H_2O$$

a fine yellow product results (yellow mercuric oxide, yellow precipitate, precipitated mercuric oxide). When prepared by calcination a comparatively coarse red product results (red mercuric oxide, red precipitate). The red oxide becomes yellow if finely powdered; (trituration color change: red → orange → yellow) the yellow becomes red on heating. Except for color and particle size the oxides are identical. They have identical crystal structures, orthorhombic, and water solubility, K_{sp}, 3.0×10^{-26}. Both forms are light sensitive. Because of its finer state of subdivision the yellow form is more reactive. Mercury(II) oxide has a mild antibacterial action. Yellow Mercuric Ointment is a 1% preparation used chiefly for treatment of the eyelids and conjunctiva. Ointments up to 5% are prepared and used for topical treatment of chancres, chronic ulcers, and fungus infections of the skin. Mercury(II) oxide preparations are relatively nonirritating.

MERCURY OLEATE: This compound is prepared by reaction of mercury(II) oxide (25 Gm.) and oleic acid (75 Gm.) at temperatures under 50°C. The reaction mixture must be stirred constantly during heating. It must not be used or dispensed if mercury globules have separated.

Mercury oleate is used as a suspending agent for insoluble mercury salts, both in ointments and in aqueous media. A 2.5% aqueous suspension is used as a germicide and fungicide.

MERCURY(I) CHLORIDE: Mild mercurous chloride, calomel, mercurous chloride, mercury protochloride, mercury monochloride, mercury subchloride, *precipité blanc* (French Codex). This compound is used as a local antibacterial for skin infections, principally as an ointment, Mild Mercurous Ointment (Calomel Ointment). Calomel is available in particles of colloidal size. If the colloidal form is substituted for the ordinary crystalline form, an ointment of higher antibacterial power is obtained. Black Lotion (*N.F. VIII*) is prepared by trituration of calomel (9.0 Gm.) and finely divided acacia (1.0 Gm.) and gradually adding freshly prepared lime water (1.0 L.). The resulting mixture of mercury(I) oxide (mercurous oxide) and mercury has a germicidal and local stimulant action. Since mercury(I) chloride is more soluble than mercuric oxide (K_{sp}, $[Hg_2^{2+}][Cl^-]^2$, 1.3×10^{-18}), a higher concentration of ionic mercury is present in calomel preparations.

Calomel is subject to several incompatibilities. Oxidizing agents must be avoided since mercury(II) ion forms. Reducing agents, e.g., sucrose lactose, acacia, alkaloids, iodide, cause darkening owing to formation of finely divided metallic mercury.

MERCURY(II) CHLORIDE: Mercury bichloride, corrosive sublimate, mercuric chloride, mercury perchloride, corrosive mercury chloride. This compound is an extremely toxic soluble mercury salt. Its LD_{50} is 1 Gm. Because of danger of mistaken ingestion of mercury(II) tablets, special requirements were set up as to color and physical shape for Mercury Bichloride Large Poison Tablets (*N.F. XII*).[3]

One Large Poison Tablet per pint of water gives a 1:1000 solution, useful as a disinfectant of inanimate objects. However, steel is corroded at this concentration and rubber deteriorates. The addition of sodium chloride reduces this corrosiveness. As a disinfectant on the unabraided skin 1:1000 to 1:2000 solutions may be used. However, due to strong irritating effects, dilutions in the range 1:6000 to 1:10,000 are preferable. Alcoholic solutions (1:1000-1:2000) were useful for treatment of seborrhea of the scalp. A 1:4000 solution was used as a diaper rinse in the management of napkin dermatitis, but surface active substances are equally effective and less dangerous.

Solutions more concentrated than 1:1000 were used as irritants. The powdered dry salt was used as an escharotic, but its action is inferior to silver nitrate or potassium hydroxide.

Mercury Bichloride Ointment is a 1:3300 preparation, in white petrolatum. It was used as an antibacterial. The *B.P.* lists a similar preparation, Strong Mercuric Nitrate Ointment, as a parasiticide for various skin disorders.

Mercury(II) chloride was used as a wood preservative in paints and as a dust or spray for certain fungus diseases in seeds. It is a valuable catalyst.

Mercury(II) chloride is prepared by several classical methods:

$$HgO + 2HCl \text{ (concentrated)} \rightarrow HgCl_2 + H_2O$$

$$HgSO_4 + 2(Na^+Cl^-) \xrightarrow[\text{small amount of } MnO_2]{\text{heat}} HgCl_2 + 2Na^+ + SO_4^{2-}$$

$$Hg \text{ (near boiling point)} + Cl_2 \rightarrow HgCl_2$$

In the two latter reactions the compound is separated by sublimation.

---

[3] *N.F. XII, pp. 245, N.F.X.* also included Mercury Bichloride Small Poison Tablets.

Numerous incompatibilities are possible. Precipitation as insoluble compounds takes place in the presence of hydroxide, carbonate, soap, phosphate, tartar emetic, borate, sulfide, iodide, and protein. Honey, sucrose, formates, aldehydes, some gums, iodide, bromide will result in reduction of the mercury(II) ion, usually to the metallic state. The solubility and stability of mercury(II) chloride solutions are improved by addition of sodium chloride or ammonium chloride.

AMMONIATED MERCURY: Hydrargyrum ammoniatum, mercury amido chloride, white precipitate;[4] $HgNH_2Cl$. This compound is used externally only, as a lotion or ointment. The *U.S.P.* currently cites an ointment for topical use and an ophthalmic ointment. A 1:2000 ointment is usual for topical application; 1:3300 is suitable as an antibacterial on the eye lids. It is also used as a dusting powder in eczemas and parasitic skin disorders, e.g., crab-louse infestations, impetigo, psoriasis, ringworm. The need for ammoniated mercury in treatment of bacterial infections has been greatly reduced by more effective antibiotics. However, ammoniated mercury is effective and is widely used for parasitic infections and fungal conditions. It must not be used in conjunction with sulfur or iodine.

Ammoniated mercury is used for bleaching hyperpigmented areas of the skin. It is potentially a sensitizer, contact dermatitis may arise under the conditions required for this purpose. The possibility of systemic poisoning also exists.

Ammoniated mercury is highly poisonous. Zinc sulfate as an emetic together with milk is an antidote for accidental ingestion.

The incompatibilities existing for this compound are those of mercury(II) compounds in general. Ammoniated mercury is stable in air, but darkens on exposure to light.

MERCURY(I) IODIDE: Yellow mercurous iodide, mercurous iodide, mercury protoiodide. This compound was used as an antibacterial ophthalmic ointment and as an antifungal agent. It was formerly used orally (Yellow Mercurous Iodide Tablets, *N.F. X)* for treatment of late stages of syphilis. This use precluded the simultaneous prescription of potassium iodide; the resulting autoxidation would yield the highly toxic mercury(II) ion.

Mercury(I) iodide is prepared by precipitation or by direct reaction of its elements. If an excess of mercury is used in the latter reaction, a green product, green mercury iodide, results. This is an intimate mixture of yellow mercury(I) iodide and the excess mercury in a finely divided state. Mercury(I) iodide is a compound of exceptionally high density.

---

[4] Do not confuse with French term for calomel, *precipité blanc.*

It is light sensitive, darkening or becoming green on exposure.

$$2HgI \xrightarrow{h\nu} Hg^0\downarrow + HgI_2\downarrow$$

MERCURY(II) IODIDE: Red mercuric iodide, mercuric iodide, biniodide of mercury, HgI_2. This compound resembles mercury(II) chloride in its activity and therapeutic uses. It was used in germicidal soaps (1-2%). It has been used in 4.0 mg. doses as an intestinal antibacterial. It was formerly used in treatment of syphilis.

Mercury(II) iodide is prepared by precipitation or by direct reaction of the elements in the presence of a small amount of ethanol. The compound is slightly soluble in water but readily soluble in excess potassium iodide.

$$HgI_2 + 2K^+I^- \rightarrow 2K^+ + [HgI_4]^{2-}$$

Mercury(II) iodide is light sensitive. On heating it undergoes an interesting series of transformations:

$$\text{red} \underset{}{\overset{127°C.}{\rightleftharpoons}} \text{yellow rhombic} \overset{250°C.}{\rightleftharpoons} \text{melts} \overset{350°C.}{\rightleftharpoons} \text{sublimes}$$

POTASSIUM TETRAIODOMERCURATE(II): Potassium mercuric iodide, mercuric potassium iodide; K_2HgI_4. This compound is a powerful germicide. It is soluble in water and polar organic liquids. Aqueous solutions are prepared in the range 1:10,000 to 1:100. The solutions must be stabilized with about 20% excess potassium iodide to prevent reprecipitation of mercury(II) iodide. The dilute solutions were used for skin disinfection and irrigations, the concentrated solutions for disinfecting instruments, excreta, and discharges. The compound is the essential ingredient in Mayer's Solution for detection of alkaloids, and Nessler's Solution for the detection of ammonia.

Anhydrous potassium tetraiodomercurate(II) has a sulfur yellow color. It can crystallize with one or two molecules of water. It is deliquescent in moist air. The salt is light sensitive. It does not precipitate albumin.

MERCURY(II) CYANIDE: Mercuric cyanide, *N.F. VIII.* This compound was first official in 1890 in the *U.S.P.* as a substitute for mercury(II) chloride. The compound presents an apparent contradiction to the generally accepted idea that the concentration of free mercury(II) ion is the effective determinant of the germicidal and irritant action of mercury(II) compounds. Mercury(II) cyanide is a covalent compound; it does not respond to the tests for mercury(II) ion. It does not precipitate protein from solution. Being nonionic, it may be employed to sterilize metallic instruments without causing mercury replacement.

The coumpound is extremely toxic. It is nonirritating. It was employed in 1:10,000 dilution for use on the mucous membranes, e.g., as a gargle in croup. As a skin disinfectant a 1:2000 dilution is used. A 1:200

solution was used for disinfecting instruments. It has also been used as a diuretic and antisyphilitic.

MERCURY(II) OXYCYANIDE: Mercuric oxycyanide, mercury subcyanide; approximately $HgO \cdot 5Hg(CN)_2$. This compound is a white crystalline powder soluble in 35 parts water. The solution is alkaline. It was used 1:10,000 for irrigation of the urinary bladder, 1:5000 as a local antibacterial, and 1:200 for disinfecting surgical instruments. As with the normal cyanide, steel is not corroded. It has been used intravenously, particularly in France, and occasionally intramuscularly for syphilis. The injections were given daily and were very painful.

OTHER MERCURY SALTS: Mercury(II) succinimide, monohydrogen phosphate $\frac{1}{2}H_2O$, and lactate have been used in treatment of various stages of syphilis. Mercury(I) acetate has been used for treatment of skin lesions of this disease. Mercury(II) salicylate has been used as a 10% dusting powder for treatment of parasitic and fungus skin diseases. Mercury(II) thiocyanate is an interesting compound which when ignited burns with a great increase in volume. It is the active ingredient of "Pharoah's Serpents."

GROUP IIIB—THE EARTHS

General group properties. This group includes boron and aluminum from the short periods, and gallium, indium, and thallium from Group IIIB proper. These elements are characterized by the valence electron structure, ns^2np^1. Except for thallium and, possibly, indium, they do not form simple trivalent cations. The bonding in ternary compounds is predominantly covalent. The hybridizations possible and the strong tendency for octet completion have been discussed earlier (page 44). Because of the electron pair acceptor capacity of the ternary boron and aluminum compounds, these elements act as Lewis acids; as such they act catalytically in the Friedel-Crafts synthesis. Pertinent structural data have been assembled in Table 17.3.

Of the five elements, only aluminum and boron have application in pharmacy and medicine. Gallium is interesting as a possible substitute for aluminum in some of its applications. It has two unusual physical properties: its melting point, 29.75°C., is lower than all metals except mercury, but its boiling point is 2070°C. As with water, its liquid phase has a higher density than its solid phase.

Thallium is one of the most toxic elements; its action is somewhat similar to that of arsenic. Sodium Thiosulfate Injection is recommended as an antidote, but there is no real proof of its efficacy. The element is absorbed from the intestine if ingested and through the skin from ointments and creams.

TABLE 17.3 ATOMIC CHARACTERISTICS OF THE TYPICAL ELEMENTS OF GROUP III

Element	Boron	Aluminum	Gallium	Indium	Thallium
Symbol	B	Al	Ga	In	Tl
Atomic weight	10.82	26.98	69.72	114.82	204.39
Atomic number	5	13	31	49	81
Valence electron configuration	$2s^2 2p^1$	$3s^2 3p^1$	$4s^2 4p^1$	$5s^2 5p^1$	$6s^2 6p^1$
Atomic radius (Å)	0.80	1.25	1.26	1.44	2.0
Ionic radius (Å), M^{3+}	0.20	0.50	0.62	0.81	0.95
Ionization potential (ev.), I	8.30	5.95	6.0	5.8	6.1
II	25.15	18.82	20.4	18.8	20.3
III	37.92	28.44	30.6	27.9	29.7
Electrode potential (v.), M/M^{3+}	−0.73	−1.67	−0.52	−0.34	−0.72
M/M^{+}	—	—	—	−0.25	−0.34
Electronegativity	2.0	1.5	1.8	1.5	1.4

Some years ago thallium sulfide (5%) appeared in depilatory preparations. After several deaths caused by its toxicity this use was discontinued. The pharmacist must be alert for possible reintroduction in this or other equally dangerous applications. Despite the availability of less dangerous and equally effective rodenticides, thallium is employed for this purpose. It is also used in ant poisons.

Thallium occurs in the +1 and +3 oxidation states. In the +1 state it is ionic. The +1 salts resemble those of silver(I) in appearance and solubility, e.g., TlF is soluble; TlCl is insoluble. There are also some similarities between thallium(I) and potassium ion, e.g., thallium hydroxide is a strong base. Many thallium salts are isomorphous with the corresponding potassium salts. Thallium(III) resembles both aluminum(III) and gold(III) in its chemical and physical properties.

Boron and its compounds. Boron occurs in the earth's lithosphere up to 0.005% usually as the minerals borax, $Na_2B_4O_7 \cdot 10H_2O$, kernite, $Na_2B_4O_7 \cdot 4H_2O$, and colemanite, $Ca_2B_6O_{11} \cdot 5H_2O$. The element is always covalently bonded in its compounds. Boron phosphate, BPO_4, is a covalent compound in which boron(III) and phosphorus(V) are tetrahedrally surrounded by commonly shared oxygen atoms.

Boron has coordination numbers of three or four in its compounds. The first requires sp^2 hybridization. The usual planar triangular arrangement results, e.g., the ortho boric acid monomer, $B(OH)_3$. These compounds tend to have layer structures in the crystalline state, e.g., boric acid (see Fig. 3.3). Because of the availability of an unoccupied *p* atomic orbital, the ternary boron compounds readily undergo hydrolysis and complexation, (Fig. 17.1). Rearrangement resulting in sp^3 hybridization takes place, giving rise to tetrahedral structures—e.g., the borate

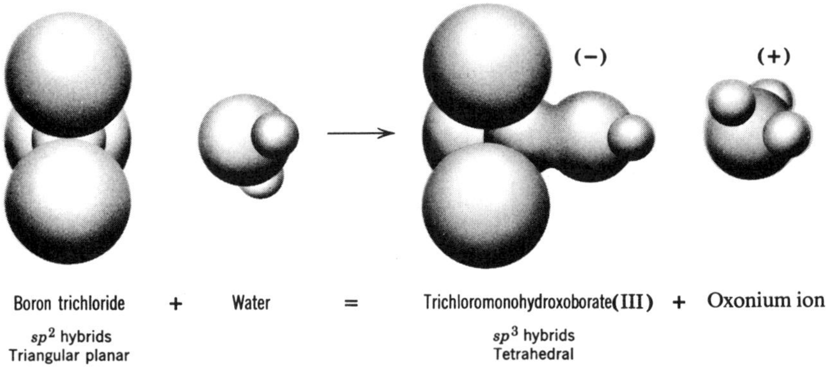

FIGURE 17.1 Solvolysis of electron-pair-deficient boron compounds. In aqueous media the three chlorides are replaced by hydroxo groups to give the tetrahydroxoborate(III) ion.

ion of aqueous solution, $[B(OH)_4]^-$. Unlike aluminum, boron does not have the *d* orbitals required for formation of octahedral complexes.

Boron(III) oxide (boron trioxide, boron sesquioxide, fused boric acid), B_2O_3, may be considered the parent of the various borates. The relationships existing may be shown by stepwise dehydration of ortho boric acid.

$$(H_3BO_3)_n \xrightarrow{105\text{–}108°C.} (HBO_2)_n + nH_2O\uparrow$$

$$2nHBO_2 \xrightarrow{185°C.} nB_2O_3 + nH_2O\uparrow$$

Phase rule studies fail to give evidence for the existence of tetraboric acid (pyroboric acid), $H_2B_4O_7$, which frequently appears in equations involving borax. Apparently this latter acid exists only in the form of its salts.

Metaboric acid and the metaborate ion do not exist as monomers. In the crystalline state polymeric ring and chain structures form, as shown in Fig. 17.2. The metaborate chain of the calcium salt is an analog of the pyroxene chain of silicon chemistry. True crystalline orthoborates are unknown; $NaBO_2{\cdot}4H_2O$ may be written, $Na[B(OH)_4]{\cdot}2H_2O$. Evidence indicates that the tetraborates should be formulated, e.g., borax, $Na_2B_4O_5(OH)_4{\cdot}8H_2O$. The structure is shown in Fig. 17.2. In the crystalline state these tetraborate units are probably hydrogen bonded to

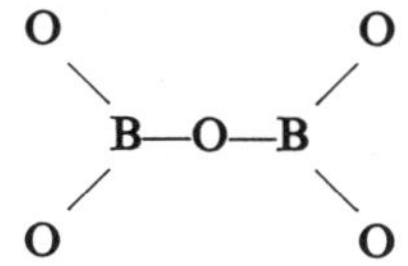

Diborate ion

$[B_2O_5]^{4-}$
(crystalline)

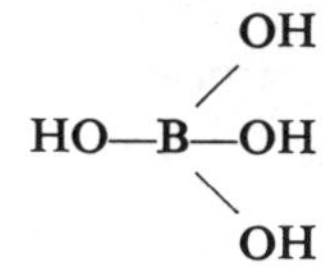

Borate ion

$[B(OH)_4]^-$
(aqueous solution)

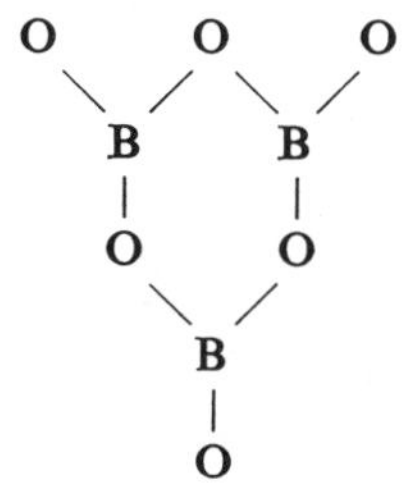

Metaborate ion

$[B_3O_6]^{3-}$
(e.g., $Na_3B_3O_6$)

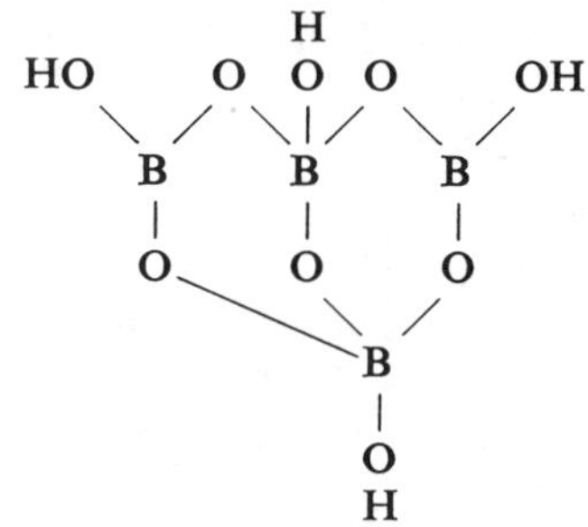

Pyroborate ion (borax)

$[B_4O_5(OH)_4]^{2-}$

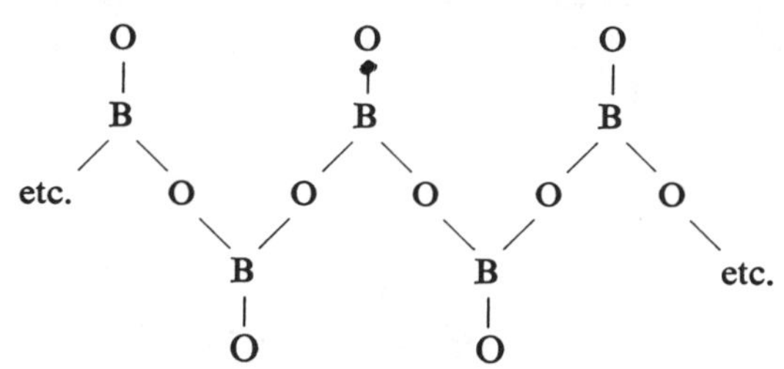

Metaborate chain

$[(B_3O_6)_n]^{3n-}$
(e.g., calcium metaborate)

FIGURE 17.2 Structure of borate ions.

give infinite chains. In addition to ortho-, meta-, and tetraborate, $[B_2O_5]^{4-}$; $[B_5O_9]^{3-}$, and $[B_6O_{11}]^{4-}$ have been identified in the crystalline state. Only the diborate exists as discrete ions; the others are ring structures.

In dilute aqueous media the various borate anions of soluble salts slowly react with water to become the orthoborate ion, $[B(OH)_4]^-$. The metaborates, e.g., $K_3B_3O_6$, simply hydrate on solution.

$$[B_3O_6]^{3-} + 6H_2O \rightarrow 3[B(OH)_4]^-$$

Tetraborates hydrolyze without the need for the hypothetical tetraboric acid as an intermediate, as follows.

$$[B_4O_5(OH)_4]^{2-} + 5H_2O \rightarrow 2[B(OH)_3] + 2[B(OH)_4]^-$$

The alkalinity of the solutions is due to ionization of the orthoborate anion.

$$[B(OH)_4]^- \rightleftharpoons [B(OH)_3] + OH^-$$

The tetrahydroxoborate(III) anion, as the borate ion of aqueous media, is supported by spectral studies of solutions of the ortho acid and its salts. Indirect evidence lends further support.

Except for the alkali metal compounds, all borates are insoluble. The degree of insolubility is a function of pH and the extent of hydration of the parent acid. As pH increases, the solubility of a given borate salt decreases. As the hydration of the parent acid increases the solubility increases; the order of solubility of a given cations's borate salts are orthoborate, most soluble, metaborate, least soluble, and tetraborate, intermediate.

Borates are stable to all but the strongest reducing agents; redox incompatibilities are for practical purposes nonexistent in pharmaceutical systems.

Although orthoboric acid exists as a polymeric hydrogen-bonded layer structure in the crystalline state, discrete molecules exist in the gaseous state and in solution. The ionization should be written

$$B(OH)_3 + 2H_2O \leftrightharpoons H_3O^+ + [B(OH)_4]^-$$

rather than the conventional

$$H_3BO_3 + H_2O \rightarrow H_3O^+ + H_2BO_3^-.$$

The limited extent of ionization is indicated by its pK_a, 9.2. The pH of a 0.1*M* solution of orthoboric acid is 5.3.

One gram of boric acid dissolves in 18 ml. of water at room temperature and 4 ml. of boiling water. The corresponding figures for ethanol are 18 ml. and 6 ml., respectively. Hydrochloric, citric, and tartatic acid increase its solubility in water. Because of a low heat of sublimation, boric acid is volatile in steam and ethanol vapors.

Boric acid is also quite soluble in polyhydroxy compounds, e.g., propylene glycol. In aqueous media it esterifies readily with cis diols, e.g., catechol, mannitol, glycerol. Some typical reactions are given in Fig. 17.3. These compounds are strong acids and are valuable for the analytical determination of boric acid, which, because of its very weak acidic character, cannot be directly titrated.

Two different reactions are possible for glycerine. In aqueous media glyceroboric acid forms (Fig. 17.3). However, it is presumed that in anhydrous media only one glycerine molecule reacts with a boric acid molecule. The resultant product, though not definitely proven, is considered to be that shown in Fig. 17.3. The compound is known as glycerylborate or boroglycerine glycerite. Its preparation is described in *N.F. XI.*[5] The resultant product is a sweet, syrupy, hygroscopic liquid. When dissolved in excess water, boric acid separates. Its solutions are slightly acid.

The boron hydrides, or boranes, illustrate the bridge element character of this element. These compounds closely resemble the silanes. The borohydride ion, $[BH_4]^-$, is a valuable reducing reagent for both inorganic and organic synthetic reactions. Sodium borohydride, $Na^+[BH_4]^-$, is commercially available in the crystalline form, or in highly alkaline aqueous solution (12% w/w). In neutral or alkaline solution the salt slowly hydrolyzes to borate ion and hydrogen gas.

$$(BH_4)^- + 4H_2O \rightarrow [B(OH)_4]^- + 4H_2\uparrow$$

Boron compounds are toxic.[6] Although it is not absorbed through the whole skin, boron readily enters the system through denuded and inflamed skin. Absorbed boron accumulates in the hair, liver, and body fats. Fatal doses in adults range from 15 to 30 Gm. In infants 3–6 Gm. are fatal; one fatality resulted from the liberal use of powdered boric acid for treatment of diaper rash. Although the reason for toxicity of boron compounds has not been established, it is suggested that a competitive esterification interfering with phosphorylation may be a factor (in view of the ease of esterification of this acid with alcohol groups). Precipitation of essential trace metals as insoluble borates may also be a contributing factor.

Boric acid and the borates have no germicidal action; at best they are feeble bacteriostats. On the basis of their toxicity and doubtful antiseptic value, the widespread use of these substances is unwarrated.

ORTHOBORIC ACID: Boric acid, boracic acid; H_3BO_3 or $B(OH)_3$. This compound has been used over the years. It has been official in the *U.S.P.* since 1882, currently listed as a local anti-infective and ophthalmic irrigant.

Boric acid is most widely used as the saturated solution, 5.3%, Boric Acid Solution, *N.F. XII* (Saturated Boric Acid Solution). Saturated solutions deposit crystals of the acid on chilling or evaporation. If, on

[5] *N.F. XI*, p. 56.

[6] J. J. Sciarra, "Review of Boric Acid Toxicity," *J. Am. Pharm. Assoc., Pract. Ed.*, **19**, 494 (1958).

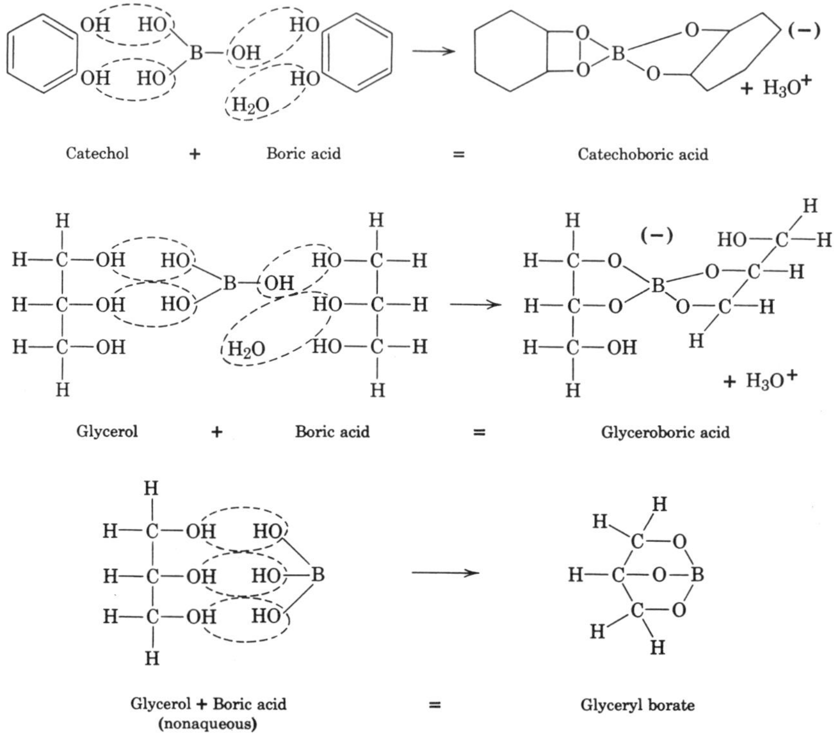

FIGURE 17.3 Typical esterifications of boric acid.

use, crystals are carried along, e.g., instillation in the eyes, mechanical irritation may result. For this reason slightly undersaturated solutions, e.g., 5%, are desirable. *N.F. XI* recommends half saturation. The 2.2% solution is iso-osmotic with the lachrymal fluid. *N.F.* Antiseptic Solution, *N.F. XII,* intended as an external or oral antibacterial, includes 2.5% boric acid. Cottony growths of ***Fuserium*** and ***Torula*** form in boric acid, borate, and zinc sulfate solutions. The organisms may be introduced by the distilled water used for preparation. Boric acid is incompatible with benzalkonium chloride.

Boric acid was popular as a 10% ointment, Boric Acid Ointment. Although of doubtful value as an antibacterial in fatty base ointments, boric acid is bacteriostatic in hydrophilic bases, especially if large amounts of water are present. Gels of methyl cellulose and sodium alginate are most effective.

Boric acid is also used as a dusting powder, usually mixed with varying quantities of inert fillers. Because of the danger of inhalation of the dust, the practice should be discontinued, especially on infants. Zinc Sulfate

Powder, *N.F. XI*,[7] is predominantly boric acid, 86.6%. It is used for hot compresses and is a frequent ingredient of cosmetics. Its use as a food preservative is to be discouraged.

When trituration is required during preparative procedures, the impalpable powder rather than crystalline boric acid should be used. Crystalline particles are extremely resistant to size reduction because of their platy character. The impalpable powder is obtained by precipitation from solution rather than crystallization.

SODIUM TETRABORATE DECAHYDRATE: Sodium borate, borax, sodium pyroborate; $Na_2B_4O_7 \cdot 10H_2O$ or $Na_2[B_4O_5(OH)_4] \cdot 8H_2O$. This compound is usually intended when sodium borate is specified. It is also available as the anhydrous salt, fused sodium borate.[8] The hydrate is efflorescent in dry air, becoming coated with a fine white powder. Borax dehydrates in stages.

$$Na_2B_4O_7 \cdot 10H_2O \xrightarrow{100°C} Na_2B_4O_7 \cdot 5H_2O \xrightarrow{150°C} Na_2B_4O_7 \cdot H_2O \xrightarrow{320°C} Na_2B_4O_7$$

One gram dissolves in 16 mL. water at room temperature and 0.6 mL. at boiling temperature; a 0.1 M solution has a pH of 9.5—10.0. One gram dissolves in 1.0 mL. glycerol at room temperature.

Borax is used for the preparation of orthoboric acid. On addition of a strong mineral acid, e.g., HCl, to a saturated solution of the salt orthoboric acid precipitates:

$$[B_4O_5(OH)_4]^{2-} + 2H_3O^+ + H_2O \rightarrow 4[B(OH)_3]\downarrow$$

Borax is toxic; 15–30 Gm. are fatal. It is a fair bacteriostat. Borax is used in the preparation of eye washes (1–2%), mouth washes and gargles, and solutions for dressing of wounds. Two of the better known oral preparations are Alkaline Aromatic Solution and Compound Sodium Borate (Dobell's Solution) both in *N.F. XI.*[9] Petrolatum Rose Water Ointment, *U.S.P. XVI,* renamed Cold Cream in *U.S.P. XVIII*[10], includes 0.5% sodium borate.

Borax has been used as a food preservative, but this use is prohibited in the United States. Borax is widely used for softening small volumes of water and for control of pH in soap powders.

OTHER BORON SALTS: Sodium perborate has been discussed (page 447). It exhibits the properties of both borate and peroxide. Boroglycerin

[7] *N.F. XI*, p. 399.

[8] Molten borax is an excellent solvent for metallic oxides. A borax bead has a color which is characteristic of the dissolved metal oxide.

[9] *N.F. XI,* pp. 35, 331.

[10] *U.S.P. XVIII,* p. 143.

glycerite (glycerylborate) was used as an antibacterial.

Aluminum and its compounds. Aluminum, unlike boron, is an abundant element, constituting 7.5% of the earth's crust. Only oxygen and silicon are more abundant. Naturally occurring aluminosilicates are extremely stable. Only bauxite, impure aluminum oxide, is an economical source of the element and its compounds.

Aluminum, like boron, forms electron-pair-deficient sp^2 planar hybrids, e.g., AlF_3. Like boron, the aluminum atom seeks its full inert gas quota of electrons; it achieves sp^3 hybrids readily by coordination with the solvent if necessary, e.g., $[AlCl_4]^-$, $(C_2H_5)_2O{:}AlCl_3$. Unlike boron, however, aluminum has unoccupied $3d$ orbitals available for hybridization and may achieve sp^3d^2 hybridization, with resultant typical octahedral arrangements, e.g., $[AlF_6]^{3-}$, $[Al(OH)_4(H_2O)_2]^-$, $[Al(H_2O)_6]^{3+}$.

Aluminum metal is a strong reducing agent. Thus it will react vigorously with even mild oxidizing agents. A medicinal application, Aluminum Paste, (10% aluminum, 5% liquid petrolatum, and 85% zinc oxide ointment) takes advantage of this reducing action. The aluminum prevents undue irritation of the skin around an intestinal fistula.

Powdered aluminum is available in flake form or in irregular particles. The flake form is superior for use in aluminum paste, having the required smooth and unctuous feel when rubbed between the fingers. Ordinary powdered aluminum has particle sizes down to the 5–10 μ range. Recently aluminum powders less than 0.1 μ in diameter have been prepared.

As a very weak oxidizing agent, aluminum(III) may be considered stable to the redox conditions ordinarily met during pharmaceutical operations. Aluminum metal and its hydroxyl bases are amphoteric. Because of its high charge, small diameter, and electron pair deficiency, aluminum(III) ion is incapable of independent existence in water and other polar solvents. Complexation always takes place. The particular structure attained is a function of the ligands available and their relative concentrations.

The complexity and properties of these solutions may be demonstrated by considering preparation of a $0.5M$ solution of aluminum chloride from its anhydrous salt. Initially reaction with a water molecule forces sp^3 hybridization.

$$AlCl_3 + H_2O = [H_2O{:}AlCl_3]$$

Diffuse field chloride ions are successively displaced as a result of the stronger dipole field of the water molecule.

$$[AlCl_3(H_2O)]^0 + H_2O = [AlCl_2(H_2O)_2]^+ + Cl^-$$
$$[AlCl_2(H_2O)_2]^+ + H_2O = [AlCl(H_2O)_3]^{2+} + Cl^-$$
$$[AlCl(H_2O)_3]^{2+} + H_2O = [Al(H_2O)_4]^{3+} + Cl^-$$

At this time, or earlier, 3*d* orbitals become available for sp^3d^2 hybridization.

$$[Al(H_2O)_4]^{3+} + 2H_2O = [Al(H_2O)_6]^{3+}$$

The hexahydrate is the final product.

The acid-base properties of the ions present must now be considered. The chloride ion is a very weak base, and for practical purposes may be neglected. The hexahydrate, however, is a mild acid ($pK_a = 4.85$). At ordinary concentration levels about 4% ionization takes place; as a result the pH of the solution falls to about 3.5.

$$[Al(H_2O)_6]^{3+} + H_2O \rightleftharpoons [Al(H_2O)_5(OH)]^{2+} + H_3O^+$$

If the solution is treated with a strong base in order to adjust the pH to about 5.0, the reaction

$$[Al(H_2O)_6]^{3+} + OH^- = [Al(H_2O)_5(OH)]^{2+} + H_2O$$

will go to practical completion. Or, conversely, it may be said that a solution of a pentaquahydroxoaluminum salt will have a pH of about 5.0. If additional base is added, adjusting the pH to about 6, the pentaqua ion is neutralized.

$$[Al(H_2O)_5(OH)]^{2+} + OH^- = [Al(H_2O)_4(OH)_2]^+ + H_2O$$

Again, conversely, it may be said that a solution of tetraquadihydroxoaluminum chloride is barely acid. No precipitation takes place during either neutralization process, both the penta- and tetraquaaluminum cations described are soluble.

If the solution is adjusted to neutral or slightly alkaline pH, insoluble aluminum hydroxide precipitates.

$$[Al(H_2O)_4(OH)_2]^+ + OH^- = [Al(H_2O)_3(OH)_3]^0\downarrow + H_2O$$

This series of reactions demonstrates why astringents and deodorants such as dihydroxyaluminum chloride, dihydroxyaluminum aminoacetate, and preparations such as Aluminum Subacetate Solution are less irritating to the skin than normal aluminum salts; the equilibrium pH of their solutions is much less acid. The irritating action of normal aluminum salts and their destructive action on clothing is largely due to the acid set free on hydrolysis. It may be noted that these reactions are the reverse of those taking place during antacid action of aluminum hydroxide and the partially hydrated cations (page 407).

If aluminum hydroxide, $[Al(H_2O)_3(OH)_3]^0$, is maintained at a weakly or even moderately alkaline pH, no further reaction takes place. Its acid ionization constant ($pK_a = 12.2$) precludes reaction with water. A strongly basic medium, pH $\sim$ 12, is required for further reaction.

$$[Al(H_2O)_3(OH)_3] + OH^- + [Al(H_2O)_2(OH)_4]^- + H_2O$$

The structure of aluminate ion and some of the aqua aluminum complexes are given in Fig. 17.4. For aqueous systems the formulas assigned

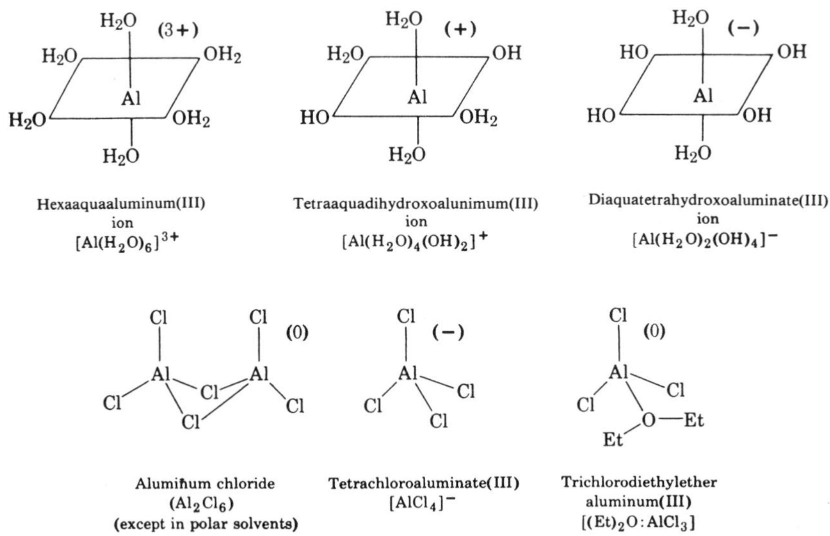

FIGURE 17.4 Aluminum complexes.

to the aluminate ion, $[AlO_2]^-$ and $[AlO_3]^{3-}$, are incorrect. Osmotic measurements on sodium aluminate solutions indicate the presence of only two ions. This rules out as improbable the ionization

$$Na_3AlO_3 \rightleftharpoons 3Na^+ + AlO_3{}^{3-}$$

Conductivity measurements and pH studies rule out the ionization

$$NaAlO_2 \rightleftharpoons Na^+ + AlO_2{}^-$$

Structural studies, coordination behavior, and chemical characteristics of aluminum indicate the ionization

$$Na^+[Al(H_2O)_2(OH)_4]^- \rightleftharpoons Na^+ + [Al(H_2O)_2(OH)_4]^-$$

as the correct one.

In concentrated solutions of aluminum chloride, or if a source of chloride ion is dissolved in solution, e.g., sodium chloride, one or more chloride ions may remain coordinated with the aluminum complex ions. The number coordinated is directly related to the chloride concentration.

Omitting water of hydration, these reactions for moderate chloride levels may be represented:

$$AlCl_3 \rightleftharpoons AlCl_2^+ + Cl^-$$
$$AlCl_2^+ \rightleftharpoons AlCl^{2+} + Cl^-$$

For high chloride concentrations, e.g., added sodium chloride or hydrochloric acid, the tetrahedral tetrachloraluminate ion forms.

$$AlCl_3 + Na^+Cl^- \text{ (excess)} = Na^+ + [AlCl_4]^-$$
$$AlCl_3 + H_3O^+Cl^- \text{ (strongly acid)} = H_3O^+ + [AlCl_4]^-$$

The foregoing discussion of aqueous systems involving aluminum ions has been summarized in Table 17.4. Although the table is based on aluminum chloride, other aluminum salts undergo corresponding reactions.

TABLE 17.4 ALUMINUM STRUCTURES AND CONDITIONS FOR STABILITY

Stable ion (or molecule)	Conditions for stability
$[Al(H_2O)_6]^{3+}$	Dilute solution, pH 3–4
$[Al(H_2O)_5(OH)]^{2+}$	Dilute solution, pH 4–5.5
$[Al(H_2O)_4(OH)_2]^+$	Dilute solution, pH 5.5–6.5
$[Al(H_2O)_3(OH)_3]$	pH 7.0–12.0
$[Al(H_2O)_2(OH)_4]^-$	pH 12.0
$[AlX_4]^-$	Large excess of anion (Cl^-, Br^-, CN^-, etc.)
$[AlF_6]^-$	Excess F^- ion

Aluminum ion has no known physiological role. Its medicinal applications depend on an excellent astringent action over a wide range of concentrations. Aluminum preparations have an excellent deodorizing action. This action is probably twofold: first, a conversion of volatile amines to nonvolatile amine salts; second, adsorption of odorous compounds by traces of hydroxide

$$[Al(H_2O)_4(OH)_2]^+ + RNH_2 = [Al(H_2O)_3(OH)_3] + RNH_3^+$$
$$[Al(H_2O)_3(OH)_3]_n^0 + xRSH = [Al(H_2O)_3(OH)_3]_n(RSH)_x$$

An aluminum salt is almost invariably the active ingredient in antiperspirants.

ALUMINUM CHLORIDE: $AlCl_3 \cdot 6H_2O$. This compound is used topically in 10–25% solution as a mild antiseptic, astringent, and anhidrotic. Because of high acidity these solutions may be irritating to a sensitive skin. For this reason a partially neutralized product is usually used (aluminum hydroxychloride, aluminum chlorhydrol, chlorhydrol, aluminum chlorhydrate).

$$[Al(H_2O)_6]^{+3} + OH^- \rightarrow [Al(OH)(H_2O)_5]^{+2} + H_2O$$

$$[Al(OH)(H_2O)_5]^{+2} + OH^- \rightarrow [Al(OH)_2(H_2O)_4]^{+1} + H_2O$$

The neutralization is stopped before complete conversion to the dihydroxy form. The combination of the two compounds has the desired properties without the irritation.

Aluminum chloride has a sweet astringent taste and a slight hydrogen chloride odor. It is deliquescent. In the solid, liquid, and gaseous states and in solution in nonpolar solvents it exists as a dimer, Al_2Cl_6. Two chlorines are shared between two aluminums, effectively permitting a tetrahedral arrangement about each aluminum atom. This structure is typical for electron-pair-deficient compounds. In polar solvents coordination between solvent molecules and the aluminum, e.g., $(C_2H_5)_2O:AlCl_3$, results in monomers.

One gram aluminum chloride dissolves in 0.9 ml. water; in 4.0 ml. ethanol. It is soluble in glycerine.

ALUMINUM SULFATE: Concentrated alum, cake alum, patent alum, pearl alum, pickle alum, $Al_2(SO_4)_3{\cdot}18H_2O$. This compound is known as alum in the chemical industry. It has a sweet astringent taste. It is very soluble in water (1 Gm. dissolves in 1 ml.), but insoluble in ethanol. The pH of a 1–20 aqueous solution is 2.9. The crystalline salt dehydrates at 250°C. On long boiling of its aqueous solutions an insoluble basic salt precipitates. For industrial purposes large quantities are shipped as concentrated aqueous solutions.

In pharmaceutical and allied preparations 5–25% aluminum sulfate solutions are used topically. A 5% solution is used for its detergent action on foul ulcers and leukorrhea. A 10% solution is used for sterilization of dental cavities.

ALUM: $KAl(SO_4)_2{\cdot}12H_2O$. Alum may be either the potassium or the ammonium compound. The label must clearly state which is dispensed. The two compounds are also available as exsiccated alum (burnt alum, dried alum).

Alum solutions are valuable for their astringent action. The 1–5% solution is used to check local sweating and to harden skin. The 0.5–1% solution is used as a skin lotion; this concentration has been used as a urethral injection. A 0.5% solution is bacteriostatic. Alum solutions have been used as gargles but their high acidity is injurious to the teeth. Powdered alum is occasionally used as a dusting powder and as a styptic.

Potassium alum is cast in pencil form and used as a styptic for minor cuts.

Exsiccated alum is used in powder form to absorb moisture. It is also used as a mild caustic.

Solutions of alum have a sweet, strongly astringent taste, and a pH of 4.0–4.2. At room temperature ammonium and potassium alum are equally soluble in water. Ammonium alum is slightly less soluble at boiling temperatures. Both are soluble in glycerol.

The alums are double salts and have been discussed (page 221). Owing to variation in the size of monovalent cations involved, three slightly different structures exist. Ammonium and potassium alums crystallize as α-alums; cesium and monoethylammonium alums crystallize as β-alums; and sodium alum crystallizes in the γ structure. The hydration water is utilized in two ways. Six molecules of water are in octahedral coordination with the trivalent cation. The remaining six serve to link these octahedra to neighboring sulfate ions. The growth of alum crystals can be interesting. Multicolored crystals may be grown.

ALUMINUM SUBACETATE TOPICAL SOLUTION: This preparation is used as an astringent wash, usually diluted with 5-10 parts of water. It is used as a wet dressing diluted with 20–40 parts water. It was at one time used as a gargle. Aluminum subacetate (basic aluminum acetate, printer's acetate) is recovered by evaporation of aluminum subacetate solution. It is a white powder, only slowly soluble in water. It has been used as a desiccant and deodorant dusting powder.

A study of the preparation of Aluminum Subacetate Topical Solution[11] reveals that the hexaqua aluminum cations have been converted to the non-irritating $[Al(H_2O)_5(OH)]^{2+}$ and $[Al(H_2O)_4(OH)_2]^+$ cations. A solution of aluminum sulfate is reacted with dry precipitated calcium carbonate. The pK_a values of the acids involved, (bicarbonate, 10.2; carbonic acid, 6.4; hexaquoaluminum, 4.85) indicate

$$[Al(H_2O)_6]^{3+} + CO_3^{2-} = [Al(H_2O)_5(OH)]^{2+} + HCO_3^-$$

as the primary reaction. During the initial stages, while hexaquo ion concentration is high, bicarbonate actively participates in the reaction

$$[Al(H_2O)_6]^{3+} + HCO_3^- = [Al(H_2O)_5(OH)]^{2+} + H_2O + CO_2\uparrow$$

On depletion of the hexaqua cation the continuing reactions may be written

$$[Al(H_2O)_5(OH)]^{2+} + CO_3^{2-} \rightleftharpoons [Al(H_2O)_4(OH)_2]^+ + HCO_3^-$$
$$[Al(H_2O)_5(OH)]^{2+} + HCO_3^- \rightleftharpoons [Al(H_2O)_4(OH)_2]^+ + H_2CO_3$$

[11] *U.S.P. XXI,* p. 31.

At this stage the pH of the solution favors active precipitation of dissolved calcium ions as the insoluble sulfate.

Upon completion of these reactions acetic acid is added to the reaction mixture. The bicarbonate, formed during the final stages, is removed as carbon dioxide. The slight excess of acetic acid has little or no effect on the pH of the finished product, which is obtained on filtration. Boric acid, 0.9%, is sometimes added as a stabilizer.

Aluminum Subacetate Solution may be described as a solution of pentaquahydroxo- and tetraquadihydroxoaluminum(III) acetates and sulfates dissolved in an aqueous medium saturated with calcium sulfate, and including a small quantity of acetic acid. Its pH is a mildly acidic 5–6.

ALUMINUM ACETATE TOPICAL SOLUTION: Burrow's Solution. This preparation is prepared by treating Aluminum Subacetate Solution with glacial acetic acid.[12] A pH shift to about 4.0 indicates neutralization of most of the coordinated hydroxyl. The finished solution may be considered aluminum acetate (5%) in excess acetic acid (4.5%).

Undiluted Burrow's Solution has some germicidal action, but is ordinarily used for its astringent and detergent properties. Before use the solution is diluted to give the desired degree of astringency. It is used in the form of wet compresses, immersions, and soaks. The solution provides a protective acid mantle for treatment of cracked, scaly hands, housewives eczema, soap and alkali dermatitis, occupational dermatitis, and prickly heat.

Aluminum Acetate Solution (former official name) was at one time prepared by reaction of lead acetate with aluminum sulfate.

OTHER ALUMINUM SALTS: Aluminum(III) salts such as the acetotartrate (70% basic aluminum acetate plus 30% tartaric acid), naphtholdisulfonate, citrate, and methionate have been prepared and used as astringents, antiperspirants, etc. Aluminum hydroxide is occasionally used as a dusting powder in treating skin diseases, largely for its drying effect.

Iron(III) compounds as astringents. Although iron(III) is not a member of Group III, its ionic radius, charge, and general characteristics closely resemble aluminum(III). Iron(III) compounds are protein precipitants; they have astringent action over a wide range of concentrations. Its solutions are generally equivalent to aluminum except for their color. Laundry and cosmetic problems inherent in this color are responsible for the limited acceptance of its compounds in this category.

FERRIC CHLORIDE SOLUTION: Iron perchloride solution, iron(III) chlo-

[12] *U.S.P. XXI*, p. 29.

ride solution. This preparation is a 5% solution of iron(III) chloride dissolved in approximately 5% hydrochloric acid. Its constitution is well represented by $[H_3O]^+[FeCl_4]^-$. The solution is too acid for topical use; it was a pharmaceutical necessity for the preparation of other iron(III) solutions.

OTHER IRON SALTS: Ferric Chloride Tincture (Iron Tincture) is prepared by dilution of Ferric Chloride Solution (350 ml.) to 1000 ml. with ethanol. It is used as an astringent (and hematinic). Ferric Subsulfate Solution (Monsel's Solution, Basic Ferric Sulfate Solution) is prepared by the oxidation of a sulfuric acid solution of iron(II) sulfate with nitric acid. It is used as an astringent in diluted form, undiluted it is useful as a styptic. Iron(III) salts, e.g., iron(III) chloride, are useful as styptics in staunching the flow of blood in difficult to control wounds.

The final medicinal ferric compounds and preparations, of the many once officially cited, appeared in *N.F. XI.* Today, Yellow Ferric Oxide and Red Ferric Oxide, *N.F. XVI,* remain as pigments, (page 536), and Sodium Nitroprusside, *U.S.P. XXI,* remains as a vasodilator (page 583).

TYPICAL ELEMENTS OF GROUP IV

General group properties. The structural properties of these elements have been summarized in Table 17.5. Carbon and silicon are included for comparison purposes. Germanium, tin, and lead resemble carbon

TABLE 17.5 ATOMIC CHARACTERISTICS OF THE TYPICAL ELEMENTS OF GROUP IV

Element	Carbon	Silicon	Germanium	Tin	Lead
Symbol	C	Si	Ge	Sn	Pb
Atomic weight	12.01	28.09	72.60	118.70	207.21
Atomic number	6	14	32	50	82
Valence electron configuration	$2s^22p^2$	$3s^23p^2$	$4s^24p^2$	$5s^25p^2$	$6s^26p^2$
Atomic radius (Å)	0.77	1.17	1.22	1.41	1.75
Ionic radius M^{2+} (Å)	—	—	—	—	1.32
M^{4+}	—	0.39	0.53	0.71	0.84
Ionization potential (ev.), I	11.3	8.15	8.13	7.32	7.41
II	23.4	16.3	15.86	14.5	14.96
Electrode potential (v.), M, M^{2+}	—	—	—	−0.136	−0.146
M^{2+}, M^{4+}	—	—	—	0.15	1.5
Electronegativity	2.5	1.8	1.8	1.7	—

and silicon in their ns^2np^2 ground state electron configuration and the tendency to form sp^3 hybrids with characteristic tetrahedral arrangements. Metallic character increases with increasing atomic weight. Germanium is a nonmetal, tin is amphoteric, and lead is metallic in the +2 state but amphoteric in the +4 oxidation state.

The tetrahalides show strong group similarities. All except carbon tetrachloride hydrolyze irreversibly in water. Hydrolysis probably proceeds through a preliminary hydration stage involving sp^3d^2 hybridization; e.g.,

$$SnCl_4 + 2H_2O = SnCl_4(H_2O)_2$$

The extremely strong positive electrostatic field of the cation weakens the O—H bonding of the water allowing successive stages of ionization

$$SnCl_4(H_2O)_2 + 2H_2O = 2H_3O^+ + [SnCl_4(OH)_2]^{2-}$$

accompanied by successive replacement of the weak field chlorides by water

$$[SnCl_4(OH)_2]^{2-} + 2H_2O = H_3O^+ + [SnCl_3(OH)_3]^{2-} + Cl^-$$

etc. When chloride replacement is complete, the insoluble hydrated acids precipitate.

Lead(II) and lead(IV) bromides and iodides have not been prepared. Lead(IV) fluoride and chloride are unstable, even at relatively low temperatures.

$$PbF_4 = PbF_2 + F_2\uparrow$$

Lead(IV) fluoride is an excellent fluorinating agent. The tetrahalides react with an excess of their respective anions to form stable octahedral complex ions, $[MX_6]^{2-}$.

Germanium(IV) oxide reacts with base giving rise to the germanates, usually isomorphous with silicates. The monoxides and dioxides of tin and lead are amphoteric; the dioxides dissolve in acid only in the presence of chloride or fluoride ions. Octahedral $[M(OH)_6]^{2-}$ anions form on solution in alkali bases. The hexahydroxo- lead(II) (plumbites) and tin(II) (stannites) ions are very strong reducing agents.

Lead, tin, and germanium find little application in pharmacy and medicine. There are no germanium compounds of value. Tin is germicidal to staphylococci. Tin is absorbed from the intestinal tract on ingestion. Unlike lead, tin ions do not accumulate in the body but are slowly eliminated through the intestinal tract and to a very small extent in the urine. Authenticated cases of tin poisoning are rare. It is a CNS stimulant.

At present the only tin compound of medicinal importance is tin(II) fluroide (page 368). Tin(IV) oxide (stannic oxide, tin dioxide) has been used as a 5% ointment for external staphylococci infection. Recently bis(*n*-tributyl) tin oxide has been used experimentally in hospitals as a disinfectant. Treatment of linens, floors, walls, and air conditioning filters is apparently successful, even against antibiotic-resistant bacteria. Tin organics are used as anthelmintics for chickens.

Lead and its compounds. At this time there are no important medicinal uses of lead compounds. However, owing to the high toxicity of its compounds some discussion is necessary. Lead(II) chloride, usually considered insoluble, is sufficiently soluble at body temperature (21.85 mg. Pb^{2+} per liter of 0.063*N* HCl) to allow considerable absorption from the intestinal tract. The hydrochloric acid of the stomach solubilizes most insoluble lead salts. Although lead(II) is not absorbed through the whole skin, it is readily absorbed through broken skin. Lead dust is readily absorbed in the lungs.

Lead is a cumulative poison. It is deposited within the bone structure in sites ordinarily occupied by calcium ions. Retention in bone deposits is related to the pH of the body fluids. Ingestion of acidifying salts, NaH_2PO_4, NH_4Cl, and $CaCl_2$, or phosphoric acid, favors its release; alkalizing salts, $NaHCO_3$ and Na_2HPO_4, favor its retention. Lead(II) is eliminated in the feces and urine.

Large quantities of sodium or magnesium sulfate should be administered immediately to individuals who have accidentally ingested a lead compound. The stomach should then be emptied. Milk, eggs, and coffee should be taken after emptying, together with a cathartic to speed elimination of any lead(II) ion in the intestinal tract. Calcium disodium ethylenediaminetetraacetate may also be administered as a complexing agent.

Lead(II) ion acts as a protein precipitant. It has excellent astringent action and negligible irritant or corrosive actions. Its high toxicity precludes its widespread adoption as an everyday astringent.

LEAD(II) ACETATE: Lead acetate, sugar of lead; $Pb(OAc)_2 \cdot 3H_2O$. This compound was used as an astringent on unbroken skin and mucous membranes. It has been used as an intestinal astringent for diarrheal inflammation. The salt is efflorescent and absorbs carbon dioxide from the air, becoming coated with a layer of insoluble basic carbonate.

LEAD SUBACETATE SOLUTION: Goulard's Solution. This preparation, essentially a solution of lead basic acetate, is prepared by reaction of lead acetate and lead monoxide. Insoluble basic carbonate precipitates from solution on exposure to carbon dioxide. Diluted with four volumes of recently boiled water it was used as a local astringent. A diluted form,

Diluted Lead Subacetate Solution (Lead Water), was used as an astringent.

OTHER LEAD SALTS: Lead(II) nitrate and lead(II) phenolsulfonate (sulfocarbolate) pentahydrate have been used as external astringents and antibacterials. Lead(II) tannate has been used for chronic inflammation and external ulcers. Lead(II) chloride has been used as an astringent dusting powder. Its former use in ointment form for burns or inflammed skin was ill advised.

TYPICAL ELEMENTS OF GROUP V

General group properties. Nitrogen and phosphorus serve as prototypes of the Group VB elements—arsenic, antimony, and bismuth. The group as a whole is nonmetallic in character; even bismuth is not a true metal.

TABLE 17.6 ATOMIC CHARACTERISTICS OF THE TYPICAL ELEMENTS OF GROUP V

Element	Nitrogen	Phosphorus	Arsenic	Antimony	Bismuth
Symbol	N	P	As	Sb	Bi
Atomic weight	14.008	30.98	74.91	121.76	209.00
Atomic number	7	15	33	51	83
Valence electron configuration	$2s^2 2p^3$	$3s^2 3p^3$	$4s^2 4p^3$	$5s^2 5p^3$	$6s^2 6p^3$
Atomic radius (Å)	0.74	1.10	1.21	1.41	1.52
Ionic radius (Å), M^{3+}	—	0.34(V)	0.69	0.90	1.20
Ionization potential (ev.), I	14.48	10.9	9.8	8.6	8.0
Electrode potential (v.), M/M^{3+}	—	—	0.25	0.21	0.32
Electronegativity	3.0	2.1	2.0	1.8	1.7

The Group VB triad differ mainly in their greatly reduced electronegativity. The atomic characteristics of the members of the family have been compiled in Table 17.6.

Examination of the ground state electron structure, $ns^2np_x^1np_y^1np_z^1$, explains a strong tendency toward the trivalent state. In the simple trivalent state these elements form three σ bonds. Since the bond angles are approximately 90°, pure p orbitals are involved; a nonbonding ns^2 lone electron pair remains. The presence of this nonbonding pair allows the compounds to act as electron pair donors; sp^3 hybridization takes place, giving rise to tetrahedral structures. By means of d hybridization all five valence electrons may be used for bond formation. Bipyramidal, square pyramidal, and octahedral arrangements result.

Although the ions are frequently assigned formal charges of +3 and +5, these numbers represent oxidation states rather than ionic valences. Except for a few bismuth compounds there is no tendency toward simple trivalent cations. Monoxo cations of antimony(III) and bismuth(III) are stable, however.

Arsenic and its compounds. Although this element and its compounds do not occur in large quantities, arsenic is a pervasive element; it is the twelfth most abundant in the human body. Recent evidence indicates arsenic is an essential trace element, possibly involved in the biosynthesis of glutathione. Arsenic and selenium appear to be closely interrelated as trace elements needed in the metabolism of sulfur compounds. There is some evidence that inorganic arsenic is a skin and lung carcinogen in man.

In greater than trace quantities arsenic is highly toxic; arsenic(III) is ordinarily more so than arsenic(V). However, in the human and other mammalian bodies arsenic(V) is reduced to arsenic(III); the symptoms observed are those of the latter. Arsenic is not a protein precipitant. It inactivates sulfhydryl groups of proteins by acting as a central unit for a complex structure. An inhibition of certain enzyme systems results, e.g., phosphatases, nucleases. Under experimental conditions pyruvate oxidation is inhibited. Because of the chemical and structural similarity of the oxophosphate and oxoarsenate ions, it is conceivable that phosphorylation interference occurs.

Symptoms of arsenic poisoning may be delayed as long as twelve hours after ingestion. Typically there is a constriction of the throat and difficulty in swallowing, followed by acute gastrointestinal pain and persistent and projectile vomiting. Treatment consists of copious gastric lavage and saline laxatives. Administration of a mixture of magnesium oxide and iron(III) hydroxide is helpful. The magnesium compound increases the pH of the gastric fluids to levels at which insoluble iron(III) arsenite may precipitate; it also provides laxative action. Sulfhydryl compounds, e.g., 2,3-dimercaptopropanol, may be injected. Shock treatment may be necessary to counteract fluid losses.

Arsenic is slowly excreted in the feces, urine, perspiration, milk, and epithelial tissues. Because of the slowness of excretion, it accumulates in the tissues if administered too frequently and over long periods. However, intestinal mucosa develops a resistance to arsenic absorption; increasingly larger doses are required to produce a given effect. "Arsenic eaters" are individuals who have developed this natural resistance.

During the Middle Ages arsenic(III) oxide was the "poison of choice." Since this compound is not easily detected by taste, it was common practice to invite the intended victim to a suitably prepared banquet.

It may be postulated that successful citizens found it expedient to become "arsenic eaters."

Arsenicals are useful for treatment of protozoan infestations. Certain types of leukemia respond to arsenic therapy by repression of white blood cell formation; a mild concomitant toxic effect is disregarded. Arsenic compounds are used as alteratives and tonics. Small doses produce slightly increased capillary permeability and may stimulate new tissue growth. The pharmaceuticals used, Arsenic(III) Oxide (Arsenous Oxide), Potassium Arsenite Solution (Fowler's Solution) were last officially cited in *N.F.X.*, while Arsenous Acid Solution was last cited in *N.F. IX*. The arsenates have no medicinal applications.

Sodium cacodylate, Na[$(CH_3)_2$AsOO], and iron(III) cacodylate (ferric cacodylate) are less toxic per given arsenic dose. The structure of the cacodylate ion is shown in Fig. 17.5. The action is that of inorganic arsenic, which is slowly released in the blood and body fluids. The cacodylates were used rather empirically for leukemia, serious anemias, and as a tonic. The cacodylates have an offensive odor. Their sodium and iron(III) salts were last official in *N.F.X*.

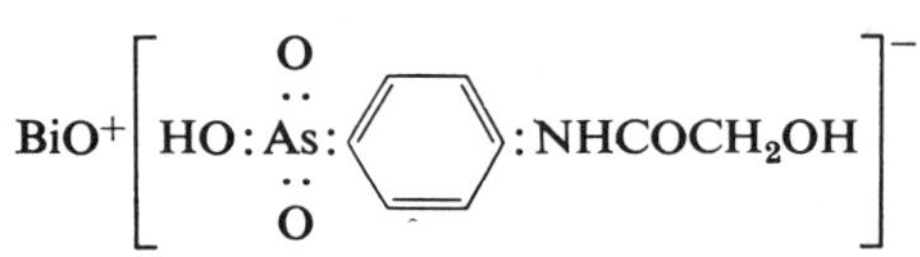

Glycobiarsol

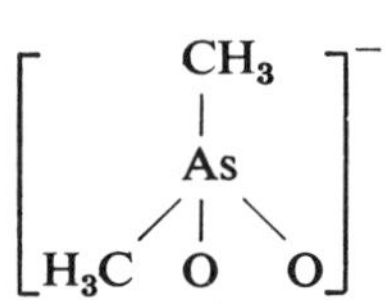

Cacodylate ion

FIGURE 17.5 Structures of arsenicals.

Glycobiarsol (Fig. 17.5) is an intestinal amebicide. It is heat labile and slightly soluble in water and ethanol. Saturated aqueous solutions have a pH of 2.8-3.5. Arsthinol is used for intestinal amebiasis and yaws. Arsphenamine and neoarsphenamine are organic arsenicals used for the treatment of syphilis before the advent of the penicillins.

Arsenic(III) oxide and sodium and potassium arsenate have been used as insecticides and hericides. The insoluble arsenates, especially lead(II) and copper(II), find application as insecticides. Stringent limits are set as to the residues of these and other insecticides remaining on foods. However, precautionary washing of fruits and vegetables is always in order. The use of the highly effective arsenate embalming fluids has been legally outlawed, for obvious reasons.

Antimony and its compounds. Antimony has physiological actions resembling those of arsenic. All antimony compounds are potentially toxic, the symptoms resembling those of arsenic poisoning. Antimony(III) compounds are more toxic than those of antimony(V). The chelating agent, 2,3-dimercaptopropanol, is an effective antidote.

Antimony compounds are useful anthelmintics. They are the reagents of choice in leishmaniasis and schistosomiasis. The latter responds to both antimony(III) and antimony(V); antimony(III) is most effective. Kala-azar (visceral leishmaniasis) responds especially well to antimony(V) therapy. Occasionally mycosis fungoides responds. Arsenic(V) compounds have superseded antimony for treatment of trypanosomiasis. In most other former applications antimony has been replaced by antibiotics. In small amounts antimony compounds act as oral nauseants and expectorants (page 583).

The antimonate(V) ion is variously represented by $[SbO_3]^-$, $[SbO_4]^{3-}$, and $[Sb_2O_7]^{4-}$. New evidence indicates that these ions do not exist. The magnesium, cobalt(II), and nickel(II) orthoantimonates have been shown to consist of hexaquo cations together with the requisite number of hexahydroxoantimonate(V) ions, $[Sb(OH)_6]^-$. Both cation and anion have the octahedral arrangement (see Fig. 17.6). The sodium and potassium salts probably have similar structures.

ANTIMONY POTASSIUM TARTRATE: tartar emetic, dipotassium bis [μ-tartrato(4-)]diantimonate(2-) trihydrate, $K_2Sb_2(C_4H_2O_6)_2 \cdot 3H_2O$. This compound and the corresponding sodium salt is used as an antischistosomal. The potassium salt is considered less irritating by some. The compound is used intravenously, the usual dose being 40 mg. in 0.5% solution three times weekly. It is too irritating to be administered by other means. Serious local and systemic side effects often outweigh medicinal efficiency.

The antimony tartrates are prepared by refluxing freshly precipitated antimony(III) oxide with the appropriate acid tartrate. The antimony(III) is tetrahedrally coordinated. The proposed structure, together with some equilibrium products of its aqueous solutions, is shown in Fig. 17.6.

The potassium salt effloresces on exposure to air; the sodium salt is hygroscopic. The sodium salt is very soluble in cold water (1 Gm. in 1.5 ml.), the potassium only moderately so (1 Gm. in 12 ml.). Aqueous solutions are acidic. Both salts are insoluble in ethanol. Antimony(III) oxide precipitates if the solution is mildly alkaline, but strong alkalis will redissolve the precipitate as the antimonate(III) ion (antimonite ion). Strong mineral acids precipitate insoluble salts. Most divalent (and trivalent) cations precipitate as tartrates. Solubility incompatibilities arise in the presence of excess citrate, tartrate, and sugar. The potassium salt is cited in *U.S.P. XXI.*

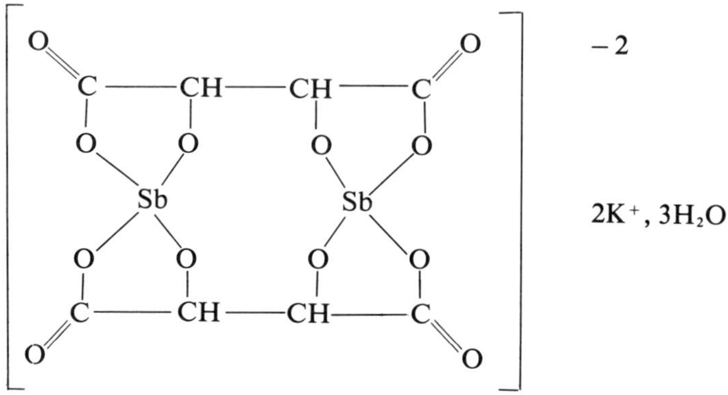

Antimony tartrate

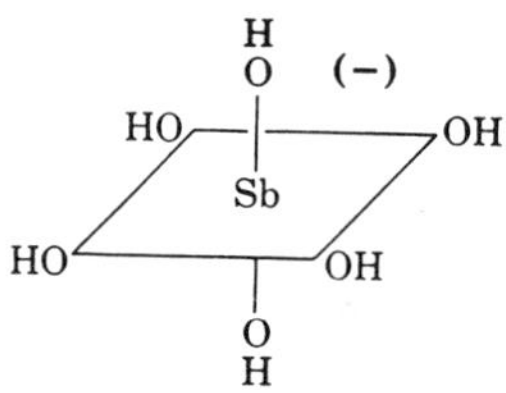

Hexahydroxoantimonate(V)

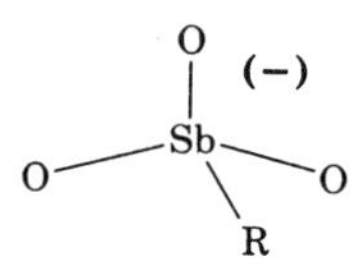

Stibonic acid derivatives

FIGURE 17.6 Structure of antimonates.

OTHER ANTIMONY SALTS: Sodium antimony(III) bis-(pyrocatechol)-2,4-disulfonate (Stibophen), *N.F. XIV,* is less irritating and less toxic than tartar emetic and may be injected intramuscularly. Tartar emetic has a cost advantage. Sodium antimony(III) thioglycollate (antimony sodium thioglycollate), $NaSb(SCH_2COO)_2$; antimony(III) thioglycollamide, $Sb(SCH_2CONH_2)_3$; and lithium antimony thiomalate (antimony lithium thiomalate) have been used as substitutes for tartar emetic. Various stibonic acid derivatives have also been used. Antimony(III) iodide (antimony triiodide) has been used intramuscularly for upper respiratory infections.

Bismuth and its compounds. The soluble inorganic salts of bismuth are protoplasmic poisons. They have a treponemicidal action similar to that of arsenic. Other than their use as antisyphilitics, in which they have been largely replaced by penicillin, the soluble salts are rarely used in medicine. The insoluble salts (basic salts) are used as mild astringents with some germicidal action, as dusting powders, and ointments. Adherence is excellent. They are used internally as astringents and mild, non-irritating anti-infectives for gastritis, diarrhea, and hyperacidity (page 426). They were formerly used as radiopaques (page 588). The nomenclature and general chemistry of bismuth compounds and the preparation of bismuth subcarbonate and bismuth magma have been discussed (page 426).

POTASSIUM BISMUTH TARTRATE: Bismuth potassium tartrate, potassium bismuth tartrate, *N.F. X.* This compound and its sodium counterpart sodium bismuthyl tartrate (bismuth sodium tartrate)—are used as anti-syphilitics. They are light-sensitive, water-soluble compounds of indefinite composition.

BISMUTH SUBCARBONATE: This salt is probably the most widely used insoluble bismuth compound. Although all insoluble bismuth salts are more or less similar in their action, this compound is generally preferred because it is less toxic. It is used for its astringent and protective properties. Good adherence to the skin and mucous membranes make it especially valuable as a protective. It is widely used in the cosmetic industry in lotions, face powders, etc.

Bismuth subcarbonate (bismuthyl carbonate, bismuth oxycarbonate, bismuth carbonate) is a product of variable composition prepared by adding an acid solution of the normal nitrate to a warm solution of sodium carbonate. After washing the precipitate is dried at moderate temperatures ($<60°C.$).

BISMUTH SUBGALLATE: This compound is similar in action to bismuth subcarbonate, but is more astringent. It is used chiefly in dusting powders for eczema and other skin diseases. It probably acts mainly as a protective, but there is also some bacteriostatic action. Bismuth subgallate (other nomenclature analogous to that given for the subcarbonate) is prepared by adding a warm solution of gallic acid to a glacial acetic acid solution of bismuth(III) nitrate. It is a bright yellow, odorless, tasteless powder. It is light sensitive.

BISMUTH SUBNITRATE: This salt is used as a mild anti-infective and astringent in dusting powders. It is used for its protective action in gastritis, enteritis, etc. It is inferior to the carbonate because of side effects arising from bacterial reduction of nitrate to nitrite. It is also used in cosmetics. Traces of bismuth ion appear in the skin and urine following ingestion of this compound. It should not be applied to denuded areas of the skin; sufficient absorption to cause serious poisoning may take place. Bismuth Paste (Beck's Bismuth Paste) is a sterile bismuth subnitrate preparation once injected into chronic fistulae in bone tuberculosis.

Bismuth subnitrate is prepared by adding a solution of bismuth(III) nitrate[11] to boiling water. The formula usually assigned, $BiO(NO_3)\cdot H_2O$,

[11] Bismuth(III) nitrate is prepared by dissolving the pure metal in nitric acid. It crystallizes from the reaction mixture as the pentahydrate.

is approximate. Aqueous suspensions have an acid reaction. It is incompatible with iodides ($BiI_3\downarrow$), alkali metal carbonates and bicarbonates, and gum tragacanth. The powder is slightly hygroscopic and is light sensitive.

BISMUTH SUBSALICYLATE: This compound is used in lupus erythematosis, as an antisyphilitic, and as a mild germicidal agent. Intramuscular injections as suspensions slowly but constantly release bismuth to the body fluids. The compound was originally introduced for treatment of enteritis. An initial protective action by the insoluble bismuth moiety was believed aided by a slow release of salicylic acid as a mild anti-infective. The feebly basic medium of the intestine, however, mediates against this decomposition.

Bismuth subsalicylate is prepared either by reacting stoichiometric quantities of freshly precipitated bismuth(III) hydroxide and salicylic acid, or by reacting a glycerine solution of bismuth(III) nitrate with concentrated aqueous sodium salicylate. The precipitated salt must be washed until its composition is constant. If a sample assays less than 60% bismuth(III) oxide, it is unsuitable for preparation of oil suspensions; on heat sterilization a hard yellow deposit forms in the vial.

OTHER BISMUTH COMPOUNDS AND PREPARATIONS: Of the previously named bismuth compounds, Bismuth Subnitrate and Bismuth Magma (Milk of Bismuth) are currently cited in the *U.S.P.* The last official citation of the others is: Bismuth Subcarbonate, *U.S.P. XVIII,* Bismuth Subsalicylate, *U.S.P. XVI,* Bismuth Subgallate, *N.F. XI,* and Potassium Bismuth Tartrate, *N.F. X.*

Glycobiarsol, bismuth [[4[hydroxyacetyl]phenyl]arsonato(1 –)]oxo, (hydrogen N-glycolyarsanilato)oxobismuth, *U.S.P. XXI*, finds use as an antiamebic. Sodium Bismuth Triglycollamate, *U.S.P. XVI,* was employed as a suppressant for lupus erythematosus.

Bismuth Magma is also used for its astringent action (page 426). Bismuth oxychloride has much the same application as the subnitrate, but without the possibility of nitrite poisoning. Bismuth(III) iodide has been used in treatment of amebiasis. Bismuth albuminate, bismuth and ammonium citrate (*U.S.P. IX*), bismuth citrate (*U.S.P. VIII*), bismuth naphtholate (*U.S.P. IX*), bismuthyl iodide, and bismuth(III) oxide (*B.P.* 1898) have been used as local topical medicants.

SUMMARY—GERMICIDES AND RELATED ACTIONS

Space considerations do not permit a thorough review of all substances studied and the uses of each. Table 17.7 gives a partial listing of some of the more important substances and their principal uses.

TABLE 17.7 Compounds Having Germicidal or Related Activities

Use	Germicide	Bacteriostat	Fungicide	Specialized	Algicide	Disinfectant	Astringent	Deodorant	Irritant	Corrosive	Bleach	Preservative
Substance												
NaCl												x
HCl	x						x					
HOAc	x						x					x
H_2O_2	3%							x		30%	6%	
ZnO_2	x						x	x				
Boric acid and borates		x										
Hypochlorites	x					x		x		x		
$NaClO_3$		x										
$KMnO_4$	x		x			x		x				
Iodine	x		x			x			x	x		
Silver compounds (sol.)	x						x		x	x		
Silver compounds (insol.)	x											
$CuSO_4$			x	x								
$ZnSO_4$							x					
$ZnCl_2$							x	x		x		
Mercury compounds (sol.)	x					x			x	x		
Mercury compounds (insol.)	x	x										
Aluminum compounds (sol.)		x					x	x				
Iron(III) (sol.)		x					x					
Lead(II)							x					
Arsenic and antimony[a]				x								
Bismuth compounds		x		x			x					

[a] Specifics for parasites, e.g., anthelmintics, amoebicides.

STUDY AND SELF-APPRAISAL AIDS

IMPORTANT TERMS. Define or identify each of the following in a single concise sentence if possible.

Amoebicide, antibiotic, anti-infective, argyria, collyrium, cumulative poison, dialysis, double salt, enteritis, expectorant, gargle, gastritis, LD_{50}, leishmaniasis, nauseant, oligodynamic, orthorhombic, poison, protective, schistosomiasis, sublimation, tetradentate, urethritis.

NOMENCLATURE OF INTEREST. Identify each of the following.

Ammoniated mercury, ammonium alum, antimony(V) oxide, antimony(III) sodium thioglycollate, antimonyl potassium tartrate, Benedict's Solution, bismuth subcarbonate, blue ointment, blue vitriol, boracic acid, borax, Burrow's Solution, calomel, chrome alum, corrosive sublimate, dichloroargentate(I) ion, Dobell's Solution, Fowler's Solution, glyceroborate ion, glycerylborate, Goulard's Solution, hexahydroxoantimonate(V) ion, Howe's Solution, Hydrargyri Oxidum Flavum, litharge, lunar caustic, Mayer's Solution, mercury bichloride, milk of bismuth, Monsel's Solution, Nessler's Solution, nitronium perchlorate, Ointment of Mercury, precipitated bismuth, *precipité blanc*, quicksilver, red mercuric oxide, Silver Nitrate Ophthalmic Solution, sodium dithiosulfatoargentate(I), sodium pyroantimonate, subchloride of mercury, sugar of lead, tartar emetic, tetramminecopper(II) sulfate monohydrate, white precipitate.

SELF-APPRAISAL QUESTIONS

1. Acetic acid can act as a mild germicide, astringent, irritant, and caustic. Explain.

2. Why is the yellow to light orange concentrated nitric acid solution a better oxidizing agent than the colorless?

3. What is the composition of the mother liquors remaining after filtration of chromium(VI) oxide from its preparative mixture?

4. Explain the relationship between the solubility of a base and its ability to act as a corrosive.

5. Why is it necessary to dispense freshly prepared silver medicinal agents? In brown bottles?

6. A student attempts to prepare a solution of 0.5 Gm. $AgNO_3$ in a buffer prepared by mixing 50 mL. of the acid solution and 10.0 mL. of the base solution of the reformulated Gifford Buffer (Table 7.8). Explain what occurs.

7. Glucose is never used to render a silver nitrate ophthalmic solution isotonic. Why?

8. What is the role of the stabilizing agent in suspensions such as silver iodide?

9. Calculate and approximate concentration of free silver ion present in 2% mild silver protein suspension. How does the free silver ion concentration present in a 10% suspension compare?

10. Discuss the high volatility of Group IIB metals and the order of volatility within the group on a theoretical basis.

11. Zinc sulfate is buffered to a pH of about 6.5 when using a borate or phosphate buffer, but at about 7.0 with a citrate buffer. Explain.

12. Explain why zinc chloride can be used as an escharotic, but is also used on delicate tissues such as the eyes.

13. Explain why a mercury(II) compound would be expected to have a higher per cent of covalent character than the corresponding mercury(I) compound.

14. Explain why tetrachloromercurate(II) is less stable than tetraiodomercurate(II).

15. On the basis of electrode potentials, explain why both mercury(I) and mercury(II) reduce to elemental mercury if a solution of either contacts a metal such as iron.

16. Why does calomel ointment have a stronger antibacterial action if colloidal mercury(I) chloride is used for its preparation.

17. Three mercuric chloride large poison tablets were accidentally ingested. Assuming a volume of 500 mL. for the gastric juice and a hydrochloric acid concentration of 0.1*M* calculate the free mercury(II) ion which can exist on solution. (1×10^{-3} M Hg^{2+})

18. What special precautions are taken to prevent accidental ingestion of mercury(II) chloride tablets?

19. How is the finish point of a mercurial ointment determined? Explain why this is suitable.

20. Theoretically the elements of Group III should exhibit both +1 and +3 oxidation states. Explain why. In practice only thallium forms stable monovalent cations. Explain.

21. What are the pharmaceutical uses of boric acid? Borax? Discuss the problems involved with each.

22. Distinguish between metaborate, pyroborate, orthoborate, and tetraborate as to structure, properties, and occurrence.

23. What are the advantages achieved by using alum for an astringent solution rather than aluminum sulfate or aluminum chloride?

24. Why are aluminum astringents and deodorants frequently in the form of the dihydroxo cation?

25. Why is dry precipitated calcium carbonate used in the preparation of aluminum subacetate solution? Explain why CO_2 is evolved when acetic acid is added to the solution. (Assume the added calcium carbonate has dissolved completely in the previous step.) Why is the finished solution saturated with respect to calcium sulfate rather than with calcium bisulfate?

26. Distinguish between the structure of the active aluminum complex present in aluminum subacetate solution and that present in aluminum acetate solution.

27. How would you prepare crystalline tetraaquo dihydroxo aluminum bromide?

28. Explain on a chemical basis the purpose of each step in the treatment of accidental ingestion of a soluble lead salt.

29. Why were lead(II) rather than lead(IV) compounds used in preparation of lead astringent solutions?

30. What are the medicinal uses of arsenic, antimony, and bismuth compounds?

31. Discuss the mode of action of arsenic and antimony.

32. Write the ground state electron structure of bismuth. Explain how this structure can give rise to ternary, quaternary, pentanary, and hexa-coordinated bismuth compounds.

33. Distinguish between bismuth(III) nitrate and bismuth nitrate oxide. How is each synthesized?

34. Classify those compounds and preparations studied as to use: germicide, anti-infective, fungicide, algaecide, astringent, irritant, corrosive, deodorant, protective, anthelmintic.

35. Give the usual medicinal uses of compounds of the following: Cu^{2+}, Ag^{+}, $[Al(H_2O)_6]^{3+}$, BiO^{+}, Zn^{2+}, Hg_2^{2+}, AsO_2^{-}.

36. Give the electronic structure, bond type, and the atomic arrangement of each of the following: CrO_3, Hg_2Cl_2, HNO_3, CrO_2Cl_2, copper(II) citrate (complex), H_3BO_3, and the diaquatetrahydroxoaluminate ion.

37. List, together with category, those compounds and preparations studied which are official in the current compendia.

38. Identify the names from which the symbols of the coinage metals, mercury, antimony, tin, and lead, are derived.

39. List those of the elements studied whose compounds are virulent poisons. Give the chemical antidotes for each.

PRACTICE IN EQUATION WRITING

40. Write and balance equations for the preparation of: copper(II) sulfate pentahydrate, Toughened Silver Nitrate, Fehling's Solution, Ferric Subsulfate Solution, Aluminum Acetate Topical Solution, Howe's Solution, Bismuth Magma, Fowler's Solution, lead acetate, mercury(II) sulfate, zinc chloride (anhydrous), boric acid, powdered bismuth, silver nitrate, tartar emetic.

41. Write equations for reactions taking place between the following pairs of substances: silver(I) ion and glucose; sodium chloride and mercuric chloride; Benedict's Solution and glucose; Howe's Solution and formaldehyde; cupric sulfate and phosphorous; magnesium sulfate and lead chloride; BAL and arsenite; ethylene glycol and boric acid in aqueous solution; Ferric Chloride Solution and a metallic closure; calomel and ammonia water; arsenite ion and iodine.

42. Write the equations for the reactions taking place when the following are dissolved in water (distinguish between equilibria and reactions which for practical purposes go to completion): borax, anhydrous aluminum sulfate, antimony(III) chloride.

43. During preparation of Unguentum Hydrargyri Oxidi Flavi an iron spatula was used. Show the reaction taking place.

44. Show equations for reactions by which zinc chloride can be distinguished from cadmium chloride.

45. Write the equations for the reactions taking place during the removal of silver stains by using iodine followed by hypo.

46. Write equations for the equilibria existing in a solution of aluminum chloride maintained at a pH of 5.5.

47. Examine each of the following prescriptions for possible chemical incompatibilities. Write the equation(s) if such an incompatibility exists.

a. Rx

Zinc Sulfate	0.2
Isotonic Gifford Buffer	90.0

b. Rx

Argenti Nitratis	600 mg.
Cocaine Hydrochloride	700 mg.
Aqua	30 ml.

c. Rx

Silver Nitrate	4 Gm.
Rose Water	30 ml.

d. Rx

Mercury(II) chloride	200
Arom. Sp. Ammonia	30

e. Rx

Sod. Nitrate	gr xi
Alum, pulv.	gr ix
P. water, qs ad	f℥ i

EIGHTEEN · GERMICIDES AND RELATED SUBSTANCES III: *topicals—protectives, dermatologicals, miscellany*

Although a few of the substances useful in these applications have been studied in Chapters 16 and 17, for the most part the compounds of interest at this time involve insoluble oxides and sulfur, polysulfides, and polythionates. This necessitates a preliminary study of the typical elements of Group VI. In addition to the application of oxides and sulfides as protectives, and dermatologicals, this group of compounds provides the inorganic pigments of pharmaceutical interest.

GROUP VIB—THE CHALCOGENS

Group properties. The characteristic structural properties of these elements are given in Table 18.1. The elements show the classical familial gradation of properties. There is no known stable isotope of polonium, the fifth member of the family.

These elements form molecules with two σ molecular bonds, leaving two nonbonding electron pairs. The ground state electronic structure indicates the possibility of an ionic state with a -2 charge, $[ns^2np^6]^{2-}$. In practice only oxygen and sulfur can achieve this state.

Because of their high negative charge, oxide and sulfide ions cannot exist independently in aqueous media. On solution of an oxide, the oxide ion, having a particularly small radius, reacts vigorously to give a strongly basic solution of hydroxyl ion.

$$O^{2-} + H_2O \rightarrow 2OH^-$$

Similarly, soluble sulfides undergo hydrolysis to give rise to highly basic solutions.

$$S^{2-} + H_2O = HS^- + OH^-$$

These two reactions illustrate the leveling effect of water on bases stronger than the hydroxyl ion.

Only the oxides and sulfides of the inert gas cations, e.g., Na^+, Ca^{2+}, and a few of the monovalent cations with an underlying $18e^-$ shell, e.g., Tl^+, Ag^+, are sufficiently soluble to give these basic solutions. Despite its high charge, the relatively nonpolarizable oxide ion (and, to a lesser extent, sulfide ion) is unable to bring about the strong polarization necessary to hold these cations in the crystal lattice on contact with water. On the other hand, the transition metal cations, the very small divalent magnesium and beryllium inert gas ions, and most $18e^-$ shell cations react strongly with the intense negative field of the oxide and sulfide ions, leading to stable crystal structures and insolubility. These latter insoluble structures are for the most part the compounds useful in the medicinal applications under discussion in this chapter.

TABLE 18.1 ATOMIC CHARACTERISTICS OF THE CHALCOGENS

Element	Oxygen	Sulfur	Selenium	Tellurium
Symbol	O	S	Se	Te
Atomic weight	16.00	32.066	78.96	127.61
Atomic number	8	16	34	52
Valence electron configuration	$2s^22p^4$	$3s^23p^4$	$4s^24p^4$	$5s^25p^4$
Atomic radius (Å)	0.74	1.04	1.17	1.37
Ionic radius (Å), M^{-2}	1.40	1.84	1.98	2.21
Covalent radius[a] (Å)	—	0.34 (+6)	0.40 (+6)	0.89 (+4)
Ionization potential (ev.), I	13.61	10.36	9.75	9.0
Standard potential (v.)	+1.23	−0.51	−0.5	−0.92
Electronegativity	3.5	2.5	2.4	2.1

[a] In oxo anion.

Both the oxide and the sulfide ions are excellent ligands. In this capacity oxygen forms many extremely stable complexes. Some, e.g., sulfate, phosphate, are so stable that they have historically been considered simple ions rather than complex ions. The thio complexes are relatively stable.

Except for its −2 oxidation state, oxygen differs radically from the other members of the family. Unlike oxygen, sulfur, selenium, and tellurium assume positive oxidation states, +2, +4, and +6, when acting as the central unit of complexes. In these complex structures some *d*-hybridization takes place; oxygen has no *d* orbitals.

Oxygen and its compounds. Oxygen is the most plentiful element, constituting about 50% of the lithosphere. Of this, 23% by weight exists as atmospheric molecular oxygen. Normal oxygen is composed of oxygen-18, 0.2%, oxygen-17, 0.04%, and oxygen-16, the remainder. Oxygen-18 can be concentrated and is a useful nonradioactive tracer; it was used to prove that the oxygen gas given off during the reaction of hydrogen peroxide and permanganate originates entirely from the peroxide moiety.

The gas is slightly soluble in water, its solubility coefficient is 0.029 at 20°C. In aqueous solution it forms a monohydrate and an unstable dihydrate.

Only fluorine is more electronegative. Oxygen reacts chemically with all the elements except the inert gases. Owing to its high electronegativity, metallic oxides tend toward simple ionic structures, with varying degrees of covalent character. Manganese(II) oxide is an ionic solid, whereas manganese(VII) oxide is a covalent liquid.

OXIDES: Oxides may be classified as basic, acidic, and amphoteric. The basic oxides are those soluble oxides giving rise to alkaline solutions, e.g., K_2O, and water-insoluble oxides, e.g., FeO, which dissolve in acidic media. All are metallic oxides. The binding tends to be electrostatic. The acidic oxides are those which dissolve in water to give acidic solutions, e.g., SO_2, or those insoluble oxides which dissolve in alkaline media, e.g., V_2O_5. The acidic oxides are the nonmetallic oxides and the higher oxides of the transition elements. The binding is covalent in the acidic oxides; the small difference in the electronegativity of oxygen and the nonmetallic elements does not allow the polarization required for ion formation. A few nonmetallic oxides appear to be neither acidic nor basic, e.g., CO, N_2O. They may be considered acid anhydrides even though they do not react with water.

$$HCOOH \xrightarrow{\text{heat}} H_2O + CO$$

There are two types of amphoteric oxides: oxides of elements lying on or near the diagonal line drawn through the periodic table, e.g., Al_2O_3, SnO, and oxides of transition elements in intermediate oxidation states, e.g., Cr_2O_3 (page 105).

Oxides are usually prepared by calcination of the hydroxide or carbonate.

$$Mg(OH)_2 \rightarrow MgO + H_2O\uparrow$$

$$ZnCO_3 \rightarrow ZnO + CO_2\uparrow$$

INSOLUBLE HYDROXIDES: When a soluble hydroxide is added to a solution of a metal cation other than the alkalis and the heavier alkaline earths, a precipitate forms. This precipitate may be a true hydroxide

$$Mg^{2+} + 2OH^- = Mg(OH)_2\downarrow$$

an oxohydroxide

$$Al^{3+} + 3OH^- = Al(OH)_3\downarrow = AlO(OH)\downarrow + H_2O$$

or a hydrous oxide

$$Hg^{2+} + 2OH^- = HgO \cdot nH_2O\downarrow$$

The latter precipitate as oxides, but occlude water during the process.

The difference between these precipitates is best shown by dehydration curves, as depicted in Fig. 18.1. The vapor pressure of a true hydroxide is relatively constant with increasing temperature until its decomposition temperature is reached, at which point it falls sharply. An oxohydroxide shows an initially high but slowly falling vapor pressure with increasing temperature, followed by a steep drop in vapor pressure over a narrow temperature interval. A hydrous oxide is characterized by a gradual drop of vapor pressure with increasing temperature.

The preparation of oxides for specialized applications requires a knowledge of the type of "hydroxide" used as the raw material. The residual water, crystal structure, and particle size of the product are affected by the structure of the hydroxide as well as the dehydration process used. The age of the "hydroxide" precipitate is also important. Freshly prepared insoluble hydroxides are usually quite reactive; on aging, changes in the crystal structure usually result in decreased reactivity.

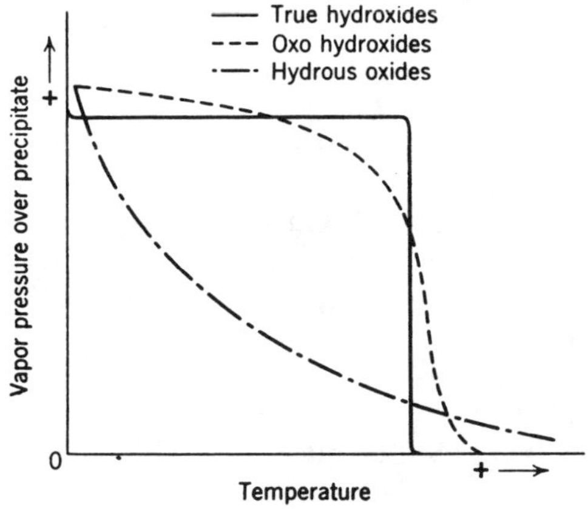

FIGURE 18.1 Dehydration of hydrous precipitates.

Freshly precipitated zirconium(IV) hydroxide reacts readily with mineral acids at room temperature; after aging it does not.

The assumption that insoluble hydroxides or oxides precipitate from basic media is too simple and too frequently erroneous. For exact work the solubility product must be considered; it allows calculation of the hydroxyl ion which will just initiate precipitation. Higher hydroxyl concentrations will act by common ion effect and reduce the solubility. Hydroxyl ion concentration is best considered in terms of pH. A few examples of the minimum pH for initiation of precipitation are: magnesium hydroxide, 10.5; manganese(II) hydroxide, 8.5; mercury(II) hydrous oxide, 7.6; zinc hydroxide, 7.0; and cobalt(II) hydroxide, 6.8.

In addition to pH, amphoterism and complexation must be considered when working with systems actively precipitating hydroxides and oxides. Amphoteric hydroxides (or hydrous oxides, etc.) will redissolve in excess hydroxide ion to give soluble anionic species, e.g., $Na^+[Al(OH)_4(H_2O)_2]^-$. High-charge cations form soluble complex ions, with hydroxy groups acting as ligands, e.g., $Na^+[Sb(OH)_6]^-$.

Basic salts, e.g., BiO^+Cl^-, represent an intermediate stage of neutralization. They occupy the same position with respect to their base that an acid salt occupies with respect to its acid. Whereas acid salts are more soluble than the corresponding normal salts, the basic salts are less soluble than the normal salts. Many of the substances used topically as protectives, antibacterial dusts, etc. are in fact basic salts, e.g., the bismuthyl compounds.

Basic salts are classified into two types, depending upon structural arrangement. The first are those in which the basic anion is associated with the cation; the second are those in which the base is present as an independent entity in an ionic lattice. The former are salts derived from polyhydroxo bases. Some examples of these ions and the ionic lattices have been compiled in Table 18.2.

Sulfur and its compounds. ELEMENTAL SULFUR: Sulfur[1] occurs in nature largely in the free state and as sulfide and sulfate minerals such as pyrites, FeS_2, zinc blende, ZnS, galena, PbS, gypsum, $CaSO_4 \cdot 2H_2O$, anhydrite, $CaSO_4$ and barytes, $BaSO_4$. Elemental sulfur, as produced in the Gulf Coast region of the United States, has a 1.2% hydrocarbon content if unrefined.

Elemental sulfur exists in several allotropic forms. At room temperature α-sulfur, or *rhombic* sulfur, is the stable form, the familiar amber or "sulfur yellow" form. Structurally it consists of eight-membered-puckered rings, with 105° valency angles and bond distances of 2.12 Å

---

[1] British spelling is sulphur.

TABLE 18.2 STRUCTURAL TYPES OF BASIC SALTS

Structural Type	Distinguishing Features	Examples
Associated with cation	Ionic structures	Basic aluminum compounds $[Al(OH)(H_2O)_5]^{2+}SO_4^{2-}$ Uranyl Compounds $(UO_2)^{2+}Cl^-_2$
Ionic lattice compounds	Three-dimensional structures	Basic calcium phosphate, $Ca_5(OH)(PO_4)_3$ Basic copper(II) carbonate, $Cu_2(OH)_2CO_3$
	Layer structures with oxo groups	Bismuth oxohalides BiOCl Hydrous hydroxides FeO(OH)
	with hydroxo groups	Basic magnesium halides Mg(OH)Cl Basic zinc halides $ZnCl_2 \cdot 4Zn(OH)_2$ Layer silicates $(HO)_2Ca_2Mg_5(Si_4O_{11})_2$ $(HO)Al_2(Si_2O_5)$

(Fig. 18.2). At 95.5°C. α-sulfur slowly changes into β-sulfur, or *monoclinic* sulfur. As the transition takes place, the crystal becomes opaque. β-Sulfur occurs in pale yellow, needle-shaped crystals. It too is probably an eight-membered ring. It is stable up to its melting point, 118.7°C. When α-sulfur is heated rapidly, it melts at 114°C. the so-called natural melting point; the conversion of α-sulfur to β-sulfur is very slow. If β-sulfur is slowly cooled to room temperature, its structure persists for some time.

When molten sulfur is heated to above 160°C. and rapidly chilled by pouring into cold water, soft, rubbery, *plastic sulfur* results. The atoms are arranged in parallel spiral chains. On standing, plastic sulfur slowly hardens to form *amorphous sulfur*, a mixture of α- and μ-sulfur. The latter is a polymeric helical structure, each loop including eight sulfur atoms (Fig. 18.2).

Precipitated sulfur may be prepared by one of several methods: acidification of a polysulfide or polythionate solutions (page 536), oxidation of an alcoholic solution of ammonium sulfide at low temperatures, or dilution of an alcoholic sulfur solution with a large volume of water.

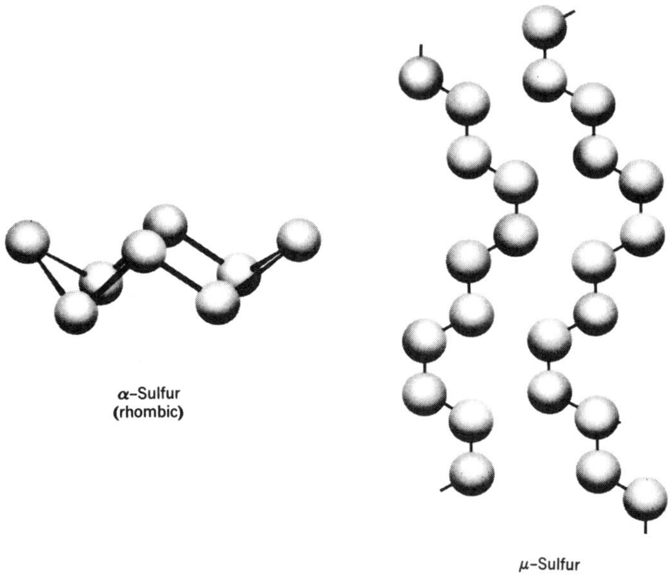

FIGURE 18.2 Elemental sulfur.

Irrespective of preparative method α-sulfur results; only particle size varies. The structure of *sublimed sulfur* has not been established.

Liquid sulfur is probably a mixture of open-chain helical molecules similar to μ-sulfur. Molten sulfur displays a viscosity maximum in the range 160–200°C., probably owing to cross-linking between adjacent chains. Liquid sulfur boils at 444.5°C. Gaseous sulfur is a mixture of S_8, S_4, and S_2 molecules. At 900°C. the vapor density corresponds to S_2. Above this temperature free sulfur atoms appear.

At temperatures somewhat below its ignition point sulfur develops a glow similar to phosphorus. A vapor-phase, flameless oxidation is involved; sulfur dioxide is the principal product.

α-Sulfur is insoluble in water and slightly soluble in ethanol and ether. μ-Sulfur is insoluble in carbon disulfide; α- and β-Sulfur are soluble in this solvent. Sulfur dissolves readily in strong alkali, carbonate, sulfide, and cyanide solutions, yielding polysulfides, thiosulfate, polythionates, with the first three substances, and thiocyanate with the last.

Sulfur reacts readily with the halogens (except iodine) and many metals, e.g., mercury, copper. Its ignition temperature in air is relatively low, 261°C.; it burns to sulfur dioxide.

HYDROGEN SULFIDE: Sulfurated hydrogen. This gaseous substance may be prepared by direct union of the elements, but is more commonly prepared in the laboratory by reacting a strong, nonoxidizing acid with a metallic sulfide, usually iron(II) sulfide.

$$FeS + 2H_3O^+ = H_2S\uparrow + 2H_2O + Fe^{2+}$$

The gas is quite soluble in water (437 ml. in 1 ml. water at 0°C.). Hydrosulfuric acid is a very weak acid. The substance is an extremely strong reducing agent. Oxygen of the air reacts slowly with a hydrogen sulfide solution giving a fine, almost white deposit of sulfur.

$$H_2S + O_2 \rightarrow H_2O + 2S\downarrow$$

Hydrogen sulfide is an extremely poisonous gas, above hydrogen cyanide in toxicity. Fortunately, the nose is more sensitive to hydrogen sulfide than to most gases. Hydrogen sulfide is an insidious poison, the familiar odor of rotten eggs is characteristic only of extremely low concentrations. At higher concentrations a progressive paralysis of the olfactory nerves takes place, and its odor is no longer detected. Instead a sweetish taste is noted. Concentrations of 600 p.p.m. are fatal in thirty minutes; the undiluted gas is fatal almost immediately. Death results from paralysis of the respiratory center. There also is a reducing action on hemoglobin.

Like carbon dioxide, hydrogen sulfide is heavier than air. As a result it tends to settle toward the floor and to accumulate in the lower levels of a building.

The compound has few medicinal applications. Solutions, and lotions incorporating hydrogen sulfide, have been employed for skin disorders.

SULFIDES: The solubility of a sulfide parallels that of the corresponding hydroxide. Aqueous solutions are strongly alkaline. Sulfides are active electron donors. They act as ligands forming many stable complex ions, e.g., CS_3^{2-}, AsS_4^{3-}.

Soluble sulfides add sulfur avidly. If sulfur is suspended in a sodium sulfide solution, the following represent possible reactions.

$$2S^{2-}(aq) + 3S = S_3^{2-} + S_2^{2-}$$
$$S_3^{2-} + 2S = S_5^{2-}$$

The polysulfide ion is usually written noncommittally, S_x^{2-}, where x may have values from 2 to 8. The reaction mechanism and some polysulfide structures are shown in Fig. 18.3. The polysulfides may be considered the analogs of the polyhalides.

Sulfides of all cations with large ionic radii, e.g., Na^+, NH_4^+, Ca^{2+}, form stable polysulfides. Those of cesium are most stable; the alkaline earth polysulfides follow. Certain alkaloids, e.g., brucine, strychnine, form stable disulfides.

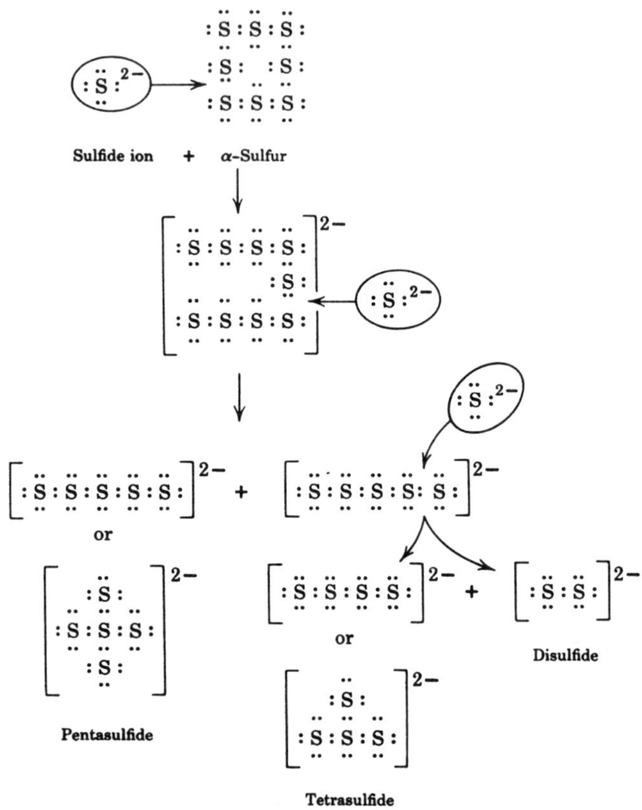

FIGURE 18.3 Polysulfide structures. Some possible mechanisms of formation are also shown.

Because of ease of oxidation, colorless solutions of hydrogen sulfides and the normal sulfides are very difficult to maintain.

$$H_2S + \tfrac{1}{2}O_2 = H_2O + S\downarrow$$

This elemental sulfur reacts as formed, giving yellow disulfide ion. As oxidation proceeds the solution's color gradually deepens to the dark orange of the penta- and hexasulfide ions. The polysulfides too are subject to oxidation.

$$S_x^{2-} + 1\tfrac{1}{2}O_2 = SO_3^{2-} + (x - 1)S^0\downarrow$$
$$SO_3^{2-} + S^0 = S_2O_3^{2-}$$
$$SO_3^{2-} + \tfrac{1}{2}O_2 = SO_4^{2-}$$

The free hydrogen polysulfides are prepared by adding a cold concentrated solution of sodium polysulfide to cold excess hydrochloric acid.

The order of addition is important; the reaction mixture must be strongly acid at all times. If the acid is added to the polysulfide solution, free sulfur and hydrogen sulfide separate as neutral pH's are approached. The hydrogen polysulfides separate as an unstable, pale yellow oil. This liquid may be fractionated in quartz apparatus at 2 mm., yielding disulfide, trisulfide, and possibly tetrasulfide acids. It is thought that the lower polysulfides are not present in the original oil, but are breakdown products of the tetra-, penta-, and hexasulfides. Glass apparatus may be used if the surface alkali is first neutralized by treating the thoroughly dried glass with dry hydrogen chloride gas. All hydrogen polysulfides are unstable at room temperature, decomposing into hydrogen sulfide and sulfur. Water and alkali accelerate the decomposition. Experimental evidence indicates that the disulfide is structurally analogous to hydrogen peroxide, HS—SH vs. HO—OH.

The polysulfides are stronger acids than hydrogen sulfide. This is demonstrated by comparing the degree of hydrolysis of their salts, as shown graphically for the sodium polysulfides in Fig. 18.4.

The polysulfides of interest in pharmaceutical chemistry are sulfurated potash and sulfurated lime. These polysulfides find application in dermatological practice and in the preparation of dermatologicals. The hydrogen polysulfides are lachrymators. It is possible that the polysulfides of other cations may have properties of value in medicinal application.

OXOSULFUR COMPOUNDS: Sulfur dioxide is an angular molecule (bond angle of 119.5°) owing its shape to the sp^2 hybrid orbitals of the sulfur; there is a small d orbital contribution. One hybrid orbital is occupied

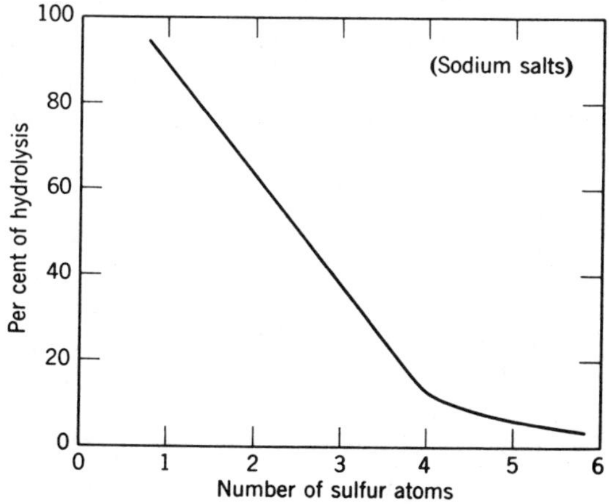

FIGURE 18.4 Degree of hydrolysis of polysulfides.

by a lone pair. The gas is prepared industrially by combustion of sulfur in air or by roasting of a sulfide ore, e.g., pyrites. In the laboratory it is prepared more conveniently by action of a strong nonvolatile mineral acid on a hydrogen sulfite (bisulfite) or sulfite.

$$HSO_3^- + H_2SO_4 = H_2SO_3 + HSO_4^-$$
$$H_2SO_3 = H_2O + SO_2\uparrow$$

Sulfur dioxide is an easily liquefied, colorless gas with a characteristic choking odor. It is a strong reducing agent, but can also act as an oxidizing agent in the presence of strong reducing agents:

$$SO_2 + 2H_2S \rightarrow 2H_2O + 3S\downarrow$$

Its application as preservative and disinfectant has been discussed (page 237).

Sulfur gives rise to two other oxides: a monoxide, SO, and a trioxide, SO_3, neither of which is of direct importance in pharmaceutical chemistry.

Sulfur gives rise to an extensive series of oxo- acids. Many are known only as their salts. They have been tabulated in Table 18.3. The structure of sulfoxylic acid has not been established for certainty. It acts as an oxidizing or reducing agent. Sodium dithionate is an important industrial reducing agent; the acid has not been isolated.

Sulfurous acid has not been isolated, known only in aqueous solution and as its salts. It is a moderately strong acid; the pK_a^1 is 1.77. Solutions of its soluble normal salts have an alkaline reaction. Strong bases form both normal and acid salts (hydrogen sulfites). However, the latter are not known in the solid state. If crystallization is attempted by evaporation of a solution of an acid sulfite salt, sulfur dioxide is evolved and the normal salt crystallizes,

$$2Na^+ + 2HSO_3^-(aq) \xrightarrow{\text{evap.}} Na_2SO_3\downarrow + SO_2\uparrow + H_2O\uparrow$$

Theoretically, a sulfur dioxide atmosphere should repress this decomposition. However, if this is done, metabisulfite rather than hydrogen sulfite crystallizes.

$$2Na^+ + SO_2 + HSO_3^-(aq) = Na_2S_2O_5\downarrow + H_3O^+$$

When a metabisulfite is dissolved in water, the resultant solution acts as a solution of the corresponding hydrogen sulfite.

TABLE 18.3 The Oxo Acids of Sulfur

Acid Name	Formula	Salt	Notes
Sulfoxylic		Sulfoxylate	Known only as salts
Sulfurous		Sulfites	Known only in solution and as salts
Disulfurous	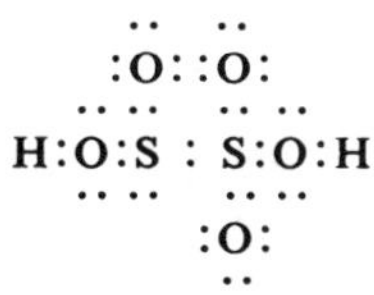	Metabisulfites (incorrectly named pyrosulfites)	Known only as salts
Dithionous	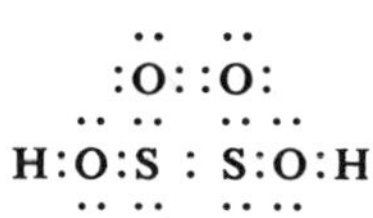	Dithionites (hydrosulfite, hyposulfite)	Known only as salts (S—S bond unusually long, 2.39 Å)
Sulfuric	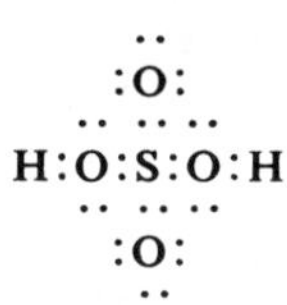	Sulfates	M. p., 10.5°C.
Disulfuric Pyrosulfuric	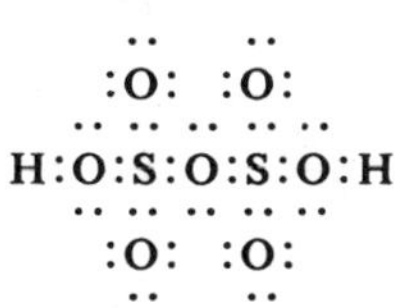	Disulfates (pyrosulfates)	M. p., 35°C.
Thiosulfuric	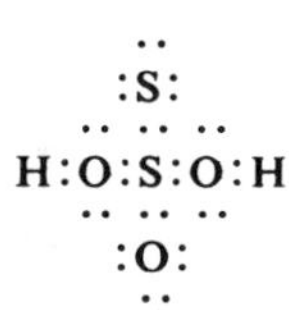	Thiosulfates (hyposulfites)	Known only as salts
Thionic acids	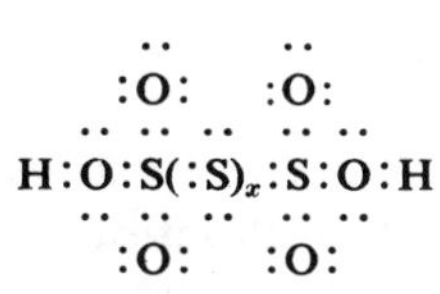	Thionates	Known only in solution and as salts (x = 1–3)

Sulfurous acid and its salts may act as oxidizing or reducing agents. The oxidizing properties are poor. The acid and its salts are excellent reducing agents.

$$SO_3^{2-} + H_2O_2 = SO_4^{2-} + H_2O$$

and

$$SO_3^{2-} + S = S_2O_3^{2-}$$

Sulfuric acid is a strong acid. The degree and mode of ionization are a function of concentration. In dilute solution both protons are available for reaction with water.

$$H_2SO_4 + H_2O \rightleftharpoons H_3O^+ + HSO_4^-$$

$$HSO_4^- + H_2O \rightleftharpoons H_3O^+ + SO_4^{2-}$$

In moderately concentrated solution only the former reaction proceeds to an appreciable extent. In the concentrated acid even this ionization becomes negligible. Pure hydrogen sulfate undergoes self ionization.

$$2H_2SO_4 \rightleftharpoons H_3SO_4^+ + HSO_4^-$$

In addition, some self dehydration takes place.

$$2H_2SO_4 \rightleftharpoons H_2SO_4 \cdot H_2O + SO_3$$

This sulfur trioxide undergoes reaction.

$$H_2SO_4 + SO_3 \leftrightharpoons H_2S_2O_7$$

$$H_2S_2O_7 + H_2SO_4 \rightleftharpoons H_3SO_4^+ + HS_2O_7^-$$

At its freezing point, 10.5°C., pure hydrogen sulfate has the following molal concentrations of the respective ionic species indicated: $H_3SO_4^+$, 0.021; HSO_4^-, 0.013; $HS_2O_7^-$, 0.0083.

Hydrogen sulfate is a useful ionizing solvent for theoretical studies. Water acts as a strong acid; 0.1 molal water in this solvent is 93% ionized. Perchloric acid, the strongest known acid in aqueous media, is a weak acid in hydrogen sulfate solution. Sulfuric acid (and phosphoric acid) is one of the least volatile of the common mineral acids. Volatilization with decomposition occurs at 340°C.

$$H_2SO_4 \rightarrow H_2O + SO_3\uparrow$$

A dense white fog forms over the heated vessel because of the reaction of micro particles of solid sulfur trioxide with moisture. Because of this low volatility, the concentrated acid is a valuable reagent for displacement of volatile acids from their salts:

$$H_2SO_4 + Na^+Cl^- \rightarrow HCl\uparrow + Na^+HSO_4^-$$

Concentrated sulfuric acid (hydrogen sulfate) is an excellent dehydrating agent. This is evident from the stability of its monohydrate and dihydrate to heat.

$$H_2SO_4{\cdot}2H_2O \xrightarrow{167^\circ C.} H_2SO_4{\cdot}H_2O + H_2O\uparrow$$

and

$$H_2SO_4{\cdot}H_2O \xrightarrow{290^\circ C.} H_2SO_4 + H_2O\uparrow$$

The avidity of sulfuric acid for water is demonstrated by placing a drop of the concentrated acid on sucrose. The removal of the OH groups as water is evidenced by immediate charring.

In hot concentrated solution sulfuric acid is an oxidizing agent.

$$SO_4^{2-} + 4H_3O^+ + 2e^- = H_2SO_3 + 5H_2O \qquad \mathscr{E}^0 = 0.20 \text{ v.}$$

However, at room temperature only strong reducing agents will elicit this property; e.g., iron(II) ion is stable in sulfuric acid solution, if protected from air oxidation.

The hydroxyl groups of sulfuric acid, $SO_2(OH)_2$, may be substituted by other electronegative groups. Chlorosulfonic acid, $HO(SO_2)Cl$, and sulfuryl chloride, $(SO_2)Cl_2$, are valuable synthetic reagents for sulfochlorinations. The former is a valuable fogging agent. Sulfamic acid, $NH_2(SO_2)OH$, is used in the form of the sodium salt as a weed killer.

Concentrated sulfuric acid readily dissolves additional sulfur trioxide to give iso-poly acids, e.g., disulfuric (pyrosulfuric), $H_2S_2O_7$; trisulfuric, $H_2S_3O_{10}$; and probably tetrasulfuric, $H_2S_4O_{13}$. The resultant *fuming sulfuric acid* (oleum) is an excellent sulfonating agent and an even better dehydrating agent than ordinary concentrated sulfuric acid. Disulfuric acid gives rise to the disulfates (pyrosulfates). In solution these salts rapidly hydrolyze to give hydrogen sulfates.

$$S_2O_7^{2-} + H_2O = 2HSO_4^-$$

The sulfate ion is tetrahedral. The four S—O bonds are equivalent. Bond length studies indicate considerable sulfur *d*-orbital hybridization, probably accounting for the unusual stability of this ion. The two negative charges may be considered to be shared equally by the four oxygens.

The thiosulfate structure (see Fig. 18.5) is analogous to the sulfate structure. Studies using sulfur-35 indicate that the two sulfurs are not equivalent.

Thiosulfates are unstable in acidic media. However, addition of an acid to a thiosulfate solution does not cause precipitation of sulfur until a pH of 4.6 is reached. This indicates that it is the $HS_2O_3^-$ ion rather

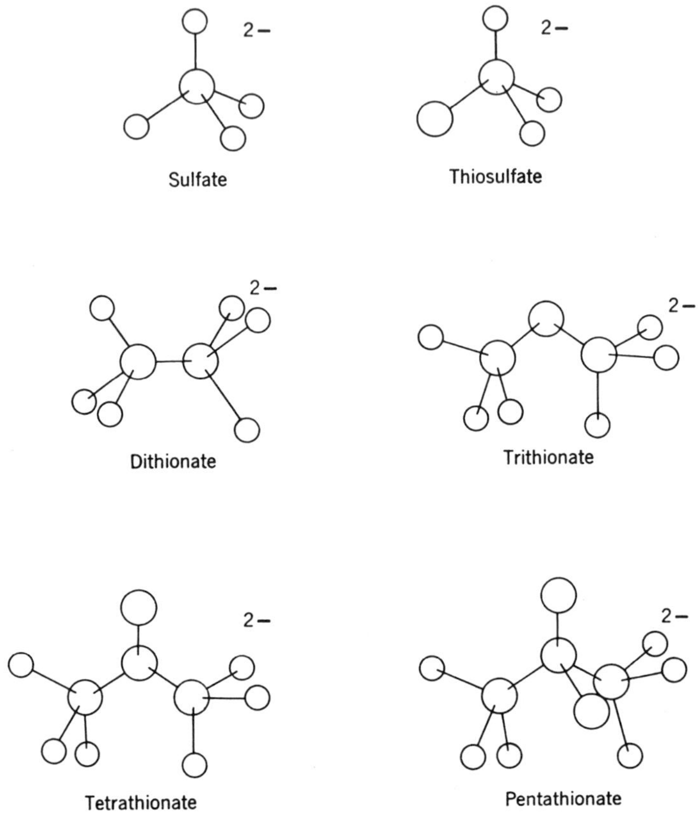

FIGURE 18.5 Sulfate and polythionate ions.

than the $S_2O_3^{2-}$ ion which is unstable, decomposition probably proceeding according to

$$S_2O_3^{2-} + H_3O^+ \rightleftharpoons HS_2O_3^- + H_2O$$
$$HS_2O_3^- = HSO_3^- + S\downarrow$$

Additional acid causes the evolution of sulfur dioxide. Crystalline acid thiosulfates have not been isolated. (Isolation of anhydrous thiosulfuric acid has been claimed recently.)

Thiosulfate reacts readily with cyanide ion. This reaction is the basis for its use as an antidote for cyanide poisoning.

$$CN^- + S_2O_3^{2-} = CNS^- + SO_3^{2-}$$

Thiosulfate ion is an active ligand. Its complexes are quite stable, e.g., dithiosulfatoargentate(I), instability constant 4×10^{-14}.

Thiosulfuric acid is the first member of a family of acids with the general formula, $H_2S_xO_3$, where x may have values of 1 to 7. These acids may be considered monosulfonic acid derivatives of sulfane.

The thionic acids have the general formula, $H_2S_xO_6$, where x has values of 2 to 6. They may be considered disulfonic acid derivatives of the sulfide and polysulfide ions. Dithionic acid, which may be considered disulfonic acid, $HO_3S{:}SO_3H$, differs markedly from other members of the family. Its sodium salt is prepared by action of iodine on sodium sulfite.

$$2SO_3^{2-} + I_2 = S_2O_6^{2-} + 2I^-$$

Some sulfate also forms. The ion, unlike the other thionates (often referred to as polythionates), is resistant to oxidation. It does not add sulfur to give higher thionates. Free thionic acid is a strong acid. Its alkali and alkaline earth salts are stable at boiling temperatures, but the salts of the other metals and the free acid decompose if heated above 50°C.

$$H_2S_2O_6 = H_2SO_4 + SO_2\uparrow$$

Oxidizing agents capable of oxidizing sulfite will oxidize polythionates. Unlike dithionates, polythionates add sulfur readily to give higher members of the series.

$$[S_3O_6]^{2-} + S^0 = [S_4O_6]^{2-}$$

Some structures are shown in Fig. 18.5. The free acids are strong acids. The acids and their salts are thermally unstable.

The polythionate salts through hexathionate have been isolated and identified; there is evidence for the existence of a heptathionate. Since the addition of elemental sulfur to a precursor gives mixtures of higher polythionates, special techniques are required for preparation of individual polythionates.

$$2K^+ + [S_2O_3]^{2-} + 4SO_2 + 3H_2O \rightarrow 2K^+ + 2[S_3O_6]^{2-} + 2H_3O^+$$

$$2[S_2O_3]^{2-} + I_2 \rightarrow [S_4O_6]^{2-} + 2I^-$$

$$SO_2 + H_2S \text{ (2:1 molar ratio as dry gases)} + H_2O \rightarrow [S_5O_6]^{2-}$$

Polythionates are unstable in both alkaline and acid solution. In alkaline solution the tetra and higher thionates first degenerate to trithionate which then decomposes accordingly:

$$[S_{4-6}O_6]^{2-} \xrightarrow{(OH^-)} [S_3O_6]^{2-} + (1\text{–}3)S^0\downarrow$$

$$[S_3O_6]^{2-} \xrightarrow{(OH^-)} [S_2O_3]^{2-} + SO_3^{2-}$$

These reactions explain why the pharmaceutical products sulfurated potash and sulfurated lime include thiosulfate but no polythionates, even though polythionates must form initially.

Decomposition of polythionates under acid conditions is complex, the products formed depending on the pH of the reaction mixtures:

$$[O_3S{-}S_n{-}SO_3]^{2-} = SO_4^{2-} + nS\downarrow + SO_2\uparrow$$

(n = 3 or 4) (weak acid)

$$2[O_3S{-}S_n{-}SO_3]^{2-} + 4H_3O^+ = (n-1)S\downarrow + 5SO_2 + 6H_2O$$

(n = 5 or 6) (strong acid)

Polythionates react with cyanide ion:

$$[S_4O_6]^{2-} + 2CN^- = 2CNS^- + SO_2\uparrow + SO_4^{2-}$$

The higher polythionates form insoluble salts with high molecular weight quaternary ammonium cations:

$$2[RN(CH_3)_3]^+ + S_5O_6^{2-} \rightarrow [RN(CH_3)_3]_2(S_5O_6)\downarrow$$

where R is an aliphatic group, e.g., lauryl. This reaction may be used for fractionation of mixtures of polythionates since a higher member of the series is less soluble than a lower member.

Selenium and tellurium compounds. Selenic acid is similar to sulfuric acid. It is a strong acid and chars organic matter. Selenates are isomorphous with the sulfates. Telluric acid differs widely from its sulfur and selenium analogs. It is a weak, insoluble acid. The tellurates are not isomorphous. Selenium and tellurium do not form acids or salts corresponding to the polythionic series of compounds. However, selenium can partially replace sulfur in these compounds, e.g., potassium selenosulfate, K_2SSeO_3.

The compounds of both elements are toxic; the toxic effects are similar to those of arsenic. Small quantities of tellurium compounds are reduced *in vivo* to the nontoxic elemental form. This is slowly eliminated, probably as the dimethyl telluride. A strong garlic-like odor is imparted to the breath. Selenium occurs as an impurity in many metallic ores. Fumes given off during the reduction of such ores are toxic. Selenium exists in comparatively high concentration in some soils, especially in areas in the Dakotas and the surrounding states. Although nontoxic to plants, it accumulates in plant tissue. Animals feeding upon such plants develop "Alkali Disease."

Selenium and tellurium have limited use in medicine. Sodium tellurate, $Na_2[TeO_2(OH)_4]$, has been used as an antidiaphoretic. Recently selenium sulfide has been successfully introduced into dermatological practice as an antiseborrheic (page 542). Even more recently it has been found that selenium acts as an essential trace element (page 396).

PROTECTIVES

Essential properties of suitable compounds. To be useful as a protective a compound must have certain well-defined properties centering about its solubility, state of division, and chemical inertness. A protective must be insoluble in the body fluids with which it comes in contact. If a compound dissolves, there can be no protective action. Some protectives dissolve to a very limited extent, thereby providing a germicidal, antibacterial, or astringent action. The protective particles must be in a state of division and have a shape inducive to good adherence on the skin or on the stomach or intestinal walls. Sieve size becomes important. If a particle is too coarse, its bulk prevents good adherence. Shape is important since a flat, platy particle will generally show better adherence.

Surface area, inversely proportional to particle size, is important in two ways. A solvate layer on the protective improves adherence to the skin; smaller particles adhere better by virtue of the greater surface area available for solvation. Solvation does not aid adherence in the intestinal tract. Increased surface area renders the protective more effective as an adsorbant for toxins, poisons, etc.

A protective should be chemically inert. It should not react with the skin or the gastric and intestinal mucosa or their associated fluids. Color may be important for cosmetic reasons. Psychologically a protective with a skin tone is more acceptable to some patients than a starkly white material.

Classification and suitable compounds. Protectives are generally classified as external or internal. Those intended for use on the skin, e.g., chalk, are known as external protectives. Those intended for use in the gastrointestinal tract are classified as internal. An internal protective is intended to coat the crater of ulcers and other lesions of the gastrointestinal tract.

Except for the insoluble zinc compounds, most substances used as protectives have been previously discussed and are compiled in Table 18.4. The peroxo compounds provide an added germicidal action. Kaolin is used as an internal protective when a strong adsorbing action is needed for inactivation of toxins. Because of the coating action of silicic acid gel, magnesium trisilicate is considered particularly valuable for treatment of ulcers. Talc is benign; applied to the skin of some individuals it can produce severe granulomatous reactions. Powdered boric acid, having a platy layer structure, has good adherence but this toxic compound is absorbed through broken skin.

ZINC COMPOUNDS: Zinc oxide (zinc white, flowers of zinc) is a white to yellowish white amorphous powder. It is insoluble in water and ethanol. Being amphoteric, it dissolves readily in dilute acids and strong bases. It is readily complexed by ammonia. Zinc oxide slowly absorbs carbon dioxide from the air. Pure zinc oxide is prepared by calcination of the basic carbonate, $(ZnOH)_2CO_3$. It sublimes without decomposition. It becomes yellow on heating, but there is no evidence of a change in structure.

TABLE 18.4 COMPOUNDS USED AS PROTECTIVES

Compounds	Common Use
Bismuth compounds	
Bismuth subcarbonate	Internal, external
Bismuth subgallate	Internal, external
Bismuth subnitrate	Internal, external
Peroxo compounds	
Medicinal zinc peroxide	External
Zinc perborate	External
Silicates	
Kaolin	Internal; external
Talc	External
Magnesium trisilicate	Internal
Miscellaneous compounds	
Basic aluminum carbonate	Internal
Boric acid	External
Chalk	External
Starch	External

Zinc oxide is a protective. It has mild astringent and mild topical antacid properties. It is used topically in the form of a dust, paste, or as a 15–25% ointment in either an oil or aqueous base. Lassar's Plain Zinc Paste is zinc oxide, 25%, starch, 25%, in white petrolatum; it is also prepared with salicylic acid, 2%. Zinc oxide is commonly used in combination with other protectives and medications, e.g., talc, kaolin, calamine, boric acid, resorcinol.

Calamine is a zinc oxide prepared by calcination of native zinc carbonate.[2] Iron compounds present in the ore as impurities are converted to hydrated iron oxides, imparting a skin-like tint to the predominant zinc oxide. This color is considered desirable for cosmetic reasons.

[2] In mining and metallurgy zinc carbonate is known as calamine. The *B.P.* defines calamine as zinc carbonate suitably colored with iron(III) oxide.

However, the tint achieved varies considerably because of the variation of the native carbonate and preparative techniques.

Neocalamine represents an attempt to control the tint of calamine. It was at one time recognized officially (*N.F. IX*). It is prepared by blending zinc oxide, 93%, red iron(III) oxide, 3%, and yellow iron(III) oxide, 4%. In practice, colors suitable for most individuals, regardless of race, can be obtained by using suitable mixtures of these and some of the darker iron(II,III) oxides.

Other than the color imparted by the iron oxides, calamine and neocalamine are physically and chemically identical to zinc oxide. However, the temperatures required for calcination of native ores produce a physically unsuitable product with a gritty quality and poor adhesion. The crude calamine must be pulverized until it passes through a 100-mesh sieve. Because of its incidence in zinc ores, arsenic limits are prescribed for both zinc oxide and calamine.

Calamine is usually applied as a dusting powder, ointment, or lotion. Turner's Cerate (Calamine Ointment) is a 1.7% calamine ointment. Calamine Lotion has probably been the subject of more studies intended to improve its stability and other properties than any other single pharmaceutical preparation. As a result, its officially recognized preparative method and composition is changed frequently. Calamine Lotion, *U.S.P. XXI,* is a zinc oxide, 8%, and calamine, 8%, suspension in lime water including a small quantity of glycerol. The suspension is stabilized with bentonite. Phenolated Calamine Lotion, *U.S.P. XXI,* includes liquid phenol, 1%.

These preparations have the protective and astringent properties of calamine. Calamine Lotion finds wide application in a variety of dermatoses, chiefly the acute exuding types. It exerts a soothing, drying, protective action. It is also useful for generalized eczemas. The phenolated version imparts an antipruritic and local anesthetic effect to the preparation.

Zinc carbonate (*U.S.P. IX*) has been used as a substitute for calamine. Its activity is similar to that of zinc oxide. Its intense white color has reduced its popularity.

Zinc stearate is a mixture of zinc salts of stearic and palmitic acids, usually with a slight excess of zinc oxide. The substance is a fine, white, soft, bulky powder with a slight, characteristic odor. Its melting point is about 120°C. It repels water and is insoluble in all ordinary solvents. It decomposes when suspended in dilute acid, the insoluble fatty acids separate. Zinc stearate is used topically as a dusting powder and in lotions and ointments for its mild astringent and anti-infective properties. It is usually used in combination with other protectives and medicinal agents. Its use as a dusting powder for infants and children is to be discouraged; inhalation of the dust causes pulmonary inflammation. It is a common constituent of cosmetics. Zinc stearate is used as a lubricant in tablet preparation. It is also used as a waterproofing agent.

Zinc gelatin (Zinc Gelatin Boot) is an aqueous preparation including zinc oxide, 10%, gelatin, 15%, and glycerine, 40%. It serves as a protective.

TITANIUM DIOXIDE: This compound, TiO_2, is an extremely inert, intensely white substance. It dissolves in only the most powerful solvents—fused alkalis, bisulfates, hydrofluoric acid, and hot concentrated sulfuric acid. In the rutile crystal structure it has a higher refractive index, 2.6, than the diamond, 2.4.[3] The anastase and brookite crystal modifications have lower refractive indices. Thus titanium dioxide is an excellent example of a substance requiring accurately controlled crystallization conditions to obtain an optimum structure.

Titanium dioxide is available commercially as a pure white or slightly cream-tinted product. The latter includes varying quantities of coprecipitated iron(III) phosphate. Titanium ores are associated with iron, which must be removed during the manufacturing process. Traces of iron(III) accumulate in the mother liquors from which titanium dioxide is precipitated. Phosphoric acid is added to these liquors to prevent coprecipitation of the iron(III) as the more highly colored hydrated oxides.

Covering power of a white pigment is directly related to its refractive index. Having a higher refractive index, titanium dioxide has the highest covering power of all white pigments. For this reason it is useful as a solar protective. It is included in suncreams to ward off undesirable light rays. It is used as a white pigment in lotions and cosmetic preparations. Small percentages of calamine or neocalamine may be incorporated for cosmetic reasons. Titanium Dioxide, *U.S.P. XXI,* is official as a solar protective.

ZIRCONIUM COMPOUNDS: Zirconyl chloride, $ZrOCl_2 \cdot 8H_2O$, is the intermediate used for the preparation of most zirconium compounds of interest in pharmacy and medicine. It is prepared by reaction of zirconium tetrachloride with water and subsequent recovery by evaporation of the solvent.

$$ZrCl_4 + 3H_2O = ZrOCl_2 + 2H_3O^+ + 2Cl^-$$

[3] Titanium dioxide can be grown into massive single crystals. Titanium dioxide jewels have a greater brilliance than the diamond. Unfortunately, they do not have the diamond's hardness. Friction erodes the facets of the cut crystal.

On solution zirconyl chloride gives rise to highly acid solutions

$$ZrOCl_2 + 2H_2O \rightleftharpoons [ZrO(OH)]^+ + H_3O^+ + 2Cl^-$$

The cation formed is probably identical to the structure present in the crystalline state (Fig. 18.6). Equilibrium is attained slowly, about 24 hours or more. The results obtained from processes requiring aqueous zirconyl chloride depend on the age of the solution; a freshly prepared solution will react differently from an aged solution.

If the free acid of a zirconyl chloride solution is carefully neutralized, a soluble but highly polymerized crystalline zirconium oxohydroxochloride, empirically $(ZrOOH)^+Cl^- \cdot xH_2O$, may be recovered. In contrast to the parent salt its solutions are only mildly acidic. When an excess of strong base, e.g., NaOH, is added to the zirconyl chloride solution, hydrous zirconium oxide (hydrous zirconia) precipitates.

$$\{H_3O^+ + 2Cl^- + [ZrO(OH)]^+\} + 2OH^- + (n - 3)H_2O = ZrO_2 \cdot nH_2O\downarrow + 2Cl^-$$

This precipitate is insoluble in water and weak acids. It will dissolve in hot hydrochloric acid. A more reactive form of the hydrous oxide is obtained by using sodium carbonate as the precipitating agent; with suitable control a product is obtained approaching the composition, $2ZrO_2 \cdot CO_2 \cdot 8H_2O$. This latter product is known as zirconium carbonate or carbonated hydrous zirconia.

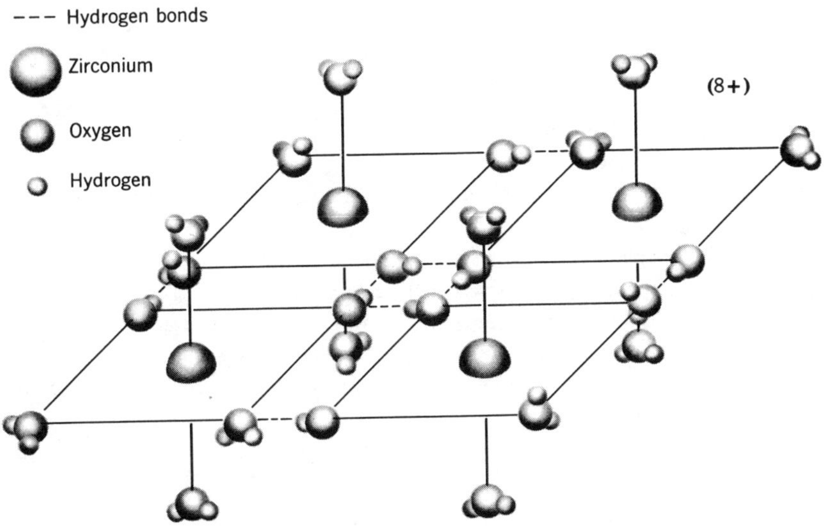

FIGURE 18.6 Structure of the zirconyl ion.

The complex oxohydroxozirconium cation, $[ZrO(OH)]^+$, of zirconium oxohydroxochloride shows a tendency to split out water and bond with certain other molecules and ions. With the organic hydroxo compounds the strength of bonding, weakest first, is monohydroxo aliphatic, 1,2-dihydroxo aliphatic, 1,2-dihydroxo aromatic, carboxylate and α-hydroxocarboxylate ions. Larger aggregates show sequestering action for cations.

Hafnium occurs in all zirconium ores. Because of extreme similarity of its properties it is carried along during the industrial processing for zirconium with virtually no separation. Since there is no significant qualitative difference between the two elements, the presence of hafnium in zirconium compounds is regarded of no consequence.[4]

Early work with test animals and humans indicated that zirconium compounds could be ingested or used topically without toxic effects. However, as with solids of all types, insoluble zirconium compounds injected subcutaneously tend to cause granulomas. Later work with aerosol sprays showed skin granulomas could occur, and tests on monkeys showed mild inflammation of the lungs.

The FDA first banned the use of zirconium compounds in aerosol deodorants and antiperspirants in 1974. Hydrous zirconium oxide and zirconium carbonate were commonly used in these applications. They were also formerly popular in treatments for contact dermatitis.

"Zirconium silicate" is used in cosmetic preparations. It has excellent covering power, good slip, and adherence (cling).

INORGANIC PIGMENTS

The inclusion of pigments in protectives and other medicinal preparations is largely for psychomedical reasons rather than direct medicinal action. Since the inorganic pigments used for this purpose and in cosmetics are insoluble oxides and some sulfides, a brief review of this class of substances is appropriate here.

The use of inorganic pigments for human decoration and otherwise goes back to antiquity. Egyptian tombs have yielded stibnite (antimony (III) sulfide) preparations intended for darkening the eyebrows. It is still used for cosmetic purposes.[5] The iron oxides and oxohydroxides are the most widely used. Depending upon degree of hydration, oxidation state, and particle size, this family of compounds provides pigments virtually of all colors. The hydrated chromium oxides give a valuable

[4] Zirconium intended for use in atomic reactors must be free of hafnium.

[5] Freshly prepared stibnite is orange-red; it becomes black on standing.

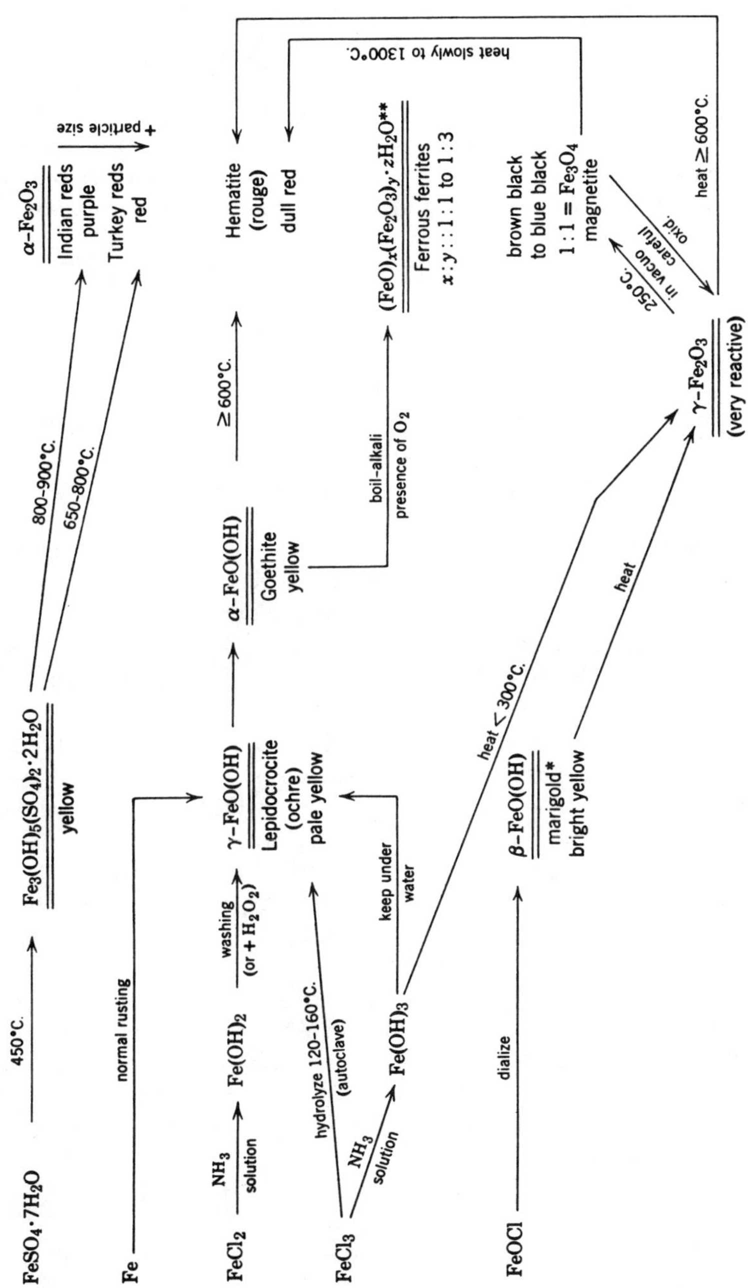

FIGURE 18.7 The iron pigments. *Traces of chloride ion are essential for stability. **May be prepared by sodium hydroxide precipitation from solutions of iron(II) and iron(III) chlorides in the desired ratio.

range of blue and green pigments. By blending various combinations of these pigments, it is possible to approximate any and all skin tones. The corresponding manganese and rare earth compounds are also available. The ultramarines have been discussed earlier (page 300).

Inorganic pigments have excellent color stability and are chemically inert under the usual conditions of use in topical preparations and cosmetics. An excellent range of colors is available. Unlike the organic pigments (or dyes), the possibility of allergic reactions and carcinogetic activity is exceedingly remote.

The iron oxide-based series of pigments is shown in Fig. 18.7. The interrelations and methods of preparation are indicated.

DERMATOLOGICAL AGENTS

This group of medicinal agents includes those substances used for treatment and prevention of skin disorders arising from causes other than infectious conditions. Sulfur and its compounds are popular in these applications.

Elemental sulfur. Sulfur (brimstone) is available in several physical forms. When molten sulfur from any recovery process is allowed to cool in large vats or is cast in cylindrical wooden molds, the commercial *rock* and *roll*[6] *sulfur,* respectively, are obtained. The sublimate obtained on heating rock or roll sulfur is known as sublimed sulfur (flowers of sulfur). (Sublimed Sulfur, *U.S.P. XXI.)* A product known as washed sulfur is prepared by pruification of sublimed sulfur. The sublimed sulfur is crushed to pass through a No. 30 sieve and washed with 1% ammonia water for three days. Traces of dissolved impurities, mainly arsenic (As_2S_3), tellurium, and sulfuric acid, are dissolved. The sulfur is then washed with water until the wash water is neutral, dried rapidly, and rescreened to pass the No. 30 sieve. Sulfur produced in the Gulf Coast region of the United States is nearly arsenic free; the preceding process is unnecessary.

Precipitated sulfur (milk of sulfur, *lac sulfuris*) (*U.S.P. XXI*), is prepared by boiling an aqueous suspension of sulfur and calcium hydroxide. The sulfur goes into solution as a mixture of polysulfide and polythionate ions (page 519). The filtrate from the reaction mixture is adjusted to pH 7 by hydrochloric acid. The polysulfide present decomposes according to

$$S_x^{2-} + 2H_3O^+ \rightarrow H_2S\uparrow + (x - 1)S^0\downarrow + 2H_2O$$

Thioarsenate(III), thiosulfate, and polythionate are stable at this pH.

[6] Reference to recently popular music is unintended.

The precipitated sulfur is washed with water until the wash water no longer gives a test for calcium. It is then dried rapidly. The product is a very fine, pale yellow, amorphous powder. Being in a finely divided state, precipitated sulfur dissolves more readily and completely in solvents and is more reactive than other forms of sulfur.

Sulfur is used topically in the form of dusts, lotions, creams, pastes, ointments, e.g., Sulfur Ointment, *U.S.P. XXI,* detergent cakes, soaps, and foams. The sulfur content of these preparations is usually within the range 2-25%, but most commonly 3-10%. When necessary a suspending agent is present; bentonite is commonly used. Most of these preparations include additional medicinal agents. Sulfur-resorcinol combinations, usually as lotions or creams, e.g., Resorcinol and Sulfur Lotion, *U.S.P. XXI,* are common; sulfur is usually held below 5%, most commonly at 3% and resorcinol is present in the range of 1-3%. Resorcinol supplements the activity of the sulfur. Salicylic acid is a frequent ingredient of sulfur preparations. Its keratolytic action supplements that of sulfur. Zinc oxide and calamine are frequently present for their protective action and the mild astringent action of the zinc ion. Talc is also a common ingredient in sulfur preparations. Iron oxides may be included for cosmetic reasons.

Sulfur is used topically for a wide range of conditions. It is used as an antiparasitic, especially in scabies. It is useful as an antifungal and as a keratolytic for a range of conditions such as acne, psoriasis, seborrheic alopecia, and some eczematous processes. A fine dust applied to the body where clothing binds controls chiggers. Some individuals are allergic to sulfur. It may cause dermatitis, headache, and fever.

Sulfur is also used as a cathartic (page 579) and as a fumigant. A sulfur candle is burned in the room to be fumigated; the effect is poor but is improved in moist air. Colored materials will be bleached if not removed. Sulfur is used in insecticide preparations, e.g., lime sulfur, principally as a fungicide.

Sulfur apparently owes its activity to a slow reaction with moisture. Both sulfide and polythionate ions probably result *in vitro.* These ions have been shown to have germicidal and fungicidal action. The keratolytic effect of sulfur as sulfide ion may be the basis for its therapeutic action in certain skin disorders not associated with infections.

Sulfurated potash. (Liver of sulfur, *hepar sulfuris*). This compound is a solid of variable composition consisting largely of potassium polysulfides and potassium thiosulfate. It is sometimes incorrectly named "potassium sulfide." Because of variable composition, the sulfur content is ordinarily used as a quantitative measure of composition; e.g., *U.S.P. XXI* specifies not less than 12.8% sulfur combined as sulfide. The usual composition ranges, expressed as sulfur, are: sulfate, 1.7–2.15%; thiosulfate, 10.0–12.5%; polysulfide, 32–35%.

The product is manufactured by controlled heating of a mixture of potassium carbonate and sublimed sulfur. While carbon dioxide is evolved, the temperature is held below 185°C. When carbon dioxide evolution ceases, the temperature is allowed to increase slowly until the reaction mixture is completely fused, but not in excess of 300°C. Above this temperature thiosulfate is converted to polysulfide and inactive sulfate.

$$4S_2O_3^{2-} = 3SO_4^{2-} + S_5^{2-}$$

The equation for the preparation of sulfurated potash is usually represented as

$$3K_2CO_3 + 8S = 2K_2S_3 + K_2S_2O_3 + 3CO_2$$

Although this equation preserves the approximate sulfur stoichiometry of the product, a closer examination of the underlying chemistry indicates a complex series of reactions. Limited quantities of sulfur are probably oxidized to the dioxide by oxygen of the air.

$$S + O_2 \rightarrow SO_2$$

Being a stronger acid, sulfur dioxide displaces carbon dioxide, leading to the formation of sulfite.

$$SO_2 + CO_3^{2-} \rightarrow SO_3^{2-} + CO_2\uparrow$$

Elemental sulfur reacts with sulfite giving rise to thiosulfate and polythionates.

$$SO_3^{2-} + S^0 \rightarrow S_2O_3^{2-}$$
$$2S_2O_3^{2-} + S^0 \rightarrow S_4O_6^{2-} + S^{2-}$$
$$S_4O_6^{2-} + S^0 \rightarrow S_5O_6^{2-}$$

As it is formed sulfide ion reacts with sulfur, giving rise to polysulfides.

$$S^{2-} + (x - 1)S \rightarrow S_x^{2-}$$

The polythionates are unstable at the temperature and high pH of the reaction medium. Using tetrathionate, the decomposition may be represented as

$$S_4O_6^{2-} \rightarrow S_3O_6^{2-} + S\downarrow$$
$$S_4O_6^{2-} + S^{2-} \rightarrow 2S_2O_3^{2-} + S\downarrow$$

Some air oxidation of thiosulfate always takes place.

$$2S_2O_3^{2-} + O_2 = 2SO_4^{2-} + 2S\downarrow$$

Additional sulfide ion may be generated through oxidation and reduction of sulfur by carbon dioxide to give carbonyl sulfide and sulfur dioxide.

$$2CO_2 + 3S = 2COS + SO_2$$

Carbonyl sulfide undergoes disproportionation,

$$2COS = CO_2 + CS_2$$

Under equilibrium conditions the latter reaction proceeds 19% at 100°C. and 27% at 200°C. Since moisture is present, hydrolysis of carbon disulfide can yield additional sulfide ion.

$$CS_2 + 2OH^- \rightarrow 2(HS)^- + CO_2\uparrow$$

By action with sulfide ion, carbon disulfide can give rise to thiocarbonates.

$$CS_2 + S^{2-} = CS_3^{2-}$$

Freshly prepared sulfurated potash and its solutions have the odor of hydrogen sulfide and a bitter acrid taste. It has a liver-brown color, but due to slow reaction with absorbed moisture (polysulfides are deliquescent), carbon dioxide, and oxygen the color changes to a greenish yellow or grey. If the decomposition reactions have not proceeded extensively, a freshly fractured crystal shows the liver-brown color on the newly exposed surface. These reactions may be written

$$2H_2O + K_2S_x = 2K^+ + H_2S\uparrow + 2OH^- + (x - 1)S\downarrow$$

$$K_2S_x + CO_2 + H_2O = K_2CO_3 + H_2S\uparrow + (x - 1)S\downarrow$$

$$2K_2S_2O_3 + O_2 = 2K_2SO_4 + 2S\downarrow$$

Sulfurated potash must be stored in tight containers. Small bottles (120 Gm.) should be used for active storage in the dispensary.

Sulfurated potash is very soluble in water (1 Gm. in 2 ml.), but there may be a slight insoluble residue. The polysulfides are soluble in ethanol, but thiosulfate is not. The substance is incompatible with even weak acids, e.g., acetic, and most metal cations.

$$S_x^{2-} + 2H_3O^+ = H_2S\uparrow + 2H_2O + (x - 1)S\downarrow$$

$$S_2O_3^{2-} + H_3O^+ = HSO_3^- + S\downarrow + H_2O$$

$$Fe^{2+} + S_x^{2-} = FeS\downarrow + (x - 1)S\downarrow$$

$$[Al(H_2O)_6]^{3+} + 3OH^- = [Al(H_2O)_3(OH)_3]\downarrow + 3H_2O$$

Sulfurated potash is used in the form of lotions or ointments, 1–5%, and aqueous solutions, up to 10%. Anise oil may be included to mask the hydrogen sulfide odor.

Alkaline sulfides dissolve the horny component of the skin. For this reason sulfurated potash is used in treatment of psoriasis. The polysulfide component has a local stimulating action and is therefore used to restore tone to dilated blood vessels in skin conditions characterized by sluggish circulation, e.g., acne rosacea. Sulfurated potash preparations are used as parasiticides for such conditions as scabies, ringworm, tinea versicolor, and others. Various chronic skin conditions are also treated.

WHITE LOTION: *Lotio Alba.* This compound is prepared by slowly adding a freshly prepared and filtered 9% solution of sulfurated potash to a 9% zinc sulfate solution (both 40 g. salt in 450 ml. purified water) with constant stirring. In order to maintain the mildly acid to mildly alkaline medium necessary to obtain a creamy white suspension, it is essential that the order of mixing be observed. The reaction may be written

$$Zn^{2+}SO_4^{2-} + K^+{}_2S_x^{2-} = 2K^+ + SO_4^{2-} + ZnS\downarrow + (x - 1)S\downarrow$$

On completion of reaction an additional 10% water is added. Bentonite magma has been used to improve the stability of the suspension.

Simplification of the preparative procedure has been attempted. Addition of zinc sulfate to the reaction mixture during the sulfurated potash preparation yields an unsatisfactory product. Freshly fused zinc sulfate and sulfurated potash when triturated in a mortar, with addition of purified water from time to time, yields a satisfactory product. However, the preparative advantage of this latter procedure is questionable. Although it has not been reported, it is conceivable that flash evaporation of the prepared white lotion could with proper controls yield a product suitable of reconstitution when needed by addition of purified water.

To be effective White Lotion must be freshly prepared. Increases in particle size and changes in crystal structure decrease its activity on standing. Loss of adhering power on aging indicates increased particle size. White Lotion is used for its astringent and protective properties. It has been popular for treatment of acne vulgaris over the past century.

Sulfurated lime topical solution. Vleminckx's Solution, Vleminckx's Lotion. Sulfurated lime chemically resembles sulfurated potash, except for the cation present and its physical state. It is prepared by slaking the necessary lime, mixing the slaked lime with pulverized roll sulfur, and adding this mixture in portions to a volume of boiling water up to two times the desired final volume. The usual proportions and procedure are found in *U.S.P. XXI.*[7] When reaction is complete, the solution is filtered hot and made up to volume. Filtration is necessary to remove insoluble

[7] *U.S.P. XXI,* p. 596.

impurities present in the lime and the calcium sulfate formed by air oxidation of thiosulfate. Except for substitution of hydroxyl for carbonate ion, the reactions taking place during preparation are essentially identical to those taking place during preparation of sulfurated potash. The equation usually given

$$3Ca(OH)_2 + 12S = 2CaS_5 + CaS_2O_3 + 3H_2O$$

expresses only the approximate stoichiometry rather than a true picture of the mechanism of reaction.

Sulfurated lime solution has a clear orange or reddish yellow color and a hydrogen sulfide odor. It has an alkaline reaction. Incompatibilities are identical to those of sulfurated potash, with the added possibility of calcium sulfate precipitation on excessive exposure to oxygen of the air.

Sulfurated lime solution (usually diluted with up to ten parts purified water) is used externally as a scabicide, for treatment of various acnes, seborrhea capitos, and other skin conditions. Its activity is largely due to sulfide itn, thitsulfate ion, and sulfur. On proper acidification, sulfurated lime yields Milk of Sulfur (page 536).

Sodium thiosulfate. Sodium hyposulfite, hypo, antichlor; $Na_2S_2O_3{\cdot}5H_2O$. Sodium thiosulfate is a colorless, odorless, crystalilne substance. It is efflorescent in dry air and slightly deliquescent in moist air. It is very soluble in water (1 Gm. in 0.5 ml.). The solution has a slightly alkaline pH. Its solutions decompose slowly when cold and more rapidly when hot. If the solution is to be stored for some time or exposed repeatedly, a small per cent of sodium carbonate is usually added to maintain the alkaline pH necessary for maximum stability. Saturation with thymol prevents mold growth.

Thiosulfate is incompatible with oxidizing agents (e.g., iodine), acids (e.g., acetic), and heavy metal ions (e.g., mercury(II)). It forms stable complexes with transition metal cations. The second sulfur atom undergoes transfer to stronger Lewis bases. The following equations illustrate these reactions.

$$2S_2O_3^{2-} + I_2 = [S_4O_6]^{2-} + 2I^-$$

$$S_2O_3^{2-} + H_3O^+ = HSO_3^- + H_2O + S\downarrow$$

$$S_2O_3^{2-} + 2Ag^+ = Ag_2S_2O_3\downarrow$$

$$Ag_2S_2O_3\downarrow + 3S_2O_3^{2-} = 2[Ag(S_2O_3)_2]^{3-}$$

$$S_2O_3^{2-} + CN^- = CNS^- + SO_3^{2-}$$

Thiosulfate is used as an antidote for cyanide and iodine poisoning and in the treatment of argyria and arsphenamine reactions. It is used topically to prevent the spread of ringworm and mange. It is effective for treatment of tinea versicolor infection of glabrous skin if infected areas are vigorously scrubbed with soap and water and thoroughly dried before application of a saturated solution of thiosulfate. Calcium thiosulfate has been used for treatment of dermatitis.

Polythionates. These compounds are employed for various skin disorders. They are bactericidal in acid medium. Pentathionic acid has been shown to be both germicidal and fungicidal. The keratolytic property of these compounds is probably due to precipitated sulfur and is of value in conditions of the skin not directly associated with infections. Solutions of predominantly tetrathionates and pentathionates are available. They are claimed of value in the treatment of various acnes and seborrheic dermatitis.

Selenium disulfide. Selenium sulfide. This compound is an antiseborrheic used in noninflammatory nonexudative seborrhea of the scalp, eyelids, and external glabrous skin. The usual treatment uses 5–10 ml. of a $2\frac{1}{2}\%$ detergent suspension left in contact for five minutes. The selenium compound must be removed by thorough washing. It is also used in a 0.5% ointment or gel on limited areas of the skin for treatment of seborrheic dermatitis and allergic dermatosis with seborrhea.

This compound is highly toxic. The patient must always be made aware of the dangers of accidental poisoning. Selenium sulfide must not be allowed to enter the conjunctival sac. It must not be used on acutely inflammed moist skin since cutaneous absorption will take place. Because it is highly toxic orally, the patient must be warned to wash thoroughly under the fingernails to remove traces lodged there during treatment of the scalp or other areas of the body. Selenium is absorbed through the whole skin only to a slight degree; if applied as directed, absorption through the whole skin is about equal to the daily selenium intake in foods.

Selenium disulfide is a bright orange powder, possessing a very faint odor. It is virtually insoluble in water and ethanol. Its actual composition has not been established. There is reasonable doubt as to the existence of a compound of the stoichiometry indicated by the name. Vapor pressure studies indicate a mixture of molecules, possibly a solid solution of selenium monosulfide in selenium and sulfur in amorphous form.

Selenium disulfide was originally prepared by passing hydrogen sulfide into selenious acid.

$$2H_2S + H_2SeO_3 = SeS_2 + 3H_2O$$

It may also be prepared by adding selenious acid to a saturated solution of hydrogen sulfide. Selenium Sulfide and Selenium Sulfide Lotion are official in *U.S.P. XXI.*

Cadmium sulfide. This compound was recently introduced as a relatively nontoxic substitute for selenium disulfide in treatment of seborrheic dermatitis. Its activity is about equal or slightly less. Neither compound is a cure, but each controls the condition if applied periodically. Cadmium sulfide is used as a 1% suspension in a detergent base; phosphoric acid is used to adjust the pH. Cadmium sulfide is prepared by precipitation from a solution of a soluble cadmium salt. The color varies from a deep orange to lemon yellow.

Carbon disulfide. Carbon disulfide is an interesting substance. It is highly toxic and for that reason has proven useful as a fumigant. Because of the explosive nature of its fumes when mixed with air, only experienced personnel should undertake this work.

Carbon disulfide is prepared by direct reaction of sulfur and carbon in the electric furnace. Its boiling point is 46.2°C. It is a colorless liquid with a pleasant odor when pure. The yellow color and offensive odor associated with this substance are due to impurities. They may be removed by treatment of the dry liquid with a mild oxidizing agent. Recontamination results from traces of moisture, or exposure to light.

When carbon disulfide is burned in air, the flame has the unusually low temperature of 150°C. Mixed with an equal volume of carbon tetrachloride to absorb the heat, cloth or paper impregnated with the solution does not ignite even though the carbon disulfide-carbon tetrachloride solution is ignited.

Other sulfides. Other insoluble sulfides may exist which possess antiseborrheic properties. Apparently an insoluble sulfide which hydrolyzes slowly under the conditions found on the skin is required. Although it has not been applied as an antiseborrheic, silicon disulfide is an interesting compound with controlled hydrolysis possibilities. It consists structurally of long fibrous macromolecules with each silicon atom sharing two sulfurs with its neighbor to the right and two sulfurs with its neighbor to the left. This macro structure allows only a slow end group hydrolysis.

Thiocarbonates deserve attention for possible dermatological application. The soluble salts are prepared by reacting carbon disulfide with a solution of the sulfide of the desired cation.

$$(2Na^+) + S^{2-} + CS_2 \rightarrow (2Na^+) + CS_3^{2-}$$

The alkali and alkaline earth salts are very stable; not all heavy metal salts are stable. Thiocarbonic acid is an oily yellow liquid and is much more stable than carbonic acid. The pure salts are a bright yellow; the aqueous solutions are red, but turn yellow on acidification. The thiocarbonates are used to destroy phylloxera (an aphid) on vines.

DEPILATORIES AND DEPILATORY ACTION

Depilatory action is essentially the reduction of the disulfide linkage in cystine. A pH of 10 is needed to facilitate this action as well as the swelling process which appears to be necessary for reaction of the sulfhydryl groups. The alkaline earth sulfides—calcium and strontium—provide both the reducing action, via sulfide ion, and the alkaline medium, via hydrolysis. Barium sulfide also supplies both actions, but the highly toxic nature of barium ion makes its use inadvisable. This is especially true since the horny layer of the skin may be dissolved sufficiently to allow systemic absorption of the barium ion. Thallium sulfide is unsuitable for the same reasons.

Calcium sulfide was once widely used in depilatory preparations but has now been replaced by thioglycolic acid.

Calcium sulfide. Calcium sulfide (erroneously named sulfurated lime, liver of lime), CaS, is prepared by igniting a mixture of gypsum and carbon (coke).

$$CaSO_4 + 2C \rightarrow CaS + 2CO_2\uparrow$$

The product is a yellow to grey-yellow powder, having the odor of hydrogen sulfide if exposed to moist air. Although calcium sulfide is ordinarily soluble, this furnace product is difficultly soluble in water.

Because of hydrolysis and air oxidation the furnace product cannot be purified by usual leaching, filtering, and recrystallization procedures. Purity is therefore a function of the purity of the raw materials used and the extent of reaction. The principal impurities are calcium sulfite, calcium carbonate, unreacted calcium sulfate, and ash. A pure product may be produced by reacting calcium metal with sulfur vapor.

Strontium sulfide. This compound is prepared by the same process as calcium sulfide. Except for good solubility, the synthetic problems involved and properties of the compound are very similar to those of the calcium compound.

MISCELLANEOUS TOPICAL PREPARATIONS

Magnesium sulfate is used locally for treatment of inflammatory conditions. The saturated solution, applied in a gauze of proper size, is used for sprains, bruises, cellulitis, and epididymites. (Boric acid is similarly used for wet compresses.) In paste form magnesium sulfate, applied to the infected area and covered with gauze, is used on boils and carbuncles.

For a satisfactory fluid paste the magnesium sulfate must be anhydrous and the glycerol used preheated at 120°C. for an hour and cooled. Morison's Paste has the proportions: magnesium sulfate, 4.5 parts; glycerine, 5.5 parts; and with or without phenol, 0.5%. The glycerine should be added to the magnesium sulfate in a mortar and triturated to a smooth paste. The phenol is added at the final stage. The mixture may be reheated to 120–130°C., with stirring. The product must be immediately packed in tight containers if it is to remain anhydrous.

Calcium hydroxide as lime water may be used as an ingredient in topical preparations. Lime water has astringent action and is used as such topically. It is claimed to have antipruritic action. Only the clear supernatant solution should be used. Lime water reacts with sodium soaps to give water-insoluble calcium soaps. These lime soaps have excellent adhesive properties which make them valuable in the preparation of ointment bases and dermatological lotions, e.g., lime liniment.

Solid carbon dioxide (dry ice) is used as a caustic for removal of warts, etc. Its action is due to the intense cold of the material, the tissue treated being destroyed by freezing. Care must be taken not to freeze healthy tissue.

STUDY AND SELF-APPRAISAL AIDS

IMPORTANT TERMS. Define or identify in a single concise sentence if possible:

Allergic, carcinogenic, covering, power, dermatology, depilatory, dye, fumigant, lachrymator, lotion, monoclinic, pigment, polysulfide, polythionate, rare earth, rhombic, seborrhea.

NOMENCLATURE OF INTEREST. Identify each of the following.

Ammonium pyrosulfate, ammonium tetraoxosulfate(VI), β-sulfur, brimstone, cadmium sulfide, Calamine Lotion, calcium trisulfide, carbon disulfide, disulfonic acid, ferrous ferrite, flowers of sulfur, fuming sulfuric acid, hypo, liver of sulfur, Lotio Alba, magnetite, milk of sulfur, neocalamine, ochre, oleum, potassium metabisulfite, potassium thiocarbonate, potassium thiosulfate, potassium trithionate, rouge, selenic acid, selenium sulfide, sodium dithionate, sodium hyposulfite, sodium pentathionate, strontium sulfide, sulfamic acid, sulfur(II) oxide, sulfurated potash, sulfoxylic acid, telluric acid, titanium(IV) oxide, trisulfuric acid, Turner's Cerate, Turkey Red, Vleminckx's Solution, White Lotion, zirconium carbonate.

SELF-APPRAISAL QUESTIONS

1. Define leveling effect. Give illustrations of the leveling effect of

water.

2. Give the structural formulae of the following ions: thiosulfate, tetrathionate, tetrasulfide, sulfate, and hydrogen sulfite. Can the structure be used to help explain the properties of a given compound? Explain.

3. Explain on a theoretical basis why silver oxide will give a strongly alkaline aqueous solution whereas silver sulfide will not.

4. The solubility of a given sulfide or oxide is related to its cation's electronic structure. Predict the solubilities of the following uncommon compounds of this family and give the basis of your answers: Tl_2S_3, MoO, Rb_2S, Cs_2S_4, PdO, Ta_2O_5, RaO, FrS, BeS, B_2O_3, and WS.

5. Distinguish between the following commercial forms of sulfur: roll, precipitated, sublimed, washed. How do they compare as to chemical reactivity?

6. Explain on a theoretical basis why metallic oxides tend to be basic in character whereas the nonmetallic oxides are acidic in character.

7. Give an explanation for the amphoteric character of the following hydroxides: aluminum(III), zinc (II), tin(II), chromium(III).

8. Give evidence indicating that the hydrogen thiosulfate ion rather than the thiosulfate ion is unstable.

9. Sulfuric acid is useful as an acid, oxidizing agent, dehydrating agent, and solvent. Explain this versatility.

10. Sulfuric acid has no reducing properties. Explain.

11. It is thought that protective action involves the slow release of the active ions from the insoluble salts. Assuming that the other substances present have no effect on the solubility of the following, calculate the theoretical equilibrium concentration of the active ion under the conditions indicated: zinc sulfide, pH 9; aluminum hydroxide, pH 7; Calmine, pH 5.5. (ZnS $-$ 1.4×10^{-5} M Zn^{2+})

12. Compare the atomic properties of zirconium and hafnium, e.g., ionic radii, and account for the extreme similarity of these elements. If you were given pure $ZrCl_4$ and pure $HfCl_4$, which properties would you expect to show the greatest difference? Explain.

13. Name types of compounds suitable as protectives and give their essential properties. Indicate their mode of action.

14. Why the preoccupation with the color of protectives?

15. Why is titanium dioxide an effective protective in suncreams? Explain.

16. Identify the different medicinal forms for topical application in which sulfur may be dispensed. Name some other compounds with which sulfur is sometimes used.

17. What is the postulated action of elemental sulfur when used in dermatological preparations?

18. Why is a "creamy" white character desirable for White Lotion?

19. Explain why solid calcium sulfide must be prepared by a roasting technique.

20. What are the important differences between: calamine and neo-calamine; α-sulfur and μ-sulfur; sodium bisulfite and sodium metabisulfite; carbon monoxide-formic acid and carbon dioxide-carbonic acid.

21. Which of the compounds and preparations discussed are official in the current compendia? Give their official categories.

22. Classify those compounds and preparations studied as to suitability in the following categories: depilatory, protective, dermatological agent (e.g., treatment of acne), cosmetic pigment, astringent, antiseborrheic, contact dermatitis.

PRACTICE IN EQUATION WRITING

23. Write the equations for the preparation of the following: White Lotion, zinc stearate, hydrous zirconium oxide, sodium ferrite, ochre, precipitated sulfur (from polysulfide), sulfurated lime solution, potassium thiosulfate, sodium polysulfide, strontium sulfide, sodium thiocarbonate.

24. Write the equations for the reactions taking place if the order of mixing is reversed during preparation of White Lotion.

25. Write equations for reactions suitable for distinguishing between ZnO, ZnO_2, ZnS, $ZnSO_3$, $ZnSO_4$.

26. Give equations for reactions suitable to distinguish between soluble salts of the following anions: S^{2-}, $S_2O_3^{2-}$, S_3^{2-}, O^{2-}.

27. Write equations for reactions taking place between the following pairs of substances: sodium trisulfate, water; trithionate, strong acid; zinc oxide, strong base; zirconyl chloride, strong base.

28. Assuming SeS_2 as its formula, write the equations for the preparation of selenium sulfide.

29. Write the equations for the reactions used for the official assay for: zinc oxide, titanium dioxide, sulfurated potash.

NINETEEN · RADIOISOTOPES IN PHARMACY AND MEDICINE

All previous studies involving the reactions of atoms and molecules have concerned themselves with events taking place at the outer volume of the atom. Only the so-called valence electrons were involved, either being shared to form covalent bonds or being transferred to form electrostatic linkages. The deeper electrons and the nucleus remained unchanged. However, the study of radioisotopes focuses on reactions that take place in the nucleus, cause the emanation of sub-nuclear particles, and result in the formation of new daughter elements.

NUCLEAR REACTIONS

Of the many types of nuclear reactions which are possible, two, fission and radioactive decay, the most important in pharmacy and medicine will be discussed in detail.

Fission. The nuclei of some of the larger elements, e.g., Uranium-235, are energetically so close to the limits of stability that the addition of even one neutron will cause them to become unstable and break apart yielding two approximately equal sized fragments as well as 2 or 3 neutrons[1], e.g.,

$$^{235}_{42}U + ^{1}_{0}n = ^{131}_{53}I + ^{102}_{39}Y + 3\,^{1}_{0}n$$

The neutrons produced by the reaction are available to initiate the decomposition of other nuclei. Under certain conditions, a chain reaction

[1]The standard notation used here indicates the Atomic Number (Z, the number of protons) as a subscript to the left of the symbol for the element and the Mass Number (A, the sum of protons and neutrons) as a superscript to the left.

can occur and tremendous quantities of energy can be released explosively as in an atomic fission bomb. However, when the conditions in the nuclear reactor are carefully controlled so that the rate of the reaction is slowed, there is no danger of explosion and a number of pharmaceutically useful fission products, such as $^{131}_{53}I$ and $^{99}_{42}Mo$, can be formed in good yield.

The products of fission reactions are usually unstable and are converted to other elements by the various processes of radioactive decay discussed below. The pharmaceutically useful isotopes, e.g., $^{131}_{53}I$, are contaminated with the reactant and with other isotopes and must be purified before use. It should be noted that, since the reactions being discussed here are taking place in the nucleus, the outer electron configuration and therefore the chemical properties of the radioactive isotopes are the same as those of the stable isotopes of the same element. Therefore, isolation of the desired substances is accomplished using the standard analytical techniques used for non-radioactive substances, *e.g.*, extraction, precipitation and chromatography.

Radioactive Decay. In this type of nuclear reaction, the nucleus of an unstable isotope spontaneously decomposes and a sub-nuclear particle is emitted. The atomic number of the resulting nucleus will be either the same as its parent or differ by ± 1 or -2 units; the mass number will be unchanged or differ by -4 units.

Alpha Decay. This process usually occurs in heavy nuclei, such as uranium and radon, which have atomic numbers of 83 or over, e.g.,

$$^{238}_{92}U = \cdot\, ^{234}_{90}Th + ^{4}_{2}He^{+2}\ (\alpha)$$

The particle emitted is a helium ion, consisting of two protons and two neutrons and having a charge of $+2$. The daughter element differs from its parent by having an atomic number lower by two units and an atomic weight lower by four mass units. Since the particle is relatively massive, it is emitted with a low speed (about .05c-.1c)[2] and therefore has a low kinetic energy. The range of the alpha particle in air is only a few centimeters and, in tissue, only a fraction of a millimeter. Therefore, it has no use as a pharmaceutical. However, alpha-emitting substances are very toxic and highly carcinogenic if ingested since the alpha particles cause ionization of molecules in the tissues which they contact. These ionized molecules then disrupt the normal biochemical pathways causing damage to the tissues.

---

[2] c is the speed of light, 3×10^{10} c.m/sec.

Beta-minus Decay. When a nucleus has a ratio of neutrons to protons which is high enough to confer instability, it can achieve greater stability by converting a neutron to a proton and an electron:

$$^{1}_{0}n = \,^{1}_{1}p^{+} + \beta^{-}$$

The daughter element produced by this process has the same atomic weight as its parent since a neutron is replaced by a proton but its atomic number is raised by one because there is one more proton in its nucleus:

$$^{32}_{15}P = \,^{32}_{16}S + \beta^{-}$$

The beta particle is identical to an electron in mass and charge and has a velocity of up to 0.96c. Because of this high speed, beta rays have good penetrating power but they have a lower ability to cause ionization than alpha particles due to their much smaller mass. Energy dissipation takes place in two ways during their passage through matter. Particles in the pathway may be ionized, or the electron may be decelerated to give up its energy as *Bremsstrahlung*-radiant energy similar to x-rays. The β^{-} particle has many diverse uses in radiopharmacy.

Beta-plus Decay. Nuclei which are "proton-rich," i.e., have a ratio of neutrons to protons which is less than that of the stable elements, are able to stabilize themselves by converting a proton into a neutron and a positron or β^{+}:

$$^{1}_{1}p^{+} = \,^{1}_{0}n + \beta^{+}$$

The daughter element produced by this process has the same atomic weight as its parent since a proton is replaced by a neutron but its atomic number is reduced by one since the proton has lost its charge:

$$^{15}_{8}O = \,^{15}_{7}N + \beta^{+}$$

The positron is identical to the electron except for its positive charge. Since it is a particle of anti-matter, it has an extremely short lifetime being destroyed as soon as it collides with a particle of ordinary matter. In the most common *annihilation* reaction, the positron collides with an electron. Both are converted to energy in the form of two photons each with an energy of 511 KeV[3] and traveling in a direction 180° opposite to the other.

Because few elements decay by this mechanism and since its lifetime is so short, the positron had not been much used in radiopharmacy until the recent development of a diagnostic technique called Positron Emission (Transaxial) Tomography or PET (PETT). In this technique, the

[3]An electron volt is the energy acquired by an electron when accelerated by a potential of one volt.

patient is administered a drug containing a positron-emitting nuclide (*e.g.,* deoxyglucose containing $^{11}_{6}C$). The patient is then passed slowly through a large ring of detectors that are sensitive to the photons emitted by the annihilation process. For any event which does not occur at the exact center of the detector ring, one of the photons reaches its detector slightly before the other. By using a computer, this difference in time can be converted to a distance and the entire organ can be mapped. PET yields information about the position and extent of metabolic events in the organ being studied whereas the more familiar CAT (Computerized Axial Tomography), which uses x-rays as its probe, yields information about the structure of the tissue. The two techniques are complementary and provide the physician with a powerful diagnostic tool.

Electron Capture (EC). Some elements which are "proton-rich" may decay by a means other than positron emission. This is called electron capture and occurs when an extranuclear electron is taken into the nucleus where it combines with a proton to form a neutron:

$$^{1}_{1}p^{+} + e^{-} = ^{1}_{0}n$$

Most often the electron is captured from the K shell because it is closest to the nucleus. The event is also called K-capture. The vacancy in the K shell is filled by transfer of an electron from an upper energy shell (L,M, or N) to the K level. The energy difference between the two levels is emitted as a photon, usually an x-ray. The result of the overall process is that the atomic number of the parent nuclide is reduced by one due to the loss of the proton but the atomic weight is unchanged since a neutron is formed.

Isomeric Transition (gamma emission). In much the same way that an electron in an excited energy state can go to the ground state by emitting radiation, a nucleus in an excited state can return to its ground state by emitting gamma rays. The excited states of the nucleus are called *isomeric states* and they have lifetimes ranging from fractions of seconds to years. When the isomeric state is long-lived, it is called a *metastable state* and is denoted by an "m" in the elemental formula, as in ^{99m}Tc.

Isomeric states frequently arise as a result of other radioactive events such as β^-, or β^+ decay, or electron capture when the daughter element formed is not in its lowest energy level, e.g.

$$^{131}_{53}I = ^{131}_{54}Xe^* + \beta^-$$

The asterisk indicates that the xenon is in an upper energy state. It may go to the ground state by isomeric transition:

$$^{131}_{54}Xe^* = ^{131}_{54}Xe + \gamma$$

The final state differs from the excited level only in energy. There is no change in atomic number or atomic weight.

The energy emitted, a gamma ray, is a photon of electromagnetic energy which differs from other photons only in that it has a very short wavelength and high energy. Its energy is proportional to its frequency ($E = h\nu$). Gamma rays have great penetrating power due to their high speed but their ability to ionize tissue is lower than that of other particles because of their lack of mass and charge. Most of the current diagnostic procedures using radiopharmaceuticals employ gamma-emitting isotopes, in particular $^{99m}_{43}Tc$.

Another mechanism which allows a nucleus to undergo an isomeric transition is called *Internal Conversion.* In this case, the gamma-ray emitted from the nucleus collides with an extranuclear electron and knocks it out of its shell, most frequently the K shell. In order to fill the vacancy in the lower electron energy level an outer electron falls into the inner shell and at the same time emits an x-ray. Internal conversion occurs most often with low energy gamma rays.

MEASUREMENT OF RADIATION

In order to use radioisotopes for therapy or diagnosis or to calculate doses for administration of radioactive drugs, it is necessary to have methods for detecting the radiation and determining its quantity. The methods most frequently used depend on the ability of the particle to cause ionization of a gas, to induce fluorescence in a suitable flurophor, or to darken a photographic emulsion.

The most common detector based on the ionization principle is the Geiger-Muller counter. Geiger counters are able to detect alpha, beta, and gamma rays, but their efficiency for alpha and gamma is low because of the short range of the alpha and the low ionizing power of the gamma rays. They are mainly used for monitoring the workplace in order to prevent undue exposure of personnel to radioactivity.

Photographic emulsions are darkened on exposure to radioactivity since the silver atoms are ionized. They are frequently used in film badges since they provide a permanent record of exposure over a period of time. X-ray films may be placed outside the body of an animal or person who has received a dose of a radioactive drug. The exposed film will then present a picture of the organ(s) in which the radioactive substance has concentrated. This technique is called *autoradiography.*

The detection method which is based on the production of fluorescence is called *scintillation counting.* Radioactive emanation causes a solid substance such as a sodium iodide crystal or a dissolved substance such as diphenyloxazole in toluene, to fluoresce. The fluorescence is then detected and quantitated by a suitable spectrophotometer. Scintillation counting is the most accurate method for determining radioactivity and is used when quantitative results are desired.

UNITS OF MEASUREMENT

The basic unit of measurement of radioactivity is the Curie (Ci). Originally, this was considered to be equal to the radioactivity emanating from one gram of radium but since this was found to be variable, the unit was redefined as equal to 3.7×10^{10} disintegrations per second. Because this number is much larger than those normally encountered in radiopharmacy, the millicurie, mCi, or the microcurie, μCi, are more convenient. The equivalent SI unit for radioactivity is the Becquerel (Bq) and is defined as one disintegration per second. It should be noted that neither unit includes the weight or volume which contains the activity. A curie of radioactivity might be found in a micromole of material or in many tons of an ore. When units of weight are included, as in curies per millimole, the combination of units is called the *specific activity.*

Because exposure of the human body to excessive quantities of radiation can cause serious damage if not controlled, a number of units have been defined to measure this exposure. The Roentgen (R) was the first unit devised to measure the exposure of a tissue to the ionizing effect of a given radiation and, indirectly, its damaging effect. It is defined as the amount of x- or γ- radiation which produces one electrostatic unit of charge per cc. of air at STP. This is approximately equivalent to the release of 931 erg/gm of water or tissue.

The unit called the rad (radiation absorbed dose) is more universal in that it applies to any material. It is defined as the quantity of radiation which releases 100 erg/gm of a specified medium. This is equivalent to 10^{-2} Joules/kg. In SI units, the Gray (Gy) is the unit of absorbed dose and is equivalent to 1.0 J/kg or 100 rads.

Another unit, the rem (roentgen equivalent man) has been defined to account for the fact that different kinds of radiation will cause different levels of damage even if given in the same dose. The rem is equal to the absorbed dose in rads times the RBE (Relative Biological Effectiveness) of the particular radiation. The RBE of x-rays, gamma rays and beta particles is one, while for alpha particles it is 20, indicating the greater danger of an absorbed dose of an alpha-emitting substance. In SI units, the dose equivalent is given in Sieverts (Sv), where one Sv equals 100 rem.

The distance of an object or person from the source of radiation is also important in controlling the exposure. Since radiation is emitted equally in all directions from a source, the inverse square law applies, *i.e.,* the radioactivity to which an object is exposed is inversely proportional to the square of its distance from the source. This can be expressed as:

$$A_1/A_2 = d_2^2/d_1^2$$

where A_1 and A_2 are the activities at the respective distances d_1 and d_2:

RADIOACTIVE DECAY LAWS

The radioactive disintegration of any particular atom in a collection of radioactive atoms is purely random and cannot be predicted. However, for a large number of atoms, the overall rate of decay follows a convenient mathematical relationship which is predictable and distinct for every radioactive substance.

The rate of disintegration or the number of disintegrations per unit time of a radioactive isotope is proportional to the total number of radioactive atoms present at a given time. Mathematically, this is expressed as:

$$-dN/dt = \lambda N$$

where N is the number of radioactive atoms, t is the time, and λ is the decay or disintegration constant which is defined as the probability of disintegration of a single nuclide in a unit time. The minus sign indicates that N is constantly decreasing. Since this is an ordinary differential equation, it can be integrated to give:

$$\ln (N_0/N) = \lambda t$$

where N_0 is the number of atoms present at time zero and N is the number present at time t.

Rates of radioactive decay are more customarily expressed in terms of the *half-life, $t_{1/2}$*, defined as the time interval during which one-half of the nuclei of a given sample will disintegrate. It may also be defined in terms of radiation as the time in which the intensity of radiation will be decreased to one-half of its value at time zero. Half-life is derived from the rate law by substituting $N_0/2$ for N and $t_{1/2}$ for t yielding:

$$t_{1/2} = \ln 2/\lambda = 0.693/\lambda$$

Since the rate law is exponential, it does not follow that for a radioisotope with a half-life of ten hours the sample has completely disintegrated after 20 hours. Instead, at the end of 10 hours one half of the original sample remains; at the end of 20 hours one quarter remains; etc.

Half-life values are required for the calculations necessary to establish the radioactivity of a medicinal sample at the time of administration from a knowledge of its measured radioactivity at some time in the past. Thus if a 10μCi dose of a given radioisotope is to be given, and the sample from which the dose is to be prepared had an activity of 250μCi

28 hours earlier, a knowledge of the half-life permits calculation of the current level of radioactivity. The calculations may be made directly on the basis of the decay law or by use of tables of *decay factors.* A decay factor is the fraction of a specific isotope which has decayed during a specified time interval after time zero, the latter being the time at which the activity was measured.

If the rate law rather than decay factors is to be used for calculations, the expression:

$$\lambda = 0.693/t_{1/2}$$

is substituted for λ in the radioactive decay law to give:

$$\log (N_0/N) = 0.693(t/t_{1/2})$$

Converting to base 10 logarithms and solving for log N gives:

$$\log N = \log N_0 - 0.301(t/t_{1/2})$$

Substitution of the known quantities, N_0, t and $t_{1/2}$ allows determination of current activity.

It should be recognized that the duration of time during which the body is exposed to radioactivity from an administered dose of an active substance is not necessarily the same as that calculated on the basis of the half-life of the nuclide present. The compound which contains the radioactive nuclide is treated as a drug by the body and eliminated by the normal biological metabolic routes. Therefore, if the substance is eliminated at a first order rate, it will have a biological half life, t_B, of $0.693/\lambda_B$, where λ_B is the rate constant for elimination. The effective rate of disappearance of radioactivity is proportional to the sum of the two rate constants and can be expressed by the following equation:

$$t_{eff} = (t_{1/2} \times t_B) / (t_{1/2} + t_B)$$

The effective half-life in the body will be shorter than either of the half lives.

RADIOISOTOPES OF IMPORTANCE IN THE HEALTH SCIENCES

Most of the radioisotopes used in pharmaceutical chemistry, pharmacy and medicine have been tabulated in Table 19.1. Of this list, technetium 99m and iodine 131 find very wide use.

Some general comment about radioisotopes and the effect of radiation is in order. Ordinary chemical and physical properties and possible incompatabilities, are not discussed; this has been done in appropriate sections dealing with the stable isotopes of these elements. Although general discussion centers about the affect on and possible damage to living tissue, inanimate substances may also be affected and undergo change on exposure. Crystals and glass may become colored. Organic and inorganic compounds undergo decomposition as a result of radiation, the energies possessed by α, β, and γ radiation being able to initiate chemical reaction. Ordinarily stable, water undergoes slight decomposition to free radicals, leading to the appearance of very small concentrations of hydrogen peroxide and hydrogen.

$$H_2O \rightarrow H. + \cdot\ddot{\underset{\cdot\cdot}{O}}:H \text{ (free radicals)}$$

$$H. + .H \rightarrow H_2$$

$$HO\cdot + .OH \rightarrow H_2O_2$$

Ionizing radiation leaves an assortment of new molecules, ions, and free radicals along its path. Reactions between these particles, solute molecules or ions, solvent molecules, etc., are possible.

Radioisotopes as therapeutics and diagnostic aids. When a radioactive substance is considered for use in these applications, factors related to its radioactivity (in addition to the usual chemical and physiological factors) must be weighed. The substance selected and the conditions of application must result in the least possible extraneous exposure of the patient to the harmful effects of ionization radiation.

The kind of radiation produced by the isotope is of great importance. Radiation for diagnostic purposes must have sufficient energy to pass through tissues between the locus of the radioactive isotope in the body and the detecting device. Radiation for therapeutic purposes must have sufficient energy to penetrate the tissues or organs being treated. Beta and gamma radiation are usually suitable.

The half-life of the istope must be considered. If it has a long half-life, exposure to radiation continues at a high level until eliminated. An isotope with a short half-life has self-limiting radiation effects; as decay proceeds radiation becomes progressively less. The desirable half-life value is considered to be about 10 days. Technical problems arise with isotopes having too short a half-life. Decay can be too fast to allow for the time required for recovery of the isotope after preparation, transportation to the clinic, and preparation into a suitable and effective dosage form. This situation has been avoided in the case of technetium by the development of a generator that permits production of technetium-99m on site and by the use of kits that permit the incorporation of Tc-99m into various carrier molecules.

The chemistry of the isotope must be considered for ease of incorporation into a carrier molecule. Technetium-99m is easily incorporated into many carriers making this useful nuclide very versatile in terms of the different kinds of diagnoses for which it is suitable. Iodine-131 is another example of an isotope that is easily incorporated into carrier molecules; it is also used as the sodium salt. The disintegration by-product of the radioisotope ordinarily need not be considered because of the extremely small number of radioactive atoms actually administered. Iodine-131 decays to a stable isotope of xenon; an average dose of iodine-131 produces $7 X 10^{-6}$ mL. xenon (S.T.P.).

Finally, the radioactive material should not remain in general circulation or achieve general distribution in all body fluids and cells. Rather, it should be selectively localized in the abnormal tissues to be diagnosed or subjected to therapy. With generalized distribution healthy tissues are subjected to excessive radiation. The concentration of iodine-131 in the thyroid provides an excellent example of selective absorption.

Methods of isotope placement in medical applications. Treatment of a patient with radioisotopes may be accomplished by physical or chemical means. Chemical placement involves administration of the isotope as a compound which will become a component of the body fluids and be transported by the blood to the desired site. Physical placement involves teletherapy or actual placement of the isotope, in metallic form, within the body cavities or interstitial spaces. Interstitial or intracavity placement may be accomplished via an arrangement of slugs of the metal or by injection of a colloidal suspension, e.g., gold-198.

Teletherapy is the name used when the source of radiation is at a distance from the body. Cobalt-60 is the usual source of the γ-radiation used in this procedure.[4] The penetrating γ-rays are intended to destroy malignancies.

Teletherapy setups capable of substituting for x-ray apparatus have many advantages. Skin shows better tolerance to cobalt-60 radiation, tumor dosage is more accurate, and, due to better control, there is less absorption of radiation by healthy tissues and bone. The units show greater flexibility than x-ray units and have a cost advantage.

RADIOISOTOPES USED IN MEDICINAL AND RELATED APPLICATIONS

The following section is divided into two parts: first, technetium is discussed since almost 80% of all radio-pharmaceuticals used are Tc-99m labeled compounds; this is followed by a discussion of iodine compounds and then by examples of some other useful nuclides. The second short

[4]Cobalt-60 rods are available with activities as high as 10 Ci/g. A pack of six small rods, having a total weight of 200 g., can have an activity of 1000Ci.

TABLE 19.1 RADIOISOTOPES USED IN THE HEALTH SCIENCES

Isotope	Radiation	Energy (Mev)	Decay Element	Half-life	Application
^{241}Am	α	5.49, 5.44	^{237}Np	433y	D
	γ	0.06			
^{198}Au	β^-	1.37	^{198}Hg	2.69 d	T
	γ	0.41			
^{82}Br	β^-	0.44	^{82}Kr	35 h	T
	γ	1.48, 1.32, 1.04 0.83, 0.78, 0.70, 0.62, 0.55			
^{14}C	β^-	0.16	^{14}N	5730 y	R
^{47}Ca	β^-	1.98, 0.67	^{47}Sc	4.54 d	D
	γ	1.30, 0.81, 0.49			
^{60}Co	β^-	0.32	^{60}Ni	5.27 y	T
	γ	1.33, 1.17			
^{51}Cr^c	γ	0.32	^{51}V	27.7 d	D
^{137}Cs	β^-	1.18, 0.51	^{137}Pr	30.1 y	T
	γ	0.66			
^{64}Cu	β^-	0.57	^{64}Zn	12.7 h	D
	β^+	0.66	^{64}Ni		
^{18}F	β^+	0.64	^{18}O	109.8 m	D
^{55}Fe^c	x-ray		^{55}Mn	2.7 y	D
^{67}Ga^c	γ	0.39, 0.30, 0.18, 0.09	^{67}Ga	78.3 h	D
^{3}H	β^-	0.02	^{3}He	12.3 y	D, R
^{197}Hg^c	γ	0.08	^{197}Hg	64.1h	D
^{203}Hg	β^-	0.21	^{203}Tl	46.8 d	D
	γ	0.28			
^{123}I^c	γ	0.16	^{123}I	13.0 h	D
^{125}I^c	γ	0.04	^{125}I	60.2 d	D
^{131}I	β^-	0.67, 0.33	^{131}Xe	8.04 d	D, T
	γ	0.72, 0.64, 0.36 0.28, 0.08			
^{113m}In	γ	0.39	^{113}In	99.5 m	D
^{111}In^c	γ	0.25, 0.17	^{111}In	2.83 d	D
^{192}Ir	β^-	0.67, 0.54	^{192}Pt	74.2 d	T
	γ	0.47, 0.32, 0.31, 0.30			

TABLE 19.1 (Continued)

Isotope	Radiation	Energy (Mev)	Decay Element	Half-life	Application
^{42}K	β^-	3.52, 1.97	^{42}Ca	12.4 h	D
	γ	1.52, 0.31			
^{24}Na	β^-	1.39	^{24}Mg	15.0 h	D
	γ	2.75, 1.37			
^{32}P	β^-	1.71	^{32}S	14.3 d	D, R, T
^{210}Pb	β^-	0.02	^{210}Bi	22.3 y	T
	γ	0.05			
^{226}Ra	α	4.78, 4.60	^{222}Rn	1602 y	T
	γ	0.19			
^{222}Rn	α	5.49	^{218}Po	3.82 d	T
	γ	0.51			
^{106}Ru	β^-	0.04	^{106}Pd	367 d	T
^{35}S	β^-	0.17	^{35}Cl	87.4 d	D
^{75}Se[c]	γ	0.40, 0.28, 0.27, 0.14, 0.12	^{75}Se	118.5 d	D
^{85}Sr[c]	γ	0.51	^{85}Sr	64.8 d	D
^{87m}Sr	γ	0.39	^{87}Sr	2.80 h	D
^{90}Sr	β^-	0.55	^{90}Y	28.8 y	T
^{99m}Tc	γ	0.14	^{99}Tc	6.02 h	D
^{201}Tl[c]	γ	0.17, 0.14	^{201}Tl	73 h	D
^{133}Xe	β^-	0.35	^{133}Cs	5.25 d	D
	γ	0.08			
^{169}Yb[c]	γ	0.31, 0.20, 0.18, 0.13, 0.11, 0.06	^{169}Yb	32.0 d	D

[a]Merck Index, 10th Ed., Merck & Co., Inc., Rahway, N.J., 1983.

[b]The following abbreviations are used: m for minutes, h for hours, d for days, y for years.

[c]The following abbreviations are used: D for diagnosis, R for research, T for therapy.

[d]In K-capture, an electron of the K shell (first quantum group) falls into the nucleus with the simultaneous emission of γ radiation.

section is a brief consideration of radioimmunoassay (RIA), an ingenious analytical procedure that is proving to be very useful in the measurement of small quantities of drugs. For more extensive information the student is referred to a textbook,[5] to an advanced textbook[6] and

[5]G.B. Saha, "Fundamentals of Nuclear Pharmacy," 2nd Ed., Springer-Verlag, New York, N.Y., 1984.

[6]M. Tubis and W. Wolf, eds., "Radiopharmacy," Wiley, New York, N.Y., 1976.

to a reference book[7] that is an excellent source of practical information and provides data on usual therapeutic doses to be administered as well as data on available products. Finally, the student is advised to be familiar with the legal standards[8] that radioisotopic pharmaceuticals must meet. For example, labeling information is specified to include: (1) time and date of calibration of the batch on hand, (2) the amount of radioactivity in mCi or μCi at the time of calibration, (3) the expiration date and (4) the statement "Caution-Radioactive Material." Furthermore, the label states that, in making dosage calculations, corrections for radioactive decay must be made; for this purpose, the radioactive half-life is specified.

Technetium-99m. Technetium belongs to Group VII A of the periodic table, together with manganese and rhenium; it has no stable isotopes and occurs in 3+, 4+, 5+, 7+ oxidation states of which 4+ and 7+ are most stable. The Tc-99m isotopes, as may be found in the pertechnetate ion with the 7+ state, is a very useful nuclide for medical diagnosis. Its relatively short nuclear half-life of 6.0 hours reduces the exposure time of the patient yet is long enough to allow its use if it can be generated on site. The 0.14 MeV γ-emission may be easily collimated to give good images and provides good penetration from distant body organs. Furthermore, since there are no electron emissions there is minimal tissue damage. If different compounds could be labeled with Tc-99m, then the radioactivity could be carried to various organs making imaging possible.

The widespread use of Tc-99m labeled compounds has been possible owing to the availability of the Molybdenum 99/Technetium 99m generator, a convenient mechanism that permits production of the nuclide at the site of use. Figure 19.1 is a drawing of the design of a closed generator. By decay of the ^{99}Mo nuclide, $^{99m}TcO_4^{1-}$ is produced as may be seen from the following equation.

$$^{99m}MoO_4^{2-} \longrightarrow \beta^- + {}^{99m}TcO_4^{1-}$$

In order to bring about a separation of molybdate from pertechnetate, an alumina column is used. Since the molybdate has a charge of 2− whereas the pertechnetate is charged 1−, the less strongly adsorbed pertechnetate may be eluted preferentially from the alumina with isotonic saline. The quantities of Mo-99 and Al that elute from the column are very small and are controlled. The closed generator produces

[7]G.D. Chase, "Medical Applications of Radioisotopes," Chapter 29, in "Remingtons Pharmaceutical Science," 16th ed., Mack Printing Co., Easton, PA, 1980.

[8]"The United States Pharmacopeia, 21st revision and the National Formulary, 16th edition," United States Pharmacopeial Convention, Inc., Rockville, Md, 1985.

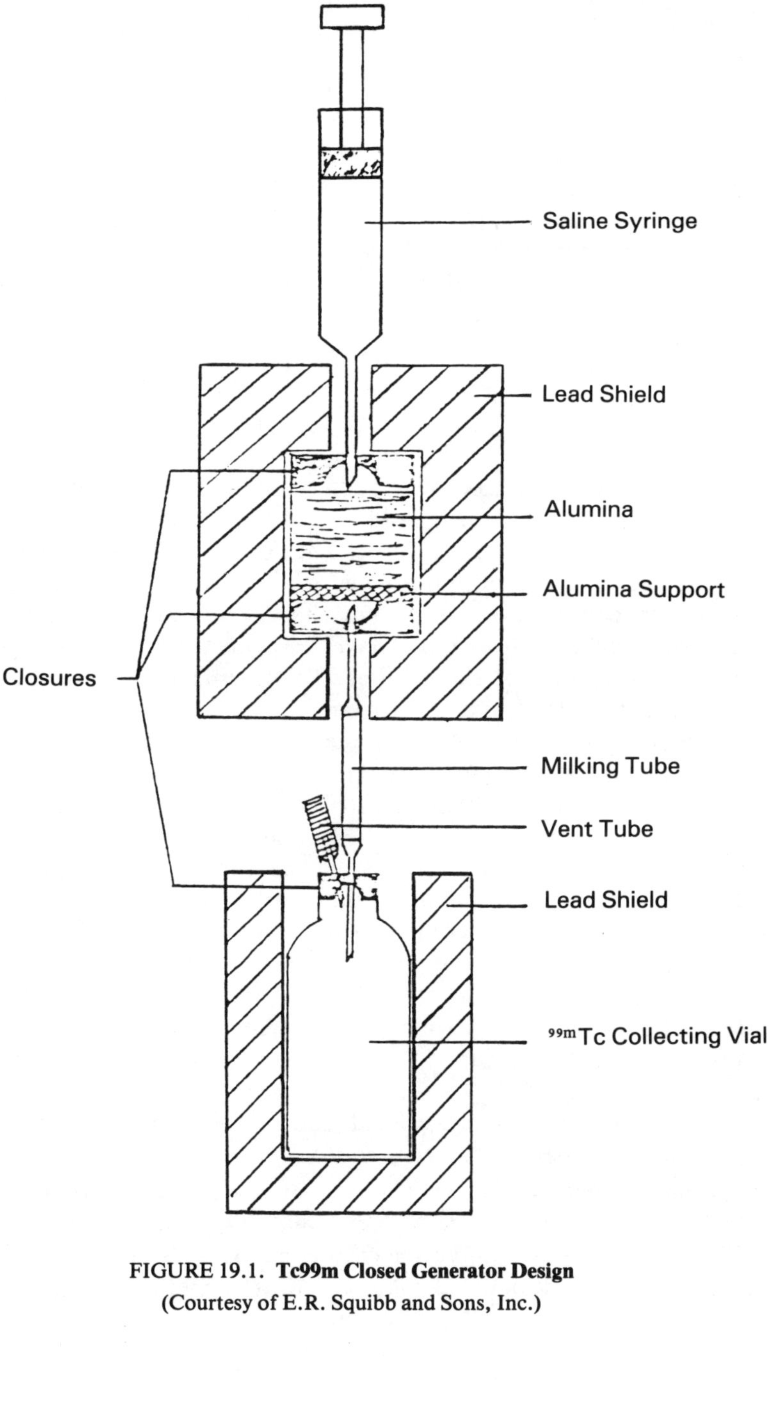

FIGURE 19.1. **Tc99m Closed Generator Design**
(Courtesy of E.R. Squibb and Sons, Inc.)

a pyrogen free, sterile product if the saline solution eluant is pyrogen free and sterile and if the collection procedure does not permit contamination.

The desirable properties of Tc-99m are responsible for its wide use since it is possible to produce many Tc-99m labeled compounds with different distribution patterns in the body. In the labeling of compounds with technetium, it is necessary to alter the oxidation state from 7+ to a less positive value, e.g., 5+, 4+, 3+. Different methods of bringing about these changes are available, e.g., stannous chloride, iron-ascorbic acid, ferrous sulfate, sodium borohydride. An equation that defines the stoichiometry in the reaction of stannous chloride with pertechnetate in concentrated hydrochloric acid solution may be written as follows:

$$3\,Sn^{2+} + 2\,^{99m}TcO_4^{1-} + 16H_3O^+ + 30\,Cl^- \longrightarrow 3SnCl_6^{2-} + 2\,^{99m}TcCl_6^{2-} + 24\,H_2O$$

The labeling of various compounds with Tc-99m is carried out in all instances except the sulfur colloid by the direct addition of $^{99m}TcO_4^{1-}$ to the compound or chelating agent. The complex may be formed using a commercial kit. Conditions are designed to ensure the formation of the maximum concentration of Tc-99m chelate; however, impurities of other Tc-99m species such as unreduced $^{99m}TcO_4^{1-}$ and hydrolysis products, e.g., $^{99m}TcO_2$, may be encountered. The chemical nature of the complex and the oxidation state of Tc-99m are not definitively known.

The list of uses of Tc-99m compounds found in Table 19.2 is evidence of the extensive usage of this nuclide owing to its favorable radioactive properties and to the easily controlled physiological distribution of its compounds or complexes. For example, the complexes formed with phosphorous compounds, e.g., the etidronate, the medronate, and the oxidronate and others, are used for bone or skeletal imaging. The pertechnetate ion behaves like the iodide ion in its biological distribution and is found in the choroid plexus (vascular network) of the brain as well as the thyroid gland, the salivary glands and the gastric mucosa. Pertechnetate ion, like permanganate, resembles the typical elements (halogens) of their group. Thus, it is not surprising that perchlorate may be administered prior to sodium pertechnetate Tc99m in order to saturate the choroid plexus and allow brain images to be recorded without undue definition of the choroid plexus.

Iodine. There are three nuclides of iodine that have use in diagnosis and treatment: iodine-123, iodine-125 and iodine-131. Iodine-123 is produced by bombardment of enriched tellurium-124 with protons or of enriched tellurium-122 with deuterons or by the decay of xenon-123;

TABLE 19.2 Diagnostic Applications of Technetium 99m Injections

Technetium Injection	Diagnostic Application
Tc 99m Albumin	Heart scan, placental localization
Tc 99m Albumin Aggregated	Perfusion lung scan, liver scan
Tc 99m Etidronate	Bone imaging
Tc 99m Ferpentetate	Renal scan
Tc 99m Gluceptate	Renal and brain imaging
Tc 99m Medronate	Skeletal imaging
Tc 99m Oxidronate	Skeletal imaging
Tc 99m Pentetate	Renal and brain imaging
Sodium Pertechnetate Tc 99m	Brain and thyroid scan blood pool examination, placental location
Tc 99m Pyrophosphate	Bone imaging
Tc 99m (Pyro- and trimeta-Phosphates)	Renal imaging
Tc 99m Succimer	Renal imaging
Tc 99m Sulfur Colloid	Liver, spleen, bone marrow scans

iodine-125 is obtained from the neutron bombardment of xenon; iodine-131 results from uranium fission or by neutron bombardment of tellurium. All three nuclides are carrier free. Iodine-123 is used as the sodium salt in solution, orally or by injection, and in a capsule dosage form. The other two isotopes are used not only as the solution and capsule but are incorporated in other systems, e.g. Iodinated I-125 Albumin Injection, Iodinated I-131 Albumin Injection, Iodinated I-131 Albumin Aggregated Injection, Iodohippurate Sodium I 131 Injection and Rose Bengal Sodium I-131 Injection. To prevent surface adsorption of the small quantities of the iodine nuclides present, the glass containers and the glassware to be used in preparative procedures must be specially prepared. These vessels are first rinsed with a solution that is approximately 0.8% sodium hydroxide, 0.04% sodium hydrogen sulfite and 0.25% sodium iodide, followed by successive rinses of purified water, repeated until the spent rinse waters are neutral to litmus.

When iodine nuclides are compared, sodium iodide I-123 is superior to I-131 for thyroid diagnosis, for both uptake and imaging studies, since I-123 has a short half life and emits only γ-radiation. However, I-131 and I-125, are useful in the treatment of hyperthyroidism and thyroid cancer.

The iodide ion is removed from the blood stream and concentrates in the thyroid. Thus, when radioactive iodide is administered, a portion of the dose is held by the thyroid whereas the rest of the radioactive iodide is excreted in the urine. The uptake of radioactive iodide is taken as an index of thyroid function. When radioactive iodide is used in imaging studies for tissues other than the thyroid, Lugol's Solution is administered first to prevent undesirable thyroid irradiation. In addition to the use of iodide, sometimes perchlorate is used to prevent the uptake of radioactive iodide by the thyroid.

Iodinated I-131 Albumin Injection and Iodinated I-125 Albumin Injection are prepared by the mild iodination of human serum albumin under conditions such that not more than one iodine gram atom is introduced per gram molecular weight (60,000) of albumin. The iodine reacts with the tyrosine moieties present in the albumin, giving monoiodotyrosine groups. The product does not differ significantly from the original albumin. It is used as a clear, colorless solution, rendered isotonic with sodium chloride, buffered to a pH of 7.0-8.5 and having an activity of not more than 1 mCi per mL of solution. Iodine is slowly released from its combination with the albumin. The solution must be stored at 2-8°C. During storage the glass of the container may darken by action of the γ-radiation released by disintegrating iodine atoms. In the body, iodinated serum albumin behaves as does the normal plasma constituent. When injected intravenously there is no transfer of radioactivity to other blood proteins. In this form the radioactive iodine remains in the circulating blood, only about 5% being found in the thyroid, probably representing iodine liberated from the albumin binding via normal metabolic processes. Blood volume determinations and cardiac output studies require doses of 5 μCi.

Various synthetic compounds incorporating iodine have been used as diagnostic aids. Iodine-131 labeled diiodofluorescein is used for location of brain tumors. Rose bengal sodium I-131 (disodium 4,5,6,7-tetrachloro-2',4',5',7'-tetraiodofluorescein) accumulates in the liver and is used in liver function tests. Iodohippurate sodium I-131 is used for diagnostic study of bilateral renal function and iodinated triolein is used to study pancreatic function.

Iodine-131 is a valuable isotope in pharmaceutical chemistry and pharmacy for tracer studies of mode of action and biological distribution of drugs and validation of analytical techniques. Amounts as small as 10^{-16}g can be detected. The microdetermination of iodine, silver (as little as 10μg can be detected when precipitated with labeled NaI), and amino acids (in microgram quantities using labeled p-iodo-phenylsulfonyl chloride), are useful analytical applications.

Phosphorus-32. Solutions of phosphorus-32 tagged sodium dihydrogen phosphate and sodium monohydrogen phosphate in ratios giving a pH of

5–6 are the usual form of administration. To prevent adsorption by glass surfaces, containers in contact with this isotope must be pretreated with 0.1% sodium hydroxide, followed by rinses with successive portions of purified water, until the rinse water is neutral to litmus. The metabolism and distribution of phosphorus-32 is identical to that of the stable isotope.

On oral administration approximately 75% is absorbed from the intestinal tract. Administered intravenously it rapidly becomes part of the phosphate pool of the body. Excretion, irrespective of administration mode, is via the urine, 25–50% appearing within the first six days.

After administration the isotope tends to concentrate in rapidly proliferating tissues and the nuclei of cells, e.g., in cancerous cells. This renders the isotope valuable for the location of brain and intraocular tumors. As a diagnostic aid doses of up to 500 μCi may be used.

Phosphorous-32 is also used in several therapeutic applications. It is administered intravenously for the treatment of polycythemia vera. Doses of 2.5-5 mCi are used depending on the severity of the disease. Chronic leukemias (myelogenous and lymphatic) are treated with intravenously administered doses of 100-175 mCi per kg over a 10 day period. The chief disadvantage is a possibility of excessive bone marrow depression.

Phosphorus-32 has been used in colloidal suspension as chromium(III) phosphate, $CrPO_4$, for injection into pleural and peritoneal cavities in treatment of intractable infusion due to carcinoma. (The stable isotope of chromium is employed.) This suspension has better stability than the gold suspension commonly used, but it is difficult to prepare in suitable form.

Chromium-51. This isotope decays by *K*-capture, being transmuted to vanadium with simultaneous γ radiation. The element is usually employed in the form of sodium chromate, chromium(III) chloride, or as a combination of the two salts.

This isotope is used to label red blood cells and/or plasma proteins for determination of circulating red blood cell volume, red blood cell survival time, and plasma volume. Administered intravenously as chromium(III) chloride, approximately 98% of injected chromium(III) is taken up by the plasma proteins, making it a useful reagent for plasma volume determinations. On the other hand, chromium(VI) is not taken up by plasma protein. If it is injected into the blood stream, as sodium chromate, it readily passes from the plasma into the red blood cells. Within the red blood cells, but not in the plasma, chromium(VI) is rapidly reduced to chromium(III), which is bound by the globin portion of the erythrocyte. It cannot leave the red blood cell, thereby making it possible to determine circulating cell volume and red cell survival time. The latter determination is valuable for diagnosis of hemolytic anemias. If chromate is injected in excess of that needed to tag the red blood cells, ascorbic acid is injected. Ascorbic acid reduces the chromate; as a result diffusion into the red blood cells ceases.

Cobalt. Cobalt-60 is used as the metal for the therapeutic effect of its γ-radiation (page 557) and as Cyanocobalamin Co-60 and Cyanocobalamin Co-57, both in capsules and solution. The radioactive cyanocobalamin is used in diagnostic procedures such as the Schilling Test in order to determine normal or abnormal absorption of vitamin B_{12}.

Miscellaneous isotopes. Some additional radioisotopes and their applications include the following: Chlormerodrin Hg-197 Injection and Chlormerodrin Hg-203 Injection-brain scan; Ferrous Citrate Fe-59 Injection-studies of iron metabolism; Gallium Citrate Ga-67 Injection-diagnosis of tumors of soft tissue and bone; Krypton Kr-81m-lung studies including function, ventilation and perfusion; Selenomethionine Se-75 Injection-studies of pancreas and parathyroid gland; Thallous Chloride Tl-201 Injection-myocardial perfusion; Xenon Xe-133 Injection-lung imaging; Ytterbium Yb-169 Pentetate Injection-brain, kidney and cisternographic (cerebrospinal fluid) studies.

Usual Doses and Radiation Doses of Some Compounds. Tables 19.3 and 19.4 provide dose information for one compound in each category when used for imaging and for non-imaging procedures. This data gives the student a concept of the dose of a labelled compound that is used for a particular purpose and of the radiation experienced by a particular organ and by the entire body. The latter is most important since the lower the irradiation of tissues the better. This data was taken from the Physician's Desk Reference for Radiology and Nuclear Medicine[9] to which the student is referred for more detailed information.

RADIOIMMUNOASSAY (RIA)

This analytical technique uses radioactivity and immunological principles in the measurement of very small concentrations of analyte materials (down to 10^{-12} *M*). For each analyte (antigen, AG, sometimes called a ligand), it is necessary to have a radioactively labeled antigen AG*. In most cases I-125 and I-131 labeled antigens are used but, in some cases, ^{3}H and ^{14}C labeled antigens are available in spite of their low specific activity.

The assay procedure is simple. A quantity of antibody (AB) that is

---

[9]L.M. Freeman and M.D. Blaufox, eds., "Physician's Desk Reference for Radiology and Nuclear Medicine," Medical Economics Company, Oradell, N.J., 1979/1980.

TABLE 19.3. **Usual Dose and Radiation Dose of Some Compounds Used in Imaging Procedures**

Procedure and Compound	Usual Dose mCi	Radiation Dose (rads) Organ	Entire Body
1. Brain Scan Sodium Pertechnetate Tc-99m	15-20	3-4 (colon)	0.2-0.3
2. Cisternography Ytterbium Yb-169 Pentetate	0.25-0.5	3.75-7.5 (cord)	0.05-0.1
3. Lung Scan Technetium Tc-99m Albumim Aggregated	1-3	0.4-1 (lung)	0.01
4. Cardiovascular System Iodinated I-131 Albumim	0.2-0.3	2.9-5 (blood)	0.2-0.4
5. Thyroid Scan Sodium Iodide I-123	0.05-0.1	1-2 (thyroid)	0.003
6. Liver Scan Rose Bengal Sodium I-131	0.15-0.30	0.2-1.4 (liver)	0.2-0.4
7. Spleen Scan Cr-51 heated RBC	0.1-0.3	4-10 (spleen)	0.05-0.07
8. Pancreas Scan Selenomethionine Se-75	0.25	3.5 (pancreas)	0.9-2.5
9. Bone Scan Technetium Tc-99m (Pyro- and Trimeta-) Phosphates	15-20	0.45 (bone)	0.15-2
10. Kidney Scan Iodohippurate Sodium I-131	0.2-0.4	0.2-0.4 (kidney)	0.006-0.012
11. Tumor and Abscess Localization Gallium Citrate Ga-67	3-5	3.7-4.5 (colon)	0.8-1.3

stoichiometrically less than the sum of the quantities of AG and AG* is added, the system is allowed to stand at constant temperature to equilibrate as is described in the equation:

$$mAG^* + nAG + pAB \rightleftharpoons qAG^*AB + rAGAB + sAG^* + tAG$$

$$\text{where } p < m + n,\ m > s \text{ and } n > t$$

If the free and bound antigens can be separated, then the ratio of bound antigen (B), actually AG*AB and AGAB, to free antigen (F), actually AG* and AG, may be determined at several different concentrations of AG with AG* remaining constant. A calibration or dose-response curve is constructed using a series of analyte solutions containing varying amounts of AG (starting with zero AG concentration)

TABLE 19.4. **Usual Dose and Radiation Dose of Some Compounds Used in Non-Imaging Procedures**

Procedure and Compound	Usual Dose mCi	Radiation Dose (rads) Organ	Entire Body
1. Thyroid Sodium Iodide I-131 (uptake)	5	6.5-9.0 (thyroid)	0.001-0.02
2. Renal Iodohippurate Sodium I-131	30	0.03 (kidney)	0.009
3. Body Space Iodinated I-131 Albumin	5	0.07 (blood)	0.01
4. Hematology Sodium Chromate Cr-51	100	0.2 (lung)	0.04
5. Blood Flow Xenon Xe-133	1000	0.006 (lung)	0.0002
6. Gastrointestinal Oleic Acid I-131	25	0.05 (intestine)	0.016

and constant levels of AG* and AB.

Data may be plotted in different ways. The B/F ratio is related to AG, the analyte. When B/F is plotted *vs.* log AG concentration, a non-linear trace is obtained; however, when the quantity logit Y (logit Y = ln Y'/1—Y' where Y' = B/B_0 and B_0 is the radioactivity of the bound AG* when the analyte AG is zero) is graphed *vs.* log AG, a straight line results.

The idea of the RIA has been applied to many biological compounds whose measurement would, in many cases, be very difficult if not impossible. Materials such as ACTH, insulin and TSH and drugs such as digoxin may be analyzed by RIA using commercial kits.

STUDY AND SELF-APPRAISAL AIDS

IMPORTANT TERMS. Define or identify in a single concise sentence if possible.

Fission, daughter element, scintillation, inverse square law, isomeric transition, background radiation, proton, Bremsstrahlung, x-ray, PET, film badge, electron volt, β^+-particle, Gray, specific activity, rem, β^--particle, metastable state, disintegration constant, Becquerel, rad, annihilation, positron, γ-ray, radioactive decay, roentgen, neutron, internal conversion, Sievert, decay factor, α-particle, Curie, photon, Geiger-

Muller counter, CAT, autoradiography, K-capture, half-life, RBE, electron capture.

NOMENCLATURE OF INTEREST. Identify each of the following.

Technetium-99m Etidronate, Technetium-99m Medronate, Technetium-99m Oxidronate, Technetium-99m Pyrophosphate, Iodohippurate Sodium I-131; sodium dihydrogen tetraoxophosphate-32(V); Polycythemia vera; chromium-51(III) chloride; cyanocobalamin; Ferrous citrate Fe-59; Gallium citrate-Ga-67; Krypton Kr-81m; Selenomethionine Se-75; Thallous Chloride Tl-201; Ytterbium Yb-169 penetate; cisternography; antibody; antigen or ligand.

SELF-APPRAISAL QUESTIONS

1. Distinguish between α, β^-, β^+, and γ rays.
2. If the following radioisotopes successively give off the indicated radiations, what is the surviving nuclide? Copper-64, $\beta-$, γ; Gallium-67, x-ray(EC); Sodium-24, $\beta-$, 2γ; Cobalt-60, $\beta-$, 2γ; Indium-111, x-ray(EC); Gold-198, $2\beta^-$, γ; Radium-226, α, γ; potassium-37, β^{+1}.
3. Explain why velocity is a useful measure of the energy of the radiation given off by a radioactive isotope.
4. Distinguish between radioactive isotope, stable isotope; teletherapy, x-ray therapy; Iodine-127, Iodine-131; electron, positron; γ-ray, x-ray; disintegration constant, decay factor.
5. Iodine-131 has a half-life of 8.14 days. In how many days will the following percentages of a given sample decompose? 75%, 50%, 25%. (25%—3.38 days)
6. Bromine-82 has a half-life of 35 hours. What percent of an initial sample remains after 20 hours? After 175 hours? (20 hrs.—67.3%)
7. A sample of Chromium-51 had a radioactivity of 500 μCi. What is the activity of this sample after 53 days?
8. A 100 ml. sample of a Phosphorus-32 solution measured 100μCi. How many ml. must be taken 20 days later if the prepared dose is to measure 20μCi? (52.8 ml.)
9. For the following radioactive elements, calculate the *decay factor* for the time indicated (Half-lives are in parentheses): Tecnetium-99m (6.0 hr.), 2 hr.; Iodine-131 (8.0 days), 19 days; Carbon-14 (5730 years), 7500 years; Gallium-68 (68 min.), 10 min.; Indium-113m (100 min.), 300 sec.
10. For Tc-99m with a half-life of 6.0 hours, calculate the percentage of radioactivity remaining after 90 minutes, after 3 hours, after 8 hours.

At what time will 74.9%, 60.6% and 39.7% of the original radioactivity remain?

11. Distinguish between chemical, physical, and interstitial placement of radioisotopes.

12. Identify each of the criteria used for judging the suitability of a radioisotope for a given application. Apply these criteria to chromium-51.

13. Why is iodine-131 considered a particularly valuable isotope? Identify some of its applications.

14. Outline conditions required for a successful precipitation of phosphorus-32 tagged chromium (III) tetraoxophosphate(V). Give the pH etc., required for a stable colloidal suspension of this solid.

15. Which of the isotopes studied are officially recognized in the current compendia? Identify the actual compound(s) present in each official listing. Give the intended use of each.

16. Glass containers used for preparing and storing solutions of radioactive isotopes require special pretreatment. Explain.

17. When an iodine nuclide is used as sodium iodide in imaging studies of tissues other than the thyroid, what is the purpose of administering Lugol's Solution prior to the diagnostic study?

18. Why is iodohippurate sodium I-131 used for diagnostic studies of the kidneys?

19. Why is the Tc-99m generator necessary and very valuable?

20. In radioimmunoassay, why is it necessary to have a concentration of antibody that is lower than the antigen concentration, both labeled and unlabeled?

TWENTY · MISCELLANEOUS INORGANIC THERAPEUTICS

In addition to the major areas discussed in the preceding chapters inorganic compounds are useful in a number of less extensive applications. Usually a specific type of compound or a specific property is of interest in each of these applications.

ANESTHETICS

Nitrogen(I) Oxide. Dinitrogen Monoxide, nitrogen monoxide, nitrous oxide, laughing gas; N_2O. Nitrogen(I) oxide is the only inorganic compound of importance in this class of medicinal agents. It is applied by inhalation for operations of short duration, e.g., dental extractions, opening of abscesses, second stage of parturition. Initially a 10–40% admixture is administered followed by pure gas until anesthesia is affected. Its use in undiluted form is contraindicated in severe cardiovascular disease and with the elderly. Nitrogen(I) oxide is not used more generally because of two drawbacks. First, the apparatus required is cumbersome to move about. Second, the anesthesia produced is inadequate without accompanying anoxia.

Nitrogen(I) oxide is prepared by carefully heating ammonium nitrate at 190°C. in aluminum vessels. Control must be exact since ammonium nitrate is explosive and may detonate at higher temperatures. The principal reaction taking place at 190°C. is

$$NH_4NO_3 = N_2O\uparrow + 2H_2O\uparrow$$

Nitrogen, higher oxides of nitrogen, and traces of oxygen and ammonia are the principal gaseous impurities. The crude gas is cooled and scrubbed to remove basic and acid materials.

$$NH_3 + H_3O^+ \rightarrow NH_4^+ + H_2O$$
$$3H_2O + N_2O_4 = 2H_3O^+ + NO_3^- + NO_2^-$$

The scrubbed gas is passed over iron filings to remove nitrogen(II) oxide (nitric oxide).

$$2NO + Fe = FeO + N_2O$$

The fixed gases are removed by fractional liquefaction; the nitrogen(I) oxide liquefies at $-89.5°C.$, considerably above the liquefying temperatures of nitrogen and oxygen. Several stages of compression are necessary to prevent excessive heating of the gas; nitrogen(I) oxide will decompose at the higher temperatures caused by rapid compression. The purified gas is supplied in blue steel cylinders under a pressure of 1000 atm.

Nitrogen(I) oxide is a colorless gas with a slight sweetish odor and taste. It is readily soluble in water; at 1 atm. pressure 1 volume dissolves in 1.5 volumes at 20°C. It is also soluble in ethanol and fats and oils. Although the gas does not react with pure water and alkaline solutions, it is the formal anhydride of hyponitrous acid, $H_2N_2O_2$. Experimental evidence indicates that nitrogen(I) oxide is a linear molecule, NNO, having the canonical forms:

$$^-{:}\ddot{N}{::}\overset{+}{N}{::}\ddot{O}{:} \quad \text{and} \quad {:}N{:::}\overset{+}{N}{:}\underset{..}{\ddot{O}}{:}^-$$

The compound is an isostere of carbon dioxide. The physical properties of these two gases are strikingly similar, and the crystalline forms are isomorphous. Although nitrogen(I) oxide is relatively inert at ordinary temperatures, it supports combustion at elevated temperatures. However, hemoglobin is unable to use it as a source of oxygen.

In addition to its use as a general anesthetic, this compound finds use as a propellant in aerosol packs (page 591). It is also used to detect leakage in pipelines; since it does not occur naturally, its appearance in an area indicates leakage.

Magnesium sulfate. When injected intravenously, this compound has powerful general anesthetic properties similar to chloroform. The anesthetic dose and lethal dose are too close together to allow its use for this purpose. However, in 10–15% solution, for intravenous or intramuscular injections this substance does find use as an anticonvulsant and as a central depressant.

Inert gases. The heavier members of this family, particularly xenon, have a general anesthetic action. However, because of high cost, they have not been exploited for this use.

ANTIRHEUMATICS

Gold compounds find their principal application in modern medicine as antirheumatics. But despite extensive experience, there is still some difference of opinion as to their value. In the early stages of active rheumatoid arthritis gold therapy is apparently effective; in advanced cases its value is questionable. It has been reported that patients under "remission" with ACTH appear to show favorable response to gold therapy. Injections of gold compounds are also used for treatment of undisseminated lupus erythematosis; it should not be used in treatment of disseminated conditions.

Soluble gold compounds are administered intravenously; oil suspensions of soluble or insoluble compounds are administered intramuscularly. They are not administered orally since they are irritating and absorption from the intestinal tract is poor. An oral dose is found almost in its entirety in the feces. When administered parenterally, the gold appears almost entirely in the urine. Efforts are being made to develop gold compounds which can be administered orally. Clinical testing is underway on one such compound, *S*-triethylphosphine gold 2,3,4,6,-tetra-O-acetyl-1-thio-β-*D*-glucopyranoside, *A*uranofin.

Gold compounds exhibit the toxic manifestations typical of the heavy metals, and to some extent those of arsenic. However, an increasing appreciation of the toxic side effects of gold therapy, e.g., gold dermatitis, and the successful use of BAL and corticotropin in controlling these toxic manifestations should encourage the use of gold compounds when applicable. It has been stated that with proper precautions gold therapy has less frequent serious complications than the long term effects of the glucocorticoids.

Chemically gold compounds are characterized by instability to heat, light, and even the mildest of reducing agents. Gold(III) (auric) compounds decompose first to the gold(I) (aurous) state and then to the free metal. Gold(I) compounds are prone to self-oxidation or "autoxidation."

$$3AuCl = 2Au^0 + AuCl_3 \qquad K \sim 10^{13}$$

Sensitivity to decomposition imposes a need for exceptional care in the handling and use of gold compounds. Because of their ease of reduction, it is best to dispense gold compounds separately.

Having a comparatively large radius and an underlying $18e^-$ shell, both gold(I) and gold(II) are highly polarizable by a wide range of ligands. Consequently gold complexes tend to be more stable than the simple gold salts. Examination of the gold compounds of interest in medicine reveals a common structural feature: the gold atom is always coordinated to the over-all molecular structure through one or more sulfur atoms. The structure of the important gold medicinals appear in Fig. 20.1.

Sodium dithiosulfatoaurate(I). Gold sodium thiosulfate, hyposulfite of gold and sodium; $Na_3Au(S_2O_3)_2 \cdot 2H_2O$. This compound is prepared by adding small successive quantities of gold chloride to a saturated solution of sodium thiosulfate. The solution assumes a reddish color after each addition; this color must be allowed to fade before the next aliquot of gold chloride is added. Upon completion of reaction the complex is precipitated by addition of ethanol to the reaction mixture. Gold(III) is reduced to

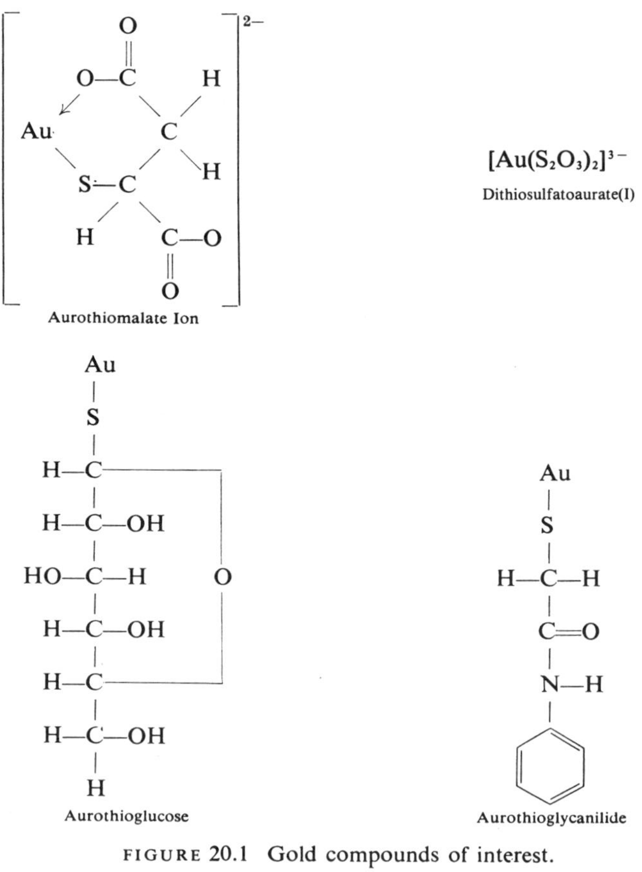

FIGURE 20.1 Gold compounds of interest.

gold(I) during the reaction. This complex is very soluble in water; resulting solutions are slightly alkaline. They decompose slowly on standing. For this reason parenteral solutions must be prepared as needed by solution of the sterile salt in sterile water for injection.

Sodium gold thiomalate. **Gold Sodium Thiomalate, *U.S.P. XXI,* sodium aurothiomalate, disodium aurothiomalate; $C_4H_3AuNa_2O_4S \cdot H_2O$. This is prepared by direct reaction of sodium thiomalate with one of the gold halides,**

$$HSC_2H_3(COO)_2^{2-} + AuI = AuSC_2H_3(COO)_2^{2-} + HI$$

Either the anhydrous compound or the monohydrate may be recovered. The compound is very soluble in water. Its solutions are slightly acid.

Aurothioglucose. Gold thioglucose; $C_6H_{11}AuO_5S$. This compound is prepared by adding gold(I) bromide to a sulfur dioxide-saturated aqueous solution of thioglucose. The reaction mixture is refluxed briefly and cooled. The compound is recovered by precipitation with ethanol. It is purified by solution in water and reprecipitation with ethanol. Aqueous solutions are slightly acid. The compound is usually stabilized by inclusion of not more than 5% sodium acetate. It is usually administered intramuscularly in a vegetable oil suspension.

Aurothioglycanilide. α-Auromercaptoacetanilide; C_8H_8AuONS. This compound is prepared by reaction of gold(I) bromide with thioglycolic acid anilide. It is usually administered intramuscularly in a fixed oil suspension.

CATHARTICS

This group of compounds facilitates the passage through and elimination of the feces from the intestinal tract, especially through the colon and rectum. The compounds are unable to relieve chronic constipation permanently. Excessive use may increase the irritability of the intestine and disrupt its normal functioning. Excessive use is likely since constipation represents a complaint which is widely treated via self-diagnosis and self-prescription.

Cathartics are usually divided into groups based on mechanism of action: irritants (e.g., calomel, castor oil); hydrophilics (e.g., magnesium sulfate, potassium sodium tartrate); emollients (e.g., liquid petrolatum); and detergent laxatives. The inorganic laxatives are either hydrophilics or irritants.

Hydrophilic laxatives. These laxatives (saline laxatives, osmotic laxatives, bulk laxatives) act by interfering with the normal fluid balance of the intestinal tract. Thus one or both of the ions comprising the salt must be nonabsorbable or nearly so. Ingestion of a laxative dose results in a hypertonic condition of the fluids in the colon. Since the ions cannot pass out of the lumen of the large intestine, water must flow from the tissue fluids to the large intestine in order to re-establish isotonic conditions. The resultant large volume of fluid acts as a mechanical stimulus for peristalsis; hence the name bulk laxative. The fluid is swept through the large intestine so rapidly that even though slow absorption of the laxative's ions is possible, there is insufficient time for absorption. Examination of certain saline laxatives will reveal that either the cation or anion present should pass freely through the intestinal wall. However, when associated with a poorly absorbed or nonabsorbable ion, as in a saline laxative, the necessity for maintenance of electrical neutrality in the intestinal fluids prevents net absorption of these ions.

Magnesium ion is the only cation which can act as a laxative; sodium and/or potassium when present in a laxative act to maintain the necessary electrical neutrality. Of the anions only sulfate is relatively nonabsorbable. Monohydrogen phosphate, monohydrogen tartrate, tartrate, and citrate are slowly absorbed, but in laxative doses their osmotic action is so effective that they are swept from the colon before appreciable absorption takes place. Insoluble laxatives, e.g., magnesium hydroxide, must be dissolved by the acid medium of the stomach before they can exert a laxative effect.

Saline laxatives are valuable for treatment of certain drug and food poisoning cases and parasitic infections. Saline laxatives are also useful for cleansing the rectum and sigmoid for proctosigmoidoscopy.

SODIUM SULFATE: Glauber's salt; $Na_2SO_4 \cdot 10H_2O$. This salt is an efficient, chemically nontoxic, and inexpensive laxative. It has an unpleasant taste. The usual 15-Gm. dose will hold about 500 ml. of water in the intestinal tract (3.2% is isotonic). Its solutions are approximately neutral.

SODIUM MONOHYDROGEN PHOSPHATE: Dibasic Sodium Phosphate, *U.S.P. XXI,* secondary sodium phosphate, disodium hydrogen phosphate, phosphate of soda. This salt is an efficient, nontoxic, saline laxative. Although it does not have the unpleasant taste of the sulfate, it is more expensive. It is available as the heptahydrate, $Na_2HPO_4 \cdot 7H_2O$, the dihydrate (Sorenson's Sodium Phosphate), and the exsiccated salt. The latter is a white hygroscopic powder, rehydrating to the dihydrate or heptahydrate. A dodecahydrate is commercially available. Since this latter salt is efflorescent, it must be kept in well-sealed containers in a cool place.

Because of its stability in air at ordinary temperatures, the heptahydrate is convenient for most pharmaceutical uses involving monohydrogen phosphates, e.g., Sodium Phosphate Solution. The usual laxative dose of sodium monohydrogen heptahydrate is 4–8 Gm. It is not recommended in veterinary medicine for dogs. The exsiccated form must be used for the preparation of Effervescent Sodium Phosphate, *U.S.P. XXI.* The hydrate water initiates a slow decomposition of this preparation (page 185). A solution of sodium monohydrogen and dihydrogen phosphate in a ratio imparting a pH of about 6 (approximately 1:3 w/w) is used as a cathartic.

POTASSIUM SALTS: The potassium salts used as laxatives are the monohydrogen phosphate (dibasic potassium phosphate, secondary potassium phosphate), K_2HPO_4; monohydrogen tartrate (potassium bitartrate, cream of tartar, crystals of tartar, potassium acid tartrate), $KHC_4H_4O_6$; potassium sodium tartrate (Rochelle Salt, Seignette Salt), $KNaC_4H_4O_6 \cdot 4H_2O$; and sulfate (sal polychrestum, *tartarus vitriolatus*), K_2SO_4. The usual cathartic doses in the order named are: 4 Gm., 2 Gm., 4–10 Gm., 1.3–3 Gm. Potassium tartrate (soluble tartar), $K_2C_4H_4O_6 \cdot \frac{1}{2}H_2O$, and a mixture of potassium hydrogen sulfate (acid sulfate, bisulfate) with sodium bicarbonate have also been used. Potassium sodium tartrate is used in Compound Effervescent Powders (page 185).

The monohydrogen phosphates and the normal tartrates give mildly alkaline solutions. The hydrogen tartrates give mildly acid solutions. The usual phosphate, tartrate, and sulfate incompatibilities will be observed. Phosphate is incompatible with alkaloids, antipyrene, chloral hydrate, pyrogallol, and resorcinol.

MAGNESIUM SALTS: Magnesium sulfate heptahydrate (Epsom Salt, Bitter Salt), $MgSO_4 \cdot 7H_2O$, magnesium chloride hexahydrate, $MgCl_2 \cdot 6H_2O$, and magnesium lactate trihydrate, $Mg(C_3H_5O_3)_2 \cdot 3H_2O$, have saline laxative action. The soluble magnesium salts are not recommended as laxatives because of the danger of absorption of significant quantities of free magnesium ion, and consequent cerebral effects. Epsom Salt is nevertheless widely used as a habitual laxative. (Its use as an aid in reducing edema has been supplanted by more efficient diuretics.) It is efflorescent, readily losing one water molecule at room temperature.

$$[Mg(H_2O)_6]^{2+} \xrightarrow{27°C.} [Mg(H_2O)_5]^{2+} \xrightarrow{75°C.}$$

$$[Mg(H_2O)_2]^{2+} \xrightarrow{100°C.} [Mg(H_2O)]^{2+} \xrightarrow{120°C.} Mg^{2+}$$

$$[SO_4 \cdot H_2O]^{2-} \xrightarrow{250°C.} SO_4^{2-}$$

As a result of its deliquescence magnesium chloride can be used only in solution. It is absorbed more readily than the sulfate.

Magnesium Citrate Oral Solution[1] and Effervescent Magnesium

[1] *U.S.P. XXI,* p. 611.

Sulfate are used as laxatives.

Insoluble magnesium compounds have a mild laxative action (in addition to the antacid effect for which these compounds are ordinarily used). The compounds of interest are magnesium hydroxide, magnesia magma, magnesium oxide, and magnesium carbonate (page 422). The laxative dose is about four times the antacid dose.

Irritant laxatives. This group exerts an irritant action on the intestinal mucosa, thereby interfering with absorption of electrolyte. Consequently enough water must remain to render the fluids in the lumen isotonic. This water, as with saline laxatives, gives bulk to the softened or fluid stool and at the same time mechanically stimulates peristalsis.

MERCURY(I) CHLORIDE: Calomel, mercurous chloride, mild mercury chloride, mercury monochloride, mercury protochloride, mercury subchloride, *precipité blanc* (French Codex); Hg_2Cl_2. This compound was at one time a very popular laxative. It was widely used to treat "biliousness," erroneously thought to be a disturbance of the liver function. Calomel has no laxative action in the acid medium of the stomach. In the alkaline medium of the small intestine some autoxidation takes place.

$$Hg_2^{2+} = Hg^{2+} + Hg^0$$

The mercury(II) ion is an extreme irritant. It is released high in the small intestine, and its laxative action sweeps most of the intestinal tract rather than only the lower portions, where water absorption normally takes place and the saline laxatives become effective. Although the action is efficient and thorough, calomel has serious drawbacks as a laxative. Because of its high density, the mercury(I) chloride is not easily swept along with the fluids of the intestinal tract, and its elimination is retarded. This raises the possibility of systemic absorption of the highly toxic mercury(II) ions. To aid in flushing the calomel from the system a saline laxative should be administered with calomel. Because of the possibility of systemic mercury poisoning, calomel should not be used. The usual dose was 120 mg.

Mercury(I) chloride has been used in various combinations. Compound Mild Mercury Pills (Compound Cathartic Pills) include colocynth extract—an intensely irritating, drastic cathartic. (Overdose of colocynth can lead to severe prostration or death. Nursing women should not use this substance. The usual dose of colocynth is 120 mg.) Santonin and Mild Mercurous Chloride Tablets (Santonin and Calomel Tablets) combine the anthelmintic action of santonin with the cathartic action of mercury(I) chloride. (Mercury(I) chloride is used in the preparation of the green Bengal Lights.)

Mercury(I) chloride is a white, tasteless, odorless powder of high density. It undergoes autoxidation in alkaline media and sunlight, as is indicated by a darkening of the powder. Chloride ion decreases its solubility through the common ion effect. Even mild oxidizing and reducing agents are incompatible.

MERCURY: In finely divided elemental form mercury has also been used. Its cathartic action depends on oxidation of the element, giving rise to traces of mercury(II). Mercury mass (Blue Mass, Blue Pills) contains 32–34% finely divided elemental mercury together with licorice, honey, althea, glycérine, and mercury(II) oleate (oleate of mercury, mercuric oleate). This latter compound is water insoluble, but soluble in fixed oils. Mercury with chalk, a grey powder with 37–39% finely divided elemental mercury, has also been used as a laxative.

Because of their highly toxic nature mercury and compounds suitable as cathartics were last officially cited in *N.F. XI.*

SULFUR: Elemental sulfur is a mild cathartic, the basis of the classical "sulfur and molasses" remedy. It is believed that its laxative action is due to a slow formation of mildly irritating hydrogen sulfide by action of reducing agents present in the intestinal fluids.

$$S + 2e^- + 2H_3O^+ = H_2S + H_2O$$

Unless a very fine powder is used, the action of sulfur is very slow and gradual. Since a mild action is desired, washed or sublimed sulfur rather than precipitated sulfur is ordinarily preferable as a laxative. Sulfur is commonly used in combination with other substances: sulfur and molasses, Compound Senna Powder.

OTHER COMPOUNDS: Sodium thiosulfate is a cathartic in large doses. Barium chloride, $BaCl_2 \cdot 2H_2O$, is used in veterinary medicine as an intestinal peristaltic, but is too toxic for human application. (Barium chloride is used in rat poisons.)

Anticathartic action. In cases of persistent diarrhea, an effect opposite to that of the cathartics is desired. Aluminum compounds, e.g., aluminum chloride, alum, are commonly employed. They have an astringent and constipating action. Kaolin is frequently included as an adsorbent. An aqueous suspension of colloidal attapulgite is also used to control diarrhea. At one time nickel pectinate was used for treatment of diarrhea.

CHOLAGOGUES

Magnesium sulfate, administered orally, is considered most effective for this action. Introduction of this salt into the duodenum is followed by stimulation of the musculature of the gall bladder and an increased bile flow. It is necessary first to determine the maximum dose which can be tolerated by the individual patient without looseness of the bowels. This quantity is then taken in concentrated aqueous solution every morning.

DIAPHORETICS

Potassium nitrate and ammonium acetate solution have been used for this purpose. The chemistry of both compounds is covered in the following section.

DIURETICS

Inorganic compounds are osmotic-type diuretics. Their elimination or the elimination of their metabolic products requires simultaneous excretion of a volume of water as the solvent. Urinary acidifiers (page 428) and urinary alkalizers (page 427) of necessity act as diuretics; the elimination of dihydrogen phosphate and bicarbonate, respectively, requires copious secretion of water to be flushed from the kidney.

Potassium salts. Potassium compounds in general are also effective diuretics. At normal potassium blood levels this ion together with chloride and water is continuously filtered from the blood plasma into the glomular fluid and subsequently reabsorbed in the tubules with the equivalent chloride and most of the water. If, however, potassium levels of the blood plasma are above normal, tubule resorption of all the potassium ions does not take place. The excess potassium ion, together with water and the necessary anions (usually Cl^-), passes from the kidney as an increased urine flow.

The order of diuretic efficiency of the common potassium salts is

nitrate > chloride > bicarbonate > acetate > citrate

The nitrate and chloride are most commonly used, especially if diuretic action uncomplicated by alkalosis is desired. With the exception of potassium nitrate the chemistry of each of these compounds, together with their sodium analogs, has been discussed previously.

Potassium nitrate (saltpeter, niter) was used as a diuretic. Because it is irritating to the mucous membrane, small doses must be used. The usual dose is 1 Gm., ordinarily as an enteric coated tablet. Although an efficient diuretic, potassium nitrate has side effects which mitigate against its use. Under the conditions existing in the intestinal tract nitrate ion is reduced to nitrite ion. Nitrite is readily absorbed, leading to methemoglobinemia (page 135). It is characterized by cyanosis, dizziness, headache, and diarrhea.

Potassium nitrate is very soluble in water; its solutions are neutral. It has a cooling, saline, pungent taste. Not being deliquescent, it is preferred over sodium nitrate.

Other compounds. Ammonium chloride potentiates the diuretic action of the organomercurials. Ammonium Acetate Solution (Spirit of Mindererus) is used as a diuretic and diaphoretic. It is prepared by dissolving 1.0 Gm. ammonium carbonate in 20 ml. dilute acetic acid (6%).

Tartrates have a mild diuretic effect. Unlike most simple organic anions used in medicine, the tartrate ions are not oxidized in the body. They are eliminated unchanged in the urine. Large doses may cause renal damage.

EMETICS

Although the only FDA-recognized emetic is ipecac syrup, several inorganic compounds produce emesis reflex by irritation of the gastric mucosa.

Zinc sulfate (heptahydrate) (page 474) activates the vomiting reflex before extreme irritation can take place. It is effective for emptying the stomach of foods and noncorrosive poisons. One to two grams in about 200 ml. water is the usual dose. Zinc phenolsulfonate (octahydrate) is also used as an emetic. Cadmium chloride is eight to nine times as potent an emetic as zinc chloride, but is much more toxic.

Copper(II) sulfate (pentahydrate) (page 468) also activates the vomiting reflex very rapidly. The usual emetic dose is 0.25-0.5 Gm. in about 100 ml. water.

Potassium antimony tartrate (tartar emetic) (page 503) is used rarely as an emetic. Its action is slow. Because it acts as a central depressant, it should not be used in this capacity; 0.2 Gm. have been fatal. Its action is not understood, but it has both a central and local component.

Cerous oxalate, $Ce_2(C_2O_4)_3 \cdot 9H_2O$, has been used to prevent the vomiting of pregnancy. Phosphoric acid is claimed to have antiemetic action, but this has not been confirmed.

ENEMAS

Evacuant enemas increase peristaltic activity in the intestinal tract by mechanical distention and stimulation. The distention is caused by the bulk of the fluid introduced. If the solution is hypertonic, the volume of fluid will be additionally increased by the osmotic effect, as with saline cathartics. Sodium bicarbonate, mono- and dihydrogen phosphate, and citrate are used to produce this osmotic effect. Sodium bicarbonate, 1–3%, is used as a cleansing enema. Sodium monohydrogen phosphate and sodium dihydrogen phosphate are commercially available in a 6%:16% solution (pH $\sim$ 6). Sodium dihydrogen phosphate, 12%, with sodium citrate, 10%, is also available. Enemas are rendered more effective by addition of irritants, e.g., soap, turpentine. Sodium Phosphates Enema is cited in *U.S.P. XXI.*

EXPECTORANTS

These substances modify the output and viscosity of the respiratory tract fluids. The more abundant and less viscous excretions aid in the removal of the secretions and exudates from the respiratory passages. The mode of action is not understood.

Steam and carbon dioxide are good expectorants when breathed. By use of a vaporizer the moisture content of air can be increased. For best results a relative humidity of at least 85% is necessary. *U.S.P. XXI* cites Sterile Water for Inhalation (page 147) for this purpose. Carbon dioxide in air air is most effective at about the 5% level. A mixture of 5% carbon dioxide and 95% oxygen is also employed for periods of 10-15 minutes three or four times daily. (Oxygen has an antiexpectorant action.)

Ammonium salts. Ammonium chloride (page 357) and ammonium carbonate (page 593) are probably the most commonly used expectorants. Both compounds are irritants on ingestion, the carbonate much more so than the chloride. Large doses of carbonate may produce nausea. The usual dose for either is 0.3–0.5 Gm. Both compounds find application in cough preparations. Stoke's Expectorant (Expectorant Mixture, *N.F. X*) includes ammonium carbonate.

Iodides. The iodides are effective and widely used expectorants. The action is rapid, iodide appearing in the bronchial fluids within 15–20 minutes after ingestion. The salts have an unpleasant metallic taste. In large and continued doses they may have unpleasant side effects. The properties and usual incompatibilities of iodides have been discussed (page 369).

Hydriodic acid is used as an expectorant, both as the dilute acid and as a syrup. Because of the ease of air oxidation, these preparations must be protected with a preservative, usually hypophosphorous acid (0.5–1.0%). They must not be dispensed if elemental iodine is present. Hydriodic acid is a very strong acid; it will attack the enamel of the teeth. The free acid must therefore be administered only in a highly diluted form, e.g., 1 ml. of the 10% acid in about 100 ml. water. Hydriodic Acid Syrup is a more pleasant tasting form of administration. It is 1-2% syrup. (*N.F. XII* specified 1.4%.) The usual dose is 5 ml.

Potassium iodide is used either in tablet form or in concentrated solution. The tablets may be enteric coated to overcome the unpleasant metallic taste. The usual dose is 0.3 Gm. The solution is prepared by dissolving 1 Kg. of the salt in sufficient purified water to make 1 L. of solution. If the solution is not to be used immediately, the iodide ion must be protected against air oxidation. Sodium thiosulfate, 0.5%, is commonly used. The usual dose of potassium iodide solution is 0.3 ml. A 0.1% aqueous potassium iodide, with 0.125% ethylhexylsulfate, is available as an aerosol.

Ammonium iodide also is used; it combines the expectorant action of both the ammonium and iodide ions. Ammonium hypophosphite (1%) is present as an antoxidant. The dose range is 0.2–0.6 Gm.

Calcium iodide is employed as an expectorant in proprietary cough preparations. Iodide ion is included in some preparations as an amine salt, e.g., diglycocoll hydroiodide.

Other salts. Potassium antimonyl tartrate (page 503), has expectorant action. However, undesirable side effects make its use inadvisable. Nevertheless, it appears in many proprietary cough and expectorant preparations. The usual dose is 1–3 mg. It is present in Brown Mixture (Compound Opium and Glycyrrhia Mixture, *N.F. XII*); the usual 5 ml. dose includes 1.2 mg. of the compound.

Potassium guaiacolsulfonate was introduced as an expectorant on the theory that guaiacol would be liberated in the body and the expectorant action made available. There is no evidence of such desulfonation. This compound is popular in proprietary cough and cold preparations. The usual dose as an expectorant is 0.5 Gm.

Potassium citrate and sodium citrate are popular as expectorants in proprietary cough and cold preparations. The expectorant action is probably empirical. The usual dose is 1.0 Gm.

FIBRILLOLYTICS

A fibrillolytic has the ability to lyse or dissolve fibrin. Sodium iodide injection is used in this capacity. The usual dose by intravenous injection is 1 Gm., either as the 10% or 20% solution. Intravenous injection of sodium iodide can be dangerous for some individuals.

HYPOTENSIVES AND VASODILATORS

Sodium Nitroprusside (Table 5.5) is the only officially listed inorganic vasodilator (*U.S.P. XXI*). It is prepared by the oxidation of potassium ferrocyanide with diluted nitric acid, followed by neutralization of the cold reaction mixture with sodium carbonate or sodium hydroxide. Ruby-red crystals of the sodium nitroprusside dihydrate form on standing. Potassium Thiocyanate (potassium sulfocyanate, potassium sulfocyanide, potassium rhodanide), KSCN, *N.F. X,* and Sodium Thiocyanate (corresponding synonyms), $NaSCN \cdot xH_2O$, *N.F. X,* were used in the management of essential hypertension.

Potentially the thiocyanates are toxic. Many patients develop mild lassitude and weakness, even on so-called safe doses. Nausea, vomiting, and striking weakness are significant at the beginning of severe toxicity. Occasionally goiter develops; this condition responds to administration of iodine.

Thiocyanate, a pseudohalogen, is readily absorbed and distributed throughout all body fluids. The sodium compound is hygroscopic, the potassium slightly so. Both compounds are light sensitive. The chemistry has been discussed (page 341).

Nitrite has the property of relaxing smooth muscle, especially the musculature of the smaller blood vessels. Sodium nitrite, $NaNO_2$, was once used as a vasodilator; because of the slowness of action, other faster-acting vasodilators (e.g., ethyl, amyl, nitroglycerine, erythrityl tetranitrate) have been preferred. Inorganic nitrite reacts with the hemoglobin to form methemoglobin (page 135). (Sodium nitrite is currently official (*U.S.P. XXI)* but as a cyanide antidote.)

Sodium nitrite is a white to very slightly yellow, hygroscopic material. It is very soluble in water, its solutions being slightly alkaline. These solutions are unstable in the presence of weak acids, nitrite ion being a relatively strong base.

$$NO_2^- + H_3O^+ = HNO_2 + H_2O$$

$$2HNO_2 = N_2O_3 + H_2O$$

The anhydride is unstable, decomposing

$$N_2O_3 = NO\uparrow + NO_2\uparrow$$

Consequently the acid solution has the yellow color of nitrogen(IV) oxide.

The nitrites are usually anhydrous or low in water of crystallization. Although nitrous acid is a strong oxidizing agent, neutral nitrites have almost no oxidizing power. The salts are mild reducing agents. The alkali metal nitrites are strong electrolytes; those of the noble metals are covalently bonded. In covalent bonding the bonds may be to the nitrogen or to one of the oxygens. This is well illustrated by the reaction of silver nitrite and methyl iodide.

$$Ag^+NO_2^- + CH_3I \begin{cases} \nearrow H_3C\text{—}ONO \text{ (methylnitrite)} \\ \searrow H_3C\text{—}NO_2 \text{ (nitromethane)} \end{cases} + AgI$$

Both products can be isolated from the reaction mixture.

Nitrites are usually synthesized from nitrates by reduction with heat or with a reducing agent.

$$NaNO_3 + Pb = NaNO_2 + PbO$$

POISONS AND ANTIDOTES

This section does not attempt to be thorough. Discussion is restricted to a few specific antidotes for inorganic poisons and the "Universal Antidote." More thorough discussions are available.[2]

"Universal Antidote" is technically a misnomer. It is a general antidote intended for use when the exact poison is unknown. The constituents are harmless and can inactivate many common poisons through physical or chemical reaction. A generally accepted composition is: powdered charcoal, 2 parts; magnesium oxide, 1 part; and tannic acid, 1 part. It is administered 15 Gm. in a glass of warm water. The intrinsic properties of the three constituents make it effective against heavy metals, acids, bases, alkaloids, and glycosides. A useful list of substances suitable as antidotes for a range of poisons is: aluminum hydroxide or magnesium hydroxide as alkalis; charcoal as an adsorbant; magnesium or sodium sulfate as precipitants and laxatives; potassium permanganate as an oxidant, and sodium chloride as an emetic.

Calcium disodium edetate (calcium disodium ethylenediaminetetraacetate, edetate) disodium edetate and dimercaprol (2,3-dimercapto-1-propanol, BAL, British Anti-Lewisite) are commonly used *in vivo* complexing agents for arsenic and the heavy metal cations—gold, mercury, lead. Edetate is administered intravenously and dimercaprol is administered intramuscularly. The cation composition of the edetate is deliberate. This tetradentate ligand is an avid complexer for most divalent cations. If the tetrasodium salt were used, calcium ion of the blood plasma would be complexed, causing a serious drop in the ionic calcium level. With the introduction of one calcium ion per edetate this avidity for calcium is satisfied, while two sodium ions are still available for exchange with heavy metal ions which may be present.

Edetate is available officially in several forms. Edetic Acid, (ethylenedinitrilo) tetraacetic acid is official in *N.F. XVI.* Edetate Calcium Disodium and Edetate Disodium and the injection for both are cited in *U.S.P. XXI.*

Potassium permanganate and copper(II) sulfate are used as antidotes for phosphorus poisoning. In the absence of a stomach pump the copper compound is preferred. Copper(II) ion not only acts to coat the phos-

[2]*Merck Index,* 9th ed., pp. misc-21 to misc-28.

phorus particles with metallic copper and copper(II) phosphide, but also quickly acts as an emetic. The usual dose is 0.5 Gm. of the pentahydrate in 120 ml. water. A 1% copper(II) sulfate solution is used for phosphorus burns on the skin.

Magnesium sulfate and sodium sulfate are used orally as antidotes for poisoning by heavy metals forming insoluble sulfates, e.g., lead(II), barium. Their laxative action speeds elimination of the precipitates. A 20% magnesium sulfate injection is used as an antidote for barium chloride; calcium gluconate injection must be available to counteract the systematic effect of excess magnesium ion if necessary.

Sodium nitrite and sodium thiosulfate are used as cyanide antidotes. Both appear in *U.S.P. XXI* as the compound and as the injection. The usual doses, in the order named, are: 10 ml. 3% solution; 0.5-2.0 Gm. as 5-10% solution. Thiosulfate is also considered an antidote for arsenic and heavy metals. Nitrites (and nitrates, page 584) are poisonous owing to formation of methemoglobin. Nitrite poisoning can be treated by injection of cyanide; cyanide converts methemoglobin to cyanomethemoglobin. Thiosulfate is then administered parenterally to convert the cyanide to thiocyanate. The thiocyanate is liberated as the free ion, reconstituting the hemoglobin.

Other compounds useful as inorganic antidotes are calcium gluconate, calcium hydroxide solution, normal saline, sodium bicarbonate, sodium lactate solution, acetic acid (5%).

RADIOPAQUES

X-rays are electromagnetic radiations of very short wavelengths (~ 1 Å). Since they are electromagnetic phenomena, interaction with the electrostatic field of the electrons of an atom results. X-ray interference is directly proportional to the atomic number of the interfering atom; an atom of higher atomic number has more electrons to interfere with the passage of the x-ray beam.

In diagnostic x-ray technology investigating the body's interior without surgical entry, only bony structures are differentiated from the uniform background of the soft tissues. The bony tissues consist of elements having a comparatively high atomic number—calcium and phosphorus. The soft tissues are almost entirely composed of elements of very low atomic number—carbon, nitrogen, oxygen, and hydrogen. The number of heavier atoms, e.g., sulfur, chlorine, present in the soft tissues is negligible. As a result only the bony tissue interferes sufficiently to be distinguished on the fluoroscope or on an exposed x-ray film.

The extension of x-rays as diagnostic aids for the study of soft tissues and organs is made possible by the use of radiopaques.[3] These high atomic number substances are introduced by suitable means into the organ or tissue to be studied. Because of the greater interference of the radiopaque with x-rays, the necessary contrast is achieved when the organ is viewed on the fluoroscope or x-ray film.

Barium sulfate. Barium has proven most useful in studies of the alimentary tract. Since soluble forms are very toxic, the highly insoluble barium sulfate must be used. Barium's 56 electrons interfere strongly with the x-ray beam and give a shadowed pattern to any space filled with the barium sulfate. The usual dose is 300 Gm. orally or 360 Gm. rectally. The alimentary tract must be empty of food. Food has a diluting effect, and the contrast obtained on exposure to x-rays will be greatly decreased.

The toxicity of barium ion necessitates extreme caution and control of all phases, beginning with the manufacture of barium sulfate by precipitation from a solution of a soluble barium compound with a soluble sulfate or sulfuric acid.

$$Ba^{2+} + SO_4^{2-} = BaSO_4\downarrow$$

A slight excess of sulfate must be used to insure precipitation of the last traces of barium ion. This also decreases coprecipitation and adsorption of soluble barium compounds. The precipitate must be washed until it no longer shows even traces of soluble barium compounds.

Other insoluble barium compounds, e.g., oxide, carbonate, phosphate, sulfide, sulfite, become soluble to some extent in the acidic medium of the stomach. Because of the similarity of the words sulfide, sulfite, and sulfate, *extreme* care must be exercised in labeling containers clearly and legibly and in *reading* the labels when a sample of radiopaque is taken. The title should always be written in full.

Because barium sulfate is a comparatively dense compound, it is difficult to maintain in a uniform suspension in the alimentary tract. Preparations are available which include suspending agents, e.g., sodium laurylsulfate, alumina gel, methyl cellulose. If present in too high concentrations, suspending agents cause a serious diminution of the interference with the x-rays; a minimum of suspending agent must be used if a maximum contrast is to be obtained. Barium Sulfate and Barium Sulfate for Suspension are official in *U.S.P. XXI.*

Iodine compounds. Iodine, atomic number 53, is another useful atom for radiopaque activity. A complex organic moiety is designed with functional groups which cause the molecules to concentrate in the organ or

[3] Another means of locating or studying a given organ involves the use of radioisotopes (Chapter 19).

body cavity to be diagnosed; as many iodine atoms as possible are incorporated into the structure. The compound is then administered in a suitable dosage form. A few examples follow. Iodized oil is used as a hysterosalpingographic radiopaque. It is prepared by reaction of iodine monochloride with vegetable oil. Iodized oil has 38-42% bound iodine. Iopanoic acid (iodopanoic acid, 3-amino-α-ethyl-2,4,6-triiodo-hydrocinnamic acid) is a cholecystographic radiopaque. The usual dose is 2-6 Gm. Sodium diatrizoate (sodium 3,5-diacetamido-2,4,6-triiodobenzoate) is used as a urographic radiopaque and administered intravenously.

Other compounds. The question may be raised why barium and iodine, atomic numbers 56 and 53, respectively, are the only elements used as radiopaques when there are 30–40 elements of higher atomic number available. Lead with an atomic number of 82, thallium, 81, mercury, 80, and bismuth, 83, certainly should give better contrast than either barium or iodine. Toxicity problems prevent safe use of all except possibly bismuth, which was at one time used as an alimentary radiopaque. The common insoluble salts of the heavy metals become sufficiently soluble in the acid medium of the stomach to give toxic levels of the ion. Although the acid solubilizing effect on sulfates is minimal, unfortunately the normal sulfates of the heavy metal cations tend to be more soluble than barium sulfate; e.g., the pK_{sp} for Tl_2SO_4, Hg_2SO_4, and $BaSO_4$ are 2.4, 6.2, and 10.1, respectively. Organically bound metals, similar to the iodine compounds, are unstable because of the strong metallic character of all heavier elements.

Water suspensions of bismuthyl nitrate and bismuthyl carbonate, 30–60 Gm., were originally used as alimentary radiopaques. The nitrate had the additional complication of causing serious nitrite poisoning. Despite their superior theoretical interference with x-rays the use of these compounds has been discontinued.

There is another aspect to the use of radiopaques. They may be used as thick puttys as shields for normal skin during roentgen irradiation (x-ray or γ ray) of tumors. A putty composed of 86% bismuthyl nitrate and 14% anhydrous wool fat applied thickly can be as effective as a lead shield 1 mm. thick.

RESPIRATORY STIMULANTS AND SYNTHETIC AIRS

Synthetic mixtures of gases intended for respiratory purposes constitute the primary use of the fixed gases in medicine. In addition to their use in administration of anesthetics, various synthetic mixtures of gases may be substituted for air to reduce dyspnea, in the treatment and prevention of hypoxia, prevention of bends, as stimulating combinations for various forms of asphyxia, for air in high altitude and space flights, and for research purposes.

Natural air is a solution of several gases, each gas mutually miscible in the others. The major constituents and their usual concentration are given in Table 20.1. Although nitrogen, oxygen, and the inert gases (primarily argon) remain reasonably constant, the carbon dioxide content varies widely. A poorly ventilated room may have a carbon dioxide concentration ten times normal; in contrast the air over the Antarctic Ocean has only 0.025% by volume. As CO_2 is appreciably heavier than the other

TABLE 20.1 MAJOR COMPONENTS OF NATURAL AIR

Component	Weight-%	Volume-%
Nitrogen	75.5	78.0
Oxygen	23.2	21.0
Inert gases	1.3	1.0
Carbon dioxide	0.04	0.03

components of air, a density fractionation of this gas can take place in poorly ventilated spaces. If undisturbed, the air in the bottom of wells or mines tends to accumulate an ever-increasing carbon dioxide content. Numerous cases of carbon dioxide asphyxiation are on record.

Water vapor, another major component of natural air, has not been included in the table. On the average it is present in less than one volume-%, but varies widely. It may reach 4% in hot, humid regions. In weather forecasts the moisture content is always given as relative humidity rather than actual concentration. Relative humidity may be defined as the per cent saturation; a 50% relative humidity means that the air is at half saturation for the temperature at the time.

TABLE 20.2 CYLINDER COLORS FOR VARIOUS PERMANENT GASES

Gas	Color of Body	Color of Top
Carbon dioxide	brown	blue
Helium	black	yellow
Nitrogen	pink	—
Nitrogen(I) oxide	yellow	aluminum
Oxygen	blue	—
Sulfur dioxide	green	blue

The atmosphere may include various contaminants especially in industrialized areas. Ozone, hydrogen peroxide, and the nitrogen oxides are present, probably for the most part formed by electrical discharges in the atmosphere. Ammonia is present, usually as a decomposition product of organic wastes. Hydrocarbons, carbon monoxide, and sulfur compounds are present as man-made wastes.

Nitrogen, oxygen, and the inert gases other than helium are obtained by the fractional distillation of liquid air. The boiling points of nitrogen, $-195°C.$, and oxygen $-183°C.$, are sufficiently different to allow a satisfactory separation; nitrogen, the more volatile, boils off first, leaving the oxygen as the residue. Carbon dioxide is recovered as a by-product during the preparation of ethanol by fermentation. It is also recovered from the flue gases of large industrial furnaces. Helium is recovered almost entirely from natural gases in certain regions in the United States. These gases may include as high as 2% helium. It is recovered by a fractional distillation process. Helium is present in air, 5.24 ml./cubic meter (S.T.P.).

The purified compressed gases are commercially available in steel cylinders under pressures of 1000 atm. and greater. Distinctive colors have been adopted to identify the cylinders used for each gas. The colors used in the United States for gases of importance in pharmaceutical chemistry and medicine are given in Table 20.2. These colors may vary slightly for different grades of a given gas; e.g., the top may have a different color or the top may have two colors. Suitable pressure-reducing valves and control valves, together with proper gauges, must be used.

Chemically the gases found in the atmosphere are characterized by inertness. At ordinary temperatures even oxygen, usually considered an effective oxidizing agent, is not reactive. Oxygen, nitrogen, and carbon dioxide owe their inertness to π bonding. A π bond shields the σ bond between the atoms concerned from the influence of reactants (See Fig. 2.4). The four gases under consideration are colorless, odorless, and tasteless, although carbon dioxide imparts the familiar acidulous taste to its solutions. The density and water solubility of the gases are given in Table 20.3.

TABLE 20.3 PHYSICAL PROPERTIES OF FIXED GASES

Gas	Density (air = 1)	Volume of Water to Dissolve One Volume of Gas at 20°C. and 1 atm.
Carbon dioxide	1.53	10
Helium	0.137	100
Nitrogen	1.25	63
Oxygen	1.43	32

Helium. As helium is the least dense of all gases except hydrogen, its purity may be determined quantitatively by weighing in a gas balloon of known volume. Impurities have a very pronounced effect on the density. Helium may be purified by passing it over activated charcoal at the temperature of liquid air; it is the only gas not adsorbed.

Helium is used as an inert diluent for medicinal gases. A synthetic air having 20% oxygen and 80% helium is of value in obstructive dyspnea. Since its density is less than that of air, much less energy need be expended to move the gas during ventilation of the lungs. This synthetic air is especially useful in severe asthma and for patients with obstructive lesions. Its use requires a tight mask. The usual oxygen tent cannot be used. Because of its low density and correspondingly high diffusion coefficient (Graham's law), helium would quickly diffuse out of an oxygen tent with natural air taking its place. Pure oxygen is also used in these cases but being denser than air it aggravates rather than relieves the mechanical distress.

Helium-oxygen synthetic airs have another valuable property. Helium is only half as soluble in blood as nitrogen and diffuses out of the blood into the aveolar spaces twice as fast, making it useful under conditions requiring pressurization. When subjected to higher than atmospheric pressure, the quantity of each dissolved gas in the blood (or any liquid) must increase according to Henry's law. If the pressure is rapidly decreased, the excess gas becomes insoluble and is released in the form of bubbles or emboli. Nitrogen is particularly troublesome in this respect. Workers in occupations requiring higher than normal atmospheric pressure, e.g., tunnelers under water, are subject to aeroembolism if the pressure is released too suddenly, a condition known as bends or caisson disease. Because of its lower solubility and higher diffusion rate, helium atmospheres are valuable in eliminating the long decompression periods ordinarily required. (The pitch of the voice is increased when breathing helium.)

Oxygen. Oxygen is used to treat hypoxia. A 100% oxygen atmosphere should not be used continuously. Hyperbaric oxygen therapy has shown promise in certain operative techniques, in control of systemic infections involving anaerobes, for shock treatment, and for reduction of radiation side effects. Pressures as high as 3 atm. are used; 100% saturation of the hemoglobin is achieved together with a 30% increase in dissolved oxygen levels. Some individuals do not tolerate high oxygen pressures.

Nitrogen. Nitrogen finds application as an inert diluent for gases, as an inert atmosphere (page 236), and as a propellant in aerosol packs.[4]

Carbon dioxide. Carbon dioxide finds employment in synthetic air

[4]In addition to nitrogen, nitrogen(I) oxide, nitrogen(I) oxide with carbon dioxide, and low boiling hydrocarbon mixtures find application as propellants for aerosol packs. Germicides and room deodorants are available in this form, as are products of subsidiary interest to the practice of pharmacy—hair sprays, shaving creams, tooth pastes, and insect sprays.

mixtures as a respiratory stimulant; it affects the vasomotor and respiratory centers. Mixed with air or oxygen, it is administered by inhalation. The preferred mixture is 5.0–7.5% in oxygen. Some common applications are to stimulate respiration in asphyxia and arrested respiration; to increase by hyperventilation the excretion rate of volatile poisons and gases, e.g., carbon monoxide, methanol; to bring about full expansion of lungs in treatment of atelectasis; to relive cough and hiccough.

Carbon dioxide-oxygen mixtures have been found preferable to pure oxygen under reduced barometric pressure, e.g., in high-altitude flights.

Respiratory stimulants. Gaseous ammonia is a reflex stimulant for the medullary respiratory and vasomotor centers,[5] probably acting through irritation of the trigeminal nerve. Because they give off ammonia gas, dilute ammonia solution, ammonium carbonate, and Aromatic Ammonia Spirit are used as reflex stimulants in conditions of syncope. Since these preparations will slowly lose volatile ammonia gas, tight closures are required.

Ammonium carbonate (ammonium sesquicarbonate, sal volatile, Spirit of Hartshorn, ammonia crystal, Preston Salt, Baker's Ammonia[6]) is a mixture of ammonium bicarbonate, $NH_4^+HCO_3^-$, and ammonium carbamate (ammonium aminoformate, anhydride of ammonium carbonate), $NH_2COO^-NH_4^+$. It is prepared by calcining a mixture of ammonium sulfate or ammonium chloride and calcium carbonate.

$$CaCO_3 + (NH_4)_2SO_4 = CaSO_4 + 2NH_3\uparrow + CO_2\uparrow + H_2O\uparrow$$

The "compound" forms by reaction of the evolved gases.

$$2NH_3 + CO_2 = NH_2COO^-NH_4^+\downarrow$$

and

$$NH_3 + CO_2 + H_2O = NH_4^+HCO_3\downarrow$$

On resublimation with a small amount of water a white, translucent, crystalline material is recovered. It is covered with the bicarbonate, as a white opaque powder. The crystals have a strong ammonia odor.

The compound is volatile even at room temperatures. It volatilizes rapidly above 60°C. It decomposes in boiling water. The decomposition reactions are the reverse of the synthetic reactions. The usual incompatibilities of ammonium salts and carbonates will be observed.

Ammonium carbonate is readily soluble in water, the carbamate reacting to form carbonate

$$[NH_2COO^-NH_4^+] + H_2O = 2NH_4^+ + CO_3^{2-}$$

[5] High concentrations in air arrest breathing and are injurious to the lungs. Death due to pulmonary edema results from breathing toxic doses of the gas. Less than 250 p.p.m. is probably harmless.

[6] A leavening agent in the baking industry.

The resultant solution is essentially a mixture of ammonium carbonate and bicarbonate. It has a sharp alkaline taste and an alkaline pH.

Ammonium carbonate is also used as an expectorant (page 582).

Aromatic Ammonia Spirit (smelling salts) is a solution of "ammonium carbonate" in diluted ammonia solution and alcohol together with aromatic oils.[7] During preparation some of the volatile oils may separate to give a cloudy solution which cannot be clarified by filtration. In older literature this material is erroneously reported to be ammonium carbonate. Aromatic Ammonia Spirit is available in single-dose, thin-wall glass vials in a strong but soft cotton envelop. Finger pressure easily shatters the vial; the cotton acts as a sponge for the released spirit and at the same time prevents harm from glass fragments.

Carbon dioxide absorbers. During administration of certain anesthetics, in some oxygen therapy techniques, in metabolic studies, in submarines during long periods of submersion and space vehicles, it is necessary to rebreathe the same air continuously. Carbon dioxide gradually accumulates unless provision is made for its removal. Condensation of moisture from expired air may also create problems.

In order to keep the carbon dioxide at near normal levels the respired air must be treated continuously for removal of the gas. A carbon dioxide absorber must be chemically inert to other gases present in the air such as the normal air constituents or added substances, e.g., nitrous oxide, ether, cyclopropane. It must absorb carbon dioxide avidly and retain it permanently in nonvolatile form. The absorber must not liquefy. If it becomes moist or liquefies, adequate contact with the respired gases is impossible and absorption becomes inefficient. Liquefaction also presents leakage and drainage problems to the apparatus and for the patient. Crystalline sodium hydroxide, an excellent, efficient carbon dioxide binder, is unsatisfactory as a carbon dioxide absorber because it liquefies in contact with moisture.

Two physical properties are important: mesh size and particle hardness. Mesh determines mean particle size, significant for two reasons which actually work in opposition to one another. The particles must be large enough to allow easy passage of the air without interference with breathing. At the same time the particles must be small enough to provide an adequate surface area for contact with the air with a reasonable volume of absorber. Because of entrainment of fine particles in the air and the resultant possibility of severe caustic burns in the respiratory tract, carbon dioxide absorbers must be dust free. Desirable limits for particle size are provided by a No. 4 mesh screen for the larger particles and a

---

[7]For preparation see *U.S.P. XXI,* p. 49.

No. 20 screen for the smaller ones, but wider limits are usually allowed. Containers should be clearly labeled as to mesh size. Particle hardness is also critical. The particles must have sufficient strength to prevent excessive breakdown under their own weight or in handling.

The ability to absorb carbon dioxide depends on the reactions

$$OH^- + CO_2 = HCO_3^-$$

and

$$HCO_3^- + OH^- = H_2O + CO_3^{2-}$$

Thus the length of time a given charge of absorbent is effective is limited by the equivalents of base present. If a carbon dioxide absorber is dyed with a suitable acid-base indicator, the change of color specified on the container will be a suitable index for determination of the end of effective life of a given charge. For absorbers not including an indicator a careful record of periods of use must be maintained; an adequate safety margin should be allowed.

Soda Lime (calx sodica) *N.F. XVI,* is prepared by fusing calcium hydroxide and sodium and/or potassium hydroxide with sufficient diatomaceous earth to impart the desired hardness to the resultant product. Magnesium ion reduces the effectiveness of soda lime as a carbon dioxide absorber. For this reason a dolomite lime cannot be used as a raw material. Soda lime contains 5-20% alkali hydroxide and 6-18% water. It can absorb 25-35% of its weight as carbon dioxide.

Barium Hydroxide Lime, *U.S.P. XXI,* is prepared similarly. Barium hydroxide octahydrate (m.p., 78°C.) is substituted for the alkali hydroxide. However, some potassium hydroxide may be present. Since barium ion is present in a dissolvable form, even in the spent crystals, this preparation is toxic if ingested.

Lithium hydroxide is used as a low-molecular-weight carbon dioxide absorber in nonpharmaceutical applications, e.g., in space craft.

Both barium hydroxide lime and soda lime are greyish white solids. If an indicator has been applied, the particles will assume the color of the indicator. Both products are powerful caustics and must not be allowed to come in contact with the skin or tissues.

The absorption reaction is exothermic, and, as seen in the equations for the neutralizations, water is a by-product of the absorption. The breathed air tends to heat up and become saturated with moisture. If uncontrolled, both heat and moisture can have deleterious effects on the patient or subject.

SCLEROSING AGENTS

Sclerosing agents are used mainly for the obliteration of varicose veins. The milder substances may be used for selected cases of hemorrhoids, but should not be employed on external prolapsed or thrombosed cases. The solutions may be combined with local anesthetics or themselves have such action.

Ionic compounds used as sclerosing agents are soluble iodides, iron(III) compounds, mercury(II) chloride, sodium chloride, sodium citrate, sodium morrhuate, sodium salicylate, and urea hydrochloride. Nonionic sclerosing agents include ethanol, dextrose, invert sugar, phenol, and quinine. A common preparation is 20% sodium chloride injected into the lumen of the varicose vein. Dextrose, invert sugar, and sodium morrhuate are less irritating than the other compounds and are less apt to produce necrosis.

SEDATIVES

Two inorganic elements, lithium and bromine, have simple ions possessing sedative properties. Lithium was first utilized by Cade when he used lithium citrate to reduce episodes of excitement encountered in manic patients and schizophrenics. Lithium can protect against depression, ameliorate unstable character disorders, and calm hyperactive children as well as aggressive and violent patients. Despite its dramatic effects in mental patients the mode of action has not been elucidated.

While Lithium Carbonate was first cited in *U.S.P. XIX,* Lithium Hydroxide, Lithium Citrate (trilithium tetrahydrate) and Lithium Citrate Syrup were first admitted to *U.S.P. XX* by supplement. All are cited in *U.S.P. XXI.*

The chemistry of lithium, a bridge element, and its toxicity have been discussed (pp. 208-209). Since the element is toxic, lithium levels in the blood should be monitored. Because it is excreted by the kidneys its use is contraindicated in cases of impaired renal function. Since toxicity increases inversely with sodium intake patients on salt-restricted diets should be monitored regularly.

Bromides were at one time widely used as central depressants. Sodium, Potassium, and Ammonium Bromides were last officially cited in *N.F. XII.* At different times in the past lithium, calcium, and strontium bromides were also used; claims for more efficient action were unsubstantiated. Preparations of mixtures were also popular; they included Bromides Syrup, *N.F. XI,* Three Bromides Elixir, *N.F. XI,* and Five Bromides Elixir, *N.F. X.* Effervescent mixtures including potassium bromide are still available as OTC preparations. Bromides have a bitter saline taste. The syrup was probably the most agreeable.

Biochemically the absorption, distribution, and excretion of bromide are similar to that of chloride. The ion can substitute for chloride ion in all intracellular fluids except in nerve tissue. But, on continued administration the concentration of bromide ion builds up, since the kidney excretes relatively more chloride ion than bromide. The overuse of bromides results in bromism, or bromide intoxication. Common reactions are severe dermatitis and urticaria. Treatment involves the use of substantial doses of sodium chloride.

Bromide depresses all nerve tissue, but there is selective action for certain centers. It is a sedative if used in proper quantities. It differs from true somnifacients such as chloral hydrate in that a single dose produces an insufficient concentration of bromide in the tissue to have significant action. Repeated administration induces sleep by decreasing general sensitivity to noise and discomfort. The sleep may not be as refreshing as that produced by barbiturates and chloral hydrate. There may be a "hangover."

STRUCTURAL REPAIR MATERIALS

Occasionally in surgical practice it is necessary to provide temporary or permanent support structures for various soft tissues of the body. Such material should be chemically inert and insoluble in the body fluids contacted. Any possible products must be nontoxic. The material must be structurally strong, since it must withstand any stress to which it may be subjected. These repairs involve the bony structures of the body. They may be permanent—replacement for shattered bone or a dental filling in a tooth—or they may be temporary—a plaster cast for a limb while a broken bone knits.

Tantalum and silver metal generally are used as a replacement or as a substitute for bony structure. Tantalum is inert in all body fluids. This property and its high tensile strength make it an ideal material. (Many of the transition metals, e.g., zirconium, have the prerequisite properties, but until recently have not become available in commercial quantities.) Tantalum has been used as a replacement and for temporary braces for the long bones, and as plates used to close openings in the skull. Tantalum in fine mesh form has been sutured with fine tantalum wires in surgical repair of hernias resistant to ordinary surgical procedures.

Silver metal has found applications similar to tantalum except for the mesh. Although silver is very slightly reactive in the body fluids, reaction is negligible and the principal product, silver chloride, is inert.

Dental surgery requires the use of various cements and mercury amalgams to fill cavities, e.g., Zinc-Eugenol Cement, *U.S.P. XXI.*[8] The mechanism of setting has not been elucidated satisfactorily. It is believed that a zinc eugenolate chelate forms as a matrix of thin elongated crystals. These enclose the undissolved and unreacted material. Zinc phosphate, $Zn_3(PO_4)_2 \cdot 4H_2O$, is also used as a dental cement.

Mercury amalgams of gold or silver are used as dental fillings. Mercury is a good solvent for metals; gold, bismuth, silver, tin, lead, zinc, and cadmium will dissolve at room temperature. Solution is hastened by using finely divided metallic powders.

Howe's Solution is used to deposit metallic silver within the exposed surface structure of freshly drilled cavities in the teeth (page 471).

Plaster of Paris (approximately $CaSO_4 \cdot \frac{1}{2}H_2O$) is used for surgical casts. The casts serve as mechanical supports or to immobilize parts of the body, e.g., support the thorax, immobilize parts of limbs. It also serves as an absorbent dressing for wounds. Plaster of Paris is valuable in preparation of dental impressions because it expands on setting. On standing it fills all impressions and depressions and thus makes an exact copy.

Plaster of Paris is for practical purposes an exsiccated gypsum having approximately one equivalent of hydrate water per two equivalents of calcium sulfate. It is prepared from the mineral gypsum, $CaSO_4 \cdot 2H_2O$, by carefully controlled calcination at temperatures not exceeding 125°C. With increasing temperatures fewer and fewer water nuclei remain within the structure. Above 204°C. there is complete dehydration; "dead burnt" gypsum results. Anhydrous calcium sulfate does not rehydrate readily, and there is no hardening of the mass.

The rate of set of plaster of Paris is influenced by the temperature of burning and the fineness. Local areas of gypsum structure must be preserved after calcination; these gypsum nuclei act as centers for the reinitiation of hydration and the subsequent hardening process. The hardening process apparently consists of a meshwork of growing gypsum crystals. The existence of local gypsum nuclei in anhydrous calcium sulfate is supported by the observation that a powdered mixture of anhydride and gypsum will set before it can be mixed with water. Fineness is important since rate of solution, necessary for growth of the gypsum crystals, will be faster than with a coarse powder. The hardening process may be described by the reaction

$$CaSO_4 + xH_2O \rightarrow CaSO_4(aq)$$

$$nCaSO_4(aq) + 2nH_2O + \underset{\text{(nucleus)}}{(CaSO_4 \cdot 2H_2O)} = (2n + 1)(CaSO_4 \cdot 2H_2O) + (aq)$$

[8]Page 1129.

The structure of gypsum has been discussed (page 319).

The rate of hardening can be influenced by inclusion of foreign substances in the plaster of Paris. The rate of set is retarded by substances such as colloids, cream of tartar, etc. The rate of set is accelerated by the inclusion of electrolytes.

STUDY AND SELF-APPRAISAL AIDS

IMPORTANT TERMS. In addition to the names of each category given in the chapter outline, define or identify each of the following in a single concise sentence if possible.

Anoxia, arthritis, autoxidation, constipation, detonation, diarrhea, dyspnea, emboli, Graham's law, Henry's law, hydrophilic, hypoxia, isostere, relative humidity, somnifacient.

NOMENCLATURE OF INTEREST. Identify each of the following.

Ammonium aminoformate, ammonium carbamate, ammonium carbonate, aurothioglucose, aurous chloride, barium hydroxide lime, Bromides Syrup, Brown Mixture, calomel, *calx sodica,* Compound Effervescent Powders, cream of tartar, dibasic sodium phosphate, Epsom Salt, freon, gold chloride, Glauber's Salt, gypsum, hyponitrous acid, hyposulfite of gold and sodium, mercury subchloride, methemoglobin, niter, nitrous oxide, plaster of Paris, potassium sulfocyanate, Preston Salt, Rochelle Salt, sal volatile, saltpeter, secondary potassium phosphate, Spirit of Mindererus, soda lime, sodium aurothiomalate, sodium dioxonitrate(III), Sodium Phosphate *U.S.P. XXI,* sodium tartrate, Sorenson's Salt, Stoke's Expectorant, tartar emetic, *tartarus vitriolatus,* tetrachlorauric acid, zinc compounds and eugenol cement.

SELF-APPRAISAL QUESTIONS

1. Explain the origin of the ammonia gas present in the crude nitrogen(I) oxide.

2. What are the disadvantages in using nitrous oxide as an anesthetic?

3. Write the canonical forms for carbon dioxide and nitrogen(I) oxide and use them to show that the two compounds are isosteres.

4. Explain the action of BAL in controlling untoward reactions arising from gold therapy.

5. What is the common structural feature of all gold compounds in use as antirheumatics? Explain.

6. Distinguish between saline laxatives and irritant laxatives. To what extent is their mode of action similar?

7. Explain why an osmotic laxative increases the fluid content of the intestinal tract. Use sodium sulfate, magnesium sulfate, and potassium monohydrogen tartrate as the basis for your discussion. Indicate the subtle differences in the actual physical behavior of these salts.

8. What untoward reactions may arise from the use of either magnesium chloride or calomel as a laxative?

9. How does the laxative dose of an insoluble magnesium compound compare with the antacid dose?

10. What are the advantages of administering a saline laxative as an effervescent mixture, e.g., Compound Effervescent Powders?

11. Explain why a finely divided sulfur will have a stronger laxative action than coarser particles.

12. What is the fundamental action of a saline enema?

13. What are the side effects of the following compounds if used as diuretics: potassium nitrate, ammonium chloride, potassium tartrate.

14. Name some of the compounds having emetic action. What is the emetic dose?

15. Give the chemical and physical properties of "ammonium carbonate." (Why is the name of the compound set off by quotes?)

16. Give the families of compounds finding application as expectorants. What advantages or disadvantages exist for each? What is the usual dose of the more commonly used compounds in each of the groups?

17. Examine the components of the universal antidote given. Name the chemical and physical property of importance for each and give an example of its action.

18. Why is EDTA used as the disodium monocalcium salt?

19. Give the theoretical basis for the use of radiopaques. What requirements must a compound meet if it is to act as a radiopaque? Illustrate your answers with actual substances.

20. Give the advantages and disadvantages of using helium as a diluent for oxygen.

21. Name some of the compounds which are useful as respiratory stimulants.

22. List the essential physical and chemical characteristics of a carbon dioxide absorber. Explain. Why is sodium hydroxide unsuitable in this application?

23. Name the lithium salts presently official as antidepressants or antimanics. Why is it advisable to monitor lithium levels in patients on these drugs?

24. Why have bromides passed out of favor as sedatives?

25. Distinguish between a nitro compound and an organic nitrite. Why is this distinction not important with ionic compounds?

26. List those compounds and preparations studied which are official in the current compendia.

PRACTICE IN EQUATION WRITING

27. Write the equation(s) for the reaction(s) taking place during the preparation of sodium nitrite, sodium dithiosulfatoaurate(I), sodium thiosulfate, sodium rhodanate, nitrogen(I) oxide, barium hydroxide lime, ammonium carbonate, magnesium hydroxide.

28. Write the equation for the reaction taking place when the first member of each of the following pairs of compounds is used as an antidote for the second: potassium permanganate, phosphorus; magnesium sulfate, barium chloride; hypo, potassium cyanide; BAL, gold(III) compounds; edetate, lead(II) ion. (Write complete structural formulas for any complexes involved.)

29. Write equations for the reactions effecting the removal of each of the following impurities from crude nitrogen(I) oxide: ammonia, nitrogen(II) oxide, dinitrogen(IV) tetroxide.

30. Show by equations why Effervescent Sodium Phosphate will decompose slowly if the heptahydrate of sodium phosphate is inadvertently used in the formulation.

31. Write the equations for all equilibria existing in Aromatic Ammonia Spirit.

32. Write the equation for the setting of plaster of Paris.

33. By means of equations show why potassium nitrite is unstable in solution at a pH of 4.

34. Demonstrate the action of barium hydroxide lime as a carbon dioxide absorber by means of suitable equations.

35. Demonstrate the antoxidant action of ammonium hypophosphite for ammonium iodide by means of equations.

36. Write the equation(s) for the reactions taking place during the official assays for: carbon dioxide, nitrogen, magnesium sulfate, potassium bromide, potassium sodium tartrate, sodium dibasic phosphate, sodium nitrite.

37. Write equations for possible reactions responsible for the laxative action of sulfur.

38. Write equations for possible chemical incompatibilities existing in the following prescriptions:

a. Rx

Codeine Sulfate	0.1
Strontium Bromide	16.0
Peppermint water	100

b. Rx

Aromatic Spirit Am.	10.0
Lime water	50.0

c. Rx

Arom. Spirit Ammonia	30
Elixir Five Bromides	90

INDEX